HAQ/BOERSMA—Introduction to Marine Micropaleontology

ERRATA

1. The caption of figure 3 on page 23 was inadvertently dropped during printing;
   it should read:

   Fig. 3. A cross-section of the protoplasm of the planktonic genus *Globigerinoides*, showing organelles and other microstructure. A. Three lunate Golgi bodies (*G*) are located around a network of endoplasmic reticulum (*ER*) which in turn is surrounded by tiny granular (black dots) ribosomes (*Rb*). Mitochondria (*M*) and lipid bodies (*L*) of stored fats are also present in this area of the protoplasm which is considered to be an active site of secretion. × 2,040. B. The nucleus is contained in this second to last chamber of the test; this is a single bi-lobed nucleus involved in the normal vegetative functioning of the test; smaller nuclei, involved in reproduction, are not present here. Note the more convolute membrane of the right lobe. × 2,240. (Courtesy Alan Bé.)

2. Magnifications on Tables III and IV (pp. 104 and 105) should be X2200.

DATE DUE

19 NOV 1981			
DEC 1 2 1983			
Jan 3			

HIGHSMITH 45-220

INTRODUCTION TO
MARINE MICROPALEONTOLOGY

CONTRIBUTORS

W.A. BERGGREN, Woods Hole Oceanographic Institution, Woods Hole, Massachusetts ●
ANNE BOERSMA, Lamont-Doherty Geological Observatory, Palisades, New York ●
KRISTER BROOD, National History Museum, Stockholm, Sweden ●
LLOYD H. BURCKLE, Lamont-Doherty Geological Observatory, Palisades, New York ●
BILAL U. AQ, Woods Hole Oceanographic Institution, Woods Hole, Massachusetts ●
YVONNE HERMAN, Washington State University, Pullman, Washington ●
LINDA HEUSSER, New York University, Tuxedo, New York ●
J. JANSONIUS, Imperial Oil Enterprises, Calgary, Alta., Canada ●
W.A.M. JENKINS, Institute of Sedimentary and Petroleum Geology, Calgary, Alta., Canada ●
STANLEY A. KLING, Scripps Institution of Oceanography, La Jolla, California ●
KLAUS J. MÜLLER, Bonn University, Bonn, Germany ●
VLADIMÍR POKORNÝ, Charles University, Prague, Czechoslovakia ●
JÜRGEN REMANE, University of Neuchâtel, Neuchâtel, Switzerland ●
GRAHAM L. WILLIAMS, Geological Survey of Canada, Darthmouth, N.S. Canada ●
JOHN L. WRAY, Marathon Oil Company, Littleton, Colorado ●

INTRODUCTION TO
MARINE MICROPALEONTOLOGY

EDITED BY
BILAL U. HAQ
Woods Hole Oceanographic Institution
Woods Hole, Massachusetts

ANNE BOERSMA
Lamont-Doherty Geological Observatory
Palisades, New York

 ELSEVIER · NEW YORK
NEW YORK · OXFORD

ELSEVIER NORTH-HOLLAND, INC.
52 Vanderbilt Avenue, New York, New York 10017

Distributors outside the United States and Canada:
THOMOND BOOKS
(A Division of Elsevier/North-Holland Scientific
        Publishers, Ltd.)
P.O. Box 85
Limerick, Ireland

Library of Congress Cataloging in Publication Data

Main entry under title:

Introduction to marine micropaleontology.

   Bibliography: p.
   Includes index.
   1. Micropaleontology. 2. Marine sediments.
I. Haq, Bilal U. II. Boersma, Anne.
QE719.I57   560'.92   78-4516
ISBN 0-444-00267-7

Manufactured in the United States of America

CONTENTS

PREFACE
Bilal U. Haq and Anne Boersma VII

1 MARINE MICROPALEONTOLOGY: AN INTRODUCTION
W.A. Berggren 1

CALCAREOUS MICROFOSSILS

2 FORAMINIFERA
Anne Boersma 19

3 CALCAREOUS NANNOPLANKTON
Bilal U. Haq 79

4 OSTRACODES
Vladimír Pokorný 109

5 PTEROPODS
Yvonne Herman 151

6 CALPIONELLIDS
Jürgen Remane 161

7 CALCAREOUS ALGAE
John L. Wray 171

8 BRYOZOA
Krister Brood 189

SILICEOUS MICROFOSSILS

9 RADIOLARIA
Stanley A. Kling 203

10 MARINE DIATOMS
Lloyd H. Burckle 245

11 SILICOFLAGELLATES AND EBRIDIANS
Bilal U. Haq 267

PHOSPHATIC MICROFOSSILS

12   CONODONTS AND OTHER PHOSPHATIC MICROFOSSILS
      Klaus J. Müller 277

ORGANIC-WALLED MICROFOSSILS

13   DINOFLAGELLATES, ACRITARCHS AND TASMANITIDS
      Graham L. Williams 293

14   SPORES AND POLLEN IN THE MARINE REALM
      Linda Heusser 327

15   CHITINOZOA
      J. Jansonius and W.A.M. Jenkins 341

      GLOSSARY OF TERMS 358

      INDEX 367

PREFACE

This textbook was conceived several years ago with the realization that there was no up-to-date, introductory-level, English language micropaleontology text. The lack of such a text has been keenly felt by instructors and students alike, but the exponential growth of micropaleontology in the last two decades and the concomitant increase in specialization has discouraged such an effort. It would have been a mammoth task for any one author to undertake a synthesis of the information now available on the various marine microfossil groups. We realized then that this vacuum could only be filled by soliciting the cooperation of specialists, each to contribute a separate chapter on a fossil group.

We envisioned a college-level introductory text, including all microfossil groups used in the study of the marine environment. We hoped to provide a source for basic information on each group, for comprehension of the type of reasoning applied to the study of microfossils and their use in (paleo)oceanography, and for locating essential background material and references necessary to pursue any group further. This text conforms closely to these concepts; only time and financial constraints have taken their slight toll.

We asked the contributors to adhere to specific outlines and to emphasize the "problem-solving" potential of each fossil group as well as the types of problems currently being investigated. Other aspects were purposely underplayed; specifically, taxonomy, the details of preparation techniques, and the burgeoning technical vocabularies associated with several of these microfossil groups. Most of this type of information is already available in the literature and references to these sources are included in the Suggested Reading section of each chapter.

All contributors have attempted to orient their writing in the directions outlined above. Nevertheless, standardization of text has been difficult. As editors, we have done our utmost to include the same sections in each chapter, to edit the language towards some sort of uniformity, and to encourage discussion of reasoning, rather than detail.

As recent advances in micropaleontology continue to transform what was predominantly a descriptive to a more interpretive science vital to (paleo)oceanographic research, the number of researchers interested in its applications, especially in the age determination of sediments has increased. Thus, in several graduate schools of Earth Sciences and Oceanography, micropaleontology is a course required of all students. We hope that, in addition to the beginning students of micropaleontology, non-micropaleontologists (e.g., sedimentologists, core-describers, shipboard geophysicists) may produce reasonable estimates of the age of sediments from the stratigraphic range-charts included with each chapter.

The biologic heterogeneity of the fossil groups included in marine micropaleontology prohibits a biologically based grouping of chapters. We have instead chosen to arrange these according to the composition of the shells; a more utilitarian approach since it combines fossil groups found together in different types of sediments. This arrangement also emphasizes that marine microfossils are composed of materials vital in the geochemical cycles of the oceanic realm as well as the cyclic processes operating between the land and the ocean and the atmosphere and the ocean.

Accordingly, we have arranged the chapters under four headings: calcareous, siliceous, phosphatic and organic-walled microfossils. Within these categories, various microfossil groups are presented in order of their present importance in historical geology and (paleo)oceanography. We have included all marine microfossil groups commonly used to interpret the marine environment, excluding terrestrial and other non-marine forms. There is a brief discussion of Spores and Pollen in the marine realm, since these commonly occur as detrital particles in near-shore sediments.

In closing, we wish to thank all the contributors for their untiring efforts, as well as all who participated in the production of this book, including the typists, graphic artists, the editorial staff at Elsevier, and the friends whose insight and encouragement (and occasional prodding) helped us complete this project.

Woods Hole, Massachusetts   BILAL U. HAQ

Palisades, New York   ANNE BOERSMA

ACKNOWLEDGEMENTS

This volume represents the joint effort of many scientists who responded to the editors' invitation to contribute introductory-level chapters in their own fields of specialization to a textbook in micropaleontology. Most contributors received help, advice and permission to reproduce published and unpublished illustrative material from their colleagues. Other colleagues critically reviewed original texts of various chapters, thereby considerably improving their quality. We gratefully acknowledge the cooperation and enthusiastic support of all these colleagues in bringing this book to fruition. Acknowledgements for individual chapters are listed below:

Foraminifera. W.A. Berggren, R. Fleischer, F. McCoy and I. Premoli-Silva for critical reading of the text; J. Aubert, A. Bé, W.A. Berggren, R. Fleischer, H. Luterbacher, D. LeRoy, W. Poag, S. Streeter, and R. Todd for illustrative material and/or permission to reproduce the same; T. Saito and D. Breger for scanning electron micrographs of planktonic foraminifera and A. Edwards and J. Weinrib for help with typing the manuscript.

Calcareous Nannoplankton. D. Bukry, K. Gaarder, K. Perch-Nielsen, R. Poore and H. Thierstein for critical reading of the manuscript. For permission to use illustrative material: A. Farinacci, S. Forchheimer, S. Gartner, S. Honjo, H. Manivit, A. McIntyre and A. Bé; H. Okada, P. Roth, K. Wilbur and N. Watabe also gave permission to use figures.

Ostracodes. H. Oertli for review of the first draft of the manuscript; A. Absolon, D. Andres, R. Benson, H. Blumenstengel, W. van den Bold, K. Diebel, M. Gramm, J. Harding, E. Herrig, N. Hornibrook, V. Jaanusson, A. Keij, R. Kesling, L. Kornicker, H. Kozur, E. Kristan-Tollmann, K. Krömmelbein, H. Malz, A. Martinsson, K. McKenzie, H. Oertli, G. Ruggieri, J. Senes, P. Sylvester-Bradley, Museum of Paleontology, University of Michigan and Senken-

bergische Naturforschende Gesellschaft, Frankfurt, for illustrative material and/or permission to reproduce the same.

Pteropods. G. Tregouboff and M. Rose, Centre National de la Récherche Scientifique, Paris, and the Publisher of Dana Reports for permission to reproduce illustrative material.

Radiolaria. W. Riedel for reviewing the first draft of the manuscript; P. Adshead, H. Forman, B. Holdsworth, H. Ling, E. Merinfeld, T. Moore, Jr., E. Pessagno, W. Riedel, and A. Sanfilippo for illustrative material and/or permission to use the same, and P. Bradley and R. White for scanning electron micrographs which were made at the research facilities of Cities Service Oil Company.

Silicoflagellates and Ebridians. D. Bukry and K. Perch-Nielsen for reviewing the manuscript and A. Loeblich, III, Y. Mandra, H. Okada, K. Perch-Nielsen and W. Wornardt for illustrative material and/or permission to reproduce the same.

Dinoflagellates, Acritarchs and Tasmanitids. E. Barghoorn, F. Cramer, E. Denton, J. Deunff, G. Eaton, A. Eisenack, W. Evitt, H. Gorka, T. Lister, A. Loeblich, Jr., F. Martin, W. Sarjeant, D. Wall, D. Williams, Cambridge University Press, Carnegie Institute, Edizioni Tecnoscienza, Harvard Museum of Comparative Zoology, *Micropaleontology, Paleontology* and *Revista Española de Micropaleontologia* for illustrative material and/or permission to reproduce the same; J. Charest and M. Trapnell for typing the manuscript and G. Cook for drafting figures.

Spores and Pollen. Y. Tsukada and D. Nicols for illustrative materials.

Chitinozoa. S. Laufeld and Exxon Production Research Co., Houston, for illustrative material.

MARINE MICROPALEONTOLOGY
AN INTRODUCTION

W.A. BERGGREN

By definition **micropaleontology**, the study of microscopic fossils, cuts across many classificatory lines. It includes within its domain the study of large numbers of taxonomically unrelated groups united solely by the fact that they must be examined with a microscope. At the same time within certain taxonomically homogeneous groups the size of some forms is such that they scarcely need be examined with microscopic aid and are more properly grouped under macropaleontology. It is not surprising then that as a discipline micropaleontology lacks a certain coherent homogeneity. Most marine microfossils are protists (unicellular plants and animals), but others are multicellular or microscopic parts of macroscopic forms. Thus, their grouping into one discipline remains essentially practical and utilitarian.

The practical value of marine microfossils in various fields of historical geology is enhanced by their minute size, abundant occurrence and wide geographic distribution in sediments of all ages and in almost all marine environments. Due to their small size and large numerical abundance, relatively small sediment samples can usually yield enough data for the application of more rigorous quantitative methods of analysis. Moreover, most planktonic and many benthic microfossils have wide geographic distributions that make them indispensable for regional correlations and comparisons, and paleooceanographic reconstructions.

Marine microfossils occur in sediments of Precambrian to Recent ages, and in every part of the stratigraphic column one or more groups can always be found useful for biostratigraphic and paleoecologic interpretations (see Fig. 1). Terminology for the major divisions of the marine realm discussed in this text are shown in Fig. 2. Fossil marine organisms lived in almost all these marine areas and can therefore be invaluable in the study of changes in the paleoenvironments.

For instance, radiolaria, silicoflagellates, calcareous nannoplankton, pteropods, and some foraminifera and diatoms (Fig. 1) are planktonic (i.e. free floating) and live in abundance from 0 to 200 m in the open ocean, but diminish rapidly near the continents. These forms are useful in monitoring past changes in the oceanic environments, particularly changes in temperature. Other groups (Fig. 1) such as the ostracodes, bryozoa, and some foraminifera and diatoms are benthic (i.e., adapted to living on the bottom of the sea), as either **vagile** (free-moving) or **sessile** (passive or attached) organisms. Since these forms exhibit distribution patterns broadly linked to depth, sediment type and various physico-chemical variables in seawater, they are useful in delineating changes in the bottom environment.

Some forms, such as the dinoflagellates, are known to contain both planktonic and benthic phases in their reproductive cycle and are particularly useful tools in paleo-

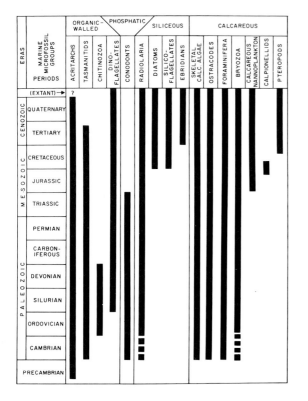

Fig. 1. Stratigraphic distribution of the major marine microfossil groups.

ecologic studies of near-shore areas and inland seas (e.g., the Red Sea, Black Sea, etc.). Others, such as the conodonts (extinct since the early Mesozoic) are presumed to have been attached to the soft body of a planktonic animal and thus their use in reconstructing Paleozoic and early Mesozoic planktonic ecology is more tenuous than the use of forms with living descendants in the Mesozoic and Cenozoic.

Spores and pollen, although derived from land plants, are strongly climate-dependent. Thus, their presence and distribution patterns in near-shore marine sediments allow interpretations of continental climates; and/or, like chemical tracers, their distribution can be used to monitor current movements.

HISTORICAL REVIEW

The earliest mention of microfossils dates back to classical times. The large benthic

foraminifer *Nummulites* — the great rock builders of the pyramids — had been mentioned by Herodotus (5th century B.C.), Strabo (7th century B.C.) and Pliny the Elder (1st century A.D.). While in the western world we usually credit Leonardo da Vinci (1452—1519) as having been the first to recognize the organic nature of fossils, it would appear that Leonardo had been anticipated by some three centuries by one Chu Hsi, who in his *Analects*, dated 1227, actually perceived their true nature.

The systematic study of microfossils awaited discovery in 1660 by Antonie van Leeuwenhoek of the microscope and it would seem that this is an appropriate date to denote the birth of systematic micropaleontology. The foraminifera were the first group of microfossils to receive the attention of early naturalists. Although illustrated in various publications as early as the 16th century and given short Latin descriptive diagnoses, it was not until the 10th edition of Linné's *Systema Naturae* (1758) that a binomial nomenclature, which was to form the basis of modern biological taxonomy, was applied to some fifteen species of foraminifera.

Alcide d'Orbigny's (1802—1875) detailed studies on the foraminifera led to the first comprehensive classification of the foraminifera in 1826 (he included them among the cephalopods) and subsequently published a large number of papers on living and fossil foraminifera. He was also first to utilize foraminifera in biostratigraphic studies of the Paratethyan Tertiary. In 1835 the French biologist Felix Dujardin demonstrated that the foraminifera could not be considered as cephalopods because they were characterized by long pseudopodia and he introduced the name Rhizopodes for the group. D'Orbigny and most of his contemporary colleagues accepted this revision. Not however C.G. Ehrenberg (1795—1876), a German micropaleontologist and contemporary of d'Orbigny. Ehrenberg was a remarkable scientist of the period and is generally credited with having made the first discovery and description of silicoflagellates, ebridians, coccoliths, discoasters, dinoflagellates, and numerous living protists, in addition to having described and illustrated numerous radio-

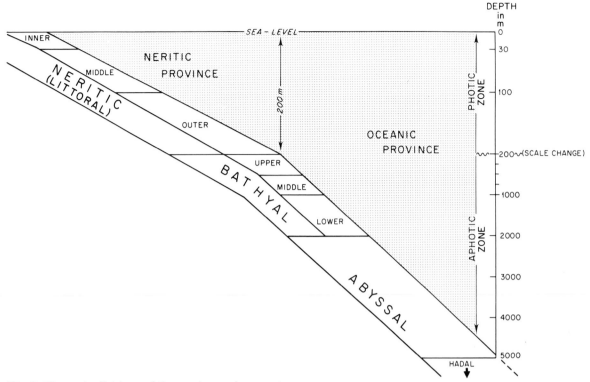

Fig. 2. The main divisions of the marine environment.

larians, diatoms and foraminiferans. Ehrenberg rejected Dujardin's views and claimed that the foraminifera belonged to the Bryozoa, a view which he staunchly maintained as late as 1858.

Basic descriptive studies were conducted on many of the major microfossil groups during the second half of the 19th century. The studies of Reuss in the 1860's and 1870's on the Cretaceous and Tertiary foraminifera of Prussia, though contemporaneous, contrast strongly with those of the so-called "English school" which included N.C. Williamson, W.K. Parker, T.R. Jones, W.B. Carpenter, H.B. Brady, and C.D. Sherborne. The latter group were distinguished by their view that foraminifera exhibited such large individual variation as to make specific differentiation virtually impossible, thus rendering the group of little use in stratigraphic studies, a view that, unfortunately, remained influential well into the present century.

At the same time the major studies by Ernst Haeckel on the Radiolaria (1862–

1887), the basic taxonomic work on ostracodes by Sars (1866), and the Schmidt diatom atlas (beginning in 1875), among others, testify to the vigorous descriptive work conducted in micropaleontology during this time. After this auspicious beginning, however, it is curious that interests subsequently waned in all these groups save the foraminifera.

The single largest impetus to descriptive micropaleontologic studies in the latter half of the 19th century was the voyage of H.M.S. *Challenger* from 1873 to 1876. In 1884 H.B. Brady published a monumental monograph of the foraminifera dredged by the *Challenger* during its voyage around the world. This work remains to this day the fundamental reference in the study of Neogene and living foraminiferal faunas.

The second, or analytic, phase in micropaleontology is concerned with such things as basic biologic structure and affinities, morphogenetic studies, creation of classificatory systems approaching a "natural system" and biostratigraphic applications.

The single major cause for a revival of interest in marine micropaleontology and which has led directly to the third or synoptic phase in most groups is the advent of oceanographic deep sea drilling. Oceanographic micropaleontology expanded greatly in the post-World War II period with the development of better techniques for obtaining deep-sea piston cores, particularly in the 1950's and 1960's. This was substantially aided by the advent of the JOIDES Program in 1965 and its successor, the Deep Sea Drilling Project (DSDP), in 1968. With the conclusion of this third phase of ocean drilling in autumn 1975, the D/V *Glomar Challenger* had drilled over 400 sites in all areas of the world oceans except the Arctic. It recovered over 41,000 m of nearly 76,000 m of drilled and cored sediments and basalt dating back to the mid-Jurassic in water depths from less than 1 to more than 4 km. The availability for study of sediment from the ocean floors covering the last 150 million years has allowed great advances in the study of the evolution of the oceanic planktonic and benthic microfossils and the utilization of these forms in deciphering the history of the evolution of the oceans themselves during this time. Refined biostratigraphic zonations and paleobathymetric and paleoecologic studies are but a few areas in which great strides have been made in the past decade. These synoptic studies were not possible until a sufficient data base was developed — a data base which had to await a sufficient amount of analytical study on the taxonomic groups themselves and the availability of a sufficient amount of material from wide areas of the marine biosphere. The oceanic region until recently remained the last great frontier for the micropaleontologist.

Commercial micropaleontology

Micropaleontology as a relatively uniform discipline is a young field of science dating back to the First World War. At that time the need for greater amounts of petroleum led to the enhanced study of foraminiferal faunas as a correlation tool. The conversion from what had previously been the somewhat esoteric descriptive studies on foraminiferal taxa to the utilitarian aspects of biostratigraphic correlation in search for greater amounts of fossil fuels was rather rapid.

Commercial, or industrial, micropaleontology is generally associated with the petroleum industry. With the recognition of the biostratigraphic utility of foraminifera in petroleum exploration, micropaleontology received a new impetus and direction. Micropaleontology continues to be widely used in the exploration and mapping of surface and subsurface stratigraphic units by geological surveys all over the world. At the same time the expansion of commercial micropaleontology has been intimately linked with its development as a part of the academic curriculum in the earth sciences. In 1911 Professor J.A. Udden of Augustana College (Rock Island, Ill.) stressed the importance of using microfossils in age determination and correlation in subsurface water well studies in Illinois. At the same time he convinced oil companies of the importance of microfossils in age determination of drill cuttings. During World War I, micropaleontology was introduced as a formal course in the college curriculum by Josia Bridge at the Missouri School of Mines and by H.N. Coryell at Columbia University and by F.L. Whitney at the University of Texas (Austin). The latter taught Hedwig Kniker, Alva Ellisor, and Esther Applin. These three women were subsequently to become the leading economic micropaleontologists in the United States and Ellisor eventually became the chief paleontologist for Humble Oil and Refining Company.

J.J. Galloway and J.A. Cushman were prime movers of micropaleontology at this time. The year 1924 was an important one in micropaleontology. Galloway began teaching micropaleontology at Columbia University. In the same year the Cushman Laboratory for Foraminiferal Research was established at Sharon, Massachusetts, and was affiliated with Harvard University. For the next quarter of a century this laboratory remained one of the major centers of foraminiferal research in North America and many of the micropaleontologists who went on to successful careers in industry and universities were at one time or another associated with J.A. Cushman. Micropaleontology was introduced at Leland

Stanford University in 1924 by H.G. Schenck and in 1933 the study of ostracodes was introduced at Columbia University by H.N. Coryell. C.G. Croneis introduced micropaleontology at the University of Chicago during the depression. In the decade immediately preceding World War II emphasis in micropaleontology and petroleum exploration remained heavily concentrated in biostratigraphic correlation, whereas in the decade immediately following the war emphasis shifted rapidly to paleoecology and paleobathymetry.

BASIC CONCEPTS AND REVIEW OF CURRENT TRENDS

Biostratigraphy and biochronology

Interpretations of earth history depend on two different systems of logic, both of which arrange geological observations into sequences of events. The first and most widely used is the logic of superposition: the ordering of events *iteratively* in a system of invariant properties simply by determining the physical relationship of features in the rocks. This is what is meant by the word **stratigraphy**. The second logical system depends on the recognition of an *ordinal* progression which links a series of events in a system of irreversibly varying characteristics. This provides a theoretical basis outside of the preserved geological record by which the nature and relationship of the events in the progression can be recognized or predicted, and according to which missing parts of the record can be identified. Geology is an historical philosophy, so the ordinal progressions we refer to are progressions in time, just as geological time is perceived by the progress in one or another ordinal series of events. This is what is meant by the word **geochronology**. A time-scale then provides a conceptual framework within which to interpret earth history; in this specific instance, phenomena related to the evolution of the marine biosphere. Only two ordinal scales are widely used today, that of **radiochronology** (based on isotope-decay rates)

and that of **biochronology** (based on organic evolution). Geomagnetic polarity reversals are non-ordinal repetitions but because of their wide applicability have been closely calibrated to the ordinal time-scale. The geologic column and its radiochronologic framework used as the basis for discussions in this text is shown in Fig. 3.

Philosophy and methodology behind the establishment of a time-scale

Stratigraphy has been succinctly defined as the "descriptive science of strata". It involves the form, structure, composition, areal distribution, succession and classification of rock strata in normal sequence. **Biostratigraphy** is that aspect of stratigraphy which involves the direct observation of paleontologic events in superposition. A biostratigraphic unit is a body of rocks which is delimited from adjacent rocks by unifying contemporaneous paleontologic characteristics.

The evolution of organisms through time has provided the framework for a system of zonations by which discrete units of time represented by material accumulation of sediments can be recognized. Biozones may generally be grouped into three categories depending on their characteristic features: (1) **assemblage zones**, those in which strata are grouped together because they are characterized by a *distinctive natural assemblage* of an entirety of forms (or forms of a certain kind) which are present; (2) **range zones**, those in which strata are grouped together because they represent the *stratigraphic range* of some *selected element* of the total assemblage of fossil forms present; and (3) **acme zones**, those in which strata are grouped together because of the *quantitative presence (abundance)* of certain forms, regardless of association or range. The latter are, qualitatively, of lesser importance than the first two. Most planktonic zones used in current biostratigraphic work are range zones and are of the following types:

(a) **Taxon Range Zone** — a body of strata representing the total range of occurrence (horizontal and vertical) of specimens of a taxon (species, genus, etc.).

(b) **Concurrent Range Zone** — a range zone

Fig. 3   THE PHANEROZOIC GEOLOGIC TIME SCALE

ERA	PERIOD	EPOCH / STAGE*		AGE Ma
Cenozoic	Quaternary	Pleistocene - Recent		1.6 — 0
	Tertiary	Pliocene		5.0
		Miocene		24
		Oligocene		37
		Eocene		54 — 50
		Paleocene		64
Mesozoic	Cretaceous		Maastrichtian	
			Campanian	
			Santonian	
			Coniacian	
			Turonian	
			Cenomanian	— 100
			Albian	
			Aptian	
			Barremian	
			Hauterivian	
			Valanginian	
			Berriasian	135
	Jurassic		Tithonian	
			Kimmeridgian	
			Oxfordian	— 150
			Callovian	
			Bathonian	
			Bajocian	
			Toarcian	
			Pliensbachian	
			Sinemurian	
			Hettangian	192
	Triassic		Rhaetian	
			Norian	— 200
			Carnian	
			Ladinian	
			Anisian	
			Olenikian	225
Paleozoic	Permian		Ochoan	
			Guadalupian	
			Leonardian	— 250
			Sakmarian	
			Asselian	280
	Carboni-ferous	Pennsylvanian	Uralian	
			Moscovian	— 300
			Bashkirian	
			Namurian	
		Mississippian	Visean	
			Tournaisian	
	Devonian		Famennian	345 — 350
			Frasnian	
			Givetian	
			Eifelian	
			Emsian	
			Siegenian	
			Gedinnian	395
	Silurian		Ludlovian	— 400
			Wenlockian	
			Llandoverian	435
	Ordovician		Ashgillian	
			Caradocian	— 450
			Llandeilian	
			Llanvirnian	
			Arenigian	
			Tremadocian	500 — 500
	Cambrian		Shidertinian	
			Tuorian	
			Mayan	
			Amgan	
			Lenan	
			Aldanian	

millions of years

defined by those parts of the ranges of two or more taxa which are concurrent or co-incident. It is based on a careful *selection* (and rejection) of faunal elements which have a concurrent, though not necessarily identical, stratigraphic range with a view to achieving a biostratigraphic unit of maximum time-discrimination and extensibility.

(c) **Oppel Zone** — a zone characterized by a distinctive association or aggregation of taxa selected because of their restrictive and largely concurrent range, with the zone being defined by the interval of common occurrences of all or a specified portion of the taxa. This is a less precise, and more restricted, relative of the Concurrent Range Zone described above. It is little used in planktonic biostratigraphy. As a biochronological concept, it is exemplified by the Land Mammal Age.

(d) **Lineage Zone** or **Phylozone** — the body of strata containing specimens representing the evolutionary or developmental line or phylogenetic trend of a taxon or biologic group defined above and below by features of the line or trend. It has been commonly referred to as a **phylogenetic zone**.

The scope of a lineage zone (phylozone) may extend from the first (evolutionary) appearance of some form in an evolutionary bioseries to the termination of the lineage, thus including the whole bioseries or lineage, or it may include only a segment of the lineage (**lineage-segment zone**). For a further discussion on the nature and application of the lineage-zone concept to biostratigraphic studies the reader is referred to Van Hinte (1969) and Berggren (1971).

(e) **Acme Zone** — a body of strata representing the acme or maximum development of some species, genus or other taxon, but not its total range. It is little used, except in local biostratigraphy in planktonic studies.

(f) **Interval Zone** — the interval between two distinctive biostratigraphic horizons but not in itself representing any distinctive biostratigraphic range, assemblage or feature. The so-called Partial-Range Zone actually corresponds to an interval zone and is commonly used in current planktonic biostratigraphic studies. For a more comprehensive discussion on the nature of biostratigraphic zones the reader is referred to the *Preliminary Report*

on Biostratigraphic Units published by the International Subcommission on Stratigraphic Classification, Report No. 5 at the 24th International Geological Congress, Montreal, 1972.

Biostratigraphic zones differ in a fundamental manner from time-stratigraphic units (see Fig. 4). The latter includes, by definition, all the rocks formed during a given interval of geologic time. Biostratigraphic zones as defined are limited to the areal distribution of strata which actually contain specific fossil forms or assemblages. No biostratigraphic zone is of mondial, absolutely synchronous extent, because no fossil form had an instantaneous mondial origin nor suffered instantaneous extinction, nor was so independent of environment as to be found everywhere in all types of sediments laid down synchronously.

In the area of marine biostratigraphy perhaps the best examples of biostratigraphic zones are those proposed for Mesozoic and Cenozoic sequences based upon calcareous and siliceous plankton. From initial studies on the early Cenozoic planktonic foraminifera of the SW Soviet Union in the decade preceding World War II, through similar studies on the Cenozoic of the Caribbean region following the war, to the advent of the Deep Sea Drilling Project in the mid-late 1960's we have witnessed the creation of a veritable flood of biostratigraphic zonations within the Cenozoic at least, and particularly for its younger part — the Neogene — local and regional biostratigraphic zonations have been created for virtually all areas of the marine realm from equator to subpolar regions reflecting the dependency of such zonations upon biogeography (i.e. climatology — paleooceanography).

Obviously, the radiochronological time-scale can be quantified since it is based on well-documented assumptions of the invariancy of isotope-decay rates through geological time; and since radiometric ages are expressed in numbers, they conform to what we are conditioned to accept as measurement itself.

The methodology and reasoning by which the marine time-scale has been developed is as follows:

(1) From 0 to 5 Ma (million years before present) the time-scale is based upon the radiometrically controlled paleomagnetic time-scale of Cox (1969) and the calibration

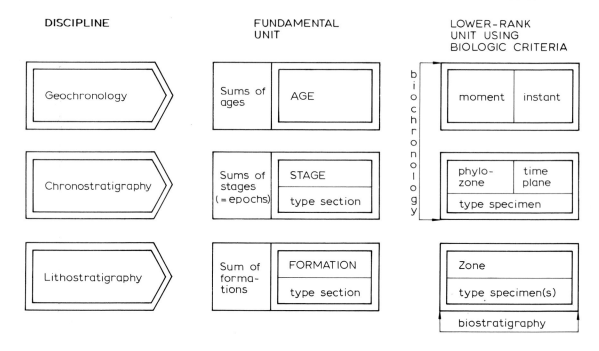

DISCIPLINE

FUNDAMENTAL
UNIT

LOWER-RANK
UNIT USING
BIOLOGIC CRITERIA

Geochronology

Sums of ages | AGE

biochronology

moment | instant

Chronostratigraphy

Sums of stages (= epochs) | STAGE / type section

phylo-zone | time plane / type specimen

Lithostratigraphy

Sum of forma-tions | FORMATION / type section

Zone / type specimen(s) / biostratigraphy

Fig. 4. Relationship between the discipline of geochronology, chronostratigraphy, and lithostratigraphy. Geochronological events are placed directly above their chronostratigraphic equivalents.

Geochronology, the measuring of time, provides an absolute, ordinal scale, or framework, for the geologic column. Geochronology further aims at depicting the time of occurrence of a given event such as an event involving a fossil organism or a magnetic anomaly. The fundamental unit for subdividing the geochronological scale is the *age*. In turn, ages may be subdivided based on some biologic criterion which is relatable to absolute time. Thus, a *moment* is the time during which a taxon is known to have existed (but due to our imperfect knowledge of many organisms, may not be its true total time of existence).

Chronostratigraphy, on the other hand, relates sequences of rock units to time. The major unit in chronostratigraphy is the *stage* which describes strata lain down during a specified period of time. A stage is based on a designated locality, its **type section**, to which all equivalent rocks can then be related. For example, the type section of the Maastrichtian Stage is near Maastricht, The Netherlands; rocks considered to be that age in England can then be compared, and if equivalent, are called Maastrichtian in age. A stage may further be charac-

terized by biologic events concerning its incorporated fauna which are related to time. Events such as the total time existence of a taxon (species), called its **phylozone**, or a single event occurring at a specific time and measurable over some distance, the **time plane**, fall into the realm of chronostratigraphic biochronology. Due to the rarity of world-wide synchroneity of events in the biosphere, just now being appreciated by biostratigraphers, time planes are difficult to prove on the basis of fossils.

Lithostratigraphy, by distinction, deals with sequences of rocks, *independently of time*. The primary unit for categorizing rock units is the **formation**, which also is based on a type section. Lithostratigraphic units describe only strata, not the time involved in their deposition. The biostratigraphic **zone** is then a biological characterization of part or all of a rock unit; such divisions may be based on one of several criteria such as the first occurrence of a taxon in a section, overlapping ranges of two taxa, etc. This type of zone does not specify the time of an event, as the same event may be diachronous (non-time equivalent) between two geographic regions, in two parts of an ancient basin, or in environments which change at different rates.

Along these lines, consider the problems involved in relating low-latitude to high-latitude zonations. (Modified after Van Hinte, 1969).

to it of biostratigraphic events which have been achieved by investigations on deep-sea cores. The calibration of this part of the time-scale is believed to have a high degree of

accuracy and reliability.

(2) From 5 to 25 Ma the time-scale is based upon calibration of biostratigraphic events to K—Ar dates, the calibration to the paleomag-

netic time-scale of selected biostratigraphic events in certain fossil plankton groups, particularly the radiolaria, and thence by second-order correlation between various planktonic microfossil groups, and to correlations between the sea-floor magnetic anomalies and paleomagnetic epochs. The calibrations within this part of the time-scale must be viewed as moderately tentative.

(3) Beyond 25 Ma the time-scale is based upon calibration of biostratigraphic events and the geomagnetic reversal record to K—Ar dates, interzonal correlations and linear extrapolations between biostratigraphic events and/or zonal boundaries in cores and outcrops. The calibrations within this part of the time-scale are considered to be tentative and subject to moderate revision as the paleomagnetic time-scale is gradually extended through the Paleogene. Current studies on the late Cretaceous and early Cenozoic (Paleocene) paleomagnetic stratigraphy of land sections and correlation with the sea-floor magnetic anomalies, as well as calibration of biostratigraphic events to K—Ar dates at the basaltic basement/sediment contact, appear to offer calibration control points upon which the time-scale may be extended over the lower part of the Cenozoic (Paleogene) and into the Mesozoic.

The major problem in creating a relatively accurate time-scale for the Mesozoic is the difficulty in matching the paleomagnetic reversal sequence in land sections with the magnetic anomaly pattern on the sea floor and their calibration to K—Ar dates, owing to the large experimental error in radiometric dates of this antiquity. Considerable progress has been made in the last few years, so that despite current controversies associated with the developing Mesozoic time-scale, we can look with optimism to a relatively reliable Mesozoic time-scale within the next few years.

The pre-Mesozoic time-scale is based solely on the correlation between biostratigraphic events and K—Ar dates with ever increasing uncertainty limits with increasing antiquity.

There is a persistent movement among earth scientists, understandably among specialists in physical geology and mineralogy but also among paleontologists who work with the more "datable" parts of the fossil record,

to do away with biochronology as a primary means of dividing up geological time and to rely entirely on arbitrary and unquestionably convenient time-lines, expressed in rounded-off and exact numbers of years, which are to be identified in the geological record by radiochronology. Biochronological age measurements, even though they are not as easily quantified, are nevertheless implicit in the fossil record itself and do not systematically lose resolution in the older parts of that record as radiometric ages do. Arguments as to the real meaning of radiometric numbers, and as to their practical utility everywhere, will eventually divide on the degree to which (numerical) certainty can be demanded in an uncertain universe, but it is our opinion that "biological time" — that is the history of the biosphere and its physical environment — is best measured biochronologically with a critically evaluated calibration assist wherever possible from radiochronology. To do otherwise is to let the radiometric tail wag the paleontological dog.

Biology and evolution

Evolutionary theory provides the conceptual framework for biological systems just as time provides the conceptual framework for (bio)stratigraphy. Patterns of fossil "behavior" such as diversity, geographic distribution, community structure, intraspecific morphologic variability, can be understood in the context of evolutionary theory. However, the path between micropaleontologic studies and evolutionary concepts is a two-way street: whereas evolutionary theory may synthesize and render more concrete (i.e. objective) the patterns of fossil "behavior", the data from micropaleontologic studies on living or fossil populations provide the raw material upon which evolutionary theory is based and continuously subjected to modification.

Evolutionary concepts applied to (micro) fossil populations are based upon studies of the biology and ecology of living organism populations which points out the strong tie between evolution and ecologic studies. Evolutionary theories are based, in turn, upon the science of genetics, the precepts of which are rarely studied by micropaleontologists. Nevertheless, a basic understanding of evolution and genetics is essential to the fields of (paleo)ecology and biostratigraphy. For instance, Simpson (1953, p. 138) has defined selection as "anything tending to produce systematic, heritable change in populations between one generation and the next". This broad, modern definition may be contrasted with the limited, restricted mid-nineteenth century Darwinian concept of natural selection as the selective reproduction by the most successful members of a population leading to adaptive changes in species and ultimately to new species. Selection is the mechanism of interaction between organism and environment and, as Simpson has pointed out, its role in evolution is the production of adaptation. The marine microorganisms discussed in this book represent some of the most successful adaptations in the history of life in terms of sheer numbers, diversity, geologic record (longevity), ecologic differentiation and evolutionary radiation.

The concept of the "species" is of critical importance to (micro)paleontology, although in fossil populations the conditions of the biologic definition cannot be met. Mayr (1966) defines a species as "groups of interbreeding natural populations that are reproductively isolated from other such groups".

In contrast the paleontologic species concept is more subjective and utilitarian: a species is defined primarily on the basis of a set of external, preservable morphologic features which can be recognized and applied by other specialists. Subjectively interpreted antecedent and descendant relationships and stratigraphic range form an integral part of the definition and recognition of a fossil taxon.

Whether a group of organisms reproduces sexually or, as most of the forms discussed in this text, asexually, further affects the definition (i.e. characterization) of a fossil species and the taxonomy used to describe it. Some asexually reproducing organisms are capable of extensive morphologic variability which will affect their classifications and nomenclature. Still others may assume distinctly different morphologies during the course of ontogeny (e.g. dinoflagellates vs. hystrichomorphs) and hence present their own specific problems of classification and nomenclature.

These and other patterns of fossil "behavior", whether it be within a given lineage, intraspecific variation due to environmental variables, large-scale adaptive radiations, or individual ontogenetic development, affect the interpretations made by the micropaleontologist.

It should be apparent that a given taxon cannot be adequately denoted by reference to a type specimen (or specimens) alone. A number of specimens should be utilized to define the concept of a species (or other lower taxonomic category) and the term **hypodigm** has been suggested by Simpson (1940) for those individuals chosen collectively from the original material which are considered to be representative of the population of the new taxon. Yet, all too often we read in past and current literature of slight differences which a specimen (or a few specimens) has from the holotype, the type specimen, of another species and of the consequent "necessity" of erecting a new species on this basis. Such static, rigid concepts cannot be applied to highly variable members of an evolutionary continuum. Only by recourse to a number of individuals (the more the better) can the intrapopulation variation of a given taxon be determined with satisfactory accuracy. The decision as to whether a given species is conspecific with another is, in the last analysis, based on the inferred range of variation for the whole taxon as a result of studying adequate comparative material, not on the morphologic similarity to any given specimen (holotype, or other). It is the population and inferences drawn from it, not the holotype specimen itself, which must serve as the ultimate criterion of identification and, ultimately, classification of a taxon. The typological approach is inextricably linked with philosophical idealism, as Simpson (1961) has pointed out, and as such has no place in modern paleontologic thought.

It is generally accepted today that classification should be based upon phylogeny. We may paraphrase Simpson (1961) in listing several criteria of strictly Darwinian taxonomy which should be utilized in attempts to erect a classification founded on phylogeny:

(1) Taxonomic groups are the results of des-

cent with modification, or phylogeny.

(2) Each valid taxon has a common ancestry.

(3) The fundamental, but not sole, criterion for ranking of taxa is propinquity of descent.

Populations, not the characters observable on the individual forms, are the things classified. The characters chosen to define the taxa are to be interpreted as showing evidence of phylogenetic affinities and to be ranked in accordance with their probable bearing on nearness of descent. The same character may in one group characterize a genus or a family, whereas in another a species or subspecies These characters are not *a priori* determinable, but their rank must be ascertained as a result of experience. The individual is merely referred, by inference, to a population of which it is a small (in the case of most marine microorganisms an infinitesimal) part. The variation which one observes in populations of different species is an inherent part of their nature and definition. Types serve merely as the legislative requirement of nomenclature. Observations of individual morphology and other somatic attributes of the species will aid in determining whether the evolutionary definition is met by a given population.

Even in instances where a strict application of quantitative data is not made, the fundamental approach of taxonomic studies is statistical in nature. This is because the observations on populations in nature can, at best, provide only partial information on the taxon of which it is a member. It is not the characters and general similarity between individuals which are of primary significance in determining membership in a given taxonomic category, but rather the *relationships* which these characters express which are of main importance. These relationships are evolutionary; in short, they are phylogenetic.

Supraspecific taxa are delimited on the principal of monophyly — all members of a taxon having a single phylogenetic origin. The pragmatic criterion of monophyly is derivation from an ancestral taxon of the same, or lower, rank (see Simpson, 1961).

Consider also the difference in material available to the specialist working on Recent faunas and the micropaleontologist working with fossil forms. Where the former has access through accurately collected plankton tow material and ocean bottom sediment cores to a suite of specimens representing all shades of morphologic development from genesis through gerontism, the paleontologist is limited to death assemblages — **thanatocoenoses**. Thus, where the Recent plankton specialist has an advantage in making observations on living populations and drawing the most logical inferences with regard to the taxonomic composition of Recent faunas, he is at the same time at a distinct disadvantage in utilizing these observations as the foundation of a coherent, logically founded classification. The reason for this is the time element which is lacking in his studies. For example, the calcareous and siliceous planktonic species in the present-day oceans are the result of a long sequence of phylogenetic events in the Tertiary, and, as classification is to be based upon phylogeny, only by taking cognizance of these events can a satisfactory classification of these organisms be formulated.

Phyto- and zoogeographic data are of further importance in providing information on taxonomy and classification. Ecologic and paleoecologic data may provide criteria whereby the difficult problem of distinguishing between convergence (which implies similar ecologic conditions among forms of unrelated phylogenies) and parallelism (which refers to the independent acquisition of similar structure in forms having a common genetic origin) may be resolved.

In the elucidation of characters distinctive of a given specific or subspecific taxon, geographic variation in the character — the **geographic cline** concept — may provide a clue as to the nature and limits of variation within that taxon. It is primarily an evolutionary and not a taxonomic concept and refers to an intraspecific gradation in measurable characters. Various types of clines have been recognized, such as the geocline (geographic), ecocline (ecologic), chronocline (successional). Examples of geoclines in the Cenozoic planktonic foraminifera include the latitudinal modification of the umbilicus and umbilical collar and the height of the conical angle (i.e. degree of convexity) within a given species and between successional species (chronocline); variation in development of intra-um-bilical "teeth" in the species *Neogloboquadrina dutertrei*. The most obvious example of an Upper Cretaceous geographic cline is the marked difference in surface ornament in the planktonic foraminifer *Rugoglobigerina* from spinose to hispid in high latitudes to meridionally oriented rugosities in low latitudes.

An evolutionary classification should be interpreted as being consistent with phylogeny. This point has been stressed by Simpson in various works and is repeated here perhaps to the point of redundancy. And one of the main problems of morphologic classification based on phylogeny is the selection of characters which are homologous or parallel in nature, not convergent, since homology is always valid evidence of affinity.

The key to understanding the evolution of microfossil groups lies in an understanding of their basic biology which can be gained from studies in both the laboratory and field and on living and dead specimens. An understanding of the conditions governing growth in microfossils is of paramount importance in the micropaleontologist's attempts at paleoecologic reconstructing. Recent success in culturing planktonic microorganisms under laboratory conditions argue well for future studies in this area. Morphologic studies suggest that phenotypic variation is related to environmental factors and that taxonomy is a function of test ultrastructure and biomineralization. We may expect morphologic studies on high-latitude planktonic forms so that they may be utilized with greater precision in paleoclimatic studies. The general potential of morphological research in paleoecology—paleooceanography is high and this generally unexplored field of research is a fertile area for creative ideas and new techniques in the years ahead.

Plankton evolution

The abundance of planktonic microfossils and the relative completeness of the stratigraphic record in the deep sea render these organisms ideal for evolutionary studies. Recent studies have been made on the factors governing temporal fluctuations in phyto- and zooplankton diversity, and rates of evolution, and extinctions. Species diversity and rates of evolution among calcareous plankton exhibit

a positive correlation with the $^{18}O/^{16}O$ paleo-temperature curve, suggesting that climate is one of the primary factors influencing the evolution of plankton (Berggren, 1969; Haq, 1973).

With recent refinements in biochronology it is now possible to make refined studies on phylogenetic trends within evolving lineages. Although these trends have usually been based upon morphogenetic changes in the external skeleton, i.e. are essentially a measure of morphologic character change, recent application of amino acid biochemistry is providing a promising area of research into the genetic relationships within and between microfossil groups (King and Hare, 1972a,b; King, 1975). This type of study should yield data allowing a more "natural" classification of taxa and thereby provide a more realistic basis upon which to conduct investigations on evolution.

Biochemical oceanography

The relationship between marine geochemistry (which until recently has been the sanctuary of the physical chemists) and micropaleontology represents a fertile area of research. Perhaps the most apparent common ground of these two fields lies in the area of carbonate dissolution. Carbonate dissolution encodes a significant message representing a potential basic environmental hazard in the form of excess CO_2 by the early part of the 21st century. Such excess CO_2 in the atmosphere should be eventually neutralized in the ocean by the production, sedimentation and remineralization of carbonate-producing planktonic organisms. Remineralization rates of biogenic carbonate, which controls the alkalinity of the ocean, depend upon residence time in the undersaturated water column and absolute dissolution rates; we may expect considerable research efforts in this field in the future.

ECOLOGY: PALEOECOLOGY

Ecology is the study of the relationships between organisms and their environment. Thus it deals with nearly all levels of organization of life, from the individual organism to the whole community of organisms living in an area to the effect of climate and geological processes on these organisms.

The marine ecologic system, the marine ecosystem, includes organisms and the factors making up the physico-chemical environment in a system similar in many ways to the organization of an individual organism. As in the case of the interrelationship between organs in a body, the arrangement of components in the marine ecosystem are not haphazard. There is a history of development, a particular spatial orientation (for example a water mass), a time factor in the operation of the system, and the involvment of specific energistic sequences, such as food chains (trophic resource regimes) or chemical cycles, etc. This organization could be summed up as four types of orderliness:

Evolutionary: Ecosystem components are products of organic evolution of individuals adapted or adjusted to that particular environment. Evolution produces biotic communities of coexistent species of plants and animals mutually adjusted (or adjusting) to each other and their milieu.

Spatial: The spatial arrangement of coexistent species in a system is determined by the **ecological niche** of each species, which simply is how and where each species "makes its living". A niche thus includes the physico-chemical habitat of a species as well as the adaptive strategy it employs to succeed in that habitat. Another means of defining spatial orderliness in an ecosystem is by **stratification**; for example marine communities are often arranged in vertical "layers", such as the microplankton in the photic zone of the ocean.

Temporal: Organisms in the marine ecosystem are not randomly active during one day, or one year. There is a periodicity, or seasonality which species obey and which often allows more species to inhabit an area as long as they have different or slightly overlapping activity periods. The seasonal blooms of phytoplankton groups vary slightly but sequentially to allow the large number of microplankton species to coexist throughout the year in the photic zone. Vertical migration of some plankton is tuned to the phases of the moon, and near-shore microbenthos is affected by tidal cycles. Reproductive seasonality is

another example of temporal orderliness in the marine ecosystem.

Metabolic or trophic: This describes the patterns of energy and material transformation in the marine environment. The chemical cycling of nutrient or other elements in the ocean (and from the land) and through the biotic community proceeds in an orderly sequence so that few chemical elements are permanently lost from the community. The $CaCO_3$ in the tests of marine microorganisms is a major contributor to the CO_2 cycling in the oceans.

From this short discussion the immense task of studying the complex ecology of any marine group is apparent. Nevertheless, using information on present ecologic relationships it is possible to examine fossils and interpret their relationship to paleoecosystems. The study of fossils and their paleoenvironment is called paleoecology; likewise the study of past distributions and geographic relations is termed paleobiogeography, which in one sense is really the charting of the distributions of ancient environments and their components. Paleooceanography, then, is really the study of the components of ancient marine ecosystems.

Until recently the principle of uniformitarianism has been the keystone of all paleoecologic interpretation. This principle states that processes and relationships operative in modern systems can be extrapolated to interpret analogous systems in the geologic past. Recent advances in benthic paleoecology, such as the studies of depth distributions of benthic ostracodes and foraminifera, have demonstrated that the spatial (depth) distributions of ancient benthic species are, in fact, not identical to the depth distributions of their Recent analogues. Such studies point out several caveats in paleoecologic studies worth mentioning:

(1) The more specific the features studied, the less likely they will be strictly analogous in past systems.

(2) The organism—environment cause-and--effect relationships may be misleading, causing one to doubt the principle of uniformitarianism, when in fact, the proper "cause" has not yet been found.

(3) The modern ocean is a geologically re-

cent feature, particularly in its temperature structure; hence, comparisons between pre-Miocene and Recent systems become tenuous if the "real" values of variable, for example temperature or salinity, in the past are not measured, but only inferred by analogy.

Despite these problems, the field of marine paleoecology or paleooceanography, is one of the fastest growing and creatively approached aspects of marine micropaleontology today.

Sea-floor distribution -- paleobiogeography

Oceanic circulation is dependent upon dynamically interrelated aspects of geography and climate. The concept that the earth is made up of a number of interlocking plates whose geometry has been subject to cyclic geographic rearrangements through time is the central hypothesis of plate tectonics — an idea which has virtually revolutionized the study of earth history. Although the exact timing of some of these interrelated events (particularly the earlier ones) during the most recent re-arrangement which began almost 200 m.y. ago is not always known, the relative sequence and general relationships are adequately known.

Reduced to its simplest scenario the earth appears to have consisted essentially of a single "supercontinent", **Pangaea**, surrounded by "superocean" **Panthalassa** in the late Paleozoic. In the early Mesozoic two major components of this single land mass, **Gondwanaland** (in the south) and **Laurasia** (in the north) were separated by a triangular reentrant of Panthalassa, the **Tethys Sea**. Northward rifting of fragment(s) from northern Australia which later became part(s) of the Asian continent began around 180—160 Ma, approximately simultaneously with the initial opening of the central Atlantic by the rifting of the North American and African continents. The subsequent fragmentation of Gondwanaland (separation of India from western Australia) and opening of the Indian Ocean) began in the late Jurassic—early Cretaceous (ca. 130 Ma) approximately simultaneously with the opening of the South Atlantic. These events herald the beginning of the evolution of the present-day oceans and signal the gradual diminution of Tethys, and

its ultimate evolution into the Mediterranean Sea — a process which was to take over 100 m.y., and was completed only recently within the life span of Man.

The transformation of biogeographic distribution patterns into fossil distribution (taphogeographic) patterns is a function of differential transport and preservation. Differential dissolution affects the preservation of planktonic organisms and has significance for biostratigraphy and paleooceanography. One of the major research areas at present is the role of selective dissolution on preservation of calcium carbonate and siliceous skeletons of microfossils (Berger, 1974). Recent studies have shown that coccolith ooze on the ocean floor and the well-preserved suspended coccoliths in the undersaturated water column are the result of accelerated sinking of coccospheres and coccoliths in the fecal pellets of small zooplankton (Honjo, 1976). Thus, the community structure of the euphotic layer is replicated with high fidelity in the depositional thanatocoenose. Destructive and non-destructive predation may be expected to play an important role in determining the nature of sedimentation patterns in the deep sea. The transfer of nutrient matter to the deep sea is also dependent upon fecal transport. Since nutrient supply is one of the major factors involved in controlling fertility in the upper layers of the ocean and thus sedimentation patterns on the bottom, we may expect to see an enhanced interest in this type of study with a view to elucidating patterns of paleofertility.

Taphogeographic sea-floor distribution studies are being conducted in most of the major planktonic groups, particularly the foraminifera, radiolaria and coccoliths, and these form the basic data by which many of the major paleogeographic, paleooceanographic—paleobiogeographic events have been dated and the history of regional and global oceanic circulation patterns have been delineated (Atlantic: Berggren and Hollister, 1974; Antarctic: Kennett and others 1975; global: Berggren and Hollister, 1977). Distinct latitudinal control on species diversity is seen within calcareous plankton whereas siliceous plankton exhibit a bimodal diversity maximum pattern, in the equatorial Pacific regions and again in polar regions.

Biogeographic distribution patterns of early Cenozoic calcareous nannoplankton and planktonic foraminiferal assemblages have been delineated for the North and South Atlantic (Haq and others, 1977). The latitudinal distribution through time allows recognition of certain assemblages which can be used as environmental indicators. On the assumptions that the latitudinal differentiation of early Cenozoic calcareous nannoplankton are related to a latitudinal temperature gradient and that the ecologic preferences of these assemblages are relatively stable through time, the latitudinal migrations which were recognized have been interpreted to have been caused by paleotemperature changes. Paleotemperature trends delineated by biogeographic migration patterns in Paleogene calcareous plankton are being correlated with paleotemperature records derived from the oxygen isotopes of marine Paleogene planktonic species, as well as trends depicted by some workers on the basis of terrestrial floras.

The role of post-depositional alteration of microfossils and the formation of deep-sea limestones and cherts through diagenetic changes is being intensively studied (Packham and Van der Lingen, 1973; Schlanger and Douglas, 1974). The diagenetic alteration of deep-sea sediments may lead to a variety of preservational states which, in turn, form the basis of acoustic stratigraphy as used by the geophysicists in mapping deep-sea reflector horizons. This forms but one example of the way in which micropaleontological and geophysical studies complement each other.

Micropaleontologic research in paleooceanography perhaps best illustrates the importance of the marriage between paleontology and geophysics. The microfossils are important as environmental indicators and as biostratigraphic tools. But the interpretation of past distribution patterns depends ultimately upon realistic paleogeographic and paleobathymetric reconstructions. The methodology exists for these reconstructions and the biostratigraphic and paleoenvironmental evidence of the microfossils places temporal constraints upon the spatial reconstructions of the geophysicist within the framework of plate tectonics. It is now possible to elucidate the history of circulation and sedimentation patterns in the ocean based on distributional patterns in various microfossil groups.

FUTURE TRENDS IN MARINE MICROPALEONTOLOGY — TOWARDS THE 21ST CENTURY

Marine micropaleontology has undergone a series of dramatic changes over the past decade, a trend that may be expected to continue into the foreseeable future. These changes have been primarily in the areas of biostratigraphy and biochronology, as well as in shifts in research emphasis to such areas as paleooceanography (including paleobiogeography and paleoclimatology) and plankton evolution. These changes have been due primarily to recent advances in technology, such as deep-sea coring, computers, and scanning electron microscopy, among others. As a result marine micropaleontology now plays a fundamental role in the interpretation of the history of the marine biosphere as may be seen from its ever increasing integration with marine geophysics and geology.

A comprehensive review of current research trends in marine micropaleontology is beyond the scope of an introductory textbook. At the same time we believe that the fact that micropaleontology is currently undergoing a "revolution" and "rejuvenation" warrants a brief overview of some of the more important trends of current research in this field. A more comprehensive, up-to-date summary (Berger and Roth, 1975) of current research trends will be found cited in the bibliography.

As the present is considered a key to the past by geologists, so too it holds the promise of the future. Where will marine micropaleontology be at the turn of the century? What are the major trends which will characterize its continued development during the last quarter of this century? It would seem that we can expect the major trends to lie within those areas of research which are at present being most actively pursued and which have been identified and briefly described above, as well as within several new areas which are only now in their infancy.

In the area of biostratigraphy and biochronology we foresee continued efforts at extending and refining planktonic biostratigraphic zonations (particularly in high latitudes), correlating them with zones in shallow-water areas based on large foraminifera, and the correlation of biostratigraphy and magnetostratigraphy and their calibration to radiochronology. Advances in this area should come primarily in the Mesozoic and early Cenozoic intervals. These studies will provide an increasingly reliable biochronologic framework within which global geologic and biologic processes may be delineated.

In the field of paleooceanography we include the diverse, but interrelated, areas of paleobiogeography and paleoclimatology. In the field of Quaternary paleoclimatology we may expect continued advances in understanding of the temperature and oceanic circulation over the last 1.6 m.y. as a clue to predicting future climatic trends. In the field of pre-Quaternary paleooceanography the use of oxygen and carbon isotope analyses will add significantly to our understanding of the biogeographic, geochemical (including productivity), climatic and circulation history of the oceans. Studies on the fluctuations in the areal distribution and relative abundance patterns of siliceous and calcareous sediments will continue to yield information on the history of productivity, dissolution and the CCD in the oceans and result in global syntheses of sedimentation and erosion patterns. Here we foresee a trend towards interdisciplinary studies involving micropaleontology, sedimentology and geochemistry.

The application of micropaleontology to studies on the geologic history of continental margins and slopes may be expected to increase in relation to increased exploration in these areas for petroleum and other mineral resources. Concomitant with such studies will be the continued use of various microfossils, particularly benthic foraminifera, in reconstructing the paleocirculation history of deep oceanic water masses and the tectonic history of various depositional basins and the subsidence history of foundered continental fragments in the oceans.

In summary, we may expect that during the course of the last quarter of this century there will be greater emphasis on quantitative analysis of assemblages which will provide greater precision in the areas of bio-

stratigraphy and paleoecology. At the same time improvements in transmission and scanning electron microscopy may be expected to yield further insight into skeletal ultrastructure. The *menage à trois* of marine biostratigraphy, geomagnetism and radiochronology should provide a continually improved geochronologic framework for studies in earth history. Finally, it would seem safe to predict that the increase in interdisciplinary studies of marine micropaleontologists with physical and chemical oceanographers will continue to place micropaleontology in the mainstream of paleo(oceanographic) research.

SUGGESTIONS FOR FURTHER READING

Berger, W.H. and Roth, P.H., 1975. Oceanic micropaleontology: progress and prospect. *Rev. Geophys. Space Phys.*, 13(3): 561—635. [Comprehensive survey of research in oceanic micropaleontology during the period 1970—1975.]

Berggren, W.A., 1971. Oceanographic micropaleontology. *EOS*, 52: 249—256. [A summary of research in oceanic micropaleontology between 1967 and 1970].

Eicher, D.L., 1968. *Geologic Time*. Prentice-Hall, Englewood Cliffs, N.J., 141 pp. [A concise treatment of the concept and growth of geologic time-scale and its application.]

Kummell, B. and Raup, D. (Editors), 1965. *Handbook of Paleontological Techniques*. Freeman and Co., San Francisco, Calif., 852 pp. [Standard reference on general procedures, techniques and bibliographies of use to (micro)paleontologists.]

Moore, R.C. et al., 1968. Developments, trends and outlooks in paleontology. *J. Paleontol.*, 42: 1327—1377. [A comprehensive survey of paleontology (including micropaleontology) until the year 1967.]

CITED REFERENCES

Berger, W., 1974. Deep-sea sedimentation. In: C.A. Burke and C.L. Drake (Editors), *The Geology of Continental Margins*. Springer-Verlag, Heidelberg, pp. 213—241.

Berggren, W.A., 1969. Rates of evolution in some Cenozoic planktonic foraminifera. *Micropaleontology*, 15(3): 351—365.

Berggren, W.A. and Hollister, C.D., 1974. Paleogeography, paleobiogeography and the history of circulation in the Atlantic Ocean. In: W.W. Hay (Editor), *Studies in Paleoceanography. Soc. Econ. Paleontol., Mineral., Spec. Publ.*, 20: 126—186.

Berggren, W.A. and Hollister, C.D., 1977. Plate tectonics and paleocirculation: commotion in the ocean. *Tectonophysics*, 38(1—2): 11—48.

Cox, A., 1969. Geomagnetic reversals. *Science*, 163 (3864): 237—245.

Haq, B., 1973. Transgressions, climatic change and the diversity of calcareous nannoplankton. *Mar. Geol.*, 15(1973): M25—M30.

Haq, B., Premoli-Silva, I. and Lohmann, G.P., 1977. Calcareous plankton paleobiogeographic evidence for major climatic fluctuations in the early Cenozoic Atlantic Ocean. *J. Geophys. Res.*, 82(27): 3861—3876.

Honjo, S., 1976. Coccoliths: production, transportation and sedimentation. *Mar. Micropaleontol.*, 1(1): 65—70.

Kennett, J.P. and others, 1975. Cenozoic paleooceanography in the southwest Pacific Ocean, Antarctic glaciation, and the development of the circum-Antarctic currents. In: J.P. Kennett, R.E. Houtz and others, *Initial Reports of the Deep-Sea Drilling Project, 19.* U.S. Government Printing Office, Washington, D.C., pp. 1155—1169.

King Jr., K., 1975. Amino acid composition of the silicified matrix in fossil polycystine radiolaria. *Micropaleontol.,* 21 (2): 215—226.

King Jr., K. and Hare, P.E., 1972a. Amino acid composition of planktonic foraminifera. A paleobiochemical approach to evolution. *Science,* 1975 (4029): 1461—1463.

King Jr., K., and Hare, P.E., 1972b. Amino acid composition of the tests as a taxonomic character for living and fossil planktonic foraminifera. *Micropaleontology,* 18(3): 285—293.

Mayr, E., 1966. *Animal Species and Evolution.* Belknap Press of Harvard University, Cambridge, Mass., 797 pp.

Packham, G.H. and Van der Lingen, G.J., 1973. Progressive carbonate diagenesis at Deep Sea Drilling Sites 206, 207, 208, and 210 in the Southwest Pacific and its relationship to sediment properties and seismic reflectors. In: J.E. Andrews, R.E. Burns and others, *Initial Reports of the Deep Sea Drilling Project, 21.* U.S. Government Printing Office, Washington, D.C., pp. 495—507.

Schlanger, S.O. and Douglas, R.G., 1974. The pelagic—ooze—chalk—limestone transition and its implication for marine stratigraphy. In: K.J. Hsü and H. Jenkyns (Editors), *Pelagic Sediments: On Land and Under the Sea. Spec. Publ. Int. Assoc. Sedimentol.,* 1: 117—148.

Simpson, G.G., 1940. Types in modern taxonomy. *Am. J. Sci.,* 238: 413—431.

Simpson, G.G., 1953. *The Major Features of Evolution.* Columbia University Press, New York., N.Y., 434 pp.

Simpson, G.G., 1961. *Principles of Animal Taxonomy.* Columbia University Press, New York, N.Y., 247 pp.

Van Hinte, J.E., 1969. The nature of biostratigraphic zones. *Proc. First Int. Planktonic Conf., Geneva,* 2: 267—272.

CALCAREOUS MICROFOSSILS

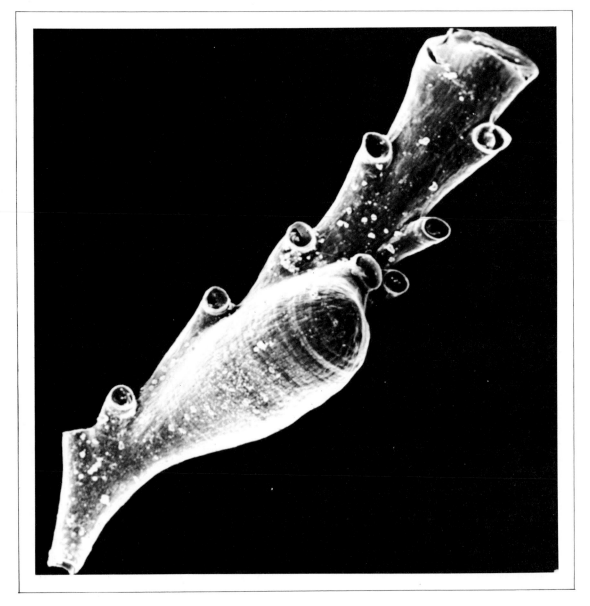

Segment of the cyclostomatous bryozoan *Crisia*

2

FORAMINIFERA

ANNE BOERSMA

INTRODUCTION

A date of 400,000 years B.P. is assigned to a Pleistocene sediment core bottom on the basis of the amino acid content in calcareous microfossils called foraminifera. The age of the opening of the North Atlantic is computed to be close to 120 m.y. B.P. from evidence lent by foraminifera. The temperature of the ocean 50 m.y. ago is estimated from the isotopes of oxygen in the shells of foraminifera. These and many other types of oceanographic information can be derived using the microscopic shells of this protozoan group.

Living both on the bottom and floating in the water column, these microorganisms presently inhabit the ocean from 5 to over 5,000 m depth. The bottom-dwelling forms have existed since Cambrian time and the planktonics since the Mesozoic. They are presently, along with the Ostracoda, the best known and most comprehensively studied of all the calcareous microfossils.

Foraminifera are roughly divided into three major groups: the planktonics, the smaller benthics, and the larger benthics which are distinguished by their larger size and by complex interiors which are visible only in thin section.

HISTORY OF FORAMINIFERAL RESEARCH

The study of any fossil group involves several steps. First the fossils are discovered and described, although their systematic position may not be determined until much later. As more descriptions accumulate, a system of organization gradually evolves. Within this system more and more taxa are described until a classification might be warranted based on new and improved information. The amassing of numbers of species descriptions then provides the groundwork for later interpretive studies. Foraminiferal research is presently at the stage where most cataloguing of species has been accomplished and researchers are in the process of interpreting this mass of data in light of recent theories of earth history.

Larger foraminifera of the genus *Nummulites*, which grow to a size of several millimeters, are abundant in the limestones used by Egyptians to build the pyramids at Gizeh, and their lenticular form led later travellers (5th to 1st century B.C.) to consider them as petrified lentils eaten by the slaves who built the pyramids.

Smaller foraminifera were first described and illustrated in the 16th and 17th centuries. Before the invention of the microscope in the late 1700's, early naturalists relied on ground magnifying lenses to view foraminiferal shells, some as small as 0.1—0.2 mm. Because foraminifera sometimes resemble the shells of gastropods and cephalopods in their coiled form and in their intricacy and beauty, early students mistook foraminiferal microfossils for tiny cephalopods or miniscule mollusks. Thus, in the 1700's foraminifera were still considered as fascinating oddities of nature, inorganic minutiae.

The French naturalist Alcide d'Orbigny revolutionized the study of foraminifera as well as several other branches of paleontology. In 1826, d'Orbigny described foraminifera as cephalopods and published an account of them in his principal work *Tableau Méthodique de la Classe des Céphalopodes*. Microcephalopods were divided into three classes, one of which we retain today as a designation for the whole group — Foraminifères, or

Foraminifera. D'Orbigny's lasting contribution was the first detailed classification of this group (1826).

After the monumental world cruise of H.M.S. *Challenger*, H.B. Brady illustrated the foraminifera dredged from the sea floor. His illustrations and a summation of the foraminiferal literature published up to that time appeared in the *Challenger* Reports in 1888. This publication has since been revised by Barker (1960) whose study remains the definitive work on Recent foraminifera of the world's oceans.

The second encyclopedic work on foraminifera was produced by R.J. Schubert, who summarized all work up to 1921.

Several workers stand out as the initiators of modern foraminiferal studies. In the U.S.A. at the end of World War I, the most prolific student of the foraminifera was Joseph August Cushman. Cushman established a research laboratory in Sharon, Massachusetts, published voluminously, trained many students, wrote one of the most influential texts in the field (*The Classification and Economic Use of Foraminifera*), and established the first journal of foraminiferal studies, *Contributions of the Cushman Laboratory for Foraminiferal Research*. Cushman's key to the identification system has been followed up to the present time, with some revisions and improvements.

A major stimulus to foraminiferal research and one that was to change the direction and nature of foraminiferal studies was provided after World War I by the oil industry. In their intensive search for oil, they came to appreciate foraminifera as invaluable aids in the determination of the age as well as the depositional environment of strata. The impetus given by oil companies accelerated the study of foraminifera, in addition to many other microfossil groups.

Modern micropaleontology really began in the 1950's. Following the lead set by the oil companies the establishment of environmental indicator faunas was pioneered in the 1940's to 1950's by Fred Phleger and Orville Bandy. These indicator faunas remain today one of the primary means of interpreting the depositional environment of ancient sediments.

Modern biostratigraphy using foraminifera also blossomed during the 1950's. Before this time biostratigraphy of sediments was based on the stratigraphic ranges of benthic foraminifera. These ranges, however, often proved to be time-transgressive and hence were not always useful when correlating from one site to another. The first zonation scheme for the planktonic foraminifera was published by the Russian worker Subbotina for sections in the Caucasus Mountains. Then the seminal work by Hans Bolli (1957) on Tertiary sections from Trinidad provided the basis for most later low-latitude zonations. Since this time several workers have refined our biostratigraphic subdivisions, notably W.H. Blow and E. Pessagno in the 1960's and W.A. Berggren in the 1970's.

Applications of micropaleontological techniques to the ocean and to deep-sea cores also began during this period of the expansion of foraminiferal biostratigraphy and paleoecology. Schott (1935) and Cushman and Henbest (1942) published some of the first studies of foraminifera from deep-sea cores. Then Ericson and Wollin (1956, and later) published extensively on deep-sea cores, proposing the first extensive scheme for zonation of Atlantic deep-sea cores and for interpreting climatic change in these cores. Emiliani and Edwards (1953) showed the usefulness of oxygen isotopes in foraminiferal tests down deep-sea cores to recognize climatic changes and the base of the Pleistocene in deep-sea cores. Since the time of these early studies, deep-sea core studies have expanded enormously and were in part responsible for the birth of the Deep Sea Drilling Project which has recovered much longer deep-sea sediment cores by drilling with an oil rig. Enormous resources, both human and financial, are now going into the study of marine cores to interpret present and past oceanic environments and continental configurations.

The work of biologists towards clarifying the nature and systematic position of the foraminifera is significant. In the 1830's Felix du Jardin concluded that foraminifera were too primitive in their cellular makeup to be cephalopods or other such biologically complex organisms. Du Jardin proposed the name Rhizopoda based on his discovery that the protoplasm of foraminifera forms branches called pseudopodia, or rhizopods;

this name was used for nearly a century afterwards.

The majority of biological work on foraminifera took place at the end of the 19th century, principally by the German and English biologists Rhumbler, Schaudin and Lister, who performed some of the first culture experiments with foraminifera and began the arduous study of foraminiferal life cycles. Their work has been continued in this century by Arnold, Jepps, Hedley, Myers, LeCalvez, and most recently by Lee and Bé, who has finally succeeded in culturing planktonic foraminifera.

BIOLOGY

Systematic position

Foraminifera belong to the Phylum Protozoa (Fig. 1). Unicellular, or acellular, organization is the single feature common to all the various members of this phylum, which is further divided into classes on the basis of the type of locomotor apparati present. Since foraminifera are non-flagellate, but possess flowing protoplasmic extensions termed **pseudopodia**, they are placed in the Class Sarcodina which includes the simplest of the Protozoa with respect to their cellular organization and specialization; but shelled sarcodines create skeletons of incredible beauty and structural complexity. (The Radiolaria, discussed in Chapter 9, are also skeletal-secreting sarcodines.)

Foraminifera are distinguishable from other Sarcodina by the possession of mineralized shells (although there are a few foraminifera that do not construct a shell but are covered by an organic material). In this chapter we will deal only with those forms that secrete or agglutinate a mineralized covering, since only these are preserved as fossils.

The cell and its contents

The cell of a foraminifer is a protoplasmic mass bounded by a limiting membrane. Some of the cell protoplasm is encapsulated in a secreted or agglutinated covering, the **test**, which may consist of one or more cavities termed **chambers**. When there is more than one chamber, a wall dividing one chamber from the next is termed a **septum**. Protoplasm is nevertheless continuous between chambers through a hole in the septum, the **foramen**, from which the name foraminifera (Latin *foramen* = hole; *ferre*, to bear) is derived. The protoplasm extends outside the test through the opening, or **aperture**, and surrounds the test in a mass of branching, anastomosing pseudopodia (Fig. 2).

Fig. 3 illustrates the contents of the protoplasm of a planktonic foraminifer. Like the Radiolaria, foraminifera may be uni- or multinucleate. The nucleus is generally round and consists of a nuclear membrane, nuclear sap, chromosomes, and the nucleoli which are the RNA-containing bodies. Although nuclei vary in size and number with different species, typically, single nuclei are larger than multiple nucleii.

The cell includes several typical protozoan structures, termed organelles. They consist of: the **Golgi bodies**, which play a role in cell secretions: **mitochondria**, sites of respiration; **ribosomes**, which contain RNA and are the sort of factory of protein synthesis; and **vacuoles**, which are fluid- or gas-filled droplets.

The protoplasm is generally colorless throughout but contains small amounts of organic pigments, iron compounds, brown and red deposits of fatty material, brown excremental particles, and green splotches produced by the presence of symbiotic algae living within the foraminiferal test. Many of these colorations can be detected in recently living specimens with a simple light microscope.

Foraminifera accomplish the life-sustaining and perpetuating activities of nutrition,

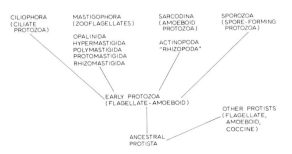

Fig. 1. Possible interrelations among protozoan groups showing the systematic position of the Rhizopoda, which include the Foraminifera.

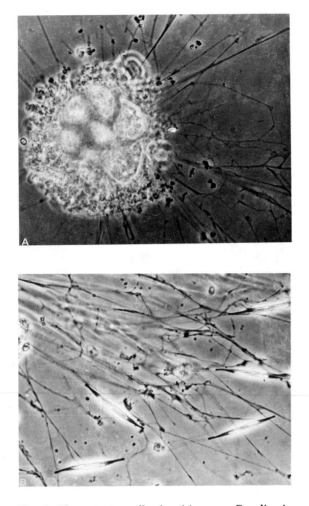

Fig. 2. The extant smaller benthic genus *Rosalina* in culture: A. The test is surrounded by food particles and debris, the pseudopodia radiate out from the test and can be seen to anastomose. × 350. B. Enlargement of the pseudopodia, showing diatoms and bacteria (dots) trapped in the pseudopodial network; these were later ingested by the foraminifer. × 420. (Courtesy D. Schnitker.)

motion, regeneration, respiration, reproduction, growth and test construction, all within the confines of this tiny protoplasmic mass.

The living animal

Nutrition

In a classic paper on protozoan motion Jahn and Rinaldi (1959) observed that pseudopodia are composed of thin filaments of a gel-like substance bent back on themselves somewhat in the manner of a conveyor belt. Like a conveyor belt, pseudopodia then move particles into and out of the inner protoplasm. This conveyor-like motion is termed **streaming** and is probably the most characteristic feature of foraminiferal protoplasm. Constantly streaming, the granular protoplasm issues from the aperture of the test; it may quickly withdraw into the test by this same streaming motion which resembles the movement of amoebas as they change their shape. The usefulness of streaming is particularly clear if we consider one activity of a foraminifer, food-gathering and excretion. Food is contacted and generally absorbed at the surfaces of the extended pseudopodia, chemically broken down into utilizable compounds, and streamed into the endoplasm. Similarly, material to be excreted agglomerates into small brown particles, is streamed out of the endoplasm through the pseudopodia, and is released into the water or dropped onto the substrate in trail-like fashion. In culture foraminifera feed several times a day, generally at the last chamber. Ingestion of food occurs outside the test except in some planktonic forms and some members of the benthic Miliolidae which have large apertures through which the food is drawn directly into the test. Several species of both planktonic and benthic foraminifera co-

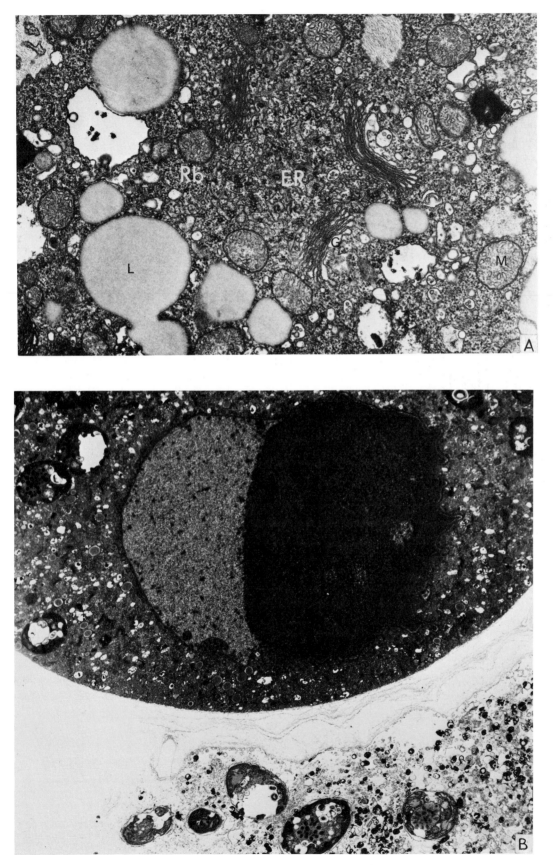

habit with symbiotic algae, which by their photosynthesis are thought to provide a source of nutrition for the foraminifer.

Some benthic foraminifera feed by filter-feeding (*Bathysiphon*), but most are deposit-feeders or grazers, and at least one genus (*Entosolenia*) is known to be parasitic. In culture benthic foraminifera feed on diatoms, algae, bacteria, and particulate organic matter, as well as capture food particles from the protoplasm of a host. Planktonic foraminifera apparently ingest planktonic diatoms and other algae, silicoflagellates, copepods, and other microplankton. Foraminifera in turn are preyed upon by microscopic gastropods, pelecypods, pteropods, crustaceans, worms, and perhaps zooplankton as large as copepods. In many cases the foraminifera are merely part of the bulk sediment another benthic invertebrate is ingesting as it grazes the ocean bottom.

Movement

Benthic foraminifera may be sessile (*Cibicidella*, Fig. 18B1) or vagile. They can move by means of their pseudopodia on ocean bottom sediment, on algal fronds, or other substrates. Indeed it is often difficult to keep track of them in culture dishes as they have the annoying habit of moving up the edges of these dishes and out. Their movement may average close to 1 cm per hour, a respectable rate for an organism of approximately 0.5 mm in length. Several species of Recent planktonic foraminifera migrate through the water column up into the surface zone. This motion is not accomplished by any sort of pseudopodial flapping, but probably by changes in the chemistry, such as the gas content, of the protoplasm.

Reproduction

One of the most puzzling aspects of the foraminifera is a life cycle which most closely resembles that of certain plants. It will facilitate understanding the different modes of reproduction if we first consider a generalized life cycle of a foraminifer (Fig. 4). A life cycle is termed heterophasic when it characteristically contains two different phases, or types of reproduction and maturation. In many plants an asexual phase, or generation, alternates with a sexual generation. Among some

foraminifera there is this same alternation of an asexual followed by a sexual generation. As seen in Fig. 4 the two principal phases are termed **schizogony**, the asexual phase; and **gamogony**, the phase involving sexuality. There are variations on this general cycle, the most frequent of which is absence or suppression of the phase involving sexuality. The benthic genus *Rosalina*, for instance, reproduces only asexually in culture. Most researchers assume that sexuality is a secondary addition to the life cycle and that foraminifera were initially asexually reproducing organisms which sometime in the late Paleozoic acquired the ability to reproduce sexually.

Individuals resulting from schizogony characteristically have a larger initial chamber, the **proloculus**, than the schizonts resulting from gamogony. The young gamonts with the larger proloculus are termed the **macrospheric generation**, while the individuals with smaller proloculi are called the **microspheric generation**. Exceptionally the microspheric generation may have a proloculus equal to or larger in size than the macrospheric generation. We call this production of two different initial morphologies in the course of a bi-phasic life cycle, alternation of generations, and the two morphologically distinct tests are termed **dimorphs**.

Dimorphic pairs are found among smaller and larger benthic foraminifera, but have not been recognized in planktonic genera. The microspheric generation with the smaller proloculus is termed the B form, whereas the megalospheric phase is called the A form. There are genera (*Cibicides*, *Triloculina*, and *Elphidium*) in which a third generation (see Fig. 4D) commonly occurs, which is called A1, the second megalospheric generation. In fossil populations the first dimorphic genera were recognized from the late Paleozoic. In most samples the A generation is more frequent and further suggests that sexuality is a secondary reproductive pattern and that asexual reproduction was the original and now the more frequent reproductive mode of the majority of foraminiferal species.

Growth

Growth has been witnessed in only a few shallow-water benthic genera. Non-periodic growth is characteristic of the single-cham-

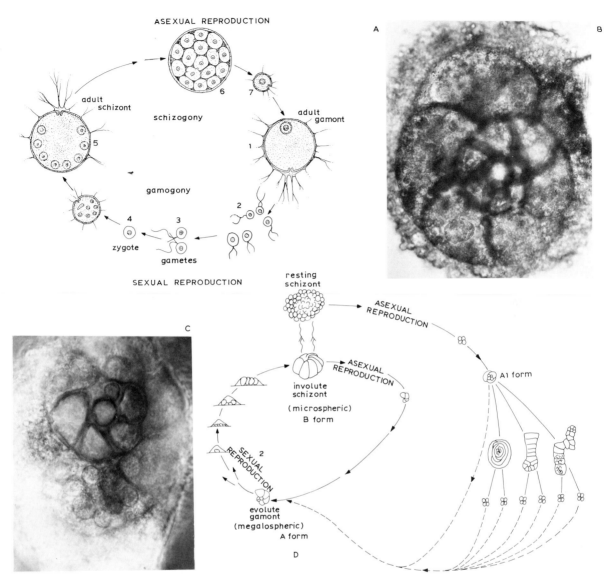

Fig. 4. Reproduction of foraminifera. A. Schematic drawing of a simple heterophasic reproduction cycle includ-ing the asexual phase (schizogony) and the phase involving sexuality (gamogony). Briefly, an adult (*1*) produces many bi-flagellate reproductive cells, gametes (*2*); when two join (*3*), the resulting individual, the zygote (*4*) then matures into an adult (*5*). Division of the protoplasm of this adult into several hundred offspring (*6*) is called schizogony. The emerging young (*7*) then grow into adults (*1*) and the cycle may be repeated. The round inclu-sions in the foraminiferal cells are the reproductive nucleii which divide during reproduction so that offspring re-ceive some original material. B. Within the benthic genus *Rosalina*, reproducing in culture, can be seen the devel-oping, well-defined offspring (lower left). × 400. C. Enlargement of the spiral region of the test (B) shows multiple offspring in the last whorl of the maternal test; this corresponds to phase *6* (schizogony) in the preceding diagram. × 420. (Courtesy D. Schnitker.) D. The production of several different morphologies during the reproductive cycle of the benthic genus *Cibicides* in culture; in this case, several different morphologies can be produced, and these multiple morphotypes can present problems to the taxonomist.

bered genera. The growth process generally takes place as depicted in Fig. 5. This is a case of periodic growth, as it involves a periodic increase in size by precipitation and accretion of mineral matter in the building of a new chamber. When a new chamber forms, the aperture of the last chamber generally be-comes the foramen of the new chamber and permits connection of the protoplasm be-tween the two chambers.

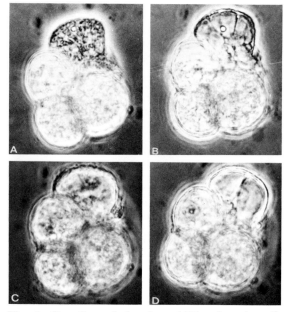

Fig. 5. Growth and chamber addition in a juvenile *Rosalina*. A. Juvenile individual, four hours old; naked cytoplasm has extended out of the test and now outlines the shape of the new chamber to be built. × 1440. B. One hour later, a membrane has formed at the surface of the cytoplasm. × 1440. C. One hour later, the calcification of the wall at the membrane has begun; the new chamber is partially filled with protoplasm. × 1360. D. Two hours later, calcification of the wall is nearly complete, and the new chamber is filled with protoplasm. × 1360. (Courtesy D. Schnitker.)

During chamber formation the foraminifer may encyst by gathering around itself a protective mass of sediment and/or the tests of other organisms. Foraminifera have been observed to encyst during chamber formation, feeding, mating, or as a resting body. Many living forms collected from the sea bottom come back to the lab still camouflaged in this mass of debris and protoplasm.

Test mineralization

We do not completely understand the process of test mineralization by the foraminiferal protoplasm. Test mineralization, at least in the major group of calcareous foraminifera, is apparently controlled by a proteinaceous organic template, a sort of anchor layer at the surface of the protoplasm which directs the growth of the calcite or aragonite crystallites. Attraction of the calcium ion from sea water by the amino acids in the protein template, is probably the key to mineralization, as the Ca ion then attracts the carbonate cation.

Although each planktonic and benthic foraminiferal species has a unique amino acid composition, the marked difference between amino acids in the calcitic and aragonitic species may help to explain the precipitation of aragonite in some foraminifera.

TEST MORPHOLOGY

Foraminifera are animals which build a shell; and for paleontologists the characteristics of the shell are the primary features which can be used to distinguish one species from another and hence to use these distinctions to form interpretations of time or environment.

Wall structure

The most readily obvious feature distinguishing one foraminifer from another is its wall type. Whether the foraminifer builds its test walls by cementing together exogenous grains, by carbonate mineralization, or by some combination of these two processes separates the three primary foraminiferal groups, the **agglutinated**, the **calcareous**, and the **microgranular** foraminifera. Original wall structure is considered to be a genetically stable feature of the test and hence to be relatively unaffected by environmental fluctuations.

Agglutinated wall structure

The geologically oldest method of test construction is the agglutination of particles together to form an external covering. Agglutinating foraminifers cement particles onto a layer of **tectin**, an organic compound composed of protein and polysaccharides. Extralocular protoplasm, calcite, silica, or ferruginous material are used to cement particles causing grains to be loosely bound in place or permanently cemented within this mineralized matrix. The grains involved may include assorted mineral particles such as sand grains, the tests of other microorganisms, distinctive sedimentary particles such as oolites, or microgranules of calcite scavenged off the deep-sea floor (see Fig. 6).

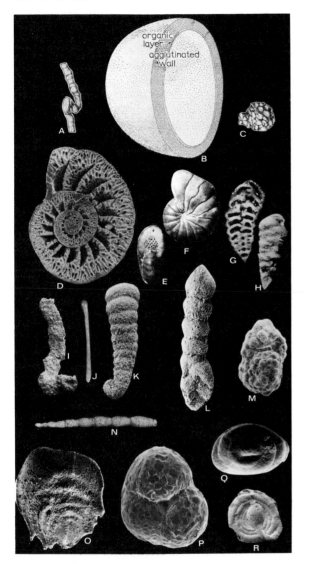

Fig. 6. Agglutinated foraminifera and wall structure: A. *Tolypammina* ✕ 25). There are two primary groups of agglutinated foraminifera, the simple, non-septate forms belonging to the superfamily Ammodiscacea, and the more complexly coiled and septate forms in the superfamily Lituolacea; compare the simple structures of figures A, B, C, I, J, Q and R to the more complexly coiled forms in this figure. B. Schematic drawing of the structure of a simple agglutinated wall; a basal organic layer is overlain by particles which grade from finer inside to coarser at the outside. C. *Lagenammina* (✕ 19) from the early Cretaceous of the North Atlantic. (Courtesy H. Luterbacher.) D. Cross-section of the planispirally coiled *Cyclammina*, showing the spongy, convoluted infolding of the chamber walls, called labyrinthic structure; Recent, Gulf of Mexico. ✕ 15. (Courtesy D. LeRoy.) E,F. End and front view, respectively, of *Cyclammina*, also from the Recent of the Gulf of Mexico. ✕ 15. (Courtesy D. LeRoy) G. Cross-section through the wall of *Gaudryina* showing the simple wall structure typical of agglutinated forminifera; both major superfamilies contain species with simple and complex interiors; Recent, Gulf of Mexico. ✕ 15. (Courtesy D. LeRoy.) H. Whole specimen of the above. I. *Rhizammina* (✕ 50), early Cretaceous, North Atlantic. (Courtesy H. Luterbacher.) J. *Hyperammina* (✕ 15), Recent, Gulf of Mexico. (Courtesy D. LeRoy.) K. *Lituola* (✕ 15), Recent, Gulf of Mexico. (Courtesy D. LeRoy.) L. *Ammobaculites* (✕ 50), middle Cretaceous, South Atlantic. M. *Recurvoides* (✕ 15). This coarsely agglutinated test is composed largely of the smaller tests of foraminifera; Recent, Gulf of Mexico. (Courtesy D. LeRoy.) N. *Reophax* (✕ 15), Recent, Gulf of Mexico. (Courtesy D. LeRoy.) O. *Vulvulina* (✕ 70), late Eocene, Rio Grande Rise, South Atlantic. P. *Trochammina* (✕ 135), Recent, from the sub-Antarctic core RS12-5. (Courtesy R. Fillon.) Q. *Glomospira* (✕ 76), middle Cretaceous, South Atlantic. R. *Ammodiscus* (✕ 15), Recent, Gulf of Mexico. (Courtesy D. LeRoy.)

Some species will select grains of a specific size and composition to affix to the test; for example the genus *Psammosphaera* places a single elongate sponge spicule across the center of its test. Other agglutinated forms are non-selective and will employ any particle from a substrate as long as it lies in the appropriate size range.

In cross-section the wall of simple agglutinated forms is composed of a simple layer where particle sizes grade from finer on the inside to coarser on the outside (Figs. 6B, G). These walls may be pierced partially or wholly by tubules considered as pores.

In more complex agglutinated genera an outer smooth microgranular wall may cover a complex inner layer. This inner wall appears spongy and where portions of the inner wall are folded inward into the chamber of the test, the result is an intricate subdivision of the chamber interior. This type of wall structure has been termed **labyrinthic** (alveolar) in the agglutinated foraminifera (*Cyclammina*, Fig. 6D).

Microgranular walls

Microgranular walls evolved during the Paleozoic and are considered the link between the agglutinated and the precipitated tests in foraminifera. Microgranular particles of calcite cemented by a calcareous cement characterize this wall type and give it a sugary appearance.

Calcareous walls — hyaline type

Calcareous walls may be composed of either low- or high-Mg calcite, or aragonite which is confined to only two foraminiferal families. Hyaline calcareous tests are characterized by the possession of minute perforations in the test wall (Fig. 7A).

When the calcite or aragonite is arranged in prisms with their *c* axis normal to the test surface, the type of wall structure is termed **radial** hyaline. When crystallites are oriented randomly, the test is **granular** hyaline. Less common are hyaline tests composed of a single calcite crystal (*Carpenteria*).

In a cross-section of a hyaline wall (Fig. 7B), remnants of the organic layers are visible. These layers were organic and contained no visible structuring. Such layering occurs in all wall types. The second feature of some hyaline wall microstructure are the calcareous laminations, or **lamellae**. In non-lamellar foraminifera, a new chamber is accreted to the preceding chamber and separated from the previous chamber only by a single thickness of septum. By distinction, in lamellar groups such as the Rotaliidae, newly accreted material may form the new chamber as well as cover the entire outside of all previous chambers (Fig. 7). Thus lamellar tests are easily recognized by the thicker early chambers, the thinner later chambers and the single-thickness final chamber. This pattern varies, of course, when the foraminifer is enrolled.

Planktonic foraminifera belong to the calcareous lamellar foraminifera and are further subdivided into two families on the basis of the surface texture of their walls. The smooth, non-spinose Globorotaliidae are thus distinguished from the spinose Globigerinidae (Fig. 8).

Calcareous walls — porcelaneous type

The term porcelaneous derives from the shiny, smooth appearance of the tests and is the result of the orientation of submicro-

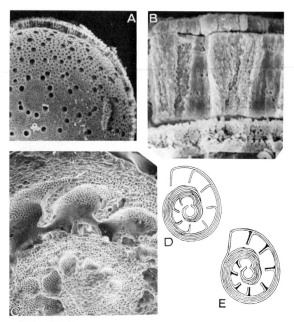

Fig. 7. Hyaline wall structure. A. Cross-section of the wall of the planktonic genus *Orbulina* (× 80). Circular holes are pores. Note the varying size and location of pores; these have been related to varying environmental conditions and can be used for environmental interpretations. B. Cross-section of the wall of the smaller benthic genus *Uvigerina*, showing the calcitic lamellae and the former position of the organic layers. × 750. C. Cross-section of the large benthic genus *Amphistegina*, showing the multi-lamellar outer wall and the characteristically multi-lamellar septae. × 105. (Courtesy R. Sherwood). D, E. Schematic drawing of the lamellar structure of foraminifera, showing how the sequential lamellae are added resulting in multiple lamellae on earlier chambers, and single-lamellar final chambers.

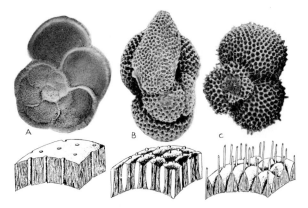

Fig. 8. Wall textures of the three main groups of planktonic foraminifera. A. Smooth-walled *Globorotalia* and associated genera are highly resistant to solution (see Fig. 25). B. The cratered wall of *Globigerinoides* is apparently most susceptible to solution, as these forms are the first to be lost from an association undergoing dissolution. C. *Globigerina* and associated forms in well-preserved samples will possess elongate spines, believed in some cases to be calcified cores of rhizopods; these forms show medium resistance to solution.

scopic crystallites of calcite that form the chamber walls. These crystallites may be randomly arranged or organized in a brick-like pattern, but both patterns give the test a smooth, opaque appearance in polarized light. Under crossed nicols the calcite crystallites

Fig. 9. Representative porcelaneous foraminifera (Miliolidae). A. *Articulina* (×40), Recent Bahama Banks; miliolids are separated primarily on the basis of the type of coiling arrangement each possesses; thus, the milioline rectilinear *Articulina* belongs to a separate family from the solely milioline coiled *Triloculinella* (D) or the partially planispiral *Peneroplis* (B). Milioline coiling is also called streptospiral coiling, as successive coils are added in different planes; apertural type is also a feature important in distinguishing different miliolid groups. B. *Peneroplis* (×10), Recent, Bahama Banks. C. *Quinqueloculina*. (×40), Recent, Bahama Banks. D. *Triloculinella* (×40), Recent, Bahama Banks. E. Miliolids in a carbonate beach sand from Puerto Rico. F. *Dentostomina* (×10), Recent, Bahama Banks. Notice the agglutinated outer covering composed of fine carbonate particles; on many forms this agglutinated covering is originally present, but lost with transport and/or burial. G. *Quinqueloculina* (×40), Recent, Bahama Banks. H. *Peneroplis* (×12), Recent, Bahama Banks. I. *Sorites* (×13) same as H. (Figs. A—D, F—I, Courtesy S. Streeter.)

appear as tiny, multi-colored flecks, so that this type of wall test is considered granular.

Porcelaneous walls are generally two-layered, consisting of a proteinaceous inner layer and a calcitic wall; some genera such as *Dentostomina* (Fig. 9F) form a third layer of agglutinated grains on top of the calcitic layer.

Both in shallow-marine (*Archaias*) and in deeper environments (*Pyrgo*) porcelaneous tests are often composed of calcite with high proportions of Mg.

Chamber shape and chamber arrangement

Foraminiferal tests may possess one or more chambers. The initial chamber is most often spherical or oblate with an aperture. Later chambers range in shape from tubular, spherical, ovate to the several others shown in Fig. 10. Additional chambers are added in a variety of patterns termed chamber arrangements. Chambers may be arranged in a single row, or **uniserially**; if it forms a curved row, it is termed **arcuate**; if a straight series, it is termed **rectilinear**. When arranged spirally around an axis of coiling and the spiral lies in a single plane, the arrangement is termed **planispiral**. When the spiral does not lie in one plane, but progresses up the axis of coiling, the chamber arrangement becomes helicoidal and is termed **trochospiral**. Trochospiral coiling in several planes of coiling is termed **streptospiral**. Both the type of coiling and the plane of coiling are the primary criteria used to subdivide the porcelaneous foraminifera into genera.

When a series of chambers is arranged spirally or coiled about an axis, the chambers involved in one complete revolution are termed a **whorl**, or **coil**. The degree to which one whorl covers, or hides a previous one, is known as the degree of involution. Where the majority of previous coils are hidden, a species is termed **involute**, while **evolute** if the majority of previous coils are visible. Like the gastropods and ammonites, foraminifera show pronounced differences in their degree of involution.

On a coiled test the side of the foraminifer showing the trace of the coil, or spiral, is termed the **spiral side**. The opposite side is termed the **umbilical side**. The umbilicus, the

Fig. 10. Terminology for chamber shapes and arrangements (bottom row) of foraminifera.

axial space between the inner wall margins of the chambers belonging to the same coil, may not necessarily be present.

If two chambers instead of one are added

in each whorl to a serial arrangement the series is said to be **biserial,** and **triserial** if two or three chambers are added in each whorl. Combinations of two or three different chamber arrangements are common in foraminifera (see Fig. 12).

The area where one chamber meets another is the **suture area** and represents the line of junction projected to the surface of the test. The sutures themselves may be simple lines flush with the test surface, indented, ornamented, or raised and subsequently reinforced with calcite in which case they are termed **limbate.**

Where they meet, chambers may be closely appressed. In forms that are streptospirally coiled, it is common for one chamber to cover substantial parts of preceding chambers.

The shape of individual chambers, their arrangement and orientation determine the overall shape of the foraminiferal test (Fig.

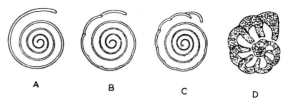

Fig. 11. Schematic drawing of the development of septation. The test without septation could belong to the simple agglutinated family Ammodiscidae (A). The gradual development of the septae (B, C) marked the change to the family Tournayellidae (D).

12). Test shape, chamber shape, and arrangement are assumed to be species-specific, although culture experiments have produced individuals with anomalous chamber arrangements showing that chamber arrangement may be altered by ecologic variables. When a morphologic change becomes fixed in the genetic information of a population (Fig. 12), a new genus is established.

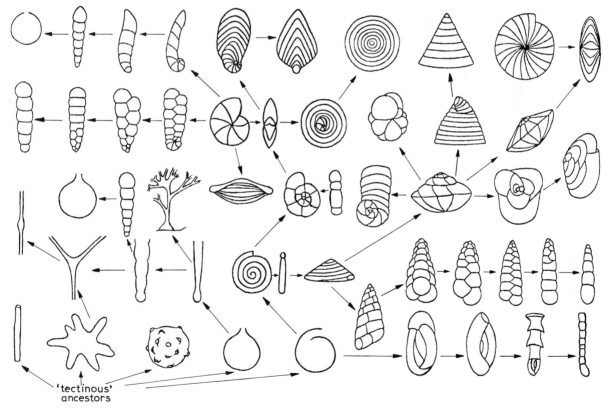

Fig. 12. Schematic diagram of some of the major trends in coiling, chamber arrangement, and test shape in the Foraminifera. Although many trends are reversible, for example from single- to multichambered to single-chambered, or from high-spired to low-spired to high-spired, others, such as the trend from triseriality to uniseriality do not seem to be reversible. The reversible trends have occurred within the Foraminifera as a group (phylomorphogenetic trends) but may occur also within a species primarily as environmental variability (ecophenotypic variation). Redrawn from Pokorný (1963).

Apertures and openings

The aperture is the primary opening of the test to the outside environment. Apertures vary in size and shape (Fig. 13) and the shape is most often a function of the shape of the chamber on which they are located.

Characteristically one aperture termed the **primary aperture** distinguishes the apertural side of the foraminifer. The aperture of a chamber that was once the final chamber of a test may remain visible after another chamber has been added. Such apertures are termed **relict** apertures (*Rotalipora*). If two or more apertures occur on one chamber, as in the Eocene planktonic genus *Globigerinatheca*, they are termed **supplementary apertures.**

We know that the size or shape of an aperture may vary with environment. For instance, the Recent planktonic species *Globigerinoides ruber* possessed a small, low-arched primary aperture during glacial times, but a large high-arched aperture during warmer interglacial conditions. The final chamber also became larger during the warmer period.

There are various structures that may modify the internal part of an aperture. A **toothplate** connects one aperture with the preceding one and in many forms is a simple sort of tube or trough (**siphon**) through which protoplasm supposedly flowed. Other modifications such as teeth, tubes, and lips are shown in Fig. 13.

External modifications of the aperture take the form of covering plates or bulbs which separate the aperture from direct contact with the external environment. These may be in the form of perforate plates, **sieve plates,** or additional chambers such as the **bullae** shown in Fig. 13. These features are regarded as evolutionarily convergent because they have similar shapes and perhaps functions although appearing at different times in different families.

Canal systems and **stolons** fall into the category of openings within the test. In the Rotaliacea and Orbitoidacea canals are tubular cavities within the shell material, often between the two layers of a lamellar septal wall. They are presumed to be the non-calcified passages through which protoplasm flowed during the formation of shell laminae. When the canals connect two chambers through a septum they are termed stolons.

Pores

Pores are round, slit-like, or irregular openings approximately 5 to 6 μm in size. They perforate the test wall as shown in Fig. 7. Pores are typical of both agglutinated and hyaline Foraminifera, and are one criterion differentiating these groups from the porcelaneous Foraminifera. Recent research has demonstrated that pores are plugged at their inner bases and thus do not allow protoplasm to extend through to the outside of the test. Whatever their function they apparently vary, like apertures, in size, shape, and distribution over the test according to environment

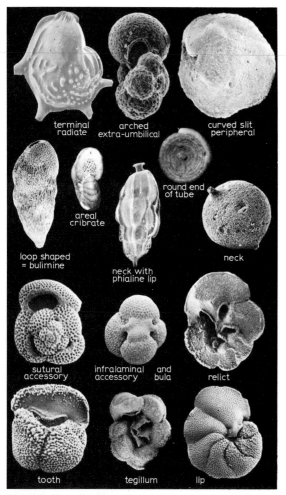

Fig. 13. Terminology for various chamber and apertural shapes and modifications.

and latitude and thus are useful as paleo-environmental indicators.

Ornamentation

Changes in wall texture, pore density, inflational thickenings of carbonate resulting in protrusions, thickening or sharpening of chamber peripheries forming keels; these comprise the ornamentation of a foraminiferal test. Among hyaline foraminifera ornamentation may take the form of ribs, ridges, furrows, spines, etc. (see Figs. 10, 13). Porcelaneous foraminifera are usually ornamented either with striae or ridges (Fig. 9), but agglutinated foraminifera have little visible ornamentation.

Ornamentation is a highly variable test characteristic and may change within a species from one environment to the next. This variation, called ecophenotypic variation, complicates our slightly rigid taxonomic concepts. Taxonomy, like stratigraphy, is linked to the concept of a type; even when some range of variation is illustrated, the limits of illustration limit our understanding of the vast genetic flexibility of asexually reproducing organisms and thus their ability to vary just slightly from one environment to the next. Hence categorization, whether taxonomic or stratigraphic, must take into account the flexibility of these animals and their inherent tendency to transcend our attempts to categorize them.

MAJOR MORPHOLOGICAL GROUPS

As a result of 500 m.y. of evolution in numerous independent lineages, there now exists a vast array of foraminiferal morphotypes.

Loeblich and Tappan (1964a) estimated that there are close to 100 families, over 1200 genera, and some 27,000 species of foraminifera described in the literature on foraminifera. To organize and categorize this vast variety of forms, over 35 schemes of classification have been proposed, some dating back to the very first publications on foraminifera. These classifications are summarized in Loeblich and Tappan (1964b), itself one of the most ambitious and complete classifications yet proposed. An abbreviated

form of this classification system is included here (Table I).

Early classifications were necessarily based on gross features of test morphology, as well as on limited faunas. More recent classifications attempt a "natural" classification based on as many criteria as possible, usually in the following hierarchical order:

1. wall composition and microstructure
2. chamber arrangement and septal addition
3. apertural characters and modifications
4. chamber form
5. life habits and habitats
6. protoplasmic characteristics
7. ontogenetic changes
8. reproductive processes
9. geologic ranges

A new approach to foraminiferal classification was published by King and Hare (1972) who analyzed the amino acid compositions of the tests of sixteen species of Recent planktonic foraminifera. They hypothesized that the amino acids in the calcified tissue vary in a systematic manner which parallels morphology and thus reflects any classification based on morphology. Their studies demonstrated that each species has a distinct amino acid pattern that differs from other species. When amino acid compositions were subjected to statistical analysis, the species did fall into two main groupings, one of which included the non-spinose species of the family Globorotaliidae and the second, the spinose species classed in the family Globigerinidae (Fig. 14).

ECOLOGY

The physical environment of the ocean basins, the chemical constitution and dynamics of sea water, and all of the organisms dwelling in the ocean comprise the marine ecosystem. Within this ecosystem are the individual habitats, local environments within which the foraminifer lives and to which it is adapted. Most foraminifera are marine and benthic, although a few genera are planktonic and some from the families Allogromidae and Lagenidae inhabit fresh water.

Ecology is generally defined as the study of the relationships between organisms and their environments. We attempt to understand such

TABLE I

Outline of classification

Suborder ALLOGROMIINA	
1. Superfamily LAGYNACEA	single-chambered; tubular, round, or flask shaped; test tectinous; agglutinated material in some genera; Paleozoic—Recent; benthic sessile and vagile (*Allogromia*)　　　　　(Fig. 4A)
Suborder TEXTULARIINA	
2. Superfamily AMMODISCACEA	multi-chambered; non-septate to protoseptate; serially or plani-spirally arranged; some branching forms; test tectinous with agglutinated outer layer, sometimes without such a layer; Paleo-zoic—Recent; benthic sessile or vagile (*Ammodiscus*)　　(Fig. 6R)
3. Superfamily LITUOLACEA	multi-chambered; septate, serially or spirally arranged; single or multiple apertures, in some groups with simple or multiple tooth-plates; test microgranular calcitic with or without agglutinated material; sometimes with pseudopores; single-layered or primarily double-layered; complicated interiors; Paleozoic—Recent; benthic vagile, a few genera sessile (*Haplophragmoides, Ammo-baculites, Textularia, Vulvulina, Clavulina, Kurnubia, Orbitolina, Lituola, Cuneolina, Choffatella, Cyclammina*)　　　　(Fig. 6)
Suborder FUSULININA	
4. Superfamily ENDOTHYRACEA	multi-chambered; septate; spirally or serially arranged, single or multiple apertures; test calcitic microgranular, often pseudo-fibrous, primarily double-layered; some genera with agglutinated material in addition; Paleozoic—Triassic; benthic sessile or vagile (*Endothyra, Climacammina*)　　　　　　(Fig. 29B)
5. Superfamily FUSULINACEA	multi-chambered; septate; planispirally arranged, fusiform; mul-tiple apertures; fluted septa or with complex inner partitions; test microgranular calcitic, primarily multilayered; some partly pseu-dofibrous; Carboniferous—Permian; benthic vagile (*Fusulina, Neoschwagerina, Triticites*)　　　　　　　(Fig. 29B)
Suborder MILIOLINA	
6. Superfamily MILIOLACEA	multi-chambered; most septate, some genera protoseptate; spiral-ly or cyclically arranged, with or without internal partitions; single or multiple apertures; some with apertural tooth or multiple toothplates (buttresses); test calcitic porcelaneous, bi-layered, with agglutinated matter; Triassic—Recent; benthic vagile (*Quinqueloculina, Triloculina, Pyrgo, Peneroplis, Archaias, Orbitolites, Marginopora, Alveolina*)　　　　(Figs. 9, 18)
Suborder ROTALIINA	
7. Superfamily NODOSARIACEA	multi-chambered; septate; serially or spirally arranged, aperture single and terminal, often with apertural chamberlet; test calcitic, radial; lamellar; late Paleozoic?/Triassic—Recent (*Nodosaria, Lenticulina*)　　　　　　　　　(Fig. 30F)
8. Superfamily BULIMINACEA	multi-chambered; septate; trochospirally arranged; comma-shaped aperture or (derived from it) terminal aperture on a neck, all with single toothplates; test calcitic, radial; lamellar; Jurassic—Recent; benthic vagile, one small group planktonic (*Bulimina, Uvigerina, Bolivina*)　　　　　　　(Figs. 20A4, 20C2, 4)
9. Superfamily CASSIDULINACEA	multi-chambered; septate; plurilocular, trochospirally arranged; comma-shaped aperture and toothplate; test calcitic with granular appearance; lamellar; Cretaceous—Recent; benthic vagile (*Cas-sidulina, Gyroidina, Oridorsalis*)　　　　(Fig. 20C5, 10)
10. Superfamily NONIONACEA	multi-chambered; septate, plani- or trocho-spirally arranged; interiomarginal or areal aperture; no toothplates; test calcitic with granular appearance; lamellar; Cretaceous—Recent; benthic vagile (*Nonion, Alabamina*)　　　　　　(Fig. 20B6)

TABLE I (*continued*)

11. Superfamily DISCORBACEA	multi-chambered; septate, trochospirally arranged; interiomarginal aperture; with or without toothplate; test calcitic with radial appearance; lamellar; Cretaceous—Recent; benthic vagile (*Discorbis, Asterigerina*) (Fig. 19A8)
12. Superfamily ANOMALINACEA	multi-chambered; septate, trochospirally arranged; interiomarginal aperture; with or without supplementary apertures; test calcitic with granular appearance; lamellar; Cretaceous—Recent; benthic vagile (*Gavelinella, Stensioeina, Anomalina*) (Fig. 20A5, 6)
13. Superfamily ORBITOIDACEA	multi-chambered; septate; trochospirally, annularly or cyclically arranged; single aperture, rarely with toothplate; or with multiple apertures; with or without lateral chamberlets; and with or without secondary chamberlets; test calcitic, radial; lamellar; Cretaceous—Recent; benthic vagile or sessile (*Cibicides, Planulina, Discocyclina, Lepidocyclina, Amphistegina*) (Figs. 18A1, 19A3)
14. Superfamily GLOBIGERINACEA	multi-chambered; septate; spirally and/or cyclically arranged; interiomarginal aperture; with or without accessory and supplementary apertures; test calcitic, radial; lamellar; Jurassic— Recent; planktonic (*Globigerina, Globigerinoides, Hedbergella, Rugoglobigerina, Globotruncana, Globorotalia, Orbulina, Heterohelix*) (Appendices)
15. Superfamily ROTALIACEA	multi-chambered; septate; trochospirally, planispirally or cyclically arranged; single or multiple apertures with toothplate and septal flap; canals, grooves and fissures; with or without lateral chamberlets; with or without secondary chamberlets, test calcitic, radial; lamellar; Cretaceous—Recent; benthic vagile (*Ammonia, Operculina, Nummulites, Miogypsina, Elphidium*) (Fig. 19A4, B2)
16. Superfamily SPIRILLINACEA	multi-chambered, protoseptate; spirally arranged; test calcitic, radial; Triassic—Recent; benthic vagile (*Spirillina*) (Fig. 30N)
17. Superfamily ROBERTINACEA	multi-chambered, septate; trochospirally arranged; single aperture, toothplate and secondary foramina; test aragonitic; lamellar; Triassic—Recent; benthic vagile (*Ceratobulimina, Lamarckina*) (Fig. 19A5)

relationships by cataloguing foraminiferal distributions in Recent environments, and then, where possible, measuring the parameters of that environment.

Indicator faunas have become one of the several indices that can be used to characterize a particular environment. Other indices now used include the planktonic to benthic (P/B) ratio, the ostracode to foraminifera ratio, the calcareous to agglutinated ratio, the percentage of various families present, diversity indices, and indices derived by more sophisticated methods of univariate and multivariate analysis (factor analysis, canonical analysis, cluster analysis, etc.).

Typically the composition of modern benthic faunas is represented by a species list and three useful diagrams: one denoting diversity; the second, generic percentages; and a triangular plot of miliolids vs. rotaliids vs. textulariids. On a larger scale, we map the distribution of foraminifera around the world and attempt to relate their occurrence to some large-scale factor such as circulation or climate. Laboratory culture work has been done to determine the parameters affecting foraminifera, but it is difficult to reproduce an organism's natural environment in the lab.

In 1955 Bradshaw published the results of a series of laboratory experiments in which he attempted to determine the relationship between foraminifera and the critical environmental parameters temperature, salinity, alkalinity, food availability, and hydrostatic

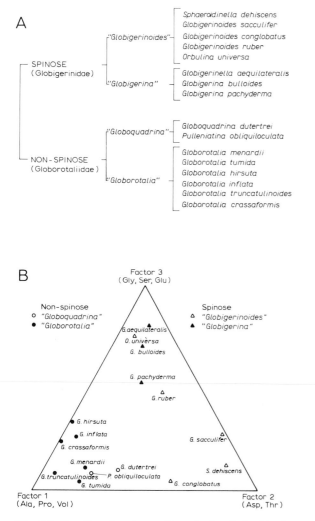

Fig. 14. Results of amino acid analysis of planktonic foraminifera. A. Sixteen species of Recent planktonic foraminifera to be analyzed for the amino acid composition of their calcareous tests; species are grouped according to the taxonomic classification based on morphological features. B. Classification of these same sixteen species analyzed for their amino acid composition and subjected to Q-mode factor analysis. The three-factor solution shown resulted from analysis of the eight most abundant amino acids: each vertex of the triangle represents 100% of the indicated factor. This diagram demonstrates that the classification based on geochemical content exactly parallels that based on morphologic criteria alone and indicates that geochemical analysis of foraminifera can be yet another tool in their classification. After King and Hare (1972).

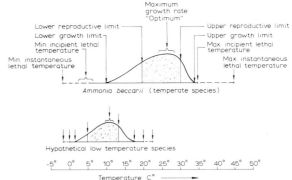

Fig. 15. In a culture situation the temperate smaller benthic foraminifer *Ammonia* was subjected to a wide range of temperatures while other environmental variables were held constant. This diagram depicts the response of the foraminifera both to high and low temperatures; beyond the optimum zone in either direction, first reproduction ceased, then growth, then life itself. This same result was produced by varying other environmental parameters such as salinity, food supply, and oxygen content of the water; a species with a slightly different preference, such as the "cold" water form shown below will demonstrate the same response, only the optimum zone will be located at a different temperature. Redrawn from Bradshaw (1955).

the principles of the curve apply whether the parameter involved is temperature, salinity, or any of several other variables. Obviously some species are more sensitive to one parameter than another, and their lower or higher tolerances will alter the shape of the curve.

The implications of Bradshaw's experiments are especially interesting in the case of a species living below its lower reproductive limit, but above its lower growth limit. Thus, it may continue to grow, but reproduce infrequently. For example, some deep-dwelling agglutinated foraminifera are small under maximum environmental conditions but grow to enormous sizes at high latitudes and in deep, cold water — environments that would seemingly be inhibitory to growth.

Physical variables

Although water depth used to be considered the major factor that influenced the distribution of benthic foraminifera, it is now thought that there is a combination of variables that controls the distribution of individual foraminifers. Temperature is one of the most important and easily determined vari-

pressure. Fig. 15 depicts qualitatively the effect of temperature on the vitality, growth, and reproductive viability of a geographically widespread benthic foraminifer. Bradshaw's and later experiments have demonstrated that

ables affecting benthics. Foraminifera are found living at temperatures from 1° to over 50°C. "Cold water" faunas inhabiting only the cooler waters of the ocean bottoms and high latitudes are fairly uniform between both hemispheres today and were in the geologic past. "Warm water" faunas occur in warmer, tropical, low-latitude areas and are also geographically uniform.

One of the least understood though fundamental parameters affecting benthic foraminifers is hydrostatic pressure. Pressure influences foraminifera indirectly through its effect on the rates of reactions in sea water and on the solubilities of gases, particularly CO_2 which is necessary to the formation of calcium carbonate. Pressure no doubt directly affects chemical processes within the foraminiferal protoplasm itself. Our lack of knowledge of pressure effects on foraminifera is best evidenced by the fact that the only benthic foraminifera that have been successfully cultured come from less than approximately 150—200 m of water.

Light intensity is not thought to control the distribution of foraminifera directly, though it may exert an indirect control, since symbiotic algae living together with foraminifera need light for photosynthesis.

There is no direct evidence to indicate a relationship between foraminiferal distributions and water turbidity, although turbidity will limit the amount of light penetration and thus be inhibitory to photosynthetic organisms. Mud-dwelling agglutinated foraminifera seem to be tolerant of turbidity as this is a main feature of their habitat.

Current systems affect the distribution of sediment and the mechanical action of currents doubtless inhibits foraminifera from areas where sediment is in active motion. In particular, currents affect the post-mortem transport of foraminiferal tests. In view of the cosmopolitan distributions of many foraminifera, it is necessary to postulate some means of migration and Adams (1967) hypothesized that larger foraminifera could be dispersed by current movement of the plant materials to which the foraminifera were attached. Smaller benthics may be swept up into surface currents by vertical convective currents.

Each type of sea bottom, whether sediment-covered or not, contains a distinctive and different population of foraminifera. Just what role the substrate itself plays in controlling foraminiferal distributions, though, is not known.

Chemical variables

Salinity

Foraminifera inhabit environments with salinities ranging from a typical open ocean value of $35^0/_{00}$ to as high as $45^0/_{00}$. The genus *Discorbinopsis* was found to tolerate salinities up to $57^0/_{00}$. At the other extreme, a river and its estuary may have salinities varying from as low as $15^0/_{00}$ to $0.5^0/_{00}$ and still contain foraminifera. (Salinity is measured as grams of dissolved solids per thousand grams of sea water and should be read as so many "parts per thousand".) However, the lower the salinity of an environment, the lower the diversity of the faunas there.

Bradshaw (1955) studied the effects of changed salinities on *Ammonia* growth rates and discovered a direct relationship between salinity and size. Very low salinities correlated with reduced size of tests as well as reduced secretion of calcium carbonate and thus thinner shells.

Alkalinity

Alkalinity is an expression of the capacity of sea water to dissolve calcium carbonate. As a function of the concentration of CO_2 in the water, alkalinity is governed chiefly by temperature, pressure, and biological respiration. The top 500 m of sea water are said to be saturated with respect to calcium carbonate which reflects the high alkalinity in this region. It also means that both Ca and CO_3^{2-} are available and can readily be precipitated.

Below 500 m water is considered undersaturated with respect to calcium carbonate and thus carbonate may begin to go into solution. Below this depth the lower alkalinities tend to cause calcium carbonate to dissolve. Since in general alkalinity decreases with increasing depth (= increasing CO_2 pressure, but decreasing temperature) there is a limiting depth below which little carbonate can remain undissolved.

When describing solution and precipitation of foraminiferal carbonate we use a geochemical term, the carbonate compensation depth, or **CCD**. The CCD is the depth of effective solution of calcium carbonate sediment on the sea bottom.

Resistance to solution in foraminifera varies according to species and varies between planktonic and benthic foraminifera, as solution of planktonic tests occurs in sediments at much shallower depths than for benthics. But at depths where the benthics are still living they are protected by their protoplasm. Most carbonate goes into complete solution below 4,500—5,000 m, although dissolution may occur at depths significantly less than this. Below the level of complete carbonate dissolution, only some agglutinated, but no calcareous foraminifera are preserved.

Trace and nutrient elements

Little is known about the trace element or vitamin requirements of foraminifera, other than that high levels of any one element, for example S_2, may be lethal. Foraminifera apparently use the elements Ca, Fe, Si, Mg, Sr, and Ti in their tests, but it is not known how many of these are actively precipitated by the foraminifera, and how many enter the test as impurities.

Nutrient elements occur in sea water in the form of dissolved bicarbonate, phosphates and nitrates. Little is known of the nutrient requirements of foraminifera, and their distributions do not seemingly correlate with the distribution patterns of these elements in the oceans. Some genera do characteristically occur in areas of high nutrient concentration (*Bulimina*, *Bolivina*), such as marine sewer outfalls where nutrient levels are raised to unnatural highs. Such faunas are generally low in species number, but very high in number of individuals.

High nutrient environments are, however, also areas low in oxygen which has been used up in the decomposition of the organic materials providing the nutrients. There is some indication that the distribution of certain benthics such as *Bulimina*, *Bolivina* and *Uvigerina* (and some planktonics) may be actually adapted to low oxygen environments and not to the nutrient content itself which

may instead have a greater affect on the general community structure on the bottom.

Biotic variables

The study of foraminifera as members of marine communities falls into the realm of **autecology**, the dynamics and ecology of communities of organisms. Such an approach seeks to relate the foraminifera to the food chain of which it is a part, as well as to understand the types of relations foraminifera have among themselves and with other members of the marine communities.

Figures for the density of living benthic foraminifera vary from 1,000 to 2,000,000 individuals per square meter of sea bottom. When the density of individuals becomes great, foraminifera have been observed to migrate away from the crowded areas.

Among foraminifera the main relationships that have been recorded are those of commensalism and parasitism (*Entosolenia*). Other then these, the interrelationships between individual species, their competition or other interactions, are unknown.

Lipps and Valentine (1970) made a detailed analysis of the role of foraminifera in marine food chains (Fig. 16). Their conclusions, as well as the recognition that planktonic foraminifera are both herbivores and carnivores give us a fairly complete picture of the significance of foraminifera in the marine trophic structure.

Parasitism by gastropods and nematodes on foraminifera was demonstrated by Sliter (1971). These invertebrates bore holes into the foraminiferal test in order to extract the protoplasm. The type of invertebrate parasite can be determined by the type of hole.

Symbiosis between foraminifera and many tiny algae, called **zooxanthellae**, is apparently common, even among the planktonic foraminifera which may live at depths up to 200 m. Apparently foraminifera can flourish in the absence of food as long as they are accompanied by their algal symbionts. Foraminifera are known to become more abundant during seasonal algal blooms. Whether this is merely a matter of increased food supply or of the symbiotic relationships is not known.

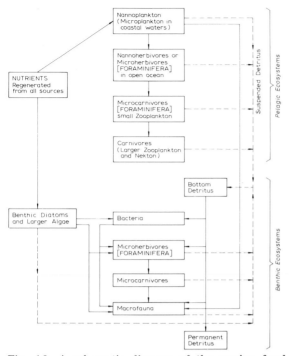

Fig. 16. A schematic diagram of the marine food chain, both benthic and planktonic, and the role of the foraminifera in that food complex. Apparently foraminifera may be microherbivores, microcarnivores, or both (omnivores); because of the strict dependence of each level in this chain to the other, a significant change in one group will have a profound effect on the entire system. It is interesting that the shell-bearing microplankton all appeared at about the same time in the Jurassic, implying some major change in the marine food chains of that time. (From Lipps and Valentine, 1970.)

Planktonic foraminiferal ecology

The distribution and ecology of Recent planktonic foraminifera is essentially similar to that of other zooplankton and is primarily governed by the availability of food, which in turn is a function of the distribution of the complex system of variables termed a **water mass**. By definition, a water mass consists of a set of physical (temperature) and chemical (salinity) variables which are distinct from other adjacent sets.

Planktonics live in the water column from the surface zone down to depths of over 1,000 m. Their density is estimated at approximately ten individuals per cubic meter of water. Distribution of taxa through the water column may change diurnally or seasonally and is considered to be a function of the water density to which the foraminifer is

adapted. In cooler seasons or at higher latitudes a species may live nearer the surface than it does in warmer waters or at lower latitudes. In the Recent juveniles and spinose species live closer to the surface, while the non-spinose or smooth species live deeper. However, the depth distribution of spinose and smooth groups in the Paleogene ocean was not identical to today, and thus is a case where analogy between the Recent and the past oceans does not strictly apply.

Geographically there are close parallels between the distribution of planktonic foraminifera in modern oceans and in the past (Fig. 17). In general smaller species are found in colder water masses or at high latitudes (small bullate globigerinids, *Globigerina*

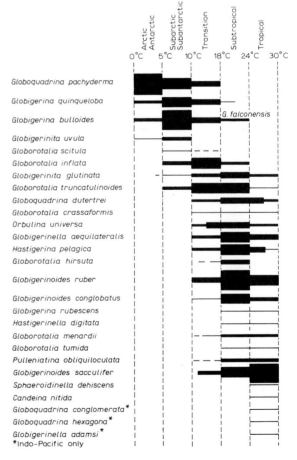

Fig. 17. Biogeographic patterns of Recent planktonic foraminifera. With some modifications, patterns derived from the study of recent forms can easily be applied to planktonic foraminiferal distributions during the Mesozoic and Tertiary (see Figs. 24 and 27). (Courtesy A. Bé.)

pachyderma) and larger species in warmer water bodies or at low latitudes (*Pulleniatina*, *Globotruncana*). Diversity is lower at high latitudes and increases toward the equator. Such generalizations can easily be applied to Paleogene or Cretaceous planktonic foraminiferal faunas.

That oceanic circulation patterns affect the distribution of planktonic foraminifera is generally accepted; there are some local patchy distributions, however, which cannot be explained by this mechanism.

The relationship between planktonic foraminifera and trace element concentrations, nutrient elements, oxygen, turbulence, or turbidity is not known. Planktonic foraminifera live throughout the oceans, even in the ice in the Arctic Ocean. They do not, however, generally live in coastal waters less than 100 m and diversity drops markedly in waters less than 300 m. As some planktonic foraminifera contain photosynthetic symbionts, it is assumed that these forms are restricted to the photosynthetic zone in the water column. Several planktonic species (*Chiloguembilina*, *Heterohelix*) characterize the oxygen minimum zone of the water column and thus must prefer or tolerate low oxygen levels.

Distribution of living foraminifera

Foraminifera have been reported from marine environments extending from tide pools in a marsh to the abyssal plains. Each environment is characterized by its particular species, their diversity and densities. We consider that past environments may have contained many analogous components and hence modern environmental indicator faunas are carefully applied to the understanding of both Recent and past environments.

One curious feature of benthic foraminifera is the similarity of faunas in geographically widespread environments characterized by many similar chemical or physical parameters. And many benthic foraminifera have essentially cosmopolitan distributions, both in the Recent and in the past. Thus, it is possible to look at shelf environments around the world's oceans and find many of the same species in marshes of England and the northeastern United States. The endemism, common among the larger invertebrates, is relatively rare in foraminifera.

Carbonate platforms, reefs, and back reefs

The carbonate platform and reef environment is one of the geologically oldest and most complex of marine ecosystems. Modern reefs occur geographically between approximately 30°N and 30°S latitude, in areas with high light penetration, warm waters, and high dissolved calcium carbonate. They are characterized by waters with high salinities and turbulent conditions. Reefs are often adjacent to or on the edges of shallow-water platforms, such as the Bahama Banks.

Foraminifera occur in coral reef environments either as adherent forms (*Homotrema*, *Miniacina*) which may contribute to the construction of the reef framework, or as epifauna in niches developed within the reef framework (*Calcarina*, *Amphistegina*, *Marginopora*, Fig. 18).

Smaller benthic foraminifera are one of the primary contributors to the sediments of shallow carbonate platforms, second in importance only to the calcareous algae. They attach to sea weeds and grasses, algal and coral fragments; this foraminiferal epifauna may even occur on coral sands which are exposed to air during low tide. Larger foraminifera inhabit these shallow waters in association with the

---

Fig. 18. Typical foraminifera found on carbonate platforms and in proximity to reefs; morphologic analogues in ancient sediments indicate similar environments. A. Reef-associated foraminifera and their locations relative to the reef structure: *1, Amphistegina; 2, Miogypsina; 3, Peneroplis; 4,* miliolids; *5, Alveolinella; 6, Cycloclypeus.* B. Carbonate platform foraminifera from the Bahama Banks, Recent: *1, Cibicidella; 2, Acervulina; 3, Triloculina; 4, Articulina; 5,* platform sediments, including foraminiferal, algal, and pelecypod debris; *6, Dentostomina; 7, Peneroplis; 8, Dentostomina; 9, Planorbulina.* (All photographs × 16. Courtesy S. Streeter.)

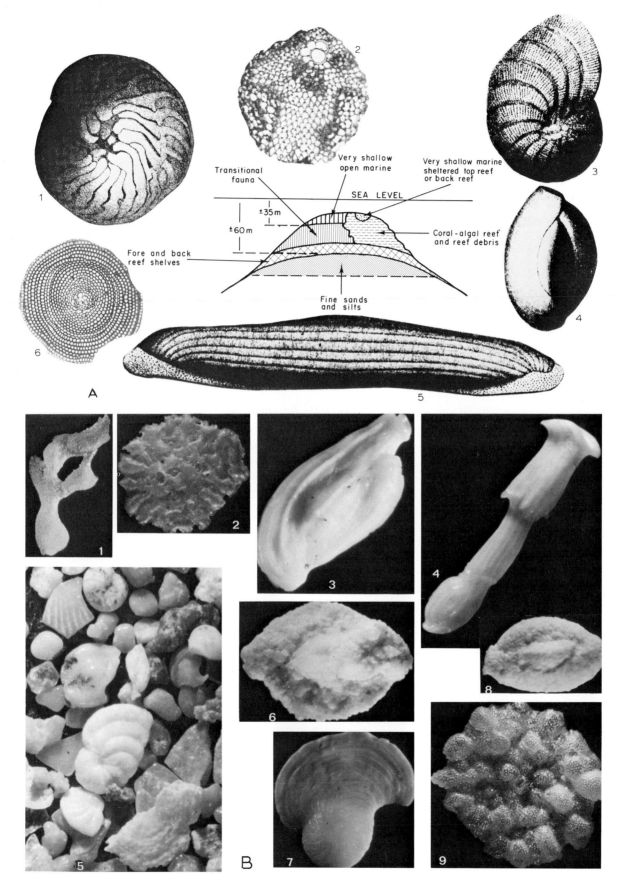

Very shallow
open marine

Very shallow marine
sheltered top reef
or back reef

Transitional
fauna

SEA LEVEL

±35 m

±60 m

Coral-algal reef
and reef debris

Fore and back
reef shelves

Fine sands
and silts

A

B

macroflora which the foraminifera use for protection, and the microflora which the foraminifera use for food (Murray, 1973).

Brackish environments

Historically foraminifera have been considered predominantly marine organisms, with primitive or aberrant types inhabiting freshwater ecosystems. There is a group of foraminifera that occur in brackish environments and this brackish-water fauna is geographically very uniform. Brackish environments are typified by finer-sized sediments containing abundant plant detritus. The critical controlling factor in this environment is apparently the low salinity (Fig. 19).

Marshes

Foraminifera live from the deepest tidal channels to shallow ephemeral tide pools in the marsh grass (Fig. 19). Marshes and bays are characteristically areas of high daily and seasonal fluctuations in temperature, salinity, water depth, turbidity, and water chemistry. In addition marshes are high in organic matter and nutrients and thus often support large biomasses low in diversity. Sediments range from clean well-sorted sands to organic-rich silt and clays with occasional pyrite attesting to the chemically reducing environment below the sediment/water interface.

There is often a marked difference between living faunas and faunas recovered from fossil marsh sediments. Hyaline, agglutinated and a few porcelaneous genera characteristically form the living faunas. The tests of the calcareous and porcelaneous genera are frequently thin. The number of calcareous genera in the sediments, however, is significantly lower or they are altogether absent; while the number of agglutinated forms is generally the same in both sediment and living populations. Dissolution of the calcareous tests has occurred in the chemically reducing zone below the sediment/water interface.

Continental shelf and open marine

One of the most frequently used methods to understand foraminiferal ecology and determine foraminiferal distribution patterns on the continental shelf, has been to obtain samples along a traverse across the shelf, slope, and out into the deep marine basins (Fig. 20).

The shallow shelf is characterized by a numerically small fauna dominated by a few species, very few of which are agglutinated and none of which are pelagic. The agglutinated species construct tests with simple interiors.

The inner shelf is often characterized by coarse-grained, clean, well-sorted sands containing abundant rounded shell fragments. The benthic faunas are usually highly dominated by a few species. Tests are small and not strongly ornamented. A few pelagic species, usually of the genus *Globigerina*, may be present.

The deep inner shelf contains fine- to medium-grained sand, silt, clay, with common glauconite and mollusk and echinoid remains. There is an increase in the number of specimens per species and a concomitant decrease in the dominance by one or a few species. Pelagic types are more numerous and agglutinated foraminifera increase in abundance, but still have simple interiors.

Middle shelf sediments are composed of clay, silt, poorly sorted sands, and abundant glauconite. Species are often highly ornamented, large and robust. Foraminifera are abundant, with pelagic types comprising from 15 to 30% of the total microfauna. Species dominance is low and the number of species is high. Agglutinated forms have more complex interior structures.

The outer shelf is characterized by fine-grained sediments such as clays and some glauconite. Species number is high and ornamentation is strong. Planktonics constitute approximately 50% of the faunas. Arenaceous foraminifera have complex interiors.

Fig. 19. Typical shallow marine foraminifera. A. Carbonate-rich shallow to middle shelf benthic foraminifera: *1, Hanzawia; 2, Robulus; 3, Cibicides; 4, Ammonia; 5, Lamarckina; 6, Hoeglundina; 7, Cibicides; 8, Discorbis; 9, Guttalina.* Notice in *4* the associated invertebrate fossil debris, particularly pelecypod fragments which help characterize this environment of deposition; all fossils × 40. (Courtesy S. Streeter.) B. Shelf and lagoonal foraminifera, characteristic of more detrital environments: *1, Ammobaculites* (× 152); *2, Elphidium* (× 147); *3, Ammonia* (× 95). Recent, Galveston Bay, Texas. (Courtesy C.W. Poag.)

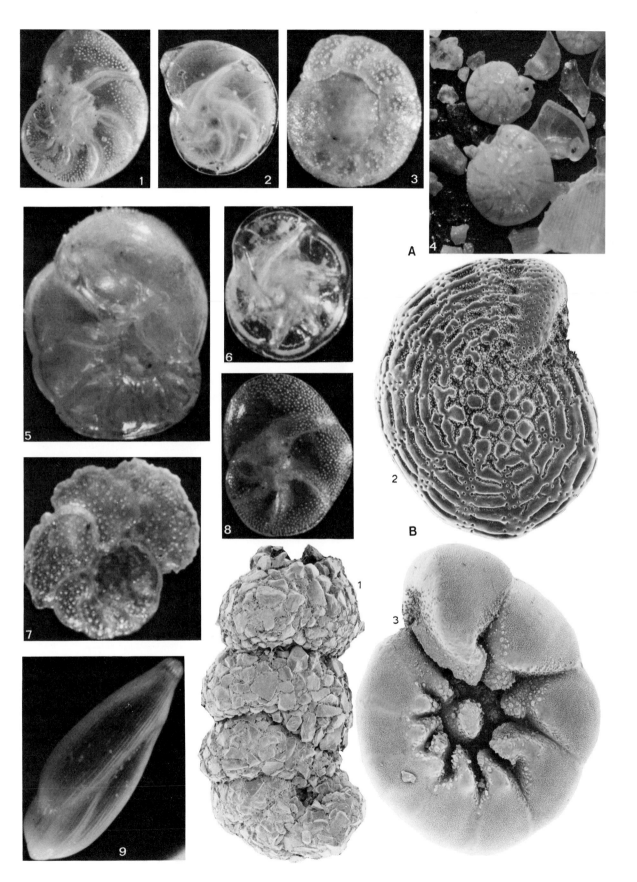

The upper continental slope strongly resembles the outer shelf. Lithologies and structures of the slope are quite complex, varying from smooth slopes to submarine canyons, so that allogenic, or transported, material is common there. Planktonic foraminifera comprise from 50 to 85% of the microfaunas. The lower slope on the other hand is more like the abyssal plain in its fine-grained calcareous marls and clays. The number of benthic species is large on the abyssal plain, though there is a dilution effect from dead planktonic tests. Planktonic foraminifera range from 75% to the more than 90% of the microfauna. However, as depth increases to the CCD, dead foraminiferal tests are dissolved and the deep abyssal areas are covered by non-calcareous sediment. Deep-sea agglutinated foraminiferal tests are composed of fine calcareous particles and appear sugary in texture. Their interiors are often complex.

The deepest-dwelling, abyssal agglutinated foraminifera are simple tube-like structures, surrounded by detrital particles held together loosely by protoplasm. Because of their fragility, these forms rarely reach the laboratory intact. The lack of carbonate in their tests reflects the absence of carbonate particles in the abyssal "red clay" environment.

PALEOECOLOGY

In a sense, paleoecology is ecology pro-

jected backward in time. Not only can paleoecologic studies provide answers to the basic paleontologic questions of why particular groups of organisms evolved as they did and what environmental pressures they were adapting to, but it can provide a basis and an approach to interpreting past oceans and climates. As a relatively new field, paleoecology relies on empirically derived concepts, an increasing multiplicity of investigative techniques, and the diversity of ideas and imagination of the paleontologist involved. It is perhaps the least "absolute" aspect of paleontology.

The primary data for paleoecologic interpretations come from the application of an understanding of present-day patterns of sedimentation, chemical cycling, circulation, and water mass dynamics of the ocean as they relate to the organisms which are also a part of this whole system. The utility of criteria derived from Recent sediments and faunas depends on what sort of observations can be made from ancient rocks as well as the reliability and persistence of faunal characteristics through geologic history. Thus, in attempting paleoecologic interpretations it is useful to keep in mind several caveats; (1) that the more specific and limited the criteria derived from Recent faunas, the lower the probability of finding analogous criteria in ancient assemblages; (2) that the present configuration of ocean basins, as well as the char-

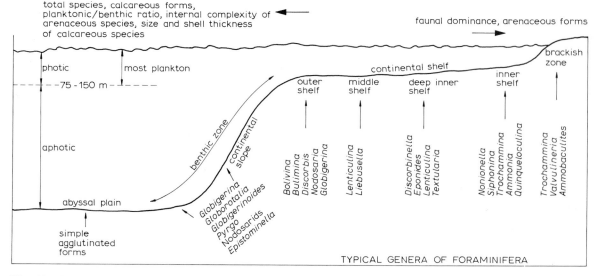

Fig. 20. Trends in bathymetry and fossil content of sediments from the shelf to the abyss. (See also pp. 45—47.)

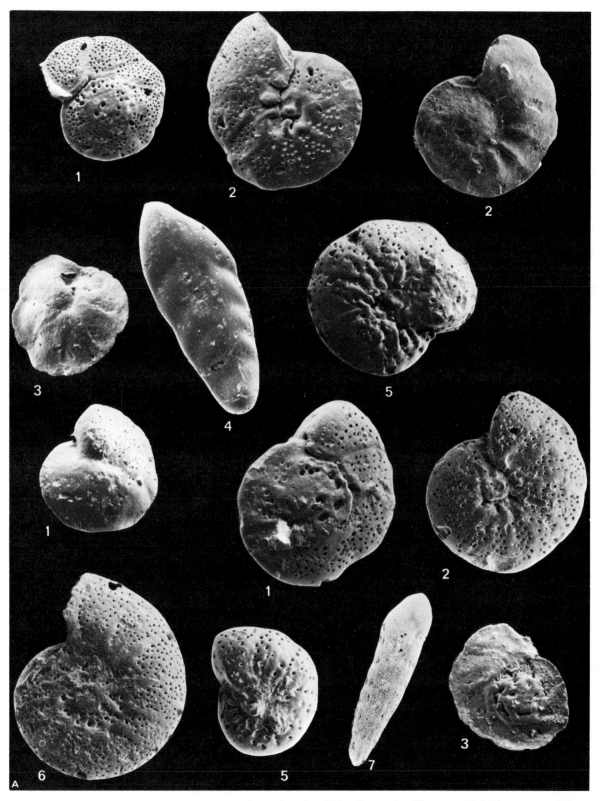

Fig. 20A. Early Tertiary deeper shelf fossils from Angola; all fossils, × 45. (Courtesy W.A. Berggren and J. Aubert: *1, Cibicides; 2, Anomalina; 3, Osangularia; 4, Bolivina; 5, Gavelinella; 6, Anomalina; 7, Loxostomoides.*

Fig. 20B. Early Tertiary outer shelf to slope fauna from Tunisia; all figures, × 45, except 9, × 33. (Courtesy W. A. Berggren and J. Aubert): *1, Gavelinella; 2, Dentalina; 3, Gyroidina, 4, Vaginulinopsis; 5, Tritaxia; 6, Alabamina; 7, Anomalina; 8, Anomalina; 9, Tappanina; 10, Marginulinopsis; 11, Osangularia.*

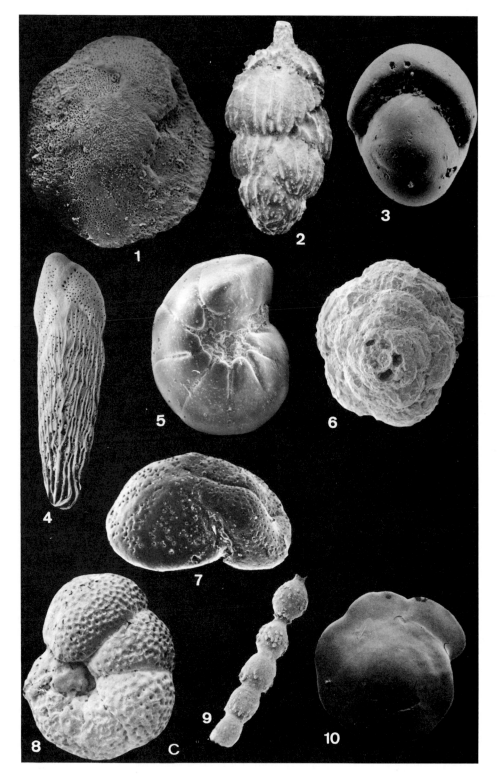

Fig. 20 C. Tertiary benthic foraminiferal association characterizing the slope to the abyss; all figures × 100: *1, Osangularia; 2, Uvigerina; 3, Pullenia; 4, Bolivina; 5, Gyroidina; 6, Trochammina; 7, Cibicides; 8, Anomalina; 9, Stilostomella; 10, Oridorsalis.*

acteristics of each ocean, are geologically new phenomena; the older the fauna observed, the less likely it is to be analogous with present-day faunas.

Approaches to paleodepth determination

Classically the paleodepth of a site has been determined on the basis of its benthic foraminiferal faunas and on the assumption that criteria derived from the Recent can be applied to past, analogous faunas. This assumption has been modified to the point that depth estimates must take into account the up- or downslope migrations of faunas during times of environmental change; and the changed adaptation of certain benthic foraminifera to environmental conditions between the Tertiary and the Recent and perhaps between the Cretaceous and the Tertiary.

Berggren and Haq (1976) used what is called the uniformitarian approach to interpret the depths of late Miocene sediments (called the Andalusian) in southern Spain. By choosing relatively young strata, they increased the probability that benthic foraminiferal depth distributions would be close to those of present-day faunas. By logging the faunas of the Andalusian and the depth distributions of Recent analogues of those taxa, they proposed a paleodepth of close to 1000 m at the base of the section. Changing faunas were then used to delineate sequential shallowing, eventually to water depths of less than 100 m (Fig. 21).

Within the Upper Andalusian faunas there also occurred specimens of the neritic species *Ammonia beccarii*; the other benthics were indicating depths of several hundred meters. Faunas containing both shallow and deeper water foraminifera are mixed and may contain species contemporaneous in age, but not in depth of deposition; or they may have contaminants of significantly different ages and lithologies. Turbidite deposition is a common mechanism for the displacement of these faunal elements.

During the Andalusian, the mixing of shallow-water faunas reflects tectonism and erosion of adjacent landmasses and turbidite deposition of large amounts of land-derived sediment downslope. Such deposition characterized this basin until the latest Miocene

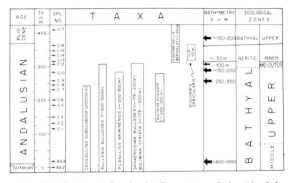

Fig. 21. Estimated paleobathymetry of the (Andalusian) Upper Miocene—Lower Pliocene stratigraphic sequence at Carmona—Dos Hermanas, southwestern Spain based on benthic foraminifera (after Berggren and Haq, 1976). The uppermost depth limits of several abundant foraminifera in these sediments were plotted against time; after considering their combined information it was possible to estimate a minimum depth of deposition for each sample and to show that these minima decreased through time as the area became more and more shallow; this shallowing coincided with a world-wide drop in sea level which occurred at this time. Although there are many problems involved in using the present depth limits of benthic species, application of this technique to later Miocene foraminifera is considered appropriate since: (1) many Recent species evolved at this time and hence it is the same species which is being interpreted; and (2) oceanographic conditions in the later Miocene were closer to those of today than at any other time in the Tertiary; hence, the closer to the Recent, the greater the probability that forms considered "analogues" will indeed be analogous. As ocean basins have changed through the Tertiary and circulation patterns have been strongly modified by changing tectonic settings, fossils from the same basin through time, will not necessarily be responding to analogous oceanic conditions; hence the idea that it is safer to compare forms only in the same ocean is not strictly accurate.

by which time a major worldwide fall in sea level took place; simultaneously, the Mediterranean Basin became isolated from the world ocean and experienced a major regression associated with desiccation.

Paleodepths may be assigned to deep marine benthic faunas by an independent method called back-tracking, which is not based on a uniformitarian approach, but independently provides an estimate of the paleodepth of a site where a fauna resides. The back-tracking technique derives from the observations of McKenzie and Sclater (1971) that as new material is formed at a mid-oceanic ridge and moves away from that ridge by sea-floor spreading, the newly accreted sea floor contracts as it cools and sinks as it contracts. Thus, sites farther from the ridge crest will be consistently deeper than those close to the ridge crest. To back-track a site is merely to move it back to the time when a fauna was deposited and then estimate its depth based on its distance from the ridge.

This method is illustrated in Fig. 22 showing the history of Site 358 in the southern Atlantic Argentine Basin. The site was formed near the Mid-Atlantic Ridge in the late Campanian. As it moved away from the ridge it sank deeper until during the Maastrichtian it came close to the level of the paleo-CCD. The highly dissolved foraminiferal faunas, containing no planktonic, but only the more resistant benthic foraminifera, attest to the proximity of this site to the late Cretaceous CCD. Although the CCD was falling during

the later Cretaceous and Tertiary, the sinking rate of Site 358 brought it below the CCD by the middle Eocene, and no calcareous sediment is found after the late Eocene at this site.

New approaches to the prediction of paleodepths of faunas both in the Pleistocene and the Tertiary are based on the application of factor analysis to benthic faunas. A faunal matrix consisting of the abundance of benthic taxa at a site or level in a deep-sea core is analyzed. The resulting factors are then correlated statistically to some measurable parameters (for example, temperature, salinity, or oxygen content) of the site where the fauna occurred. In the case of Pleistocene and Recent benthics, these parameters may be directly measured. In the case of Tertiary faunas, direct measurement is not possible so that latitude or a back-track depth may be used. The factors with their correlated physical or chemical variables can be compared with factors from analysis of another fauna, and the degree to which the variable applies to this second fauna can be predicted. The Factor Analytic Method thus can produce estimates of depth for a site where it cannot be directly measured, or temperature for a fauna whose depositional temperature cannot be measured. The possibilities of this method are just being explored.

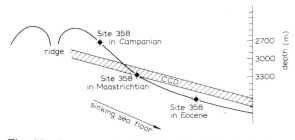

Fig. 22. An example of the back-tracking technique for estimating the paleodepth of a site, and hence the depositional depth of a foraminiferal fauna. Site 358, DSDP Leg 39 is located in the northern Argentine Basin, South Atlantic; the site is currently at a depth close to 5,000 m. This part of the sea floor apparently formed at the Mid-Atlantic ridge in Campanian time and has spread away from the ridge and sunk since then. By backtracking the site it is possible to interpret the later Cretaceous and early Tertiary calcareous faunas there. At the bottom of the section the samples contain moderately well-preserved planktonic foraminiferal oozes with abyssal benthic foraminifera. Higher in the section the faunas become more and more dissolved, eventually the planktonic foraminifera are entirely absent from the samples, and only dissolving abyssal benthic foraminifera remain; by middle Eocene time all foraminifera are gone from the sediments which can then be called a non-calcareous "red clay". Knowing the degree of solution of the samples and the back-tracked depth of the site allows us: (1) to speculate that the site was continually near the paleo-CCD and sank below the paleo-CCD in the middle Eocene; and (2) to estimate roughly the depth of the paleo-CCD both in the later Cretaceous and the early Tertiary from the depth of this site. When all calcareous sediment is gone, it is no longer possible to locate the CCD depth in this area.

Approaches to paleotemperature estimation

To determine past oceanic temperatures it is possible to use the oxygen isotope method which involves the determination of the two stable isotopes ^{16}O and ^{18}O from the calcitic tests of foraminifera. Many foraminifera incorporate both isotopes of oxygen into the carbonate of their tests in the same proportions as found in sea water and are said to precipitate their tests in equilibrium with sea water. By converting foraminiferal carbonate to CO_2 and measuring the ratio of the heavy to the light isotope, we can relate the isotopic ratio in sea water to its paleotemperature through empirical equations evolved by the geochemist Harold Urey.

Oxygen isotope ratios provide us with several sorts of information. Prior to 5 m.y. ago, the $^{18}O/^{16}O$ ratios in foraminiferal tests primarily reflect isotopic abundances in sea water and thus can be directly related to water temperature.

First Emiliani and Edwards (1953) and later Savin and others (1975) have shown that there has been an overall decrease in oceanic temperatures over the past 65 m.y. However, detailed analysis of one South Atlantic Tertiary site (Fig. 23) demonstrates the intricate complexity of the ocean's thermal structure. Using both benthic and planktonic foraminifera in detailed sequences such as these should eventually allow a substantial understanding of the oceans as they change through the Tertiary and evolved toward the glaciated world of the Pleistocene.

Imbrie and Kipp and their associates in CLIMAP have used statistical analyses of Recent planktonic foraminifera to produce estimates of oceanic temperatures during the maximum glacial episode of the Pleistocene. Imbrie has pioneered the Factor Analytic Method which is based on three main assumptions: (1) that associations of planktonic foraminifera characterize an oceanic water mass and that if the temperature or circulation of that water mass changes, a different assemblage will then be present at that locality; (2) that present types of relationships between planktonic foraminiferal associations and the environmental variables temperature, salinity and oxygen isotopic incorporation were operative in the past; and (3) that the tops of deep-sea cores which contain sediments less than 1,000 years old are a reflection of present-day water chemistry and circulation patterns.

Imbrie uses several multivariate techniques including Q mode factor analysis of a matrix of planktonic foraminiferal abundances to find those combinations of numbers that best explain the variation in his data matrix. These combinations, or factors, are composed of one or more planktonic taxa. Imbrie first studied planktonic foraminifera in core tops from the North Atlantic where he also had information on the temperature, salinity, and the $^{18}O/^{16}O$ content of the surface waters. After deriving his biotic factors, he then correlated them statistically with the measured values of the water. The result was a set of biotic factors with their correlated temperature, oxygen and salinity values.

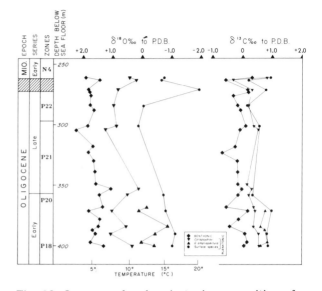

Fig. 23. Oxygen and carbon isotopic composition of planktonic and benthic foraminifera of Oligocene age from the central eastern Atlantic, DSDP Site 366. The measurement of these isotopic values has provided several types of information about the foraminifera themselves, as well as the temperatures and some water mass characteristics of the waters they inhabited. For example, by measuring the isotopic composition of several species of planktonic foraminifera, it is possible to derive their relative order of depth stratification as well as the temperatures at which they lived. This sort of work has shown that the depth habitats of early Tertiary planktonic foraminifera were not strictly similar to those of their modern "analogues". There is a significant correspondence between the temperature record of the benthic foraminifera (far left) and the deepest dwelling planktonic, *Catapsydrax*. Knowing that at present this site lies at intermediate depths and probably did in the early Tertiary allows us to suggest that the temperature structure of intermediate waters in this part of the Atlantic was particularly uniform and was controlled by high-latitude processes independent of tropical surface conditions; and by comparing the carbon values of benthic foraminifera at this site with those from middle latitudes in the South Atlantic, it is possible to suggest that the higher-latitude waters are older and contain more carbon, while the equatorial waters are "younger" and somewhat depleted in oxygen. If oxygen depletion occurs as a water mass travels farther from its source, then it is possible to speculate that in the early Tertiary intermediate waters were generated solely at high southern latitudes, while by contrast in the Recent, there is a significant effect of intermediate water production from the northern North Atlantic which flows into the equatorial zone. These and many other types of information can be read from the isotopic composition of foraminifera.

Imbrie next turned to Pleistocene deep-sea cores and conducted a Factor Analysis every 10 cm down these cores. He then could correlate his new core factors with core top factors and their associated measured temperature values, and thereby posited temperature values at 10-cm intervals down the cores. Combining analyses from many cores in the North Atlantic, and subsequently the Pacific, Imbrie and others have been able to construct models for thermal conditions in the oceans and to produce a map of estimated sea surface temperature ~18,000 years B.P. (Fig. 24). These sea surface temperature estimates may then be used by others to model atmospheric circulation and climate during a peak glaciation.

Approaches to paleochemistry of sea water

Since calcium carbonate and carbonate-secreting organisms are major components of marine sediments, chemical oceanographers, sedimentologists, and foraminiferologists alike have focused attention on the relationships between calcium carbonate and sea water. Carbonate concentrations in deep-sea sediments are a function of two processes: accumulation and removal, primarily by solution. Since we base many conclusions on death assemblages, **thanatocoenoses**, we would like to know whether solution is altering the information they contain.

Wolfgang Berger (1971) conducted a set of experiments to determine the solution effects on fifteen species of Recent planktonic foraminifera. He suspended samples of sediments and planktonic species for four months in the Pacific Ocean to assess the amount and nature of solution to each species. After four months Berger was able to quantify the amount of solution that had taken place and rank species according to their solution susceptibility. In the process of his research Berger also proposed a model for the relationship between foraminifera and solution in the Pacific which has since been extrapolated to all the oceans. Berger found that there is a level in each ocean, the foraminiferal **lysocline**, below which significant solution of foraminiferal carbonate begins to occur. This level is significantly shallower than the CCD

and varies within and between oceans (Fig. 25).

Using this information on the solution of foraminiferal tests and the resultant sediment types, Berger and Winterer (1974) among others have plotted the depth of the CCD in the modern oceans and have attempted to locate the paleo-CCD during the Cretaceous and Tertiary. These studies have shown that the CCD has dropped from close to 3,200 m in the Cretaceous to its present average depth of 4,500 m.

Understanding of the selective solution of foraminiferal species was begun in studies of cores taken on DSDP Leg 2. Cita and others (1970) noted corrosion of foraminiferal tests and commonly, fragmentation. Fragmentation was attributed to solution and not destruction by burrowing organisms, since fragmentation is known to be common in tests approaching the CCD in Recent sediments and the fragments belonged only to planktonic foraminiferal tests which appeared very thin and highly perforate, the pores possibly having been widened by solution. Accompanying benthic tests were unaltered.

At Site 10, Leg 2 Cita found that the assemblages of middle Eocene age planktonic foraminifera were far less diversified than would be expected for that time and latitude in the Atlantic and consisted of three genera: *Globigerinita*, *Orbulinoides*, and *Globigerinatheca*, foraminifera which have spherical shapes, bullae, and the latter two, secondary thickening of calcite. Very rarely, rounded globigerinids (*Subbotina*) were present. However, the typical middle Eocene genera *Globorotalia* (*Morozovella*) and the thin-shelled *Hantkenina* were both absent. By analogy with Berger's solution ratings for Recent planktonic foraminifera, Cita showed that among Eocene species also thin and spinose forms are the first to be dissolved and that massive tests are more solution-resistant.

Approaches to paleogeography

The study of the paleobiogeography of foraminifera is, in a sense, the study of paleogeography itself; foraminifera characterize specific environments and by plotting foraminiferal distributions through time we

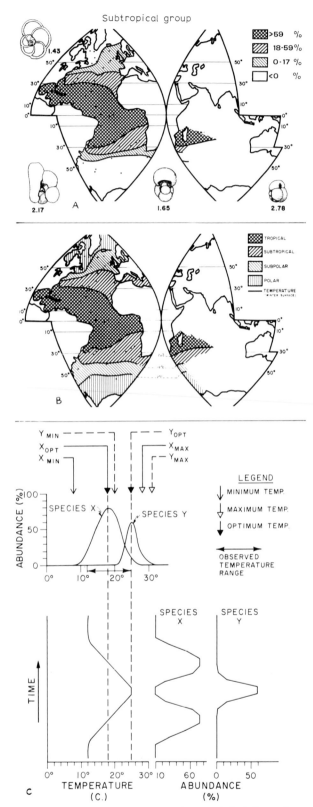

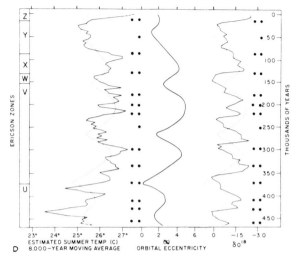

the four numerically dominant species. Note that the distribution of this assemblage defines geographic "belts" roughly parallel to the equator; compare this distribution with the Jurassic/Cretaceous paleobiogeographic maps of Fig. 27. B. Regions dominated by each of the five major assemblages derived from Factor Analysis of the core tops. Comparing these regions with the lateral distribution of surface temperature isotherms showed that the limits of each successive region coincided with the 2°, 12°, and 20° C winter surface isotherms in both hemispheres. C. To apply this information down cores, a model for the temperature response of species down cores during climatic variations was designed. This hypothetical model shows the abundance—depth curves resulting from the response of two species (X and Y) having different temperature optima to an interglacial temperature increase. Simply speaking, the basic concepts of Bradshaw (Fig. 15) are here being applied to two associated species. Minimum, maximum and optimal temperatures for each species are indicated in the upper left. The contrasting stratigraphic records (lower right) during the hypothetical interglacial temperature increase then, are caused by the different species' optima. D. Factor Analysis down core V12-122 produced a set of factors every 10 cm. The measured winter and summer temperatures at each core-top locality was associated with the dominant factor there and core-top factors and their indicated temperatures were compared with down-core factors by regression analysis which then could predict the winter and summer temperature for each down-core sample. Plotting these temperatures down the core, it was possible to compare the estimated summer surface temperatures with other climatic indices, such as the oxygen isotope record down this same core, as well as the calculated curve for the perturbations in the earth's orbit around the sun. This orbital eccentricity curve is repeated twice (dotted) to facilitate comparisons. The heavy dots represent peaks in 'Milankovitch' curves for the periodicity in earth—sun relationships. The U—Z zonation signifies the classic Ericson faunal zones based on the presence and absence of *Globorotalia menardii*. The time scale for these zones assumes an average sedimentation rate of 2.35 cm per 1000 years. After Imbrie and Kipp (1971).

Fig. 24. Factor analysis technique of Imbrie and Kipp. A. Counts of planktonic foraminifera in 61 core tops from the Atlantic Ocean were submitted to Q-mode factor analysis. Five assemblages (factors) were considered significant and the distribution of one, the subtropical assemblage, is shown along with

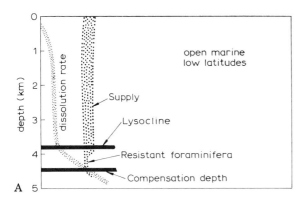

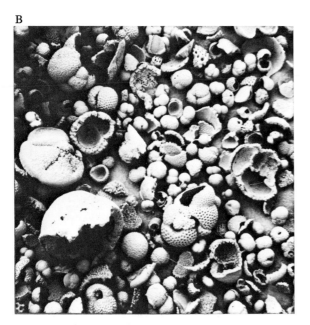

Fig. 25. Models for the dissolution of calcium carbonate with depth. A. This diagram represents the increasing dissolution of CaCO$_3$ with depth, given a constant supply from the oceanic water column. The foraminiferal lysocline is the level in the water column in a specific geographic region where significant dissolution begins. This dissolution is recognized by etching of the test surface, enlargement of holes in a test, loss of apertural bullae, "peeling" of the outer wall of a test, and eventually fragmentation. The CCD represents the level below which no carbonate tests should remain undissolved (although occasionally carbonate may be preserved at these depths by rapid burial). The depth of the foraminiferal lysocline is slightly different from that of the calcareous nannoplankton which are preserved in fecal pellets. The CCD is apparently the same for both groups, but varies from one geographic area to the next depending on water mass properties in that area. B. The highly resistant species *Globorotalia menardii*, strongly dissolved and almost totally fragmented, indicates deposition very near the CCD, about 4,500 m, in the eastern Pacific Ocean.

are plotting the distributions of those environments through time.

Historical biogeography necessarily involves both geological and biological evolution, so that within the framework of geographic evolution (tectonism, sea-floor spreading, etc.) we may see ecologic evolution (salinity and temperature shifts, chemical changes, changing sedimentary components). And within the framework of this environmental evolution we look for the phyletic evolution (and morphotypic adaptation) of the foraminifera themselves.

Paleozoic fusulinids

During the 100 m.y. of their existence the fusulinids evolved into more than 100 genera and 5,000 species, many of which were giants close to 10 cm long. To date fusulinids have been reported from every continent except Australia and Antarctica. During the later Paleozoic major synclines formed in America, Africa, and Europe and there were intermittent connections between them until latest Permian time when they were destroyed by orogenesis.

Paleozoic faunal realms are designated by key fusulinid genera and species (see Fig. 29 B). Connection between two faunal provinces is indicated by a high measure of similarity between faunas in the same environment in two separate areas. Loss of connection between provinces is then evidenced by: the earlier evolutionary transition from one genus to another in one particular province; the lack of a cosmopolitan genus in one particular province; or differing stratigraphic ranges of key species in similar environments in different provinces.

Tectonism of the Caribbean

The evolution and distribution of the two Pliocene planktonic foraminifera, *Globorotalia miocenica* and *Globorotalia multicamerata*, have been instrumental in understanding the tectonic history of the Caribbean Basin (Fig. 26) in the Pliocene. *G. multicamerata* evolved about 4.5 m.y. ago and lived in the Pacific, Atlantic, and Caribbean. Typical *G. miocenica* evolved in tropical latitudes about 3.5—3.6 m.y. ago and is found in deep-sea sediments from the Caribbean and Atlantic Oceans, but has not yet been found in the Pacific. Because

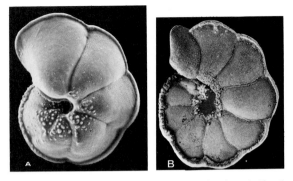

Fig. 26. The two Pliocene planktonic forminiferal species *Globorotalia miocenica* (A) and *Globorotalia multicamerata* (B) used to date the tectonic closure of the Pacific from the Caribbean by the Central American Isthmus. (×45.)

of the modern cold water masses and the northerly flowing currents along the eastern coast of South America, we presume that planktonics spread between the Atlantic and Pacific through the Caribbean, which is known to have been a seaway throughout the earlier Cenozoic. But some barrier to migration had apparently evolved and prevented *G. miocenica* from entering the Pacific.

Evidence from geophysics and from vertebrate paleontology tells us that the Middle American passageway between North and South America was closed sometime during the Pliocene, but their data could not provide a more accurate date for the time of closing. The foraminifera, however, allow us to date very precisely the major movement of land mass at between 3.5 and 4.5 m.y. Central America separates Caribbean and Pacific faunas to the present day.

Approaches to describing paleoclimates

Foraminifera today, as in the past, are distributed in belts which roughly parallel latitude. The poleward changes in faunas is a reflection of the latitudinal changes in temperatures, moisture, wind systems, continental positions, and oceanic circulation patterns on a spinning planet. The thermal gradation from warmer temperatures and more equable climates at the equator to cold waters and severe climates toward the poles has changed in intensity, but not in direction through geologic time. We are today in a period of strong equator-to-pole thermal gradients and this is reflected in the extreme provinciality of modern faunas.

Cretaceous and Jurassic faunas, though demonstrating provincialization, show the more poleward extent of equable climatic conditions.

After investigating the biogeography of Jurassic foraminifera Gordon (1970) determined two major assemblages: (1) a neritic shelf assemblage dominated by nodosariids, simple agglutinated species, or shelf calcareous species; and (2) a Tethyan assemblage, containing planktonics in some areas and complex arenaceous foraminifera in shallow waters bordering the circum-global Tethys. These assemblages typified the entire Jurassic, although planktonics were not abundant until the late Jurassic. Plotting Gordon's data on paleoreconstructions of the continents in the Jurassic demonstrates that foraminifera were, with local exceptions, distributed in belts roughly paralleling the Jurassic equator and that the Tethyan (tropical) faunas were restricted towards the equator as compared to the late Cretaceous (Fig. 27).

By plotting the distribution patterns of Cretaceous foraminifera studied by Dilley (1971) on similar reconstructions for the Cretaceous, two global patterns emerge among the larger foraminifera. The larger, shallow-dwelling, or reef-associated benthics, both agglutinated and calcareous, occurred in restricted zones roughly paralleling latitudes and extending to near 40° N and S. Although in the early Cretaceous these zones paralleled the extent of the Tethys, in the late Cretaceous these same forms had developed a high degree of provincialism. This increased provincialism probably reflects the opening of the Atlantic Ocean, followed by the subsequent connection between the northern and southern Atlantic into one large ocean which could form an effective barrier between the Eurasian and American continents.

During the Cretaceous the smaller benthic foraminifera and the planktonics can be divided into lower and higher latitude, boreal or "austral", assemblages. As shown in Fig. 27 the characteristics of assemblages including types of general morphologies, amount of ornamentation, and diversity changed markedly toward the polar regions as they do today.

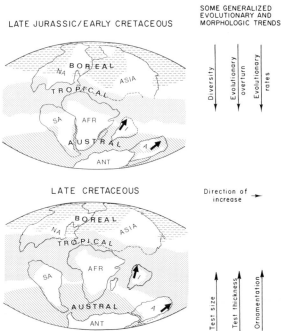

Fig. 27. Major Jurassic and Cretaceous bioprovinces based on the distribution patterns of planktonic, larger, and smaller foraminifera. Increased climatic equability and latitudinally widespread warm temperatures are reflected by the longitudinal distribution of species, paralleling the entire extent of the Tethys. Post-Cretaceous tectonism gradually closed off Tethyan circulation and resulted in increased north—south orientation of currents, oceanic basins, and increased latitudinal faunal bioprovincialism (see Fig. 24B). Besides variations in species and species dominance, other features of morphology and evolution change between the tropical and higher latitude regions. Such trends are common not only in the Mesozoic, but also in the Cenozoic and can be recognized in many other fossil groups. There are many local controls on these trends, but the general direction of each is depicted here.

In addition it is clear that the tropical belt at this time stretched as far as 40—50°N paleolatitude, compared to its restriction between 30°N and S today.

Nearly all planktonic foraminifera became extinct at the Cretaceous—Tertiary boundary. Subsequent rediversification of planktonic foraminifera took place slowly during the Paleocene and more rapidly during the early Eocene. At the end of the Eocene many species again became extinct. Relative to the late Cretaceous, the Paleocene and particularly the Eocene are periods of increased climatic zonality and increased equator-to-pole thermal gradation of ocean waters. This increased zonality is reflected in the increased differentiation of latitudinally characteristic planktonic foraminiferal assemblages. Recently I. Premoli-Silva has determined the latitudinal preference of four primary and several secondary planktonic foraminiferal assemblages of Paleocene and Eocene age in the Atlantic Ocean. By noting the changing dominance of these assemblages through time at low, high and middle latitudes, she was able to speculate on the changing climatic zonality which caused the movements of their assemblages. In particular by plotting the times and extent of the poleward migration of the low-latitude keeled morozovellids, she could show distinct warming trends in climate during the Paleocene and Eocene. These climatic interpretations have been corroborated by paleotemperature estimates made on planktonic foraminifera from this same time interval.

BIOSTRATIGRAPHY AND GEOCHRONOLOGY

The intent of biostratigraphy is to subdivide strata based on their biotic content and to make finer and finer orderings of what are considered sequential events in the geologic record. When biostratigraphic subdivisions can be related to time, and particularly to some measurement of time, we are speaking of the field of geochronology. These fields are both distinct and complementary.

Researchers have erected zonations for most of the geologic record of foraminifera; however, foraminifera have proved to be most useful for the subdivision of the Upper Paleozoic, the Upper Cretaceous, and the Tertiary.

Paleozoic biostratigraphy

Biostratigraphic subdivisions of the Paleozoic are based on several microfossil groups, but until the Carboniferous—Permian, foraminifera are not particularly useful as compared to conodonts or chitinozoa. For the Carboniferous there are fourteen biostratigraphic zones based on evolutionary changes in both the agglutinated and microgranular foraminifera (Mamet and Skipp, 1970). In the Upper Paleozoic there are ten fusulinid biostratigraphic assemblage zones based on groupings of fusulinid genera (see Appendix).

The basis for both the classification and stratigraphy of the fusulinids are the internal shell structures which in later taxa reach an incredible degree of complexity.

Mesozoic biostratigraphy

Although there are assemblage zones of shelf-dwelling benthic foraminifera in the Upper Triassic and Jurassic, by the middle of the Cretaceous the planktonic foraminifera become useful for biostratigraphy since they: (1) are abundant in open marine samples; (2) generally have short ranges and thus delimit short time periods; (3) are geographically widespread and their diffusion after origination is essentially instantaneous in geologic time; and (4) occur in sediments, representing water depths from 200 m to over 3,000 m thus allowing some correlation between events on the shelf and in the deep sea. Their utility is slightly limited in those environments where assemblages are highly altered by dissolution and by the fact that not all faunal events are synchronous between higher and lower latitudes.

Biostratigraphic zonations have been based primarily on lineages of the distinctive genus *Globotruncana*. In the zonation of Van Hinte (1976) the refinement of time intervals for zones ranges from an average of 3—4 m.y. to a low of 1 m.y. in duration. A zone based on the clearly demonstrable evolution from one species to its derivative, is considered the most reliable of all biostratigraphic zones (see Appendix).

Tertiary biostratigraphy

The most detailed biostratigraphic zonations have been made on the basis of the evolution of Tertiary planktonic foraminifera; at present there are 23 Paleogene and 23 Neogene low latitude biostratigraphic zones and numerous subzones (Berggren, 1972). The average length for a Tertiary zone is 2 m.y., however resolution close to 200,000 years is possible in the Pliocene due to rapid faunal changes (see Appendix).

Zonations have been established for both high latitude and temperate latitude areas. In the case of high latitude zonations, these are locally applicable but do not correlate strictly with low latitude zones. It is slightly confusing that what may be called the Oligocene at high latitudes, correlates with the Eocene as it is defined at low latitudes. Diachroneity of faunal events between lower and higher latitudes is responsible for this confusion which cannot be resolved until the higher latitude zonations are related to some absolute time scale.

The larger foraminifera of the Tertiary have been subdivided into zones on the basis of the ranges of key genera, the evolutionary degree of complexity, and the first and last appearance of taxa. By using the planktonic foraminifera found along with the larger benthics, it has been possible to correlate this zonation to the low latitude planktonic foraminiferal zonation and hence to time, thus rendering it much more useful.

Some researchers prefer to rely on datums, which are well dated, easily recognizable first or last appearances of a taxon. For example, the first evolutionary appearance of the genus *Orbulina* from its evolutionary predecessor is called the *Orbulina* datum. This datum has been found in sediments sandwiched between volcanics which could be dated; an age of 14—16 m.y. brackets the datum. This datum has been further related to the paleomagnetic time-scale in middle Miocene sediments from Italy. Here the paleomagnetics of the sediments were measured and this datum fell within the top of Magnetic Epoch 16. Using the age of 14—16 m.y. for the event, it was possible to date the top of Magnetic Epoch 16 at 14—16 m.y.

The relating of faunal events to the paleomagnetic time-scale has been accomplished back to the Miocene, and is presently being extended farther and farther back into the Tertiary by T. Saito and his colleagues at Lamont Geological Observatory.

Several methods have been developed to use the chemical composition of foraminifera to relate them to absolute time.

Some of the amino acids found in fossil foraminifera are products of the thermodynamic change of the amino acid L-isoleucine to D-alloisoleucine through time. At a given time horizon a foraminiferal species will exhibit a unique ratio of isoleucine to alloisoleucine. Wehmiller and Hare (1971) were able to determine the empirical relationship between the A/I ratio and time for mollusks. King and Hare (1972) then measured the A/I ratio for several recent species of the planktonic foraminiferal genus *Globoquadrina* down a deep-sea core with a continuous and reliable paleomagnetic record. They found that each species exhibited a different degree of changed amino acids at each magnetic time horizon. They could then construct an empirical curve relating age to the A/I ratio (by way of the Wehmiller—Hare equation) for each species. Thus, for those species whose rate of change has been discovered, it is possible to relate the A/I ratio to time and hence to determine the age of a sample reliably from 50,000 years B.P. to 10 m.y. B.P.; that is, in the time range when other dating methods are not applicable.

EVOLUTION

Foraminifera evolved in an evolving ocean. We know that the chemical composition of both atmosphere and hydrosphere have changed through geologic time. For example, the oxygen content of the atmosphere changed within the Precambrian and probably markedly during the Devonian with the evolution of land plants. The type and amount of elements contributed to the oceans changed with repeated uplift and denudation of continental land masses. The amount of carbon dioxide available to the ocean may vary according to the extent of tropical forests where it is generated. And continental drift has altered through geologic time the amount of continental shelf present, the latitudinal situation of shelves and seas, and the number and types of ecologic niches present. These and other changing features of the hydrosphere demonstrate that we should relate the evolution of foraminifera to the constantly changing face of a dynamic earth.

Organisms evolve by adaptation, which allows them to cope with changing environmental regimes, exploit new environments, diversify, and function efficiently. The history of the Foraminifera follows this pattern (Fig. 28).

Paleozoic

Foraminifera apparently evolved the ability to encase themselves in a mass of sedimentary particles sometime in the Cambrian. Prior to this time foraminifera may have existed as naked cells or have been covered in an organic sheath composed of tectin. The earliest foraminiferal tests were single-chambered sacks or tubes, some with radiating or branching tubular extensions. These tests were all composed of poorly cemented mud or sand particles (*Astrorhiza*). By the Silurian tubular and coiled-tubular species (*Lituotuba*), attached (*Tolypammina*), planispirally coiled (*Ammodiscus*) and high-spired (*Turitella*) forms occurred in foraminiferal faunas. All these genera were single-chambered and agglutinated.

For nearly 100 m.y. after their origination, testate foraminifera exclusively inhabited shallow marine epicontinental seas. Although there is little record of deeper water Paleozoic, we assume that few foraminifera could have lived there, as they do not appear commonly in deep water sediments until the Jurassic.

A major pulse in foraminiferal radiation occurred in the Devonian. This increased diversification of late Devonian faunas was predicated on two major evolutionary steps: (1) the development of septae, resulting in multi-chambered tests and thus periodic growth; and (2) the evolution of new wall types, the calcareous microgranular and the calcareous fibrous.

Fig. 11 depicts one idea of the development of septation in tubular foraminifera. In this fashion an early ammodiscid type could have evolved to the Devonian septate family, the Tournayellidae, so that by the late Devonian septation and multilocularity were established features of foraminiferal morphology. And with multilocularity came the possibility of new coiling arrangements; biserial, triserial and multiserial coiling all evolved at this time.

Since early foraminiferal tests were presumably tectinous with cemented exogenous grains, a significant change in protoplasmic chemistry must have occurred to enable the protoplasm to align calcite particles as well as to precipitate a microgranular test. Some researchers explain this change in shell chemistry

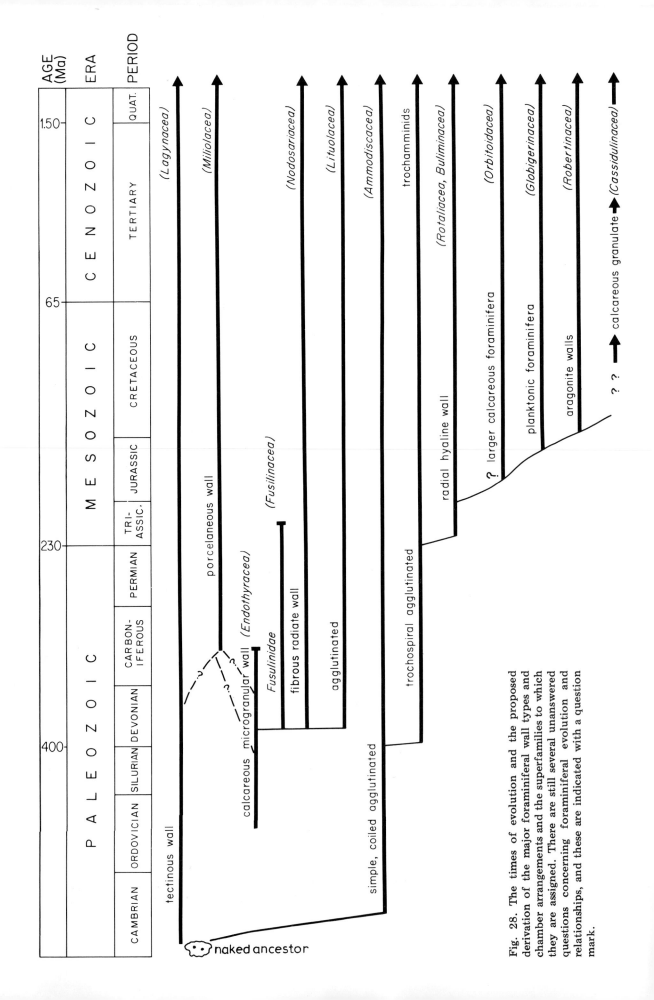

Fig. 28. The times of evolution and the proposed derivation of the major foraminiferal wall types and chamber arrangements and the superfamilies to which they are assigned. There are still several unanswered questions concerning foraminiferal evolution and relationships, and these are indicated with a question mark.

by the hypothesis that the oceans at this time finally became supersaturated with calcium carbonate and so foraminifera could use this compound to construct a test. Others suggest that microgranularity evolved as foraminifera advanced into carbonate-rich environments where they could easily use calcium carbonate as cement for sedimentary particles. Eventually, then, the foraminifera could precipitate their own calcium carbonate. The two new Devonian superfamilies with microgranular tests were the Parathuraminacea and the early Endothyracea.

Foraminiferal faunas of Devonian age have been retrieved from two facies, shallow carbonate facies and basinal facies. Each environment contains a unique fauna, although the microgranular genus *Parathurammina* apparently ranged through both environments. The Devonian is considered a period of extensive carbonate environments, as evidenced by the widespread limestones and dolomites of this age. It may be no coincidence that foraminifera first invaded and survived in carbonate environments of this time. At the end of the Devonian, tectonism along the Acadian and Uralian sutures produced environmental changes affecting the foraminifera.

By the termination of the Devonian the following families are represented in foraminiferal faunas: the Lituolidae, Nodosariidae, Endothyridae, and Parathuramminidae (all new families); as well as the preexisting Ammodiscidae, Astrorhizidae, Tournayellidae, Saccamminidae, Moravamminidae, Nodosinellidae, Colaniellidae, Ptychocoadiidae, and Semitextulariidae.

The Carboniferous was a period of radiation among pre-existing families of foraminifera, particularly the new microgranular families from the Devonian. The endothyrids, in particular, built tests of increasing complexity and size, giving rise during this time to the family Fusulinidae, probably the most important and complex of all Paleozoic foraminifera (Fig. 29). The increasingly complex interiors of the microgranular foraminifera were paralleled by the evolution of internal complexity (labyrinthic walls) among the agglutinated Lituolidae.

A significant evolutionary novelty in morphology at this time was the evolution of trochospiral coiling. Before this time there

were planispiral involute (*Haplophragmoides*) or planispiral evolute (*Forschia*) species; then trochospiral coiling appeared in both microgranular (*Tetrataxis*) and agglutinated (*Trochammina*) genera. These calcareous trochoid forms are thought to be ancestral to the very important Mesozoic and Cenozoic rotaloid foraminifera.

During the Carboniferous foraminifera became more significant members of invertebrate communities and important rock-forming elements.

In the Mississippian there are limestones composed primarily of the genus *Endothyra*. Abundant fusulinid limestones occur in the late Pennsylvanian and Permian around the world. Such limestones apparently formed on shallow tropical shelves rich in calcium carbonate.

In the late Carboniferous—Permian the first porcelaneous family, the Agathamminidae, appeared. These forms are thought to possess an imperforate porcelaneous inner wall and a finely particulate, agglutinated outer wall. Some porcelaneous genera today also possess two outer walls built onto a tectinous basal layer.

The Permian could really be termed the period of the fusulinids as this group radiated into more than 5,000 species at this time. Less important, though well represented in the Permian, were the smaller foraminifera. During the late Paleozoic seventeen new families of smaller foraminifera evolved, but this radiation was abruptly terminated by a world-wide episode of mass extinctions which eliminated the fusulinids, the endothyrids, and members of ten other foraminiferal families. Both shallow and deep living macroinvertebrate taxa also became extinct at this time.

Beginning in the early Carboniferous the earth developed glacial conditions at the south pole and later at the north pole, as well. This glaciated earth must have been significantly different from the earlier Paleozoic in terms of climate, ocean circulation, sea levels, and available carbon. In addition, the late Paleozoic was a time of active tectonism and continental movements, all of which combined to alter drastically conditions in the hydrosphere to the detriment of marine ecosystems.

Mesozoic

Diversification of foraminifera in the Triassic was slow. Rocks of early Triassic age contain rare traces of poorly preserved foraminifera, visible only in thin-section. The eventual radiation into twelve foraminiferal families took place in shallow shelf seas; however, in the Triassic we have perhaps the first record of deeper living species from the family Nodosariidae. These forms are thought to have lived at outer shelf to upper slope depths.

Members of the Nodosariidae first evolved a perforate fibrous radiate wall in the early Mesozoic (Fig. 28). Although both perforate and fibrous walls are to be found in Permian fibrous walls are to be found in Permian species, the Triassic genera *Nodosaria*, *Marginulina*, and *Astacolus* are significant because they represent the combination of perforations and a fibrous wall, called **radial laminated calcite** in Recent nodosariids.

Another important group linking Permian to later Mesozoic foraminifera are the Duostominidae. These forms are thought to be the bridge between trochoid Permian microgranular genera and the earliest Mesozoic trochoid, perforate radial-walled forms that eventually became the lamellar foraminifera. These duostominids were shallow water foraminifera often found in proximity to reefs.

From the Triassic there are some of the first truly deep geosynclinal deposits preserved. Although these deposits do contain other invertebrate fossils, they do not contain any foraminifera and thus we assume that the foraminifera had not yet entered this environment.

The Triassic is generally considered an arid period characterized by regressions of the sea. Restriction of the extent of epicontinental seas, and thus marine niches, may have been responsible for the slow recovery and diversification of foraminiferal families.

A major widespread stratigraphic break between the Triassic and the early Jurassic marks the beginning of an entirely different world ocean system. The Jurassic was the beginning of a period of extensive transgressions, equable climates, shallow and warm epicontinental seas, and circum-global equatorial warm current systems; some researchers refer to this as the period of "milk and honey". Later the seas transgressed onto land in conjunction with the suturing of the Atlantic rift system and the opening of the Atlantic Ocean.

In the epicontinental seas of the Jurassic both calcareous and agglutinated bottom-dwelling foraminifera flourished. Lituolids grew to enormous sizes. The nodosariids experienced a major period of radiation in open basins and shelves, producing a vast array of new genera. These Jurassic nodosariids may be the first "ornamented" foraminifera, as many developed striae, nodes and other surface ornamentation (*Palmula*, *Lenticulina*, *Lagena*, and *Frondicularia*; Fig. 30).

The first bathyal foraminifera are reported from the Jurassic. The nodosariid genus *Spirillina* and some small, simple agglutinated forms have been retrieved from sediments considered to represent bathyal depths. While the nodosariids dominated the slope and shelves, near-shore and back-reef habitats were filled by the first lamellar foraminifera (*Turrilina*, from the family Buliminidae), simple adherent foraminifera (*Nubecularia*), the large lituolids, and several new genera of porcelaneous foraminifera (*Triloculina*, *Quinqueloculina*, and *Opthalmidium*, of the family Miliolidae).

An entirely novel wall composition was evolved during the Jurassic in the new superfamily, the Robertinacea, which employed aragonite instead of calcite in the test. Nevertheless the majority of forms were still composed of calcite.

Probably the most revolutionary evolutionary event in the Jurassic, representing a major step in foraminiferal evolution and in the evolution of the hydrosphere, was the adaptation of foraminifera to a planktonic mode of life. The first planktonic foraminifera were tiny, trochospiral, often high-spired calcareous forms (*Gubkinella*) that are found primarily in indurated sediments (and thus visible in thin section). These minuscule planktonics appeared in the later Jurassic and continued into the Cretaceous when they underwent a spectacular radiation.

Several other groups of micro-organisms,

notably the nannoplankton and diatoms, also entered planktonic niches at this time. Some major change in ocean chemistry or food chains must have occurred to facilitate this multitude of microplankton.

The Cretaceous is considered a continuation of the period of "milk and honey", for it was during this time that foraminifera spread into the vast variety of niches that they now occupy. Thus by the Cretaceous it is possible to delimit environmental indicator faunas ranging in depth from lagoons and marshes to the abyss. Widespread equable climate, extensive shelf seas, and a circum-global warm active current system, Tethys, no doubt were responsible for the extensive migrations and geographic diversification of Cretaceous foraminifera.

Lagoonal, back-reef, and shallow neritic deposits of the Cretaceous contain primarily agglutinated foraminifera, particularly the cuneolinids, dicyclinids, orbitolinids, and valvulinids; the large Jurassic lituolids are no longer present. On carbonate platforms porcelaneous foraminifera, both the smaller forms deriving from the Jurassic and some new, large and complex taxa of the family Alveolinidae are accompanied by the larger orbitoidal foraminifera (*Orbitoides*, *Cymbalopora*) and by the larger agglutinated genus *Loftusia*. Although large agglutinated ⸤raminifera existed before the Cretaceous, larger calcareous forms, both porcelaneous and hyaline, were a novelty during this period (Fig. 28).

Outer shelf deposits contain primarily the rotaliid groups of nodosariids, buliminids, and cibicidids, while a new deeper water rotaliid fauna was characterized by the key genera *Aragonia*, *Nuttalinella*, and *Gavelinella*, usually accompanied by several genera of agglutinated foraminifera (Fig. 30).

Despite their origin in the Jurassic, there is little record of the continuation of planktonic

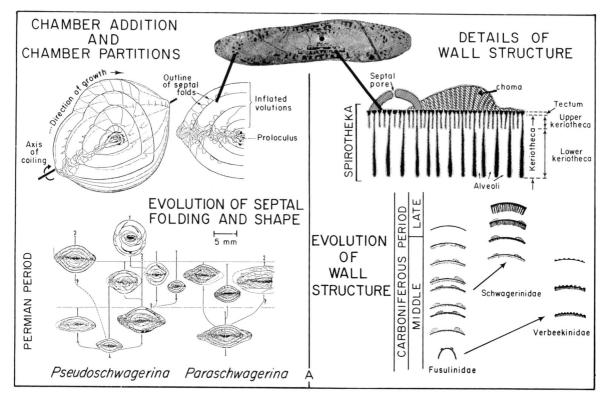

Fig. 29. A. Diagrammatic representation of the internal features of microgranular tests and walls and how these features evolved through the late Paleozoic. (Fig. 29B on pp. 62—63).

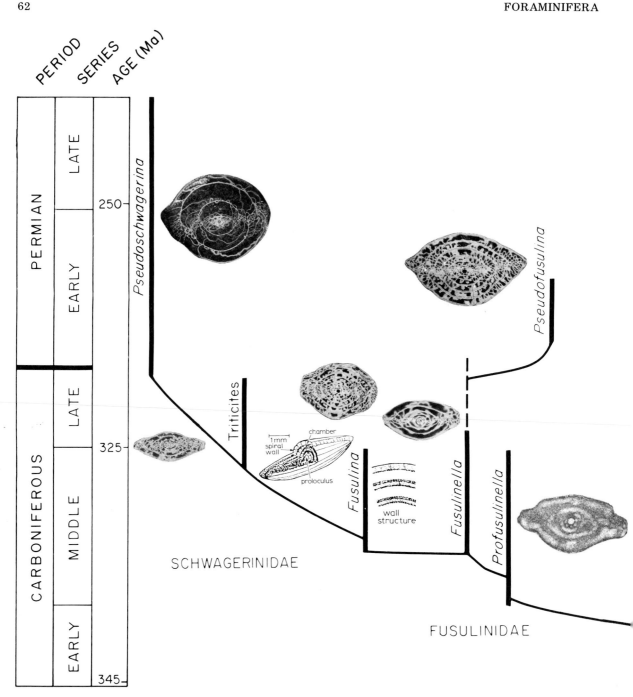

Fig. 29B. Phylogenetic relationships of some of the major microgranular benthic foraminiferal families of the late Paleozoic. Key genera are pictured; radiation into additional genera, unpictured, is indicated by arrows; photos are thin sections cut along the axis of coiling to show the major morphological features used to distinguish genera and species.

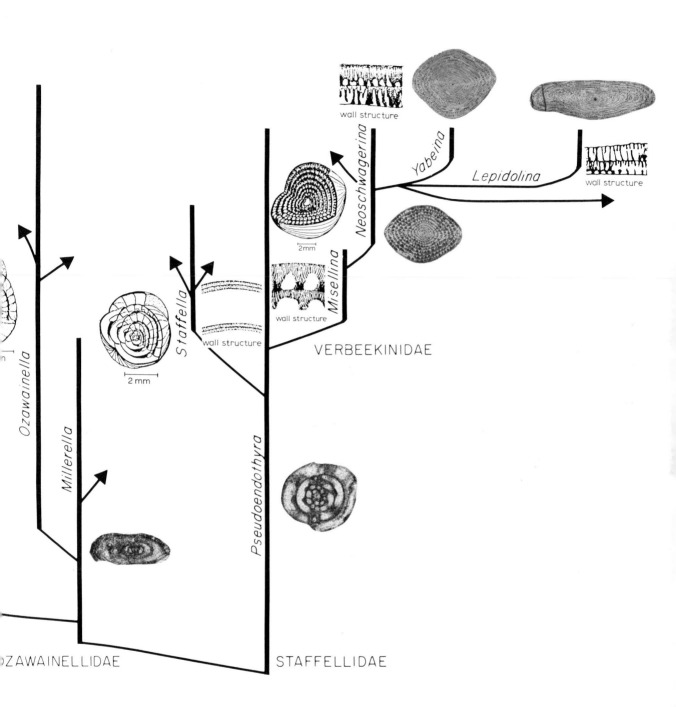

Ozawainella

Millerella

Staffella

wall structure

2 mm

Pseudoendothyra

OZAWAINELLIDAE

STAFFELLIDAE

Misellina

wall structure

2mm

VERBEEKINIDAE

Neoschwagerina

wall structure

Yabeina

Lepidolina

wall structure

2 mm

genera across the Jurassic—Cretaceous boundary. Later in the Cretaceous, however, the planktonics underwent three main episodes of radiation in which the major planktonic morphologic novelties, such as keels, relict apertures, tegillas, and double keels evolved. Cretaceous planktonic faunas contained "globigerinids", triserial and biserial planktonics (*Heterohelix*) and in the late Cretaceous the stratigraphically important genus *Globotruncana* (see Appendix).

The Cretaceous terminated with abrupt mass extinctions that destroyed the majority of planktonic genera as well as the shallower dwelling benthics. The fact that most of the abyssal and bathyal benthic fauna did not disappear in latest Cretaceous time suggests that whatever oceanic changes occurred, affected only the top 200—300 m of the water column and the shallow epicontinental seas. Controversy continues as to the causes of these massive extinctions; explanations range from high radiation due to supernovas, volcanic plumes of noxious gases poisoning the ocean, to a breakdown in marine food chains due to oxygen depletion in the surface waters.

Fig. 30. Typical Mesozoic foraminifera. All figures approximately × 62. A. *Globotruncana*, the major planktonic foraminiferal genus used for late Cretaceous biostratigraphy. B. *Aragonia*, a smaller benthic foraminiferal genus which evolved during the Cretaceous and, along with *Gavelinella*, is one of the deep--water, open-ocean environmental indicators. C. "*Globigerina helvetojurassica*", in thin section, one of the earliest planktonic foraminifera of the Mesozoic. This specimen is from the late Jurassic of the North Atlantic. D. "*Globigerina*", a rarely found free specimen from the Jurassic of France. These earliest globigerinids are also called *Gubkinella*, particularly in reported Triassic occurrences. The earliest planktonics are supposed to have been very high spired, trochospirally coiled forms which became lower spired trochospiral and evolved into the "globigerinids" of the Jurassic. (Courtesy J. Aubert.) E. *Citharina*, a smaller benthic representative of shelf deposition in the middle Jurassic (Bathonian). Note the heavy striations; many components of shelf faunas of this time are heavily ornamented. F. *Lenticulina*, an early Cretaceous bathyal benthic species from DSDP Site 105 in the North Atlantic. G, H. *Hoeglundina*, a bathyal aragonitic genus belonging to the first aragonitic superfamily, the Robertinacea, which evolved in the Jurassic. I. *Opthalmidium*, a porcelaneous genus from the late Jurassic upper bathyal zone of the North Atlantic. J, K. *Dorothia*, a triserial becoming unserial, agglutinated genus from bathyal depths in the North Atlantic. L. *Bigenerina*, an early late Jurassic upper bathyal smaller agglutinated foraminifer from the North Atlantic. M. *Saracenaria*, a smaller benthic genus found at shelf depths from the middle Jurassic of southern France. N. *Spirillina*, the smaller calcareous genus whose enrolled test is composed of a single calcite crystal. These forms are typical of the Triassic and, along with various lagenids, are the first foraminifera to characterize truly deep marine environments. This specimen is from the late Jurassic of the North Atlantic. (Figs. C, F—L, N Courtesy H. Luterbacher).

Cenozoic

The Cenozoic saw a revitalization of foraminiferal faunas and a recolonization of the upper water column and the shallow shelves. One new family of aragonitic foraminifera developed at this time, but the Tertiary is primarily characterized by the continuation of Mesozoic families. Modern species, both planktonic and benthic, are considered to have evolved during the course of the Miocene.

Larger calcareous foraminifera (*Nummulites*) were abundant in the early Tertiary but dwindled in importance after the Miocene and are now primarily found only in the Indo-Pacific and similar tropical regions. The geographic distributions of the other Tertiary foraminifera parallel their present-day distributions despite the climatic cooling which has altered the thermal structure and circulation patterns of the Cenozoic oceans.

Iterative and convergent evolution

Closer inspection of the picture presented by the evolution and adaptive radiation of the Foraminifera reveals repeated patterns of evolutionary change. Such repeated evolution of a particular morphologic type is called iterative evolution and is particularly well illustrated by the planktonic foraminifera. There were two periods of extensive diversification of planktonic foraminifera; one during the late Jurassic, but most prominent in the middle Cretaceous; and the second following the mass extinctions at the Cretaceous—Tertiary boundary. In both cases the ancestral form appears to be a tiny, high-spired trochospiral form which evolved into several low-spired trochospiral genera which in turn radiated into the vast variety of later forms. Frerichs (1971) in depicting the repeated evolution of morphologic features of Cretaceous and Cenozoic planktonics showed most clearly the iterative evolution of many morphologies in this group (Fig. 31).

Convergent evolution is really a term for iterative evolution in unrelated groups. The evolution of similar external appearance by

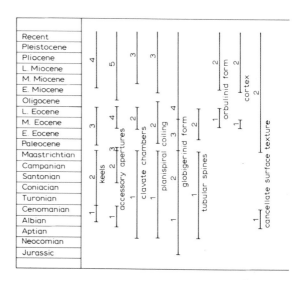

Fig. 31. Iterative evolution among the planktonic foraminifera. Repetition of morphologic traits among the planktonic foraminifera since their evolution in the Jurassic (Triassic?). Lines indicate the stratigraphic range of a morphologic feature. Continuous lines indicate that the feature has evolved in related groups; broken lines indicate that the feature evolved in one group and then later in another; numbers over the lines indicate the number of times a feature has appeared in the stratigraphic record. Thus, keels have evolved four times in the planktonic foraminifera: (*1*) in the rotaliporids, (*2*) in the globotruncanids, (*3*) in the morozovellids, and (*4*) in the globorotalids. Many of these features are shown in the Appendix at the end of this chapter.

calcareous and agglutinated foraminifera is common; for example, *Cribrostomoides—Pullenia, Pseudobolivina—Bolivina, Trochammina—Globigerina.*

Phyletic trends

The pattern of change within an evolving lineage is termed a phyletic trend. Some of the most well-known phyletic trends have been recognized through studying the microspheric and megalospheric individuals of larger foraminifera. Recall that the microspheric, or B generation, generally has a smaller proloculus and is the result of sexual recombination; while the more numerous megalospheric, or A generation, individuals result from asexual reproduction. In the larger

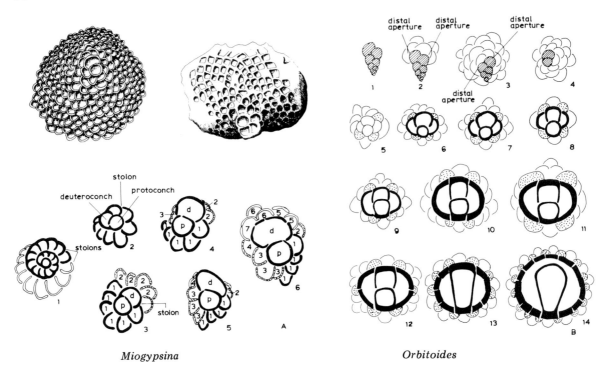

Miogypsina   *Orbitoides*

Fig. 32. Evolutionary trends in two genera of larger foraminifera. A. *Miogypsina*. Schematic drawings of the Miocene larger foraminiferal genus *Miogypsina*. Top of the test is cut away along a median layer revealing the chamber interiors. This figure depicts the evolutionary trends in chamber arrangement and shape resulting from the evolution of stolons, which are tube-like projections of chambers connecting one to another. On this diagram they are represented by holes in the outline of a chamber. As can be seen from this diagram the acquisition of a stolon allows growth in a direction secondary to the original growth direction, and in this case, in a spiral: 1, represents the original rotalid spiral; 2, stolons have formed beginning with the third chamber (at the outer periphery where it meets the second chamber) and in subsequent chambers; 3,4, protoplasm extends through the stolon and begins building new spirals composed of equatorial chambers. This trend is more obvious in the series below of *Orbitoides*. B. *Orbitoides*, a larger benthic foraminiferal genus of Cretaceous age. This series depicts the evolutionary change in the megalospheric generation of *Orbitoides* during the late Cretaceous. The original biserial stage (*Heterohelix*) evolved to the genus *Planoglobulina* by the addition of concentric chambers. The evolution of distal apertures in these concentric chambers then provided the base from which secondary series of concentric chambers could develop. This series also demonstrates alterations in the growth patterns due to the evolution of two stolons in one chamber (7). Successive cycles of chamber addition are presented by different patterns. Note that as chamber arrangement changes, so do the size and shape of the proloculus. Each of these figures corresponds to a species found in the geologic record, thus allowing the evolutionary pattern to be inferred from morphological changes in sequential populations from Cretaceous strata.

foraminiferal genus *Orbitoides* (Fig. 32) the embryonic chambers of the microspheric generation reflect the development of this genus from a probable biserial ancestor. Similarly the development from a rotaliid ancestor to the larger foraminifer *Miogypsina* is visible in the microspheric and sometimes the macrospheric generations. Through detailed statistical analysis Drooger (1963) was able to characterize the degree of phyletic development in miogypsinids and to correlate stratigraphically faunas with similar degrees of phyletic development. This degree, or level, of

morphologic development is most useful for biostratigraphy in individual basins or between basins known to have a similar ecologic succession; as environment may affect the timing of the development from one stage to the next, the appearance of developmental stages in widely separated localities *may* be diachronous.

There remains a gap in our knowledge of the comparative rates of evolution of foraminifera, particularly between shallow and deep environments and between the plankton and the benthos.

APPENDIX

Planktonic foraminiferal species or genera which typify each geologic epoch from the Cretaceous through the Tertiary and their ranges are illustrated. Benthic foraminifera used as biostratigraphic indices of the Jurassic are discussed but not illustrated in these charts. Faunas of each epoch are briefly described, particularly their general appearance, in order to facilitate their recognition. Several species, important in the faunas, but with particularly long geologic ranges, are illustrated at the bottom of the range charts. The various publications listed in the references and suggestions for further reading for biostratigraphy of the Tertiary and Cretaceous should be used in conjunction with these charts if recognition of units smaller than an epoch is desired, or for the appropriate biostratigraphic zones of each epoch.

Jurassic

Biostratigraphy of the Triassic and the early Jurassic is based primarily on invertebrate fossils, but also on the smaller benthic foraminifera. In early Jurassic faunas there is an abundance of calcareous genera such as *Lingulina*, *Frondicularia*, and *Marginulina* (see Fig. 30); and a rapid expansion of the genus *Lenticulina*, which at this time is unornamented. Typical agglutinated genera are *Ammodiscus*, *Ammobaculites* and *Glomospira* (Fig. 6). Together these genera typify the early Jurassic, although many are remnants of late Triassic faunas. Minuscule planktonic foraminifera are reported from Upper Triassic rocks of Europe, but these are very rare.

The middle Jurassic is characterized by a continuation of abundant species of *Lenticulina* (Fig. 30) many of which become strongly ornamented. Originating in the top of the lower Jurassic, the new genus *Epistomina* typifies the middle Jurassic along with other new calcareous genera such as *Paalzowella* and *Unicospirillina*, which ranges into the late Jurassic. Typical agglutinated genera include *Neocorbina* and lituolids with complex interiors, some of which grow to great size relative to the calcareous benthics.

The late Jurassic is then typified by a reduction in the lenticulinids, a continuation of the other calcareous genera, and a further expansion of the complex lituolids. Although more often found, the planktonic foraminifera are still relatively rare in Jurassic faunas.

Early and middle Cretaceous (App. I, p. 69)

Due to a lack of other suitable microfossils, the earliest Cretaceous is zoned on the basis of the calcareous calpionellids (see Chapter 6). Berriasian strata are recognized by the first abundant planktonic foraminiferal faunas which, however, are low in diversity, containing only low trochospiral, globigerinidform species of *Hedbergella*. Later in the early Cretaceous *Hedbergella* is joined by the nearly planispiral group *Ticinella*, and the planispiral genus *Globigerinelloides* (not shown). Although biserial genera evolve in the early Cretaceous, they are not important until the Albian and later. Also in early Cretaceous faunas two genera develop chamber

peripheries elongated into tubular spines (for example, *Shackoina*, not shown).

A major change in the late Albian and extending into the Cenomanian was the evolution of the first keeled genus, *Rotalipora*. This genus is characterized by a single keel and is distinguished from later, similar genera by the extraumbilical to umbilical position of the aperture. Cenomanian strata are also the locus of the final expansion of the hedbergellids which continue, but are never again so abundant in the Cretaceous. The extinction of *Rotalipora* as well as the extinction or dwindling in importance of many of the other early genera mark the termination of the Cenomanian.

Forms intermediate between *Rotalipora* and the true globotruncanids of the late Cretaceous are called marginotruncanids (*Globotruncana angusticarenata*, *G. renzi*, and *G. pseudolinneana* on this chart). These forms radiate during the Turonian and are separated from the later *Globotruncana* by the possession of apertures in an extraumbilical to nearly umbilical position. These forms are strongly keeled, with keel structures extending over the umbilical peripheries of the chambers in many species. The appearance of these marginotruncanids marks a major change in the overall appearance of Cretaceous faunas, which henceforth are dominated by individuals of this general appearance.

The faunas described above are characteristic of the tropical regions of the Cretaceous; boreal faunas contain mostly globigeriniform individuals such as are typical of all faunas of the early Cretaceous. Rotaliporids do not extend into the boreal zone through most of the middle Cretaceous, while the later marginotruncanids do.

Late Cretaceous (App. II, p. 70)

In the Coniacian biserial genera, for example *Heterohelix*, become more common in faunas and the marginotruncanids attain their greatest abundance. The Coniacian, thus, is a sort of transition period between early Cretaceous faunas and the very morphologically diverse faunas of the late Cretaceous. Typically new late Cretaceous genera first appear near the top of the Coniacian and begin to expand during the Santonian, but it is not until the top of the Santonian that a true radiation throughout the planktonic realm produces several major new groups of both biserial and trochospiral foraminifera. By the end of the Santonian the marginotruncanids are completely replaced by the important genus *Globotruncana*, which differs by the possession of a truly umbilical aperture and by the subsequent evolution of double keels. The first striate biserial species evolve in these faunas.

Several distinct morphotypes of *Globotruncana* occur during the Maestrichtian; these include the rugose globotruncanids of the lowest Maestrichtian, the very high-spired *Globotruncana contusa* and the tall, almost box-like in side view *Globotruncana gansseri* (not shown), both of the middle Maestrichtian. Very distinct forms such as *Abathomphalus mayaroensis* facilitate recognition of the late

Maestrichtian. Both *Globotruncana* and the hetero-helicids reached an acme in their diversity and abundance prior to the extinction of most of all planktonic foraminifera at the top of the Cretaceous.

Late Cretaceous faunas at middle and higher latitudes strongly resemble those from the tropics; however, certain species of *Globotruncana* are known to be restricted to the tropics and/or shallower waters (for example, *Globotruncana contusa* and *G. gansseri*).

Tertiary (App. III—VII, pp. 71—75)

The Paleocene is characterized by the gradual reappearance of the planktonic foraminifera. The ancestral form is hypothesized to be a tiny, high trochospiral, possibly triserial form which gradually at the base of the Paleocene radiated into the lower trochospiral globigerinid *Subbotina*, and eventually the keeled globorotalid genus of the Paleocene, *Morozovella*. Biserial planktonics also appear in these strata, implying that their genotype also persisted across the Cretaceous—Tertiary boundary. Thus, early Paleocene faunas are rather simple, consisting of globigerinid-form genera, somewhat reminiscent of early Cretaceous faunas in their simplicity and low diversity.

By the middle Paleocene planktonic foraminifera had evolved keels and new genera such as *Acarinina* made their first appearance. Low-latitude faunas are full of these keeled globorotalids (*Morozovella velascoensis* and *Morozovella subbotinae*) in the late Paleocene through the early Eocene which is a period of radiation in the keeled forms.

In the middle Eocene several new genera, most of which are to characterize the remainder of the Eocene, first appear; these include *Hantkenina*, *Globigerinatheka*, *Truncorotaloides* (not shown) and *Globigerina*. Keeled species become much less abundant and globigerinid-form, spherical forms, and acarininids dominate the faunas. By the late Eocene the keeled morozovellids and the acarinids have disappeared from the faunas, which are dominated instead by the globigerinids and one keeled globorotalid, *Turborotalia cerroazulensis*. Species which characterize the Oligocene first appear in the late Eocene giving the latest Eocene an "Oligocene" appearance.

In higher latitudes or in dissolved samples *Globigerina* and *Globigerinatheka* are abundant and biserial genera are more abundant than at lower latitudes. At lower latitudes the extinction of several genera including *Hantkenina* and *Globigerinatheka* characterizes the top of the Eocene.

Throughout the Oligocene faunas are dominated by several groups of globigerinid-form planktonics (*Globigerina ampliapertura*, *Globigerina ciperoensis* group (not shown) and the unkeeled globorotalids (*Globorotalia opima* group). Loss of the biserial and planispiral species (*Chiloguembelina* and *Pseudohastigerina*, respectively) characterize the middle Oligocene, while the evolution of new, Neogene genera and species typify the late Oligocene. These genera include: the new biserial genus *Streptochilus*, the unique *Globigerinoides* with supplementary apertures on the spiral side, and the toothed *Globoquadrina*.

Globorotalia kugleri, which evolves in the late Oligocene, may be the first true species of the Neogene genus *Globorotalia*.

The Miocene is characterized by the gradual evolution and expansion of modern genera and species. As compared to the later part, early Miocene faunas are somewhat less diverse; globigerinid-form, globoquadrinids and unkeeled globorotalids dominate the faunas. The genus *Catapsydrax* with its umbilical bullae, also characterizes early Miocene faunas, and becomes extinct at the top of this section. One globular genus, *Globigerinatella* (not shown) occurs towards the top of the early Miocene and is replaced in middle Miocene faunas by the distinctive globular genus, *Orbulina*. During the middle Miocene the first Neogene keeled globorotalids of the *Globorotalia fohsi* group evolved, soon to be followed by *Globorotalia menardii* and, the forms more typical of middle to high latitudes, the *Globorotalia conomiozea* group. Along with *Orbulina*, these keeled globorotalids constitute some of the best documented foraminiferal lineages for use in detailed biostratigraphy of the Tertiary. Late Miocene faunas are replete with the keeled globorotalids, *Globigerinoides*, and the distinctive group the *Sphaeroidinellopsis* lineage, which eventually in the Pliocene leads to the evolution of the characteristic Pliocene/Pleistocene tropical index species, *Sphaeroidinella dehiscens*.

The extinction of *Globoquadrina dehiscens* is an excellent marker for the top of the Miocene; Pliocene faunas then undergo a radiation among the keeled globorotalids in lower latitudes and the unkeeled globorotalids in intermediate latitudes. The keeled globorotalids evolve into several large, multi-chambered species such as *Globorotalia multicamerata*. While early Pliocene faunas retain a 'Miocene' character due to the continuation of several diagnostic species across the boundary, most of these Miocene species become extinct in the mid Pliocene so that late Pliocene faunas have an essentially modern appearance.

Loss of the multi-chambered keeled globorotalids, the strangely fistulose *Globigerinoides fistulosus*, and the evolutionary transition from the unkeeled *Globorotalia tosaensis* to the keeled *Globorotalia truncatulinoides* straddle the Pliocene/Pleistocene boundary. Pleistocene faunas are characterized by abundant keeled globorotalids and a vast variety of genera and morphologies, including a biserial genus in the Pacific. Higher-latitude faunas are characterized by the globigerinids, neogloboquadrinids, the planispiral genus *Hastigerina* (not shown); while one species, *Globigerina pachyderma*, characterizes the highest latitudes of both hemispheres.

The transition to the Holocene, though difficult to recognize paleontologically, can be located by the marker species *Globorotalia fimbriata* with its characteristically fluted keel. Some researchers use variations in the direction of coiling of some planktonic foraminifera to make detailed subdivision of Pleistocene and younger faunas. This feature can also be used in older sediments; however, each species has a unique coiling history and this feature is too detailed to be included here.

APPENDIX I   Early and late Cretaceous

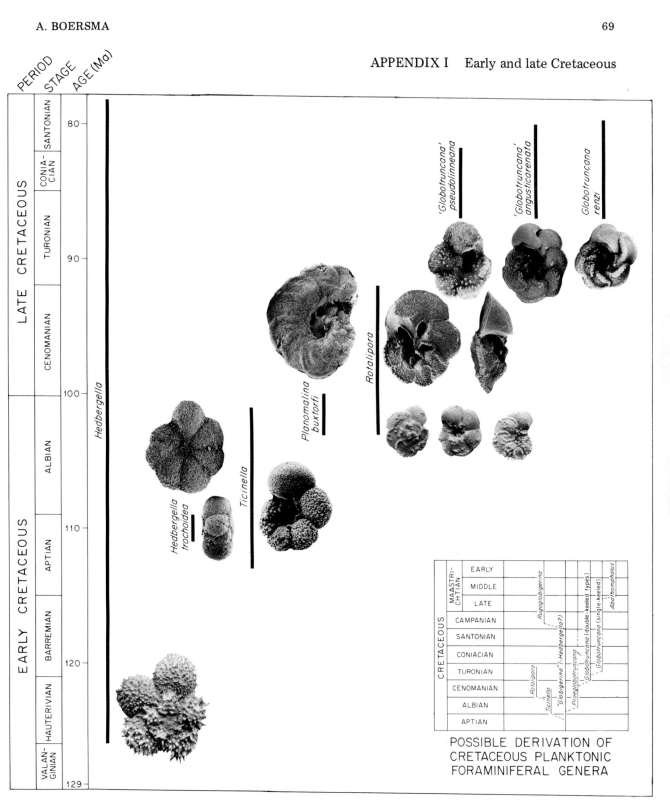

POSSIBLE DERIVATION OF
CRETACEOUS PLANKTONIC
FORAMINIFERAL GENERA

APPENDIX II   Late Cretaceous

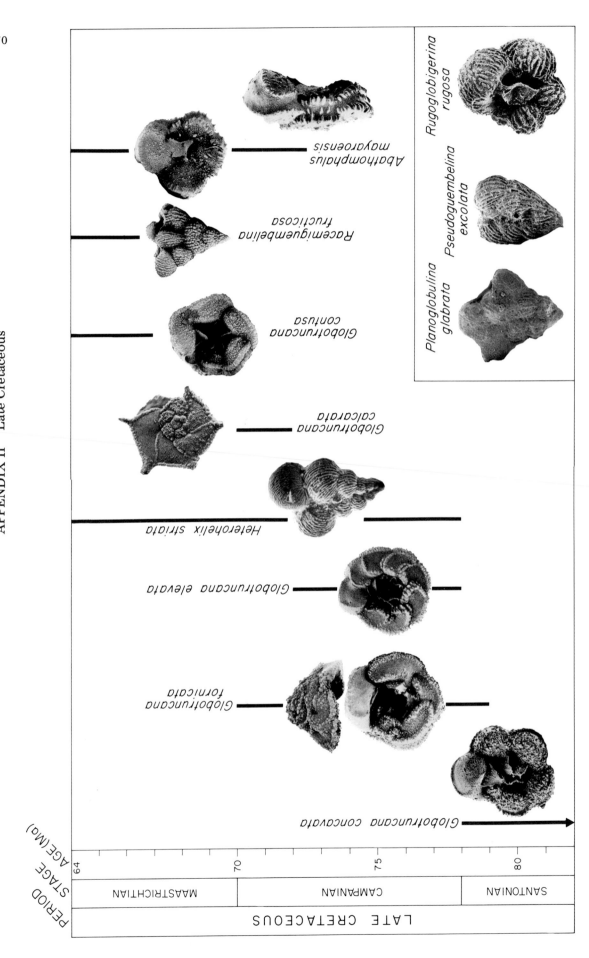

Rugoglobigerina rugosa

Pseudoguembelina excolata

Planoglobulina glabrata

Abathomphalus mayaroensis

Racemiguembelina fructicosa

Globotruncana contusa

Globotruncana calcarata

Heterohelix striata

Globotruncana elevata

Globotruncana fornicata

Globotruncana concavata

AGE (Ma)

STAGE

PERIOD

MAASTRICHTIAN

CAMPANIAN

SANTONIAN

LATE CRETACEOUS

64

70

75

80

71

APPENDIX III   Paleocene

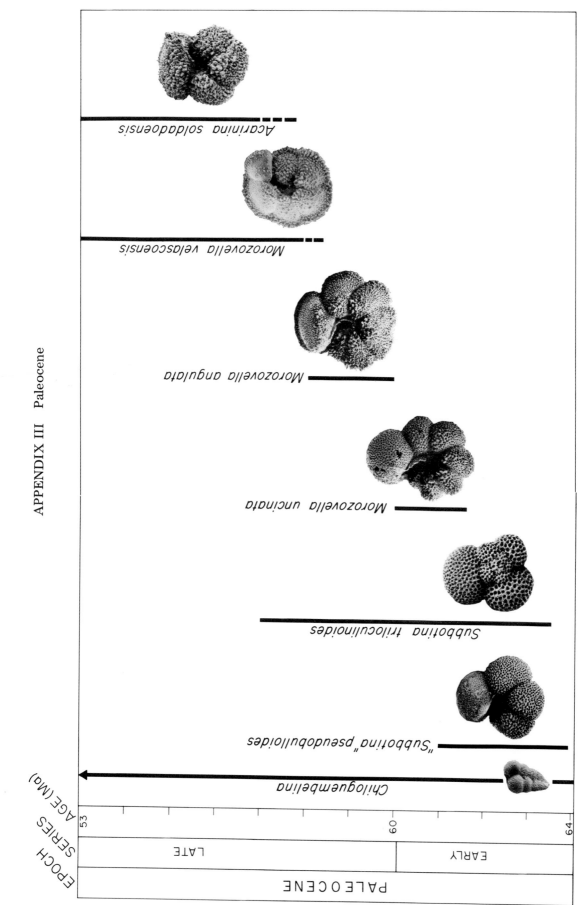

Acarinina soldadoensis

Morozovella velascoensis

Morozovella angulata

Morozovella uncinata

Subbotina triloculinoides

"Subbotina" pseudobulloides

Chiloguembelina

AGE (Ma)
SERIES
EPOCH

53
60
64

LATE
EARLY

PALEOCENE

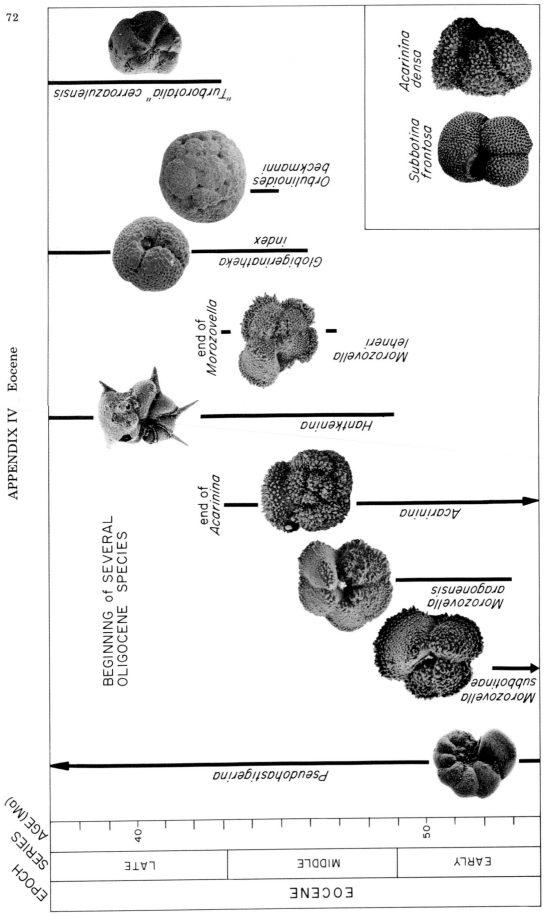

APPENDIX IV   Eocene

"Turborotalia" cerroazulensis

Acarinina densa

Subbotina frontosa

Orbulinoides beckmanni

Globigerinatheka index

end of Morozovella

Morozovella lehneri

Hantkenina

BEGINNING of SEVERAL OLIGOCENE SPECIES

end of Acarinina

Acarinina

Morozovella aragonensis

Morozovella subbotinae

Pseudohastigerina

AGE (Ma)

40

50

EPOCH   SERIES

LATE

MIDDLE

EARLY

EOCENE

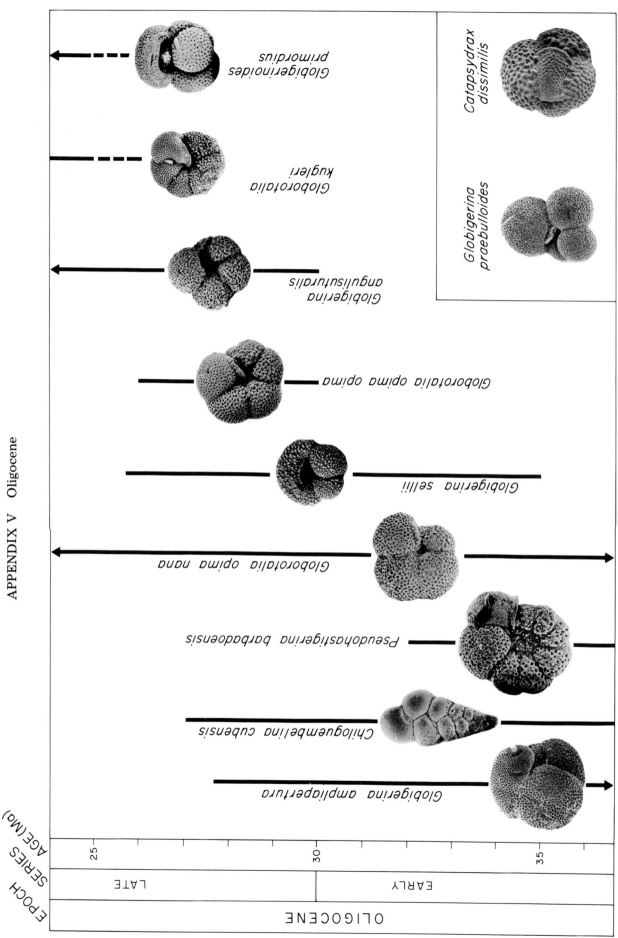

APPENDIX V   Oligocene

Globigerinoides primordius

Globorotalia kugleri

Globigerina angulisuturalis

Globorotalia opima opima

Globigerina sellii

Globorotalia opima nana

Pseudohastigerina barbadensis

Chiloguembelina cubensis

Globigerina ampliapertura

Catapsydrax dissimilis

Globigerina praebulloides

EPOCH | SERIES | AGE (Ma)

OLIGOCENE

LATE

EARLY

25

30

35

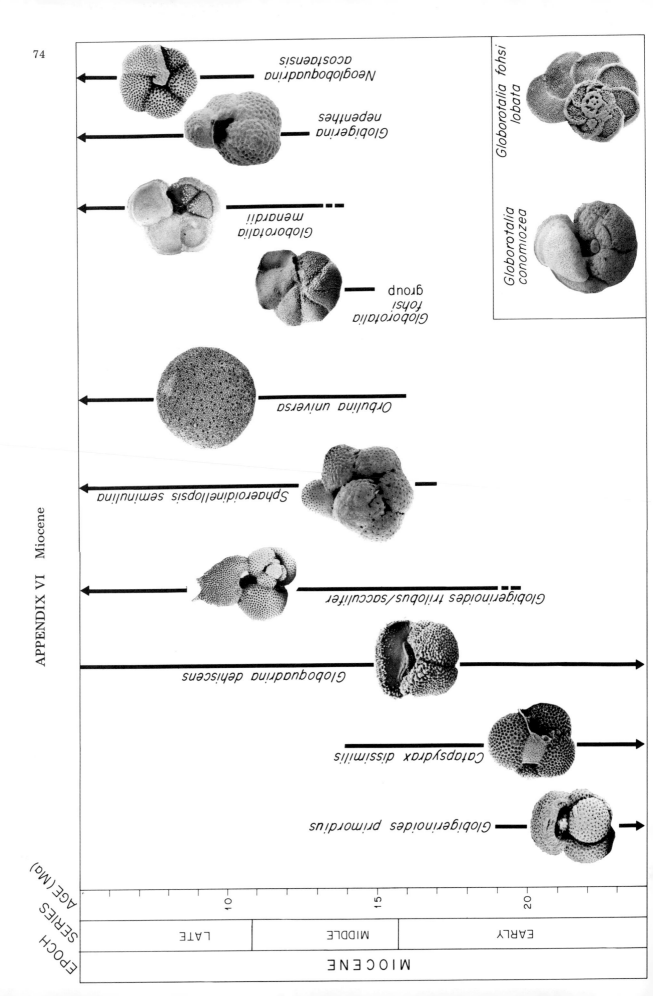

74

Globorotalia fohsi
lobata

Globorotalia
conomiozea

Neogloboquadrina
acostaensis

Globigerina
nepenthes

Globorotalia
menardii

Globorotalia
fohsi
group

Orbulina universa

Sphaeroidinellopsis seminulina

Globigerinoides trilobus/sacculifer

Globoquadrina dehiscens

Catapsydrax dissimilis

Globigerinoides primordius

AGE (Ma)

10

15

20

EPOCH   SERIES

LATE

MIDDLE

EARLY

MIOCENE

APPENDIX VII   Pliocene and Pleistocene

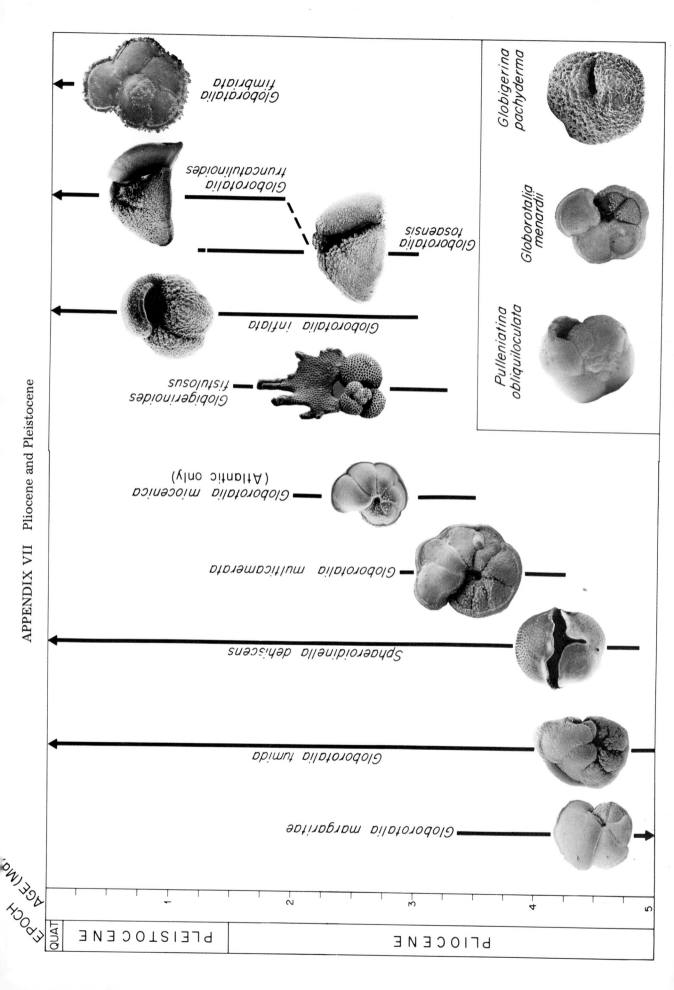

SUGGESTIONS FOR FURTHER READING

Bé, A.W.H., 1977. An ecological, zoogeographic and tax-onomic review of Recent planktonic foraminifera. In: A.T.S. Ramsay (Editor), *Oceanic Micropaleontology*, 1. Academic Press, London, pp. 1—88. [Summary of knowledge on modern planktonics, both in cores and from plankton tows. Includes all the useful references to the study of Recent planktonic foraminifera and their identification.]

Berger, W. and Gardner, J., 1975. On the determination of Pleistocene temperatures from planktonic foraminifera. *J. Foraminiferal Res.*, 5(2): 102—114. [A complete intro-duction to the subject of paleotemperature analysis and its many possibilities; constraints on the method and the problems involved in paleotemperature and climatic inter-pretation are also discussed.]

Berger, W. and Roth, P.H., 1975. Oceanic micropaleontology; progress and prospect. *Rev. Geophys. Space Sci.*, 13(3): 561—634. [Includes a brief description of almost all the work done on microfossils for a decade upto 1975. The references contain almost all major and many minor studies in the field, with an emphasis on the American literature.]

Berggren, W.A. and Phillips, J.D., 1971. Influence of conti-nental drift on the distribution of Tertiary benthonic foraminifera in the Caribbean and Mediterranean regions. *Symp. Geol. Libya, Univ. of Libya, Tripoli*, pp. 263—299. [An interesting presentation of the basics of tectonics, the reconstruction of past oceans from geophysical infor-mation, and the use of the tectonic setting in studying and understanding the distributions of foraminifera, in this case, the smaller benthic foraminifera. Also contains a summary of all the classic literature on Tertiary benthic foraminifera.]

Dilley, F.C., 1973. Larger foraminifera and seas through time. *Spec. Pap. Paleontol.*, 12: 155—168. [This paper sum-marizes the distributions of larger foraminifera and includes much useful literature on aspects not covered in any detail in this chapter.]

Douglas, R.G., 1973. Benthonic foraminiferal biostratigraphy in the Central North Pacific, Leg 17, Deep Sea Drilling Project. *Initial Rep. Deep Sea Drill. Proj.*, 17: 607—672. [One of the few updated catalogues of Tertiary smaller benthic foraminifera as well as excellent photography to facilitate their identification. Includes some intriguing ideas on benthic foraminiferal distributions through time and space which are now still being investigated by many workers.]

Hallam, A., 1973. *Atlas of Palaeobiogeography*. Elsevier, Amsterdam, 531 pp. [A collection of articles covering the paleobiogeography and paleogeographic methods applied to mega- and microfossils from the Paleozoic to the Recent, including foraminifera.]

Hedley, R.H., 1964. The biology of foraminifera. *Int. Rev. Exp. Zool.*, 1: 1—45. [An older, but good summary paper on foraminiferal biology, including synthesis of most of the classic papers on culturing techniques; does not include the planktonic foraminifera, or recent studies on larger foraminifera.]

Hedley, R.H. and Adams, C.G., 1974. *Foraminifera*. Academic Press, London, 276 pp. [This is planned to be a continuing series of volumes containing detailed synthesis papers on all aspects of foraminifera; the first volume contains updated taxonomy and an excellent detailed review of biometric approaches used to study foraminifera.]

Jenkins, D.G., 1971. New Zealand Cenozoic planktonic foraminifera. *N.Z. Geol. Surv. Paleontol. Bull.*, 42: 278 pp. [This volume is a most complete guide to higher latitude foraminiferal identification and biostratigraphy; this can be supplemented by later stratigraphic papers by Jenkins, Edwards, and Kennett in recent DSDP volumes.]

Kummel, B. and Raup, D., 1965. *Handbook of Paleontolog-ical Techniques*. Freeman and Co., San Francisco, Calif., 852 pp. [Contains short articles describing techniques for preparing all types of rocks and samples containing micro-fossils; ingenious tools used to study or manipulate fossils, both invertebrate and microfossils, are also described.]

Laporte, L.F., 1968. *Ancient Environments*. Prentice Hall, Englewood Cliffs, N.J., 115 pp. [An excellent and enjoy-able introduction to the study of environment and the principles of interpretation of past environments; the book deals in concepts.]

Luterbacher, H.P., 1972. Foraminifera from the Lower Cretaceous and Upper Jurassic of the northwestern Atlantic. *Initial Rep. Deep Sea Drill. Proj.*, 11: 561—576. [This paper is an excellent catalogue of smaller benthic foraminifera, and some planktonic foraminifera, to be found in open marine sections; it also includes some interesting paleoenvironmental analyses and comparisons between classic land sections, in this case sections in the Appenines, and the marine sections.]

Murray, J.W., 1973. *Distribution and Ecology of Living Benthonic Foraminifera*. Crane, Russak and Co., New York, N.Y., 274 pp. [It describes most of the studies done on this group in the last ten years or so and enlarges on the information presented by Phleger in 1956 in a book of similar title which is a more elementary, but interesting introduction to the subject. Many of the publications of Orville Bandy should also be consulted on this subject; much recent work has already been anticipated in his papers of the 1950's and 1960's.]

Pessagno, E.A., 1969. Upper Cretaceous stratigraphy of the western Gulf Coast area of Mexico, Texas, and Arkansas. *Geol. Soc. Am. Mem.*, 111: 139 pp. [A classic study of Cretaceous foraminifera with a good catalogue of the species; the drawings can be supplemented by more recent publications in DSDP volumes which usually provide good photography of the species, as well as the Cretaceous sections in Bulletin 215 (see Cited References). For the early Cretaceous, there are useful publications by Moullade and Van Hinte, summarized by Premoli-Silva and Boersma in DSDP Leg 39.]

Pickard, G., 1963. *Descriptive Physical Oceanography*. Pergamon Press, New York, N.Y., 200 pp. [A compact volume with a good deal of the introductory information and many useful ocean-wide charts of temperature, salinity, nutrient, and current distributions from other references are included.]

Pokorný, V., 1963. *Principles of Zoological Micropaleontol-ogy*. Macmillan, New York, N.Y., 652 pp. [There are many existing texts published in Europe which take a classical and thorough approach to the study of micro-fossils; most are not translated or have not been revised in recent years. This translated text has a large section on foraminifera. It has an excellent compilation of the literature, particularly the European literature which is often neglected in American publications. It would be a good companion to the present text for more complete taxonomy and classification of foraminifera, preparation techniques, and the more classic and basic approach to the study of microfossils.]

Postuma, J., 1971. *Manual of Planktonic Foraminifera*. Elsevier, Amsterdam, 420 pp. [A good catalogue for the identification and taxonomy of both Cretaceous and Tertiary index planktonic foraminifera; most of the im-portant species are included.]

Von Koenigswald, G.H.R., Emeis, J.D., Buning, W.L. and Wagner, C.W., 1963. *Evolutionary Trends in Foraminifera*. Elsevier, Amsterdam, 355 pp. [This is an older collection of some of the classic studies in foraminiferal evolution; it covers the larger benthic foraminifera in detail and presents a more complete approach to their study; the subject of foraminiferal evolution has been less studied in recent times and most of the useful literature on newer evolutionary concepts is to be found in the non-special-ized literature, for example in journals such as *Paleo-biology* and *Evolution*.]

Wagner, C.W., 1964. *Manual of Larger Foraminifera*. Bataafsche Petroleum Mij., The Hague, 307 pp. [This is a difficult publication to acquire as it is produced for an oil company; if available it is an excellent introduction to the types and taxonomy of larger benthic foraminifera.]

[See also various Colloquia volumes, e.g., the Planktonic Conference volumes and the International Micropaleontolog-ical Colloquia in Africa, and the Caribbean. Useful journals include *Micropaleontology*, *Revista Española di Micropale-*

ontologia, Journal of Foraminiferal Research, Special Publications of the Cushman Foundation, Utrecht Micropaleontological Bulletins, and *Marine Micropaleontology*.]

CITED REFERENCES

Adams, C.G., 1967. Tertiary Foraminifera in the Tethyan, American and Indo-Pacific provinces. *Syst. Assoc., Publ.*, 7: 195—219.

Barker, R.W., 1960. Taxonomic notes on the species figured by H.B. Brady in his report on the foraminifera dredged by H.M.S. "Challenger" during the years 1873—1876. *Soc. Econ. Paleontol. Mineral., Spec. Publ.*, 9: 238 pp.

Berger, W.H., 1971. Sedimentation of planktonic foraminifera. *Mar. Geol.*, 11: 325—358.

Berger, W.H. and Winterer, E.L., 1974. Plate stratigraphy and the fluctuating carbonate line. In: K. Hsü and H. Jenkyns (Editors), *Pelagic Sediments on Land and Under the Sea. Int. Assoc. Sedimentol., Spec. Publ.*, 1: 11—48.

Berggren, W.A., 1972. A Cenozoic time-scale — some implications for regional geology and paleobiogeography. *Lethaia*, 5: 195—215.

Berggren, W.A. and Haq, B., 1976. The Andalusian Stage (Late Miocene): biostratigraphy, biochronology, and paleoecology. *Palaeogeogr., Palaeoclimatol., Palaeoecol.*, 20: 67—129.

Bolli, H.M., 1957. Planktonic foraminifera from the Oligocene—Miocene Cipero and Lengua formations of Trinidad. In: A.R. Loeblich and others, *Studies in Foraminifera. U.S. Natl. Mus. Bull.*, 215: 97—125.

Bradshaw, J.S., 1955. Preliminary laboratory experiments on ecology of foraminifera. *Micropaleontology*, 1: 351—358.

Brady, H.B., 1884. Report on the Foraminifera dredged by H.M.S. "Challenger" during the years 1873—1876. *Rep. Voyage "Challenger" Zool.*, 9: 814 pp.

Cifelli, R., 1969. Radiation of Cenozoic planktonic foraminifera. *Syst. Zool.*, 18: 154—168.

Cita, M.B., Nigrini, C. and Gartner, S., 1970. Biostratigraphy. *Init. Rep. Deep Sea Drill. Proj.*, II: 391—413.

Cushman, J.A., 1948. *Foraminifera, Their Classification and Economic Use.* Harvard University Press, Cambridge, Mass., 605 pp.

Cushman, J.A. and Henbest, L., 1942. Foraminifera. In: *Geology and Biology of North Atlantic Deep Sea Cores. Geol. Soc. Am. Prof. Pap.*, 196: 163 pp.

Dilley, F.C., 1971. Cretaceous foraminiferal biogeography. In: F.A. Middlemiss and others (Editors), *Faunal Provinces in Space and Time. Geol. J., Spec. Iss.*, 4: 169—190.

Drooger, C.W., 1963. Evolutionary trends in the Miogypsinidae. In: G.H.R. von Koenigswald and others (Editors), *Evolutionary Trends in Foraminifera.* Elsevier, Amsterdam, pp. 315—350.

Emiliani, C. and Edwards, G., 1953. Tertiary ocean bottom temperatures. *Nature*, 171: 887.

Ericson, D.W. and Wollin, G., 1956. Correlation of six cores from the equatorial Atlantic and Caribbean. *Deep Sea Res.*, 3: 104—125.

Frerichs, W.E., 1971. Evolution of planktonic foraminifera and paleotemperatures. *J. Paleontol.*, 45: 963—968.

Gordon, W.A., 1970. Biogeography of Jurassic foraminifera. *Geol. Soc. Am. Bull.*, 81: 1689—1704.

Imbrie, J. and Kipp, N.G., 1971. A new micropaleontological method for quantitative paleoclimatology: application to a Late Pleistocene Caribbean core. In: K.K. Turekian (Editor), *The Late Cenozoic Glacial Ages.* Yale University Press, New Haven, Conn., pp. 71—181.

Jahn, T.L. and Rinaldi, R.A., 1959. Protoplasmic movements in the foraminiferan *Allogromia laticollaris* and a theory of its mechanism. *Biol. Bull.*, 117: 100—118.

King, K. and Hare, P.E., 1972. Amino acid composition of the test as a taxonomic character for living and fossil planktonic foraminifera. *Micropaleontology*, 18: 285—293.

Lipps, J.H. and Valentine, J.W., 1970. The role of foraminifera in the trophic structure of marine communities. *Lethaia*, 3: 279—286.

Loeblich, A.R. and Tappan, H., 1964a. *Protista. Sarcodina — Chiefly Thecamoebians and Foraminiferida.* Geol. Soc. Am., Univ. of Kansas Press, Lawrence, Kansas, 900 pp. (2 vols.).

Loeblich, A.R. and Tappan, H., 1964b. Stability of foraminiferal nomenclature. *Contrib. Cushman Found. Foraminiferal Res.*, 15: 30—33.

Mamet, B. and Skipp, B., 1970. Lower Carboniferous foraminifera: preliminary zonation and stratigraphic implications for the Mississippian of North America. *Int. Stratigr. Geol. Carboniferous, Sheffield*, 3: 1129—1146.

McKenzie, D. and Sclater, J.G., 1971. The evolution of the Indian Ocean since the Late Cretaceous. *Geophys. J. R. Astron. Soc.*, 25: 437—528.

d'Orbigny, A., 1826. Tableau méthodique de la Classe des Céphalopodes. *Ann. Sci. Nat.*, 7: 245—314.

Savin, S.M., Douglas, R.G. and Stehli, F.G., 1975. Tertiary marine paleotemperatures. *Geol. Soc. Am. Bull.*, 86.

Schott, H., 1935. Die Foraminiferen in dem äquatorialen Teil des Atlantischen Ozeans. *Wiss. Ergebn. Dtsch. Atlantischen Exped. "Meteor" 1925—1927*, 3(pt.3): 42—134.

Sliter, W.V., 1971. Predation on benthonic foraminifera. *J. Foraminiferal Res.*, 1(1): 20—29.

Subbotina, N.N., 1936. Stratigraphy of the Lower Paleogene and Upper Cretaceous of North Caucasus by foraminiferal fauna. *Tr. Heft. Geol. Razved. Inst.*, Ser. A, 245 pp. (in Russian).

Van Hinte, J.E., 1976. A Cretaceous time-scale. *Am. Assoc. Pet. Geol. Bull.*, 60: 498—516.

Wehmiller, J. and Hare, P.E., 1971. Racemization of amino acids in marine sediments. *Science*, 173(4000): 907—911.

3

CALCAREOUS NANNOPLANKTON

BILAL U. HAQ

INTRODUCTION

When fine particles of a marine sediment rich in carbonate are observed under a high magnification of ×500 to ×1000 one is struck by the great diversity and the quantitative prominence of tiny arrays of calcite crystallites known informally as coccoliths and formally as calcareous nannoplankton [*nano* (Greek) = dwarf]. Over 150 known species of these minute unicellular, autotrophic, marine algae (coccolithophores) are living in the present-day oceans. The fossil coccolithophores and related groups of nannofossils (collectively known as nannoliths) have been important constituents of marine carbonate sediments since early Jurassic time.

In the coccolithophores the cell secretes a skeleton of minute calcareous shields which may envelop the cell completely or partially (coccosphere) (Fig.1). Post-mortem disintegration of the coccosphere usually dislodges and scatters the shields before or after they come to rest at the ocean bottom. These individual elliptical to circular shields, or coccoliths, range in size from about 1 to 15 μm. Other related, but morphologically dissimilar groups of organisms traditionally included under calcareous nannoplankton comprise a wide array of elaborate designs (see Figs. 10—31). Of these the Tertiary lineages of asteroliths (discoasters) are more important because of their higher diversity, relative quantitative prominence in the tropical and subtropical sedimentary provinces and their great value to the biostratigrapher.

HISTORY OF THE NANNOPLANKTON RESEARCH

First reference to the nannoplankton was made by the German biologist C.G. Ehrenberg. In 1836 he reported the occurrence of small, "flat", elliptical discs of "agaric-mineral" in the chalk from the island of Rügen. He also figured the first discoasters and called them "calcareous crystal-discs" and considered both these and coccoliths to be of inorganic origin.

In 1858, T.H. Huxley reported the presence of Ehrenberg's "crystalloides" in deep-sea oozes recovered prior to the laying of the first trans-Atlantic telegraphic cable and referred to them as **coccoliths**. Huxley also regarded them as of inorganic origin. In 1861 G.C. Wallich and H.C. Sorby came to the conclusion independently that coccoliths were parts of larger spherical objects to which the former gave the name of **coccospheres**. Wallich was tempted at first to compare the

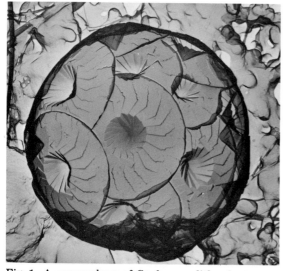

Fig. 1. A coccosphere of *Cyclococcolithus leptoporus* (Murray and Blackman) Kamptner. × 7,300. (Photo: H. Okada.)

coccospheres to the juvenile stages of Foraminifera, but Sorby continued to consider them as separate organisms because of their peculiar optical properties under polarized light. It was Wallich, however, who in a series of short papers between 1863 and 1877 clarified the true nature of the coccospheres and in 1865 reported the discovery of living coccospheres from the tropical waters of the Indian and Atlantic oceans.

In 1891, John Murray and A.F. Renard published their report on deep-sea sediments collected during the famous expedition of the H.M.S. *Challenger*. In this first attempt to understand the nature of the life in the oceanic realm, the naturalists of the *Challenger* expedition described the diverse flora and fauna in a great opus of over 35 volumes. In one of the volumes devoted entirely to deep-sea deposits, Murray and Renard recorded a wide variety of microfauna and flora, including nannoplankton. They noted that both the rhabdospheres (with funnel-shaped coccoliths) and the coccospheres were common in the tropical regions, but the former quickly diminished in numbers in the higher latitudes, rarely occurring in the regions with surface water temperatures less than 65°F (about 18.5°C). On the other hand, the coccospheres, they noted, were encountered in surface water temperatures as low as 45°F (about 7.5°C).

At the turn of the last century the importance of calcareous nannoplankton as primary link in the marine food chain was realized. This provided a strong impetus for the study of living coccolithophores. The German biologist Hans Lohmann was amongst the leading workers engaged in the recording of living nannoplankton, mainly from the Mediterranean Sea. In 1930 in L. Rabenhorst's classic work on the *Kryptogamen-Flora*, Josef Schiller presented a complete account of all known species of extant coccolithophores and this work persists as one of the standard references on the subject to this day. French researchers of this period, particularly F. Bernard, were also active in this field. Two other central European biologists who were particularly active in the field of nannoplankton biology and paleontology in the first half of this century were G. Deflandre and

E. Kamptner. These pioneers in the field of "nanno-paleontology" — a term coined by the latter — have published extensively on the biologic, structural and stratigraphic aspects of coccolithophores.

During the late 1930's the center of research on living coccoliths shifted from central Europe to Norway, where T. Braarud and his colleagues were to become the most active group in this field. These biologists began the difficult task of maintaining and studying pure cultures in the laboratory. The little that we know of the physiology and life cycle of coccolithospheres is mainly due to efforts of these workers at Oslo and Mary Parke and her colleagues at the Plymouth Biology Laboratory in England.

The usefulness of nannofossils as biostratigraphic indicators remained obscure until 1954 when M.N. Bramlette and W.R. Riedel pointed out the distinctiveness of the Mesozoic and Tertiary assemblages and suggested their usefulness, particularly that of discoasters, in worldwide correlation of pelagic sediments. This provided the necessary impetus and great strides were made in the recording of fossil assemblages by central European and American micropaleontologists during the next decade. By the late 1960's our knowledge of the fossil nannoflora and the vertical ranges of species, particularly those of the Cenozoic had reached a stage where biostratigraphic zonations could be proposed. W.W. Hay, M.N. Bramlette and their associates were specially active in this field and were able to establish a refined and practical biostratigraphic zonation which has been successfully used in tropical—subtropical and temperate regions for worldwide correlations of Tertiary strata.

BIOLOGY OF THE ORGANISM

Most living coccolithophores are known to possess the flagellar apparatus, **haptonema**, and the organic surface scales characteristic of the unicellular algal class Haptophyceae (Fig. 2). The ability of the coccolithophores to secrete calcareous plates or coccoliths on these organic scales distinguishes them from other members of the Haptophyceae and all other algae.

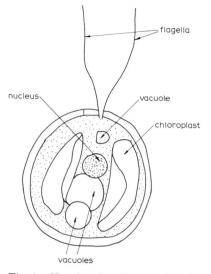

Fig. 2. Sketch of a living cell of *Cyclococcolithus leptoporus* (Murray and Blackman) Kamptner, showing flagellar apparatus (haptonema), vacuoles, chloroplasts and nucleus. × 2,000. (After Schiller, 1930.)

In spite of the importance of coccolithophores as one of the major constituents of marine phytoplankton and their value as primary producers in the food chain, culture studies on coccolithophores are rare. Like other protists, coccolithophores are exceedingly difficult to culture in the essentially restricting environments of the laboratory. Of the few species that have been kept alive, *Emiliania huxleyi* (Lohmann) Hay and Mohler, is relatively easier to culture. This is a ubiquitous species occurring both in pelagic and near-shore areas. It is distributed in tropical, subtropical, as well as higher latitudes, and warm and cool water ecophenotypes have been observed. The high tolerance of *E. huxleyi* is probably responsible for the relative ease in maintaining a culture of this species. Since this species has been more extensively studied than any other, we will take it as an example and follow its case history.

Organization at cellular level

The components of the cell of *E. huxleyi* are shown in Fig. 3. The cell is bound by two double membranes that enclose the protoplasm, containing a prominent nucleus, two chloroplasts, mitochondria, the dictyosome or the Golgi apparatus, and a vacuolar body referred to as "body-X". **Chloroplasts** contain the chlorophyll responsible for photosynthesis. The chloroplasts can change their shape and position within the cell in relation to light intensity. The process of photosynthesis is not fully understood, however, it is basic for the existence of all plant life and indirectly for all other forms of life (excluding certain chemosynthetic bacteria). In simple terms, photosynthesis is a process by which light energy is converted into chemical energy with the help of photosynthetic pigments. Subsequently the chemical energy is used in the reduction of carbon dioxide to form carbohydrates and liberate oxygen.

Mitochondria are bodies with double membranes and tubular extensions into the interior. They contain the oxidative enzyme systems which produce energy for the various cell functions. The dictyosome, or the **Golgi apparatus**, is well developed in *E. huxleyi*. It consists of parallel aligned, closely packed cisternae with vesicular expansions at the ends. The function of the Golgi apparatus is not well understood, but it is believed to participate in the secretion of cell wall material.

Strains of *E. huxleyi* normally secrete coccoliths, but strains which do not calcify have been obtained in higher nitrate culture media. The cell shown in Fig. 3 is from such a strain which does not produce coccoliths. In these naked cells an unidentified vacuolar structure somewhat larger than a mitochondrion has been observed by Wilbur and Watabe (1963). This structure (body-X) usually lies near the nucleus and the cell surface and appears to contain crystalline material. In strains that produce coccoliths, body-X is not present and is replaced by another body containing a reticular structure, composed of strands arranged roughly parallel. The function of this **reticular body** is not well understood, however, Wilbur and Watabe have shown that it is intimately connected with the formation of coccoliths.

Formation of coccoliths

In their study of cross-sections of the various stages of development of *E. huxleyi*

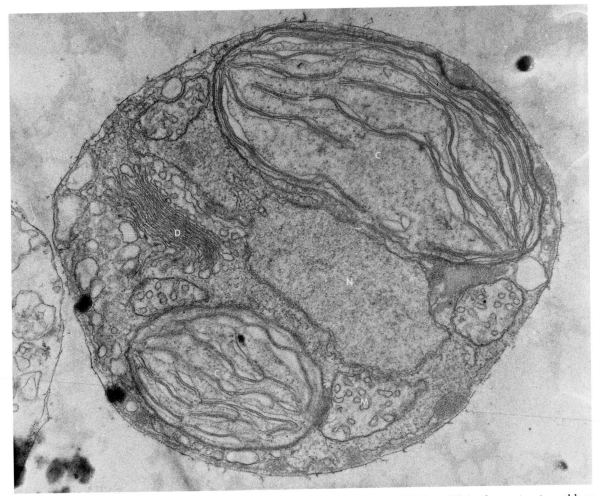

Fig. 3. Cell components of *Emiliania huxleyi* (Lohmann) Hay and Mohler. Nucleus (*N*) in the center, two chloroplasts (*C*) at top and bottom, four mitochondria (*M*) on both sides of chloroplasts, and dictyosome (*D*) or Golgi apparatus at left of nucleus. × 33,700. (After Wilbur and Watabe, 1963.)

under the electron microscope, Wilbur and Watabe (1963) noticed that the mineralization of the coccolith occurs within the protoplasm immediately next to the reticular body. Between this body and the nucleus a non-granular matrix material is differentiated and first assumes the general shape of a coccolith (Fig. 4). Precipitation of the mineral calcite starts from various central points in this matrix, advancing from the base of the coccolith upwards and then to the periphery of the shields.

After calcification the coccoliths are evidently pushed to the surface of the cell where they form into an interlocking spherical envelope around the cell. This coccolith envelope may become multi-layered

as more coccoliths are produced internally.

The process of calcification in all species of coccolithophores is evidently not identical. Manton and Leedale (1969) have observed that in *Coccolithus pelagicus* (Wallich) Schiller and *Cricosphaera carterae* (Braarud and Fagerland) Braarud the coccoliths are attached to underlying unmineralized organic scales and both of these originate within the cisternae of the Golgi apparatus.

Most coccolithophores have a definite range in the number of coccoliths around the cell (e.g., from about 10 to 30 in *Gephyrocapsa oceanica* Kamptner), but these ranges vary considerably in different species. The final shape of the coccolith envelope also shows great variety in various species, the

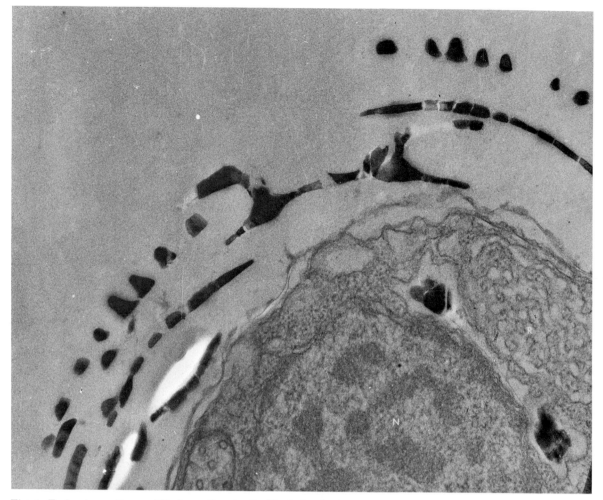

Fig. 4. Early stage of coccolith formation in *Emiliania huxleyi* (Lohmann) Hay and Mohler. Coccolith begins to form between the nucleus (*N*) and reticular body (*R*). The rim of the basal shield is formed first. Some complete coccoliths in cross-section are seen outside the cell at the top. × 70 000. (After Wilbur and Watabe, 1963).

commonest of which is a near-spherical shape. Oblong and spindle-shaped tests and other intermediate varieties are also seen in some extant species.

Life cycle (reproduction, nutrition and growth)

Some coccolithophores pass through 2 phases in their life cycle: a motile phase during which they possess the flagellar apparatus, the haptonema, alternating with a non-motile phase. The cell in the non-motile phase normally secretes coccoliths, whereas in the motile phase it is either naked or bears coccoliths of a different type than those in the non-motile phase. The only known living example of the latter is *Coccolithus pelagicus* (Wallich) Schiller.

Formerly the motile phase of *C. pelagicus* was described as a different species, namely *Crystallolithus hyalinus* Gaarder and Markali, characterized by coccoliths (also called crystalloliths) consisting of loosely bound, unmodified rhombohedral microcrystals contained in the outer hyaline layer of the cell wall. Parke and Adams (1960) were able to culture *C. hyalinus* and discovered that the cell reproduces by fission and the two daughter cells after being rounded off remain inside the outer hyaline layer before emerging as naked cells, eventually forming their own crystalloliths. The motile phase passes into the non-motile phase in 5—8 weeks when the cells begin to collect at the bottom of the flask in a dark olive brown layer. After another two weeks the conversion to the non-motile *Coccolithus pelagicus* phase is complete and large cells with coccoliths consisting of modified crystals arranged into two shields connected by a central tube (placoliths) are observed. Sexual fusion of two motile cells to form non-motile cells may be involved in this conversion from one phase to the other. According to Paasche (1968), if sexual fusion actually occurs then the non-motile phase would represent a **diploid stage** (when the nucleus contains chromosomes in complete pairs) and the motile phase a **haploid stage** (containing a single set of unpaired chromosomes of half the number of diploid stage) in the life cycle of this species. Thus coccolithophores, like Foraminifera and possibly Radiolaria, may have heterophasic life cycles.

Deviations from these modes of reproduction have also been observed in culture. For example, in one strain of *Emiliania huxleyi* studied by Braarud the cell divided only after the protoplasm had escaped the coccolith envelope of the non-motile phase as a free-floating naked swarmer. New coccoliths were gradually produced by the daughter cells produced by division of the swarmer. In yet another strain when fission divided the non-motile cell, each daughter cell retained some of the original coccoliths.

Nutrition

Coccolithophores are mostly photoautotrophic, i.e. they manufacture organic materials for their sustenance from water, CO_2, nitrates and other inorganic salts with photosynthesis of sunlight as the source of energy. However, some instances of ingestion of foreign organic material have been reported. For example, the motile cells of *Coccolithus pelagicus* (Wallich) Schiller have been observed to ingest bacteria and smaller algae. Such reports of **phagotrophy** amongst coccolithophores are rare, but some species are believed to be heterotrophic (utilizing both organic and inorganic substances from the surroundings). Their requirements of at least one vitamin and the ability of some littoral species to assimilate organic sources of nitrogen instead of inorganic nitrates in experiments, can be cited as points in favor of heterotrophic tendencies.

Growth

The growth rates of coccolithophores are relatively high, as some species multiply more than twice in one day. In the laboratory *Emiliania huxleyi* (Lohmann) Hay and Mohler has been shown to divide every 19 hours or so. In nature the cell division rate of this species varies from a low of 1.2 divisions per day in the Atlantic to a high of 4.8 divisions per day recorded in the Black Sea. Another rapid-growing coccolithophore species is *Cricosphaera elongata* (Droop) Braarud which has a rate of up to 2.25 divisions per day in cultures.

Most coccolithophore species exist within a relatively narrow temperature range. The temperature ranges of optimum growth are even narrower. In laboratory cultures *Emiliania huxleyi* seems to tolerate a temperature range of 7 to 27°C, however, in nature this species has been recorded in waters with temperatures as low as 2°C. Optimum growth of *E. huxleyi* occurs only within a temperature range of 18 to 24°C, when it grows almost four times as fast as at 7°C, and calcification of coccoliths is two- to three-fold as compared to lower and higher temperatures (Watabe and Wilbur, 1966).

In their laboratory experiments on *E. huxleyi*, Wilbur and Watabe have also shown that this species grows twice as fast in higher nitrate media than in lower nitrate media. At the optimum growth temperature of 18°C the number of normal coccoliths produced by the cell were also higher, abnormal coccoliths occurring more frequently at other temperatures. Other features such as number and shape of crystal elements of the coccolith and their dimensions also varied at different temperatures.

Mineralogy of the coccoliths

The calcium carbonate in coccoliths normally crystallizes as calcite and to a lesser degree as aragonite. In laboratory cultures minor traces of a third polymorph of lime, vaterite, have also been found. For example, in *Emiliania huxleyi* (Lohmann) Hay and

Mohler all three polymorphs have been detected, but aragonite and vaterite are present only in very small amounts (Wilbur and Watabe, 1963). Due to the relatively unstable nature of aragonite and vaterite, it is not surprising that these polymorphs have not been found in fossil coccoliths.

Hetero- and holococcoliths

Two different types of crystallization can be distinguished amongst coccolithophores: the majority of coccoliths of living and fossil species are made up of crystallites of varied shapes and sizes in which the basic rhombohedral shape of the calcite has been modified by the cell to fit into specialized morphologies. These coccoliths are informally known as the **heterococcoliths** (or coccoliths formed of crystallites of different shapes and sizes). A smaller group of species produce coccoliths composed of crystals which are minute, usually equidimensional and, more or less, maintain their original rhombohedral or hexagonal prism habits. These coccoliths are known collectively as **holococcoliths** (or coccoliths formed of similar types of crystals). Holococcoliths may consist entirely of unmodified or slightly modified rhombohedral or hexagonal crystals or a combination of both.

Coccolithus pelagicus (Wallich) Schiller which produces holococcoliths of minute, unmodified rhombohedral crystals in the motile phase of its life cycle and heterococcoliths of large, completely modified crystals in which the original rhombohedral symmetry is obscured in the non-motile phase, is a unique example of its kind. No other coccolithophore is known to produce coccoliths of both types during its life cycle and species are restricted to coccoliths of one type or the other.

As mentioned earlier, complete life cycle studies of coccolithophores are rare; however, a survey of the literature on extant species reveals that most (and perhaps all) of the species that produce holococcoliths or less modified coccoliths are those that bear coccoliths only during the motile phase of the life cycle. One is thus tempted to offer the explanation that since in motile phase most of the cell energy is used in flagellar motion, little or none is available for chemical reorganization within the cell and for crystal

modification. On the other hand, in the non-motile phase flagellar motion being discarded, this energy can be utilized for chemical changes and modifying crystal shape and dimensions.

About thirty extant species are known to produce holococcoliths. Their morphologies range from simple packings of rhombohedra, e.g. in *Sphaerocalyptra papillifera* (Halldal) Halldal, to varied arrangements of rhombohedra to produce a variety of interior and wall patterns (see e.g., *Calyptrosphaera catillifera* (Kamptner) Gaarder) (Fig. 5). The reader is referred to Gaarder (1962) and Black (1963) for a summary of the fine structure of coccoliths. Seven fossil holococcolith species have also been recorded from Cenozoic strata, all of which are composed of tightly packed, minute, equidimensional crystallites and vary in morphologies from the elliptical disc of *Holodiscolithus macroporus* (Deflandre) Roth, to the domes shape of *Daktylethra punctata* Gartner, an elongated stem surmounting a prominent base as in *Zygrhablithus bijugatus* (Deflandre) Deflandre and the crescents of *Peritrachelina joidesa* Bukry and Bramlette. The crystallites in these species are so small that they are not discernible under the light microscope, and electron microscope observation becomes necessary. Gartner and Bukry (1969) have studied the Tertiary holococcolith with both types of microscopes and

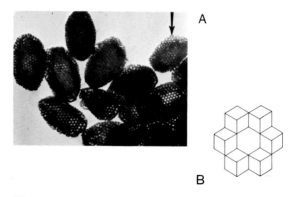

Fig. 5. Most common type of crystal arrangement in holococcoliths. A. *Sphaerocalyptra papillifera* (Halldal) Halldal showing coccoliths in slight degree of disintegration (at arrow packings of rhombohedra can be seen). B. Sketch of the original crystal arrangement in *S. papillifera*; the C-axis of rhombohedra is placed normal to the surface of the coccoliths. (Figures and data courtesy K. Gaarder.)

the reader is referred to this work for fuller descriptions.

Function of coccoliths

No experimental studies have been undertaken so far to determine the function of coccoliths or their significance to the living cell. However, there is no lack of speculations about it. One popular theory is that the coccoliths shield the enclosed cell from excessive sunlight. Equally popular is the opposing idea that coccoliths concentrate light towards the cell interior. The latter is more acceptable in view of the convexo-concave shape of most coccoliths, whose convex surfaces then face outward. Even if these convexo-concave surfaces are not true lenses, they would still be more efficient as light gatherers, rather than light dispersers. Moreover, as Gartner and Bukry (1969) point out, light energy decreases rapidly with depth, but nutrient supply increases, and a photosynthetic organism capable of concentrating the lesser amounts of light at greater depths in the photic zone would have an ecologic advantage over its competitors who are not so equipped.

Other such varied functions as floating devices, metabolic barriers, stabilizers, defensive shields, etc., have also been invoked for coccoliths. Some biologists think of them as by-products resulting from the detoxification of carbonate through the fixation of calcium (Isenberg and others, 1967).

Discoaster mineralogy

The Tertiary group of asteroliths — discoasters — became extinct at the close of the Pliocene Epoch. Although Bursa (1965) reported relict discoaster-like forms living in the Arctic, this occurrence has not been substantiated so far and Bursa's discovery can best be ascribed to contamination of his material. In the absence of an uncontroversial living analogue, we can only speculate about the manner in which discoasters calcify. Most researchers believe that, unlike coccoliths, discoasters more likely remained within the cell after calcification and were not secreted around it.

Discoasters show many complex and varied morphologies (see Fig. 23) but their ultra-structure is relatively simple as compared to coccoliths. Discoasters did not calcify rhombohedral or hexagonal crystals like the coccolithophores, but a third, tubular, form of calcite. Individual rays or arms are made of single tabular crystals.

ECOLOGY

Coccolithophores are exclusively planktonic marine organisms and are distributed from the open ocean, pelagic environment to nearshore littoral and inshore lagoonal environments.

The occurrence of coccolithophores in littoral, lagoonal and estuarine areas where salinities are either much higher or lower than the average salinity of the open ocean (35°/oo) demonstrates their tolerance of a wide range of salinities. *Emiliania huxleyi* (Lohmann) Hay and Mohler, for example, can tolerate a salinity range of 45°/oo to 16°/oo. In laboratory cultures species of *Cricosphaera* withstand salinities as low as 4 to 8°/oo in the motile phase of their life cycle and as high as 236°/oo during the non-motile phase.

Coccolithophores, being photosynthetic, live in the upper 100—150 m of the oceans, or the photic zone, where sunlight can easily penetrate. Although coccolithophores are found throughout the photic zone, they are most abundant a few meters to about 50 m below the surface of the water, and their concentration decreases rapidly at greater depths.

Biogeography

Although over 150 species of coccolithophores have been recorded living in the oceans, few detailed studies on the ocean-wide distribution of individual species exist. Quantitatively, coccolithophores show greater concentrations in the zones of high organic productivity where more nutrients are available because of the upwelling of bottom waters or the convergence of currents. These high productivity zones are situated north and south of about 45° latitudes and along a narrow equatorial belt. In the marine planktonic food chain inorganic nutrients support high standing crops of coccolithophores and other phytoplankton, which in turn sustain high standing crops of the zooplankton which feed on them.

Significant contributions to our knowledge of the biogeography of coccolithophores have been made by Andrew McIntyre and his colleagues, who have mapped the occurrences of selected species in the Atlantic Ocean. By studying both plankton samples and surface sediments from the ocean bottom, McIntyre and Bé (1967) were able to group the coccolithophores of the Atlantic into five discrete latitudinal climatic assemblages: tropical, subtropical, transitional, subarctic and subantarctic. Tropical and subtropical assemblages contain over three times more species than the subarctic and subantarctic assemblages and thus conform to present ideas on latitudinal changes in diversity (Fig. 6). From seasonal sampling data these authors also determined the temperature ranges of thirteen selected species and mapped their occurrences in both the plankton and the surface sediments. Most species thus mapped showed a slightly wider distribution in the plankton than in the surface sediments of the bottom. This discrepancy was attributed to the warming of the ocean since the end of the last glacial age approximately 12,000 years ago. A rapid warming would lead to a poleward migration of the surface

isotherms and warm-water species will thus show a broader latitudinal distribution, while the distribution of cold-water species becomes narrower.

In a biogeographic study of the Pacific Ocean, McIntyre and others (1970) delineated the geographic ranges of selected species of coccolithophores and from these distributional patterns they were able to assign temperature ranges to each taxon (Fig. 7). They found the various species associations to correspond to four Pacific water masses, each characterized by a particular temperature range. These biogeographic patterns, although admittedly of seasonal nature, nevertheless make the nannoplankton assemblages potentially of great value for paleoclimatic and paleooceanographic interpretations.

In another study of the distribution of coccolithophores in the water column along a north—south transect in the North and Central Pacific Ocean, Okada and Honjo (1973) also found surface temperatures and distribution of surface currents to be the most important factors delineating the biogeography of various species. They recognized six latitudinal assemblage zones. As in the Atlantic an increasing species diversity gradient was observed from high to low latitudes. In addition, these authors found a vertical differentiation in the coccolithophore assem-

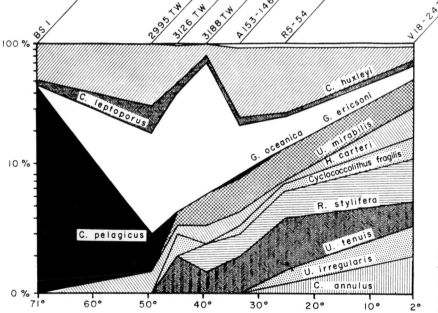

Fig. 6. Changes in species diversity from subarctic to tropical areas in the North Atlantic. (After McIntyre and Bé, 1967).

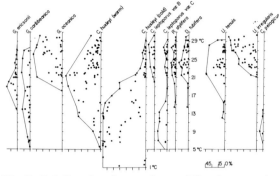

Fig. 7. Relative abundance of coccolithophore species in the Pacific Ocean plotted against mean surface temperature of water masses. Such data assign temperature ranges to taxa and enhance their paleoclimatic usefulness. (After McIntyre and others, 1970.)

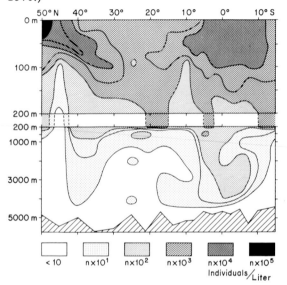

Fig. 8. Vertical distribution of coccolithophores along a north—south transect in the North Pacific. Note the wide variations in the number of individuals at various depths and high concentrations at 50°N and around the equator. (After Okada and Honjo, 1973.)

blages within the upper 200 m of the water column, which also changed with latitude. There was also a wide variation in the number of individuals with depth at various latitudes (Fig. 8).

MAJOR MORPHOLOGIC GROUPS

Calcareous nannoplankton constitutes a diverse group of morphological forms, many of which are either clearly related or show some similarity to the extant coccolithophores (forms with coccolith-like shields). Other

forms with no clear morphologic relationship to coccolithophores (e.g. discoasters) occur as calcareous microfossils within the same size fraction as coccoliths and may form a substantial part of the nannofossil assemblages. Thus both the coccolithophores and the associated non-coccolithophore nannoliths are traditionally studied together by the nannopaleontologists.

Because of the very diverse nature and at times unknown affinities of the morphologic groups, no satisfactory classification for calcareous nannoplankton has been suggested to date. Also because of the dual, plant and animal, characteristics of nannoplankton (see section on biology), they are claimed by both the botanists and zoologists, and a complicated double systematics has developed over the years. Most present-day nannopaleontologists, however, favor the plant origin, and essentially by consensus, the Code of Botanical Nomenclature is used for the description of taxa. Botanists usually include coccolithophores in Order Heliolithae, Class Coccolithophyceae and Division Chrysophyta of the Plant Kingdom.

It is beyond the scope of this text to discuss the details of taxonomic and classifactory problems. Instead we have chosen to introduce the major groups of nannoliths that most commonly occur in nannofossil assemblages by grouping them under informal epithets. The groups are arranged into three categories: (a) those that show clear relationship to coccolithophores; (b) non-coccolithophores, but common nannoliths; and (c) common *incertae sedis* genera (for a summary of major groups, see Table I). With the exception of four groups (coccolithids, zygodiscids, braarudosphaerids and thoracosphaerids) that occur in both Mesozoic and Cenozoic sediments, all other morphologic groups are either restricted to the Mesozoic or the Cenozoic. For each group the most commonly occurring genera have been illustrated. See Fig. 9 for commonly used coccolith and discoaster terms. For details of terminology of other nannolith groups, the reader is referred to Farinacci (1971).

The catalogue of genera and species of nannofossils by Farinacci (1969 and later) and the annotated index and bibliography of

TABLE I

Key to major morphologic groups of calcareous nannoplankton

A. COCCOLITHOPHORES AND RELATED NANNOLITHS

1.	Arkangelskiellids	Family Arkhangelskiellaceae Bukry (Mesozoic)
2.	Coccolithids	Family Coccolithaceae Poche (Mesozoic and Cenozoic)
		Family Prinsiaceae Hay and Mohler (Cenozoic)
		Family Helicosphaeraceae Black (Cenozoic)
3.	Podorhabdids	Family Podorhabdaceae Noël (Mesozoic)
4.	Pontosphaerids	Family Pontosphaeraceae Lemmermann (Cenozoic)
5.	Rhabdosphaerids	Family Rhabdosphaeraceae Lemmermann (Cenozoic)
6.	Stephanolithids	Family Stephanolithionaceae Black (Mesozoic)
7.	Syracosphaerids	Family Syracosphaeraceae Lemmermann (Cenozoic)
8.	Zygodiscids	Family Eiffellithaceae Reinhardt (Mesozoic)
		Family Zygodiscaceae Hay and Mohler (Cenozoic)

B. NON-COCCOLITHOPHORE NANNOLITHS

1.	Braarudosphaerids	Family Braarudosphaeraceae Deflandre (Mesozoic and Cenozoic)
2.	Ceratolithids	Family Ceratolithaceae Norris (Cenozoic)
3.	Discoasterids	Family Discoasteraceae Vekshina (Cenozoic)
4.	Fasciculithids	Family Fasciculithaceae Hay and Mohler (Cenozoic)
5.	Heliolithids	Family Heliolithaceae Hay and Mohler (Cenozoic)
6.	Lithastrinids	Family Lithastrinaceae Thierstein (Mesozoic)
7.	Lithostromationids	Family Lithostromationaceae Haq (Cenozoic)
8.	Sphenolithids	Family Sphenolithaceae Deflandre (Cenozoic)
9.	Thoracosphaerids	Family Thoracosphaeraceae Schiller (Mesozoic and Cenozoic)

C. GENERA *INCERTAE SEDIS*

1.	*Isthmolithus* (Cenozoic)		4.	*Nannoconus* (Mesozoic)
2.	*Microrhabdulus* (Mesozoic)		5.	*Triquetrorhabdulus* (Cenozoic)
3.	*Micula* (Mesozoic)			

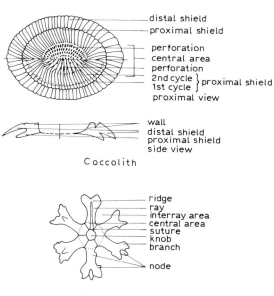

Fig. 9. Coccolith and discoaster terminology. (After Farinacci, 1971.)

Loeblich and Tappan (1966 and later) can also be consulted for description and validity of taxa.

Nannofossils can be studied rapidly for biostratigraphic analysis with the help of a light microscope at about ×1000. Smear-slides can be either made directly from raw sediment samples or from suspensions that have been cleaned chemically and short-centrifuged to concentrate nannolith-size particles. The use of normal, phase-contrast and cross-polarized light conditions are necessary to bring out finer details and the peculiar optical properties that help differentiate species. Transmission and scanning electron microscopes have been used increasingly in recent years to study the ultrastructure of nannoliths, the details of which are not discernible under light microscope. The reader is referred to Stradner and Papp (1961) and Hay (1965) for details of light microscopic methods and to Reinhardt (1972) for a discussion of the optical properties of nannoliths and a summary of both light and electron microscopic methods.

Under description of the various groups below we have included only electron micrographs to illustrate the morphologic variations and ultrastructural differences. For range-charts of biostratigraphically important taxa, on the other hand, light micrographs have been included to make these taxa useful for light microscopic determinations.

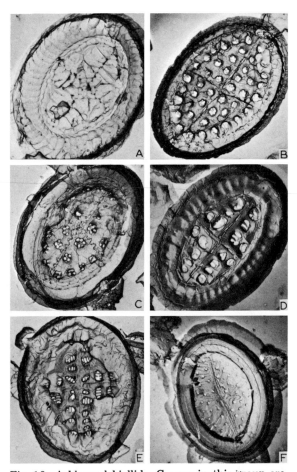

Fig. 10. Arkhangelskiellids. Genera in this group are distinguished by the difference in construction of the coccolith rim. *Arkhangelskiella* has a three-tiered rim in proximal view; *Broinsonia* an additional inner rim cycle in distal view and distinctive perforations; *Gartnerago* has a multi-tiered rim in proximal view and *Kamptnerius* an asymmetrical outer rim cycle. A. *Arkhangelskiella cymbiformis* Vekshina, proximal view. × 5,120. B, C. *A. specillata* Bukry. B. Distal view. × 4,320. C. Proximal view. × 4,640. D. *Broinsonia parca* (Stradner) Bukry, distal view. × 5,600. E. *Gartnerago costatum* Bukry, distal view. × 3,600. F. *Kamptnerius percivalli* Bukry, proximal view. × 3,360. (After Bukry, 1969.)

A. Coccolithophores and related nannoliths

Under this division we include those groups that show some similarity to the basic coccolith-like shield construction.

(1) **Arkhangelskiellids.** In the Mesozoic Family Arkhangelskiellaceae the elliptical coccoliths are composed of complex shields of two to four cycles of joined elements. Common genera are: *Arkhangelskiella, Broinsonia, Gartnerago* and *Kamptnerius* (Fig. 10)

(2) **Coccolithids.** This is one of the four groups that occur in both the Mesozoic and Cenozoic strata. This is also one of the most commonly occurring groups with wide morphologic variety. The group is characterized by two-shielded coccoliths, each shield composed of one or more cycles of crystal elements connected at their inner margins.

Of the three families included in this group, the Family Coccolithaceae occurs in both the Mesozoic and Cenozoic. Common Mesozoic genera are: *Biscutum, Discorhabdus, Sollasites* and *Watznauria* (Fig. 11). The more common Cenozoic genera of this family are: *Coccolithus, Chiasmolithus, Cruciplacolithus, Cyclococcolithus* and *Umbilicosphaera* (Fig. 12).

Fig. 11. Mesozoic coccolithids. Distinguishing criteria among genera are: shape of the coccolith and central area, and number of cycles in two shields. *Biscutum* consists of two single-cycle elliptical shields; *Discorhabdus* of two single-cycle circular shields; *Sollasites* has elliptical coccoliths with two single-cycle, narrow shields and a distinctive central area and *Watznauria* has two or three cycles of elements in distal shield. A—C. *Biscutum constans* (Gorka) Black. A. Proximal view. × 2,925. B. Distal view. × 3,825. C. Coccosphere. × 3,400. D,E. *Discorhabdus rotatorius* (Bukry) Thierstein. D. Proximal view. × 3,825. E. Distal view. × 6,075. F. *Sollasites horticus* (Stradner and others) Black, distal view. × 4,050. G. *Watznauria barnesae* (Black) Perch-Nielsen, distal view. × 4,050. (A,B,D,E, and F after Bukry, 1969.)

The second family, Prinsiaceae, is restricted to the Cenozoic. The basic construction of the elliptical to subcircular coccoliths of the Prinsiaceae is similar to that of the Family Coccolithaceae and differentiated from them only under light microscope where the former shows bright proximal shields and dextrogyre (clock-wise coiling) extinction lines under crossed-polarized light. Common genera include: *Prinsius, Toweius, Reticulofenestra, Dictyococcites, Gephyrocapsa* and *Emiliania* (Fig. 13).

The third family, Helicosphaeraceae occurs in the Cenozoic. It differs from the other two in the construction of the proximal shield of the coccoliths which are helicoid. *Helicosphaera* is the only genus in this family (Fig. 14, see also Fig. 32).

(3) **Podorhabdids.** In the Mesozoic Family Podorhabdaceae the coccoliths are composed of two shields, with a single cycle of radial elements on the proximal shield and with varied central area structures that help differentiate genera. Common genera are: *Podorhabdus, Cribrosphaera, Prediscosphaera* and *Cretarhabdus* (Fig. 15).

(4) **Pontosphaerids.** Members of the Family Pontosphaeraceae have elliptical coccoliths with the two shields closely appressed. The distal shield may have distally enlarged walls to give it a basket-shaped appearance. The central area is either closed, perforated or spanned by a bridge. Pontosphaerids are restricted to the Cenozoic. Common genera are *Pontosphaera, Transversopontis, Lophodolithus* and *Scyphosphaera* (Fig. 16).

(5) **Rhabdosphaerids.** The elliptical to circular coccoliths of members of the Family Rhabdosphaeraceae are characterized by one or two shields of simple construction, one or more cycles of minute

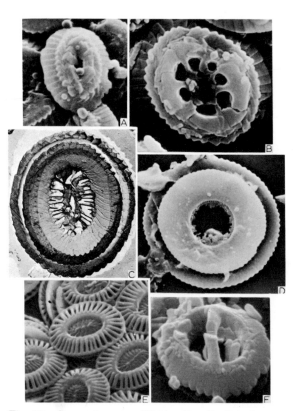

Fig. 13. Cenozoic coccolithids. Family Prinsiaceae. Genera distinguished by structure of the rim and central area. *Prinsius* has a multi-cycle distal shield but a single, non-distinct central opening; *Toweius* has a multi-cycle distal shield and multiple openings; *Dictyococcites* has a central grid of bar-shaped elements; *Reticulofenestra* has a central area of fine reticulate grille; *Gephyrocapsa* a central cross-bar and *Emiliania* a distinctive rim of I-shaped elements and a central grid. A. *Prinsius martinii* (Perch-Nielsen) Haq, distal view. × 3,300. B. *Toweius craticulus* Hay and Mohler, distal view. × 4,400. C. *Dictyococcites* sp., proximal view. × 6,875. D. *Reticulofenestra dictyoda* (Deflandre) Hay and Mohler, proximal view. × 3,300. E. *Emiliania huxleyi* (Lohmann) Hay and Mohler, distal views. × 6,600. F. *Geophyrocapsa oceanica* Kamptner, distal view. × 11,000.

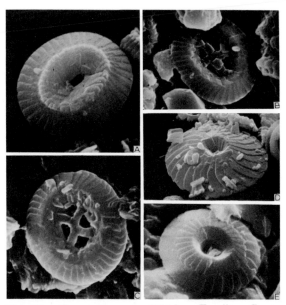

Fig. 12. Cenozoic coccolithids. Family Coccolithaceae. Generic distinguishing criteria are: shape of the coccolith and characteristics of the central area. *Coccolithus* is oval with a central pore; *Cruciplacolithus* has a +-shaped central structure; *Chiasmolithus* an x-shaped central structure; *Cyclococcolithus* is circular with a single cycle in the distal shield and *Umbilicosphaera* is circular with a central pore and two or more cycles in distal shield. A. *Coccolithus pelagicus* (Wallich) Schiller, distal view. × 3,250. B. *Cruciplacolithus tenuis* (Stradner) Hay and Mohler, distal view. × 2,250. C. *Chiasmolithus californicus* (Sullivan) Hay and Mohler, distal view. × 3,000. D. *Cyclococcolithus leptoporus* (Murray and Blackman) Kamptner, distal view. × 6,250. E. *Umbilicosphaera sibogae* (Weber-Van Bosse) Gaarder, distal view. × 5,000.

crystal elements in the central area, and a central structure that gives the coccolith a funnel or dome-shaped appearance. Rhabdosphaerids occur in the Cenozoic only. Common genera are: *Rhabdosphaera*, and *Blackites* (Fig. 17).

(6) **Stephanolithids.** Members of the Family Stephanolithionaceae have coccoliths of varied geometric shapes, with a distal cycle of crystal elements forming a low or high cone or a cylinder and a proximal cycle of smaller brick-like crystals. The central area is either open or spanned by subradial to radial bars which emerge from the proximal cycle of crystal elements. Stephanoliths are restricted to the

Mesozoic. Common genera are: *Stephanolithion*, *Corollithion* and *Cylindralithus* (Fig. 18).

(7) **Syracosphaerids.** Coccoliths of the Cenozoic Family Syracosphaeraceae have a simple to complex wall and a characteristic central area with crystal laths extending partially or completely up to the center. Syracosphaerids are most common in Recent. Common genus: *Syracosphaera* (Fig. 19).

(8) **Zygodiscids.** This is the second group that is distributed in both Mesozoic and Cenozoic strata. Zygodiscids are elliptical coccoliths with a marginal wall of one or two cycles of imbricated elements and a thin inner rim. The central area structures may be a simple bar, an X or more complicated. In the two families included here, Eiffellithaceae (Mesozoic) and Zygodiscaceae (mainly Cenozoic) the basic construction of the shields is similar, however, the former may possess a stem on the central structure. Common Mesozoic genera are: *Eiffellithus*, *Chiastozygus* and *Vagalapilla* (Fig. 20). Common Cenozoic genera are: *Zygodiscus* and *Neococcolithes*.

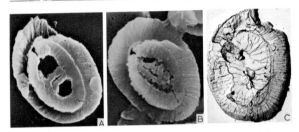

Fig. 14. Cenozoic coccolithids. Family Helicosphaeraceae. In the genus *Helicosphaera* the species are differentiated by the overall shape of coccolith and terminal flange and characteristics of the central area. A. *Helicosphaera wilcoxonii* (Gartner) Jafar and Martini, proximal view. × 2,000. B. *H. dinesenii* (Perch-Nielsen) Jafar and Martini, proximal view. × 2,000. C. *H. recta* (Haq) Jafar and Martini. × 3,200.

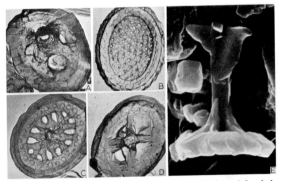

Fig. 15. Podorhabdids. Genera are distinguished by characteristics of the central area. *Podorhabdus* has four radial bars supporting a central process; *Cribrosphaera* has a closed central area occupied by a grid and no central process; *Cretarhabdus* central area is spanned by four main radial bars supporting a central process and the areas between bars with secondary structures; *Prediscosphaera* has sixteen petaloid elements in rim and central area spanned by four radial bars supporting a characteristic process. A. *Podorhabdus granulatus* (Reinhardt) Bukry, distal view. × 3,650. B. *Cribrosphaera ehrenbergi* Arkhangelsky, proximal view. × 3,650. C. *Cretarhabdus crenulatus* Bramlette and Martini, distal view. × 4,800. D,E. *Prediscosphaera cretacea* (Arkhangelsky) Gartner. D. Distal view. × 4,800. E. Side view. × 4,320. (A—D: after Bukry, 1969.)

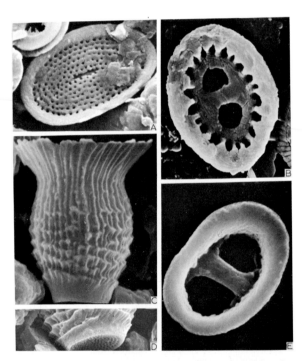

Fig. 16. Pontosphaerids. Genera differentiated by the wall and central area characteristics. *Pontosphaera* has a thin wall with multiple pores in the central area; *Transversopontis* has a central bar; *Scyphosphaera* is basket-shaped due to enlarged walls, and *Lophodolithus* has an asymmetrically enlarged wall. A. *Pontosphaera japonica* (Takayama) Haq, distal view. × 3,640. B. *Transversopontis pulcher* (Deflandre) Hay, Mohler and Wade, proximal view. × 3,120. C,D. *Scyphosphaera pulcherrima* Deflandre. C. Side view. D. Tilted proximal view showing perforated central area. × 2,600. E. *Lophodolithus nascens* Bramlette and Sullivan, distal view. × 2,860.

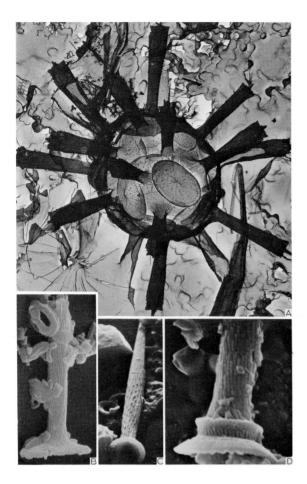

Fig. 18. Stephanolithids. A. *Stephanolithion laffittei* Noël. × 8,100. (Photo: H. Thierstein.) B. *Corollithion exiguum* Stradner. × 8,625. (After Bukry, 1969.) C. *Cylindralithus coronatus* Bukry. × 4,875. (After Roth and Thierstein, 1972.)

Fig. 17. Rhabdosphaerids. A. A coccosphere of *Rhabdosphaera clavigera* Murray and Blackman. × 3,710. (Photo: H. Okada). B. *Rhabdosphaera procera* Martini, side view. × 4,770. C, D. *Blackites spinosus* (Deflandre) Hay and Towe. C. Titled view of complete specimen. × 3,180. D. Side view of basal plate and part of the stem. × 4,770.

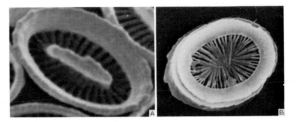

Fig. 19. Syracosphaerids. A. *Syracosphaera lamina* Lecal-Schlander, distal view. × 6,300. B. *S. pulchra* Lohmann, proximal view. × 4,500.

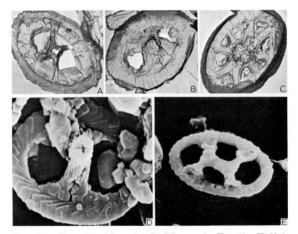

Fig. 20. Zygodiscids. A—C. Mesozoic Family Eiffellithaceae. Genus *Chiastozygus* has an x-shaped cross; *Eiffellithus* has eight large crystal elements around the central cross and *Vagalapilla* has a cross whose axes are parallel to the axes of the ellipse. A. *Eiffellithus turriseiffeli* (Deflandre) Reinhardt, distal view. × 3,430. B. *Chiastozygus inturratus* (Reinhardt) Bukry, distal view. × 3,020. C. *Vagalapilla octoradiata* (Gorka) Bukry, distal view. × 5,200. (A—C: after Bukry, 1969.) D—E. Family Zygodiscaceae. Genus *Zygodiscus* has a cross-bar parallel to short axis of ellipse and *Neococcolithes* has a thin rim and X- to H-shaped central cross. D. *Zygodiscus sigmoides* Bramlette and Sullivan, distal view. × 4,160. E. *Neococcolithes dubius* (Deflandre) Black, distal view. × 2,860.

B. Non-coccolithophore nannoliths

This division includes groups with no clear relationship to extant coccolithophores, but which occur commonly in the nannofossil assemblages.

(1) **Braarudosphaerids.** This is the third group that is common to both Mesozoic and Cenozoic. Braarudosphaerids are pentaliths (composed of five crystal units) belonging to the Family Braarudosphaeraceae. The genera *Braarudosphaera* and *Micrantholithus* are encountered in both the Mesozoic and Cenozoic, but *Pemma* is found in the Cenozoic only (Fig. 21).

(2) **Ceratolithids.** These horse-shoe shaped nannoliths occur in the late Cenozoic only. The Family Ceratolithaceae includes two genera: *Ceratolithus* and *Amaurolithus* (Fig. 22).

(3) **Discoasterids.** This is a numerically important group of asteroliths, some disc-shaped, others star-shaped with or without stems on one side of the asterolith. The Cenozoic Family Discoasteraceae includes three commonly occurring genera: *Discoaster*, *Discoasteroides* and *Tribrachiatus* (Fig. 23).

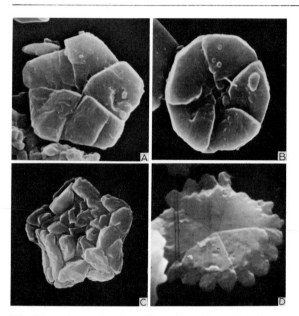

Fig. 21. Braarudosphaerids. A. *Braarudosphaera bigelowi* (Gran and Braarud) Deflandre. × 3,135. B. *B. discula* Bramlette and Riedel. × 2,850. C. *Micrantholithus obtusus* Stradner. × 3,200. (After Thierstein, 1971.) D. *Pemma papillatum* Martini. × 3,420.

Fig. 23. Discoasterids. A. *Discoaster multiradiatus* Bramlette and Riedel. × 3,000. B. *D. mirus* Deflandre. × 2,500. C. *D. lodoensis* Bramlette and Riedel. × 2,700. D. *D. saipanensis* Bramlette and Riedel. × 2,500. E. *D. brouweri* Tan Sin Hok. × 4,000. F. *D. challengeri* Bramlette and Riedel. × 3,250. G. *Discoasteroides kuepperi* (Stradner) Bramlette and Sullivan. × 1,500. H. *Tribrachiatus orthostylus* Shamrai. × 2,500.

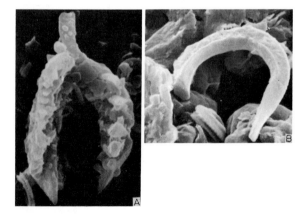

Fig. 22. Ceratolithids. A. *Ceratolithus* aff. *acutus* Gartner and Bukry. × 2,475. B. *Amaurolithus delicatus* Gartner and Bukry. × 4,125.

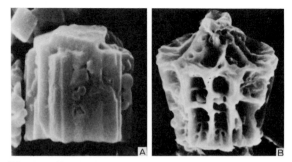

Fig. 24. Fasciculithids. A. *Fasciculithus tympaniformis* Hay and Mohler, side view. × 4,550. B. *F. schaubii* Hay and Mohler, oblique side view. × 2,100.

Fig. 25. Heliolithids. A. *Heliolithus* sp., oblique side view. × 3,000. B. *H. kleinpelli* Sullivan, top view. × 3,000. (B: photo, H. Okada.)

Fig. 26. Lithastrinids. A. *Lithastrinus grilli* Stradner. × 3,590. (After Forchheimer, 1972.) B. *Bukryaster hayi* (Bukry) Prins. × 4,450. (After Bukry, 1969.)

Fig. 27. Lithostromationids. A. *Lithostromation perdurum* Deflandre. × 3,120. B. *Polycladolithus operosus* Deflandre. × 5,580.

(4) **Fasciculithids.** Members of the Cenozoic Family Fasciculithaceae are subcylindrical bundles of wedge-shaped crystals whose thin edges meet along the center. One end of the cylinder is concave, the other pointed or flat and its surface may be ornamented. A single genus, *Fasciculithus*, is included in this family (Fig. 24).

(5) **Heliolithids.** Two abutting unequal shields lacking a connecting tube and one shield with subradial sutures are the characteristics of the Family Heliolithaceae. A single Cenozoic genus, *Heliolithus*, comprises this group (Fig. 25).

(6) **Lithastrinids.** Members of the Mesozoic Family Lithastrinaceae consist of one or more cycles of five or more imbricated crystal elements. Elements of one cycle may extend outward to form an asterolith-like nannolith. Common genera are: *Lithastrinus* and *Bukryaster* (Fig. 26).

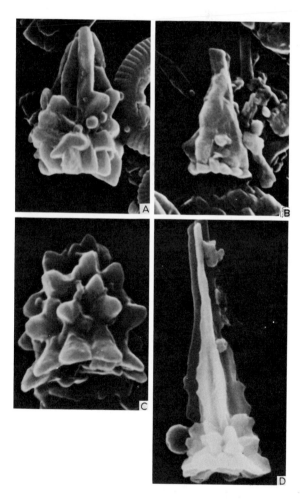

Fig. 28. Sphenolithids. A. *Sphenolithus radians* Deflandre. × 4,500. B. *S. distentus* (Martini) Bramlette and Wilcoxon. × 5,625. C. *S. moriformis* (Brönnimann and Stradner) Bramlette and Wilcoxon. × 5,625. D. *S. heteromorphus* Deflandre. × 5,625.

(7) **Lithostromationids**. These are thick, triangular tests with highly ornamented surfaces. The Family Lithostromationaceae includes only one Cenozoic genus, *Lithostromation*, but other Cenozoic genera such as *Catinaster*, *Trochoaster* and *Polycladolithus* show similarity to this group (Fig. 27).

(8) **Sphenolithids**. These Cenozoic forms have a rounded to polygonal base, surmounted by an elongated rib-cone with one or more spines radiating from the center. The Family Sphenolithaceae contains a single genus: *Sphenolithus* (Fig. 28).

(9) **Thoracosphaerids**. Members of the Family Thoracosphaeraceae are found in both the Mesozoic and Cenozoic. They are spherical shells composed of a mosaic of interlocking crystal units. The shells may be with or without an opening which may have a lid. A single genus, *Thoracosphaera*, is included in this family (Fig. 29).

C. Genera incertae sedis

Only five of the more common *incertae sedis* genera are described here.

(1) *Isthmolithus*. Nannoliths of this Cenozoic genus are shaped like elongate parallelograms with wide walls and two transverse septa dividing the parallelogram into three windows (Fig. 30A).

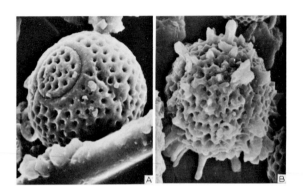

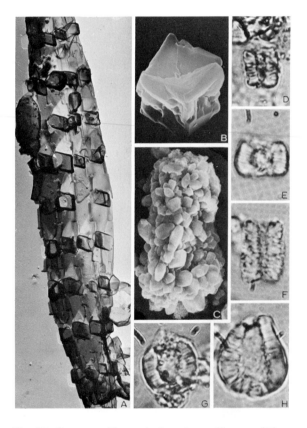

Fig. 30. Common Cenozoic *incertae sedis* nannoliths. A. *Isthmolithus recurvus* Deflandre. × 9,000. B. *Triquetrorhabdulus carinatus* Martini. × 10,000.

Fig. 31. Common Mesozoic *incertae sedis* nannoliths. A. *Microrhabdulus belgicus* Hay and Towe. × 8,000. B. *Micula staurophora* (Gardet) Stradner. × 4,950. (Photo: H. Thierstein.) C—H. *Nannoconus* spp. C. Overgrown outer surface of *Nannoconus* sp. D. *N. minutus* Brönnimann. E. *N. truitii* Brönnimann. F. *N. elongatus* Brönnimann. G. *N. globulus* Brönnimann. H. *N. wassali* Brönnimann. (D—H, × 1,100, after Manivit, 1971.)

(2) *Microrhabdulus*. These are long, cylindrical rod-like nannoliths, either truncate or tapering at ends and showing checkered extinction patterns under cross-polarized light. These forms are most probably calcareous remnants of other unknown microfossils. They occur in the Mesozoic only (Fig. 31A).

(3) *Micula*. Nannoliths of this Mesozoic genus are rectangular to subcubical with concave sides and constructed of two or more units of different calcite orientation (Fig. 31B).

(4) *Nannoconus*. Nannoconids are cone-shaped nannoliths composed of minute wedge-shaped crystal units arranged radially around and perpendicular to the long axis which is occupied by a canal open at both ends of the test (Fig. 31C—H). This genus occurs only in the Mesozoic.

(5) *Triquetrorhabdulus*. This genus has rodlike nannoliths with three edges 120° apart and pointed or rounded ends (Fig. 30B). Occurs in the Cenozoic only.

EVOLUTIONARY TRENDS

Two aspects of the evolution of calcareous nannoplankton have been considered so far by researchers: (1) evolutionary trends in individual phylogenetic lineages, i.e. phyletic changes in morphology with time from ancestral to descendant populations of groups; and (2) taxonomic frequencies in evolution, or changes in the tempo (origination and extinction) of evolution in various groups.

Lineage studies

These studies involve a close scrutiny of all taxa (usually at the specific or generic level) within a family or within another group of closely related taxa. From such detailed study trends in the modification of structural features with time can be easily discerned.

In one such study Gartner (1969a) was able to recognize two distinct lineages within the lower Tertiary coccolith genus *Chiasmolithus*. The species of this genus are characterized by an open central area which is spanned by an X-shaped crossbar. One lineage probably evolved from the early Paleocene species *Chiasmolithus danicus* (Brotzen) Hay and Mohler, developing a two-part crossbar, and the other lineage typified by a crossbar constructed of closely packed tubular prisms which are aligned parallel to the plane of the placolith, originated from another Paleocene species, *Cruciplacolithus tenuis* (Stradner) Hay and Mohler. Fig. 32 shows these lineages together with the geological ranges of the species.

The species of the Cenozoic coccolith genus *Helicosphaera* are characterized by two unequal shields, the proximal shield showing a helicoid construction. In a detailed study of this genus Haq (1973a) recognized three distinct trends in the evolution of helicosphaerids through the Cenozoic: (1) the outline of the coccoliths which was typically ovoid in early and middle Eocene species changed to rhomboid and broad-elliptical in the late Eocene, to roughly rectangular in the Oligocene and then to long-elliptical in Miocene and younger species; (2) the terminal flange of the helicoid distal shield shows a marked evolution from compressed types to sharply pointed types and then to large extended types; a return to smaller, rounded flanges is seen in two of the extant species; (3) the construction of the central areas of the coccoliths changed from generally more complex types as in the Eocene species to less complex ones in Oligocene and younger species. In Fig. 33 these evolutionary trends in the morphology of helicosphaerids are shown (numbers in Fig. 33 refer to species listed in Haq, 1973a).

As a group the discoasters show a distinct trend towards reduction in the amount of skeletal calcite resulting in less massive skeletal elements and a reduction in the number of arms (see Fig. 23). According to Prins (1971) two lines of evolution can be followed in the discoasters. One line probably originated from the Paleocene genus *Heliolithus* giving rise to the *Discoasteroides kuepperi* — *Discoaster barbadiensis* lineage, and the other developing from another early Tertiary genus, *Fasciculithus*, resulting in the lineage beginning with *Discoaster mohleri* Bukry and Percival and leading to all other discoasters.

Researchers have only recently begun to appreciate the phylogenetic evolutionary trends among calcareous nannoplankton and the above cited are a few examples of the many lineages that lend themselves to such studies. Many lineages appear to start rather

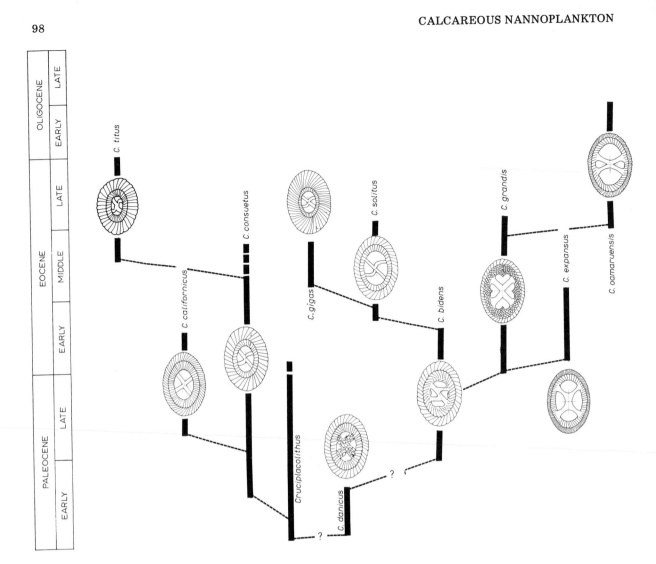

Fig. 32. Suggested lineages in Cenozoic genus *Chiasmolithus*. (After Gartner, 1969a; figs. after Perch-Nielsen, 1971.)

suddenly without an obvious ancestor or morphologically intermediate forms between the ancestor or the descendents. Asexual reproduction has sometimes been invoked as a possible reason for such sudden appearances because it can lead to repeated multiplications of a mutant without genetic exchange with non-mutants.

Evolutionary rates

The taxonomic frequency, or variation in the number of taxa with time, provides data on the tempo or the pulse of evolution. After the initial stages in the history of study of a fossil group when the majority of taxa have been described, total frequencies of taxa, plotted against a time scale, should depict changes in the total diversity of that group. In this way time-frequency curves serve as an indirect reflection of the entire evolutionary process at work during a certain time-span. From the frequency curves of first appearances and last occurrences of taxa, two important aspects of the tempo of evolution can be deciphered: the rates of origination or diversification and the rates of extinction.

The study of evolutionary rates should be undertaken with caution because a number of factors can bias the quantitative data. Differ-

Fig. 33. Suggested lineages in the Cenozoic genus *Helicosphaera*. (After Haq, 1973a).

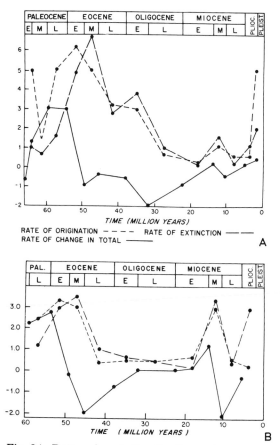

Fig. 34. Rates of evolution for Cenozoic coccolithophores (A) and discoasters (B). Vertical scale represents number of species/m.y. in case of rates of origination and extinction and a ratio of these two rates (i.e. origination/extinction) in case of change in total. (After Haq, 1973b.)

ing taxonomic concepts of individual researchers can introduce a strong bias in the taxonomic data. Other factors such as multiple names for a taxon (synonymies) and preferential study of assemblages of more popular geologic epochs can introduce further bias into the data. Among important physical factors to be considered are the selective solution of species after death and before deposition on the sea floor, and the diagenetic changes in assemblages within the sediments. These are but a few of the many factors which must be considered carefully before such a study is undertaken.

Haq's (1973b) study of the evolutionary rates in the Cenozoic coccoliths and discoasters shows that both these groups evolved rapidly from Paleocene to early Eocene (Fig. 34). In the case of the coccoliths the rates slowed down sharply during the middle Eocene, slowing even further in the Oligocene. The advent of the Miocene saw a slight change in this trend and the coccoliths evolved a bit faster as compared to the Oligocene Epoch. Another decline occurred in the late Miocene followed by a revival in the Plio-Pleistocene. Discoaster evolution proceeded in much the same manner as the coccoliths, except that in the Oligocene the discoaster rates show a plateau rather than a marked decline, and the

early Miocene revival is more pronounced than that of the coccoliths. On the average Cenozoic coccoliths evolved at the rate of one species every 30,000 years and the discoasters at the rate of one every 50,000 years, which are among the highest rates for any group of microfossils.

NANNOPLANKTON PALEOBIOGEOGRAPHY AND PALEOCLIMATIC INTERPRETATIONS

Mapping of distribution patterns of taxa within the sediments is the first step towards paleoclimatic/paleooceanographic interpretations. To what degree a calcareous nannoplankton assemblage preserved in the sediment reflects the actual life-assemblage depends on a number of factors of which carbonate dissolution is the most important.

The dissolution of coccolith-calcite begins soon after the individual dies and sinks into water undersaturated with respect to calcium carbonate. Dissolution is, however, retarded by the fact that most coccoliths drop to the sea floor within the fecal pellets of planktonic grazers. A protective membrane around the pellet protects the coccoliths against spilling out and dissolution for a major part of their descent (Honjo, 1976). This communal sinking within fecal pellets also accelerates the sinking of coccoliths (calculated to be on the order of 100 years in the deep sea for an average coccolith and on the order of only a few weeks for an average-sized fecal pellet). Accelerated sinking further protects the coccoliths by decreasing their residence time in deeper undersaturated waters. In this way a large quantity of the coccoliths from a living population may reach the sea floor relatively unaffected by dissolution. A short sinking time also ensures better correspondence between life-assemblages and those eventually deposited on the sea floor.

Coccoliths may thus avoid dissolution within the water column, but once the fecal pellets reach the sea floor they disintegrate rapidly due to bacterial action and the dissolution process is activated. In deeper waters a coccolith lysocline (a level below which all coccoliths start showing signs of dissolution) has been recognized between 3 and 4 km depth (Berger, 1973; Roth and Berger, 1975). Selective removal of less resistant species begins to occur within these depths and dissolution increases rapidly below 4 km, resulting in markedly decreased assemblage diversity. Below 5 km, which is the approximate level of the calcite compensation depth (CCD) that separates the predominantly calcareous sediments from the carbonate-poor ones, only the most resistant species, if any, are preserved.

Diagenesis can also alter the nature of fossil assemblages. Recrystallization can result in heavy calcite overgrowth on crystal faces sometimes altering the morphologic features of nannoliths beyond recognition, especially in discoasters (Wise and Kelts, 1972). In older sediments these problems are particularly acute. Thierstein (1974) found a number of Cretaceous taxa that spanned equal strati-

graphic intervals to be conspecific, distinguished from each other by diagenetically induced features. For example, perforated species of *Kamptnerius*, *Gatnerago* and *Broinsonia* were much more common in well-preserved samples, whereas the imperforate forms of some genera (many apparently conspecific with perforate forms) were more frequent in overgrown assemblages.

The effect of these problems can be minimized by considering sediment samples from relatively shallower depths and only those that contain assemblages in which relatively delicate forms are still recognizable. This procedure, although ensuring better quality of distributional data, restricts the number of samples (data points) from which meaningful biogeographic information can be obtained.

Paleobiogeographic study of the Pleistocene

The present-day coccolithophore distributional patterns are clearly related to a latitudinal climatic gradient and to water masses (see section on biogeography) which makes them excellent tools for the study of past climatic fluctuations and water mass dynamics. This kind of comparative paleoclimatic inference can, however, be made only as far back in time as the first evolutionary appearance of the assemblages with essentially the same components as the present day (i.e. within the Pleistocene).

One example of such a study is that by McIntyre and others (1972) in which they used both nannofloral and planktonic foraminiferal associations in a series of cores from the high latitudes in the North Atlantic to delineate the intensity and extent of polar front migrations in the late Pleistocene. They first defined polar, subpolar, transitional and subtropical water masses as characterized by present-day faunal and floral associations. Latitudinal shifts of these associations southwards as seen in the cores were then interpreted as polar front movements. Six southward polar front migrations were recognized in the relatively short geological time-span of 225,000 years (Fig. 35). Their evidence points to over 10° latitudinal climatic shifts during the late Pleistocene time in the open ocean of the North Atlantic.

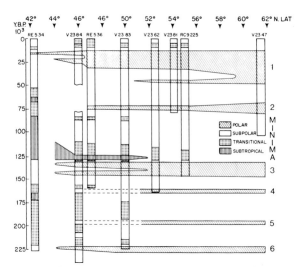

Fig. 35. Polar front migrations in the North Atlantic during the last 225,000 years B.P. (After McIntyre and others, 1972.)

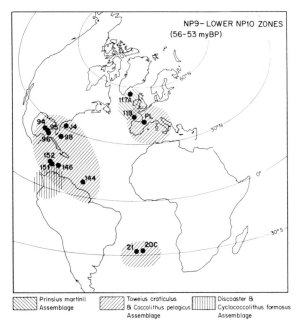

Fig. 36. Nannoplankton paleobiogeography during late Paleocene—early Eocene nannofossil assemblages, plotted on paleogeographic reconstruction of Atlantic Ocean at 56—53 m.y. B.P. Numbers refer to DSDP sites. Late Paleocene—early Eocene was an interval when low-latitude assemblages shifted towards higher latitudes (After Haq and Lohmann, 1976.)

Pre-Pleistocene biogeography and paleoclimatology

Paradoxically, the high rates of evolution in nannofossils that make them very useful in biostratigraphy, prevent a more effective use of these fossils as paleoclimatic indicators in older ages. Assemblages have evolved so rapidly that those of pre-Pleistocene age have few components in common with extant ones. This precludes climatic inferences by comparison to present-day patterns. To resolve this problem assemblages have to be first mapped for different times and their latitudinal distributions delineated. Extreme assemblages that show constant restriction to either higher or lower latitudes can then be selected as paleoclimatic indicators and by comparison to these all other assemblages can be scaled along a relative latitudinal (temperature) gradient.

Haq and Lohmann (1976) have attempted such a study for the early Cenozoic of the Atlantic Ocean. They mapped biogeographic patterns of Paleocene to Oligocene nannoplankton assemblages for twelve time intervals which allowed them to recognize the paleoclimatic indicators and to establish the relative latitudinal preferences of all major assemblages (Fig. 36). Assemblages were then grouped into relatively low-, middle- or high-latitude nannofloras and five distinct latitudinal shifts were discerned in their distribu-

tional patterns through the early Cenozoic. These shifts were then interpreted as a response to major climatic fluctuations. The early Cenozoic climatic history of the Atlantic Ocean is summarized in Fig. 37. Beginning with a relatively warm earliest Paleocene a cooling occurs in the latest early Paleocene shown by the equatorward migration of cool *Prinsius martinii* and *P. bisulcus* assemblages. This is followed by a major warming trend with peak warming in late Paleocene and early Eocene as evidenced by the poleward migration of warm low-latitude assemblages including the discoasters. A general cooling in the middle Eocene is indicated by the return of high-latitude assemblages to mid latitudes, followed by a slight warming when patterns are somewhat similar to the pre-mid Eocene. In the Oligocene two cooling episodes are indicated: first, a sharp, but relatively short-lived, cooling in the early Oligocene when higher latitude assemblages temporarily invade the low latitudes, and a second, less severe, cooling episode occurs in the middle Oligocene when the high-latitude assemblages shift up to the mid latitudes only.

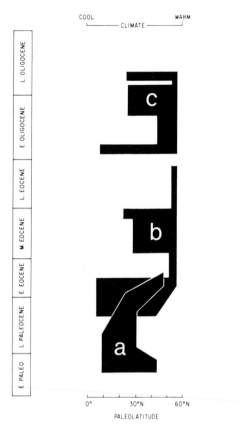

Fig. 37. Early Cenozoic climatic history of the Atlantic Ocean as interpreted by nannofloral migrations through latitudes. *a* represents migrations of relatively cool nannofloral assemblages through Paleocene—early Eocene, *b* represents migrations of relatively warm nannofloral assemblage through the Eocene, and *c* represents shifts of relatively cool nannofloral assemblages through the Oligocene. (After Haq and Lohmann, 1976.)

GEOLOGICAL DISTRIBUTION AND BIOSTRATIGRAPHY

Calcareous nannoplankton first appeared in the earliest Jurassic and species diversities remained relatively low through the early and middle Jurassic. The first significant diversification occurred during the early part of the late Jurassic (Oxfordian) and a second marked increase occurred in the Aptian and Albian, culminating in a very high abundance of nannofossils in the latest Cretaceous (Maastrichtian).

At the end of the Maastrichtian a massive extinction of marine organisms occurred which depleted the oceans of calcareous nannoplankton and only four or five species

survived into the earliest Tertiary (Danian). From this relatively small genetic pool new taxa evolved, at a relatively slow pace at first, and then rapidly during the late Paleocene and early Eocene resulting in the highest diversities of the Cenozoic during the early Eocene. This exponential increase is also partly due to the appearance of discoasters in the late Paleocene and their rapid diversification in the early and middle Eocene. Total diversity dropped gradually during the remainder of the Eocene and relatively sharply during the early and late Oligocene. A second, less pronounced, radiation of taxa occurred during the middle Miocene after which a trend towards decreased diversity continued until the Pleistocene.

The relative abundance and short stratigraphic ranges of nannofossils makes them an excellent group for the biostratigraphic subdivision of the Mesozoic and Cenozoic strata. Their planktonic habit and thus relatively rapid dispersal over large areas enhances their usefulness as tools for inter-regional correlations. Many species are ubiquitous in tropical and temperate regions and can be used for relatively refined biostratigraphic zonations in these areas. In the higher latitudes, however, the species are few and usually the hardy, long-ranging, forms which are of little stratigraphic value.

A number of biostratigraphic zonation schemes have been suggested for various parts of the Cenozoic and a fairly high-resolution zonation for the entire era is now available. Hay and Mohler (1967) and Hay and others (1967) have suggested zonations for the Paleocene—Eocene interval; Bramlette and Wilcoxon (1967) and Roth (1970) for the Oligocene; and Gartner (1969b) and Hay and others (1967) for the Neogene and Quaternary. Bukry (1971) presented a comprehensive zonation for the entire Cenozoic of the Pacific Ocean. It is beyond the scope of this text to discuss the details of these zonation schemes. The interested reader is referred, in addition to the above papers, to Martini (1971) for a summary version of the Cenozoic zonations. Martini's zonation is, however, more applicable to the near-shore and shelf, tropical to temperate areas as it relies upon many hemipelagic species not found in the

TABLE II

Stratigraphic ranges and light microscopic illustrations of selected Mesozoic nannofossil species (figures *a* under phase-contrast and *b* under cross-polarized light conditions; all figures × 2,080)

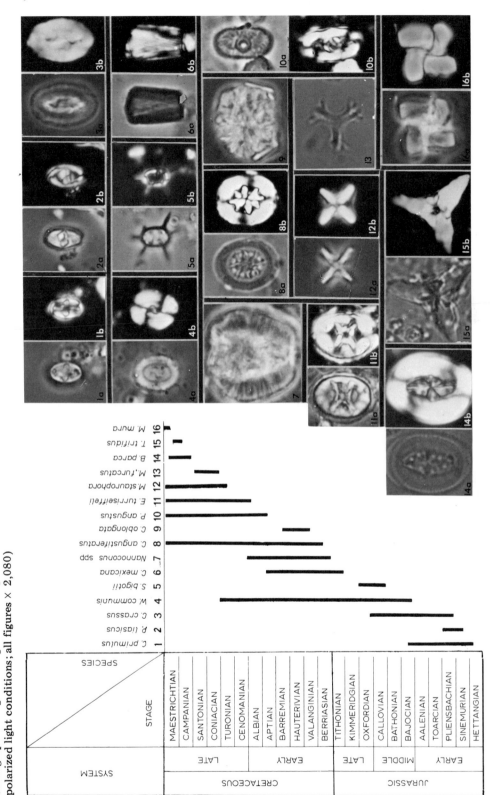

Legend: *1, Crucirhabdus primulus* Prins ex Rood, Hay and Bernard; *2, Parhabdolithus liasicus* Deflandre; *3, Crepidolithus crassus* (Deflandre) Noël; *4, Watznauria communis* Reinhardt; *5, Stephanolithion bigotii* Deflandre; *6, Conusphaera mexicana* Trejo; *7. Nannoconus bucheri* Brönnimann; *8, Cretarhabdus angustiforatus* (Black) Bukry; *9, Calcicalathina oblongata* (Worsley) Thierstein; *10, Parhabdolithus angustus* (Stradner) Stradner and others; *11, Eiffellithus turriseiffeli* (Deflandre) Reinhardt; *12, Micula staurophora* (Gardet) Stradner; *13, Marthasterites furcatus* (Deflandre) Deflandre; *14, Broinsonia parca* (Stradner) Bukry; *15, Tetralithus trifidus* (Stradner) Bukry; *16, Micula mura* (Martini) Bukry. (Photos: H. Thierstein.)

TABLE III

Stratigraphic ranges and light microscopic illustrations of selected early Cenozoic nannofossil species (figures a under phase-contrast and b under cross-polarized light conditions; all figures × 3,200)

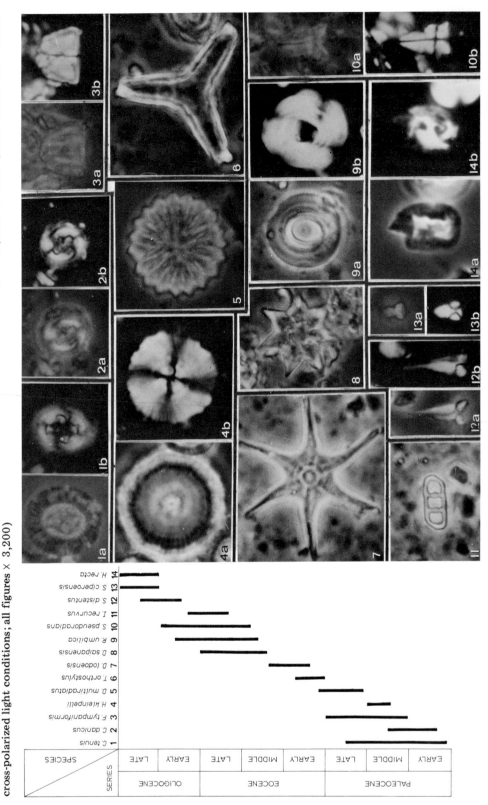

Legend: 1, Cruciplacolithus tenuis (Stradner) Hay and Mohler; 2, Chiasmolithus danicus (Brotzen) Hay and Mohler; 3, Fasciculithus tympaniformis Hay and Mohler; 4, Heliolithus kleinpelli Sullivan; 5, Discoaster multiradiatus Bramlette and Riedel; 6, Tribrachiatus orthostylus Shamrai; 7, Discoaster lodoensis Bramlette and Riedel; 8, D. saipanensis Bramlette and Riedel; 9, Reticulofenestra umbilica (Levin) Martini and Ritzkowski; 10, Sphenolithus pseudoradians Bramlette and Wilcoxon; 11, Isthmolithus recurvus Deflandre; 12, Sphenolithus distentus (Martini) Bramlette and Wilcoxon; 13, S. ciperoensis Bramlette and Wilcoxon; 14, Helicosphaera recta (Haq) Jafar and Martini.

TABLE IV

Stratigraphic ranges and light microscopic illustrations of selected late Cenozoic nannofossil species (figures *a* are under phase-contrast and *b* under cross-polarized light conditions; all figures × 3,200)

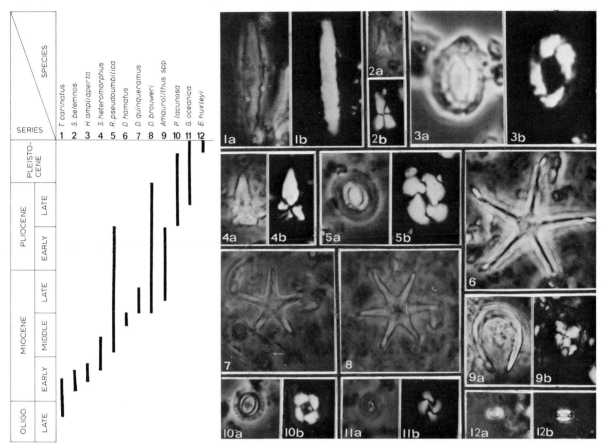

Legend: *1, Triquetrorhabdulus carinatus* Martini; *2, Sphenolithus belemnos* Bramlette and Wilcoxon; *3, Helicosphaera ampliaperta* Bramlette and Wilcoxon; *4, Sphenolithus heteromorphus* Deflandre; *5, Reticulofenestra pseudoumbilica* (Gartner) Gartner; *6, Discoaster hamatus* Martini and Bramlette; *7, D. quinqueramus* Gartner; *8, D. brouweri* Tan Sin Hok; *9, Amaurolithus delicatus* Gartner and Bukry (notice the almost complete extinction under cross-polarized light); *10, Pseudoemiliania lacunosa* (Kamptner) Gartner; *11, Gephyrocapsa oceanica* Kamptner; *12, Emiliania huxleyi* (Lohmann) Hay and Mohler.

open ocean. For a scheme more readily applicable to the oceanic sections zonations suggested by Bukry (1971 and 1973) should be consulted. For a Paleogene zonation of the relatively higher latitudes of New Zealand and adjacent Southern Ocean areas the reader is referred to Edwards (1971).

As compared to the Cenozoic rates, the nannoplankton evolutionary rates were relatively slower through most of the Mesozoic and thus the zonations for this era are not as refined. Prins (1969) suggested a zonation for the early Jurassic; Thierstein (1973) for the early Cretaceous and Čepek and Hay (1969) and Manivit (1971) for the late Cretaceous.

The reader is referred to Thierstein (1976) for a complete summary of the existing knowledge on the Mesozoic nannofossil zonation.

To make this text biostratigraphically more useful for the reader, range-charts of key species and their light microscopic illustrations are included in Tables II—IV. As far as possible those species with well-known ranges which are fairly common and can be more readily recognized have been chosen to present three- or four-fold subdivisions of each epoch (see Table II for the Mesozoic; Table III for the early Cenozoic and Table IV for the late Cenozoic).

SUGGESTIONS FOR FURTHER READING

Black, M., 1965. Coccoliths. *Endeavour*, 24: 131—137. [A short, introductory and beautifully illustrated article on nannoliths.]

Farinacci, A., 1971. Round table on calcareous nannoplankton, Rome, Sept. 23—28, 1970. *Proc. II Planktonic Conf. Rome*, 2: 1343—1369. [Summary of nannoplankton terminology in English, French, German, Italian, Russian and Spanish.]

Noël, D., 1965. *Sur les coccolithes du Jurassique Européen et d'Afrique du Nord*. C.N.R.S., Paris, 209 pp. (29 pls.).

Noël, D., 1970. *Coccolithes crétacés de la Craie campanienne du Bassin de Paris*. C.N.R.S., Paris, 129 pp. (48 pls.). [These two papers present ultrastructure of Jurassic and Cretaceous nannoliths. Numerous transmission and scanning electron micrographs of the Mesozoic species are illustrated.]

Paasche, E., 1968. Biology and physiology of coccolithophorids. *Ann. Rev. Microbiol.*, 22: 71—86. [A summary of all available data on the subject till 1967.]

Perch-Nielsen, K., 1971. Durchsicht Tertiärer Coccolithen. *Proc. II Planktonic Conf.* Rome, pp. 939—980. [Summary of Cenozoic nannofossil taxonomy; valid and invalid taxa and morphological comparisons through hand-drawn illustrations of taxa.]

Reinhardt, P., 1972. *Coccolithen, Kalkiges Plankton seit Jahrmillionen*. Die Neue Brehm-Bücherei, A. Zeimsen Verlag, Wittenberg, 99 pp. [A short introductory textbook on nannofossils in German language.]

Schiller, J., 1930. Coccolithineae. In: *L. Rabenhorst's Kryptogamen-Flora*, 10. Akad. Verlagsgesellschaft, Leipzig. pp. 89—263. [Morphology and physiology of all living coccolithophores known until the 1920's.]

Catalogues

Farinacci, A., 1969 and later. *Catalogue of Calcareous Nannofossils*. Edizioni Tecnoscienza, Rome, vols. 1 to 7. [This compilation gives the original description and illustrations of all taxa at generic and species level without comments about validity of names, etc. Volumes published intermittently.]

Loeblich Jr., A.R. and Tappan, H., 1966. Annotated index and bibliography of the calcareous nannoplankton. I. *Phycologia*, 5: 81—216. Subsequent Nos. of the same: II: 1968, *J. Paleontol.*, 42: 584—598; III: 1969, *J. Paleontol.*, 43: 568—588; IV: 1970, *J. Paleontol.*, 44: 558—574; V: 1970, *Phycologia*, 9: 157—174; VI: 1971, *Phycologia*, 10: 315—339; VII: 1973, *J. Paleontol.*, 47: 715—759. [These and future numbers index all nannoplankton taxa with comments about their non-validity under Botanical or Zoological Codes if not described properly. Also includes up-to-date bibliography on nannoplankton.]

CITED REFERENCES

Black, M., 1963. The fine structure of the mineral parts of Coccolithophoridae. *Proc. Linn. Soc. Lond.*, 174: 41—46.

Berger, W.H., 1973. Deep-sea carbonates: evidence for a coccolith lysocline. *Deep-Sea Res.*, 20: 917—921.

Bramlette, M.N. and Wilcoxon, J.A., 1967. Middle Tertiary calcareous nannoplankton of the Cipero section, Trinidad, W.I. *Tulane Stud. Geol.*, 5: 93—131.

Bukry, D., 1969. Upper Cretaceous coccoliths from Texas and Europe. *Univ. Kansas Paleontol. Contrib.*, 51: 1—79 (40 pls.).

Bukry, D., 1971. Cenozoic calcareous nannofossils from the Pacific Ocean. *Trans. San Diego Soc. Nat. Hist.*, 16: 303—328.

Bukry, D., 1973. Low-latitude coccolith biostratigraphic zonation. In: *Initial Reports of the Deep Sea Drilling Project, XV*. U.S. Government Printing Office, Washington, D.C., pp. 685—703.

Bursa, A.S., 1965. *Discoasteromonas calciferous* n.sp., an Arctic relict secreting *Discoaster* Tan Sin Hok, 1927. *Grana Palynol.*, 6: 147—165.

Čepek, P. and Hay, W.W., 1969. Calcareous nannoplankton and biostratigraphic subdivision of the Upper Cretaceous. *Trans. Gulf Coast Assoc. Geol. Soc.*, 19: 323—336.

Edwards, A.R., 1971. A calcareous nannoplankton zonation of the New England Paleogene. *Proc. II Planktonic Conf., Rome*, pp. 381—419.

Forchheimer, S., 1972. Scanning electron microscope studies of Cretaceous coccoliths from the Köpingsberg Borehole no. 1, SE Sweden. *Sver. Geol. Unders., Ser. C.*, no. 668: 141 pp.

Gaarder, K.R., 1962. Electron microscope studies on holococcolithophorids. *Nytt. Mag. Bot.*, 10: 35—51 (12 pls.).

Gartner, S., 1969a. Phylogenetic lineages in the Lower Tertiary coccolith genus *Chiasmolithus*. *Proc. N. Am. Paleontol. Conv.*, G: 930—957.

Gartner, S., 1969b. Correlation of Neogene planktonic foraminifera and calcareous nannofossil zones. *Trans. Gulf Coast Assoc. Geol. Soc.*, 19: 585—599.

Gartner, S. and Bukry, D., 1969. Tertiary Holococcoliths. *J. Paleontol.*, 43: 1213—1221.

Haq, B.U., 1973a. Evolutionary trends in the Cenozoic coccolithophore genus *Helicopontosphaera*. *Micropaleontology*, 19: 32—52.

Haq, B.U., 1973b. Transgressions, climatic change and the diversity of calcareous nannoplankton. *Mar. Geol.*, 15: M25—M30.

Haq, B.U. and Lohmann, G.P., 1976. Early Cenozoic calcareous nannoplankton biogeography of the Atlantic Ocean. *Mar. Micropaleontol.*, 1: 119—194.

Hay, W.W., 1965. Calcareous nannofossils. In: B. Kummel and D. Raup (Editors), *Handbook of Paleontological Techniques*. W.H. Freeman, San Francisco, Calif., pp. 3—6.

Hay, W.W. and Mohler, H.P., 1967. Calcareous nannoplankton from Early Tertiary rocks at Pont Labau, France and Paleocene—Early Eocene correlations. *J. Paleontol.*, 41: 1505—1541.

Hay, W.W., Mohler, H.P., Roth, P.H., Schmidt, R.R. and Boudreaux, J.E., 1967. Calcareous nannoplankton zonation of the Cenozoic of the Gulf Coast and Caribbean—Antillean area and transoceanic correlation. *Trans. Gulf Coast Assoc. Geol. Soc.*, 17: 428—480.

Honjo, S., 1976. Coccoliths: production, transportation, and sedimentation. *Mar. Micropaleontol.*, 1: 65—79.

Isenberg, H.D., Douglas, S.D., Lavine, L.S. and Weissfellner, H., 1967. Laboratory studies with coccolithophorid calcification. *Proc. Int. Conf. Tropical Oceanography, Miami, Fla.*, pp. 155—177.

Manivit, H., 1971. *Les nannofossiles calcaires du Crétacé français (de l'Aptien au Danian). Essai de biozonation appuyée sur les stratotypes*. Thesis, Paris University, 187 pp. (32 pls.).

Manton, I. and Leedale, G.F., 1969. Observations on the microanatomy of *Coccolithus pelagicus* and *Cricosphaera carterae*, with special reference to the origin and nature of coccoliths and scales. *J. Mar. Biol. Assoc., U.K.*, 49: 1—16.

Martini, E., 1971. Standard Tertiary and Quaternary calcareous nannoplankton zonation. *Proc. II Planktonic Conf.* Rome, pp. 739—785.

McIntyre, A. and Bé, A.W.H., 1967. Modern coccolithophoridae of the Atlantic Ocean — I. Placoliths and cyrtoliths. *Deep-Sea Res.*, 14: 561—597.

McIntyre, A., Bé, A.W.H. and Roche, M.B., 1970. Modern Pacific Coccolithophorida: A paleontological thermometer. *Trans. N.Y. Acad. Sci.*, 32: 720—731.

McIntyre, A., Ruddiman, W.F. and Jantzen, R., 1972. Southward penetrations of the North Atlantic Polar Front: faunal and floral evidence of large-scale surface water mass movements over the last 225,000 years. *Deep-Sea Res.*, 19: 61—77.

Okada, H. and Honjo, S., 1973. The distribution of oceanic coccolithophorids in the Pacific. *Deep-Sea Res.*, 20: 355—374.

Parke, M. and Adams, I., 1960. The motile (*Crystallolithus hyalinus* Gaarder and Markali) and non-motile phases in the life history of *Coccolithus pelagicus* (Wallich) Schiller. *J. Mar. Biol. Assoc. U.K.*, 39: 263—274.

Prins, B., 1969. Evolution and stratigraphy of coccolithinids from the Lower and Middle Lias. *Proc. I Planktonic Conf., Geneva*, 2: 547—558.

Prins, B., 1971. Speculations on relations, evolution and stratigraphic distribution of discoasters. *Proc. II Planktonic Conf., Rome*, pp. 1017—1037.

Roth, P.H., 1970. Oligocene calcareous nannoplankton biostratigraphy. *Eclogae Geol. Helv.*, 63: 799—881.

Roth, P.H. and Berger, W.H., 1975. Distribution and dissolution of coccoliths in the south and central Pacific. In: *Dissolution of Deep-Sea Carbonates. Cushman Found. Foram. Res.*, 13: 87—113.

Roth, P.H. and Thierstein, H., 1972. Calcareous nannoplankton: Leg 14 of the Deep Sea Drilling Project. In: *Initial Reports of the Deep Sea Drilling Project, XIV*. U.S. Government Printing Office, Washington, D.C., pp. 421—485.

Stradner, H. and Papp, A., 1961. Tertiäre Discoasteriden aus Österreich und deren stratigraphische Bedeutung. *Jahrb. Geol. Bundesanst. (Wien)*, 7: 1—160.

Thierstein, H.R., 1971. Tentative Lower Cretaceous calcareous nannoplankton zonation. *Eclogae Geol. Helv.*, 64: 459—488.

Thierstein, H.R., 1973. Lower Cretaceous calcareous nannoplankton biostratigraphy. *Abh. Geol. Bundesanst. (Wien)*, 29: 52 pp.

Thierstein, H.R., 1974. Calcareous nannoplankton — Leg 26 of the Deep Sea Drilling Project. In: *Initial Reports of the Deep Sea Drilling Project, XXVI*. U.S. Government Printing Office, Washington, D.C., pp. 619—667.

Thierstein, H.R., 1976. Mesozoic calcareous nannoplankton biostratigraphy of marine sediments. *Mar. Micropaleontol.*, 1: 325—362.

Watabe, N. and Wilbur, K.M., 1966. Effects of temperature on growth calcification and coccolith form in *Coccolithus huxleyi* (Coccolithineae). *Limnol. Oceanogr.*, 11: 567—575.

Wilbur, K.M. and Watabe, N., 1963. Experimental studies on calcification of the alga *Coccolithus huxleyi. Ann. N.Y. Acad. Sci.*, 109: 82—112.

Wise, S.W. and Kelts, K.R., 1972. Inferred diagenetic history of a weakly silicified deep sea chalk. *Trans. Gulf Coast Assoc. Geol. Soc.*, 22: 1—77.

4

OSTRACODES

VLADIMÍR POKORNÝ

INTRODUCTION

If asked to indicate the most useful group of Crustacea in geological sciences, every stratigrapher or paleontologist would name, without hesitation, the ostracodes. Indeed, the remains of these small, mostly microscopic, crustaceans are widely distributed in rocks of all the periods of the Phanerozoic Era. Beginning in the Cambrian their evolutionary history can be followed with a completeness exceptional among the crustaceans. This subclass is one of the best documented groups within the whole animal kingdom due to the most characteristic feature of their bodies — a bivalved, well-calcified shell which fossilizes easily.

The majority of ostracodes have a length between 0.15 and 2 mm. Recent marine swimming forms attain up to about 25 mm in length and the largest Paleozoic species up to 80 mm. Their shells possess a considerable number of morphological features allowing taxonomic and phylogenetic study as well as analysis of their functional morphology and ecology.

Ostracodes have undergone a spectacular ecological radiation. They live in fresh, brackish, saline and hypersaline waters, and rarely even in extra-aquatic environments. In the sea they are found from the shoreline down to hyperabyssal depths. By far the greatest number of fossil marine ostracodes are benthic forms. The planktonic species, due to their weakly calcified shells, are generally rare in fossil assemblages, and play a minor role in paleontology. Rapidly evolving ostracode lineages are extremely useful as markers in intraregional stratigraphy, especially in cases where foraminifera are absent. Lacking planktonic larvae, many shallow and warm-water species cannot cross physical barriers, so that they are only of limited use for intercontinental or interregional stratigraphy, especially in the Cenozoic. This disadvantage, however, is outweighed by their usefulness in paleogeography and paleoecology.

HISTORY OF OSTRACODE STUDY

Recent ostracodes

The scientific study of ostracodes began in 1776 when O.F. Müller named the first living species. One of the most important taxonomic works on recent ostracodes was written by Sars (1866) who established a firm basis for all subsequent classifications. The monograph by G.W. Müller (1894) on the ostracodes of the Gulf of Naples stands out because of its detailed morphological analysis and careful descriptions of soft parts and includes important observations on phylogeny, morphological adaptations, ecology and reproduction.

A renowned school of ostracode workers active in the nineteenth century in England was represented by Brady and his co-workers. Apart from voluminous descriptive work, these workers opened new directions for ostracode research. Brady (1868) introduced a new approach by recognizing five biogeographic ostracode provinces in Europe. These workers also initiated paleoecological studies with their comparison of Scottish Pleistocene ostracode communities with Recent ones. The important monograph by Brady (1880) describing the material of the *Challenger* expedition gives the first comprehensive insights into the global distribution of marine ostracodes.

Another outstanding monograph, compiled by Skogsberg (1920) gave, among other things, a masterly analysis of ostracode extremities, critical discussion of older systematic works and of morphology and functional analysis of adaptations to planktonic life. Outstanding work on functional morphology was also carried out by Cannon (1925 and later). The era of modern ecological studies on Recent forms was opened by the work of Elofson (1941).

Fossil ostracodes

The first fossil ostracode was described in 1813. The stratigraphical value of the group was recognized soon afterwards. In 1850 Forbes was able to divide the English Purbeckian (uppermost Jurassic to lowermost Cretaceous) into five biostratigraphic zones. A good deal of work on fossil ostracodes was done by Jones between 1849 and 1901, who also studied Paleozoic ostracodes.

Interest in fossil ostracodes was stimulated in the 1920's by increased demand for oil when the stratigraphic value of fossil ostracodes led to a rapid revival of research on this group. Ostracodes became second in importance only to the foraminifera in applied micropaleontology. This development, however, was not accompanied by a well-developed scientific basis. The taxonomic work on Recent ostracodes was based almost exclusively on soft parts and was therefore of little use to the paleontologist. For the immediate needs of practical work, this difficulty was overcome by using provisional designations for ostracode taxa by letters or numerals. Simultaneous paleontological studies aimed at the development of the necessary scientific background. It began with the study of shell morphology which has been, with few exceptions, largely neglected by zoologists. After the pioneer work by Zalányi (1929) important papers by Alexander (1933), Sylvester-Bradley (1941) and Triebel (1941) appeared, followed by a wide array of subsequent contributions. Diagnoses of the shell characters of Recent genera were completed mainly by paleontologists.

After World War II, the study of ostracodes entered the stage of neontological and pale-ontological synthesis. First manuals devoted to ostracodes, several synopses of ostracode taxa and their card catalogues, appeared in the 1950's and 1960's, stimulating the development of the entire field. The expanding taxonomic studies, chiefly by paleontologists, resulted in the publication of about 6000 papers containing descriptions of about 2000 genera.

The improved taxonomic basis greatly enhanced the development of other fields of ostracode research: ontogeny, functional morphology, phylogeny, ecology and paleoecology, biogeography and paleobiogeography. The periodical joint meetings of zoologists and paleontologists at special ostracode symposia started in Naples in 1963. These symposia volumes edited by Puri (1965), Neale (1969), Oertli (1971) and Swain (1975) are important milestones in ostracode research. The utility of ostracodes in deep-sea research has been demonstrated through the recent development of programs of oceanic research chiefly in the U.S.A. and the Soviet Union.

BIOLOGY

Ostracodes belong to the phylum Crustacea, generally characterized by a segmented body covered by a jointed external skeleton with different number of paired appendages.

The soft body of an ostracode (Fig. 1) is compactly built and more or less laterally compressed. Slight constriction divides it into the head (cephalic) region and the post-cephalic part (thorax). The existence of original segmentation is indicated by the presence of several pairs of appendages. Covering the body is a cuticle, the two folds of which hang down on each side of the body and secrete a bivalved shell, the carapace.

Ostracodes have five to seven pairs of appendages which are considerably diversified in their shapes and functions. All, however, may be derived from the basic type of arthropod extremity (Fig. 2). Four pairs of appendages are present on the head region of the body (Fig. 1). The first two have chiefly a locomotory and sensory function, whereas the third and the fourth function mainly as

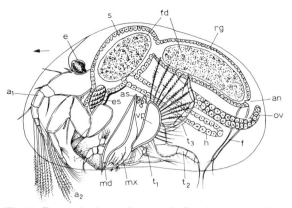

Fig. 1. *Eucypris virens* (Jurine). Left valve removed to show the position of internal organs and structures. Legend: a_1 = antennule; a_2 = antenna; *an* = anus; *as* = adductor scars; *e* = eye; *es* = esophagus; *f* = furca; *fd* = food inside the digestive tube; *h* = hepatic gland (liver); *l* = upper lip; *md* = mandible; *mx* = maxillule; *ov* = ovary; *rg* = rear gut; *s* = stomach; t_1 to t_3 = thoracic legs; *vp* = vibratory plate of the mandible. (After Vávra, 1892; terminology modified.)

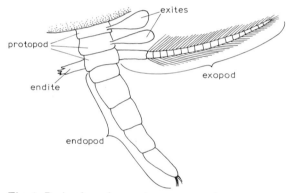

Fig. 2. Basic plan of a crustacean appendage.

food-intake organs. Three, two or one pair of variously shaped appendages are present on the postcephalic region. They are modified for feeding, locomotion, and for creating water currents; they may function as male clasping organs or are used in cleaning the carapace inside. The sixth and seventh pairs of appendages are sometimes absent.

The body of most ostracodes ends in the **furca**. This consists of two branches usually folded underneath the body. It can be protruded from between the valves and aids in locomotion. In some groups it is reduced or lacking.

The sex organs are paired. In the females they consist of ovaries, uteri, vaginae and seminal receptacles. The ovaries are situated

either in the body cavity or in the valve cavity (Fig. 3). The male genitalia are complicated and voluminous due to unusually large spermatozoa which are the largest in the animal kingdom, being up to ten times longer than the carapace. The male organs (Fig. 3) consist of testes, vasa deferentia and penes. In some ostracodes four gonads on each side lie in the valve cavity; the distal part of their ejaculatory duct is muscular and provided with chitinous platelets in some taxa. It is known as the **Zenker's organ**.

The digestive tract consists of the mouth, the esophagus, a mid-gut with the stomach located anteriorly, the intestine posteriorly, and the anus. Digestion is facilitated by salivary glands opening at the anterior end of the digestive tube and by the liver. The livers of some forms are situated in the valve cavity and their imprints may be seen beneath those of the gonads (Fig. 3).

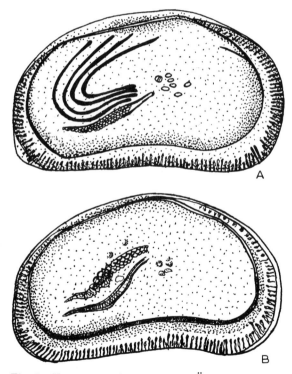

Fig. 3. *Hungarocypris madaraszi* (Örley), a Recent fresh-water cyprid; left valves. A. A male. Four curved male gonads are situated in the posterior part of the body. Below them a liver gland extending from the adductor scars in the posteroventral direction. B. A female. Above the liver gland is the ovary. In the cyprids, traces of gonads and livers are sometimes well preserved on the valve. (After Bronstein, 1947.)

The majority of modern ostracodes have no special organs for blood circulation. Among living ostracodes a heart is present only in a few groups. In the heart-possessing species a system of anastomosing blood canals radiating from the adductor muscle scar group is present between the two valve lamellae. Similar markings in fossil ostracodes have been interpreted, probably mistakenly, as indicating the presence of a heart (see under the leperditicopids, p. 129).

Respiration takes place through the soft body wall. Exchange of gases is facilitated by the movement of vibratory plates which produce a water current within the carapace. Some forms have special gills on the rear of the body.

Among Recent ostracodes only the members of one group have a pair of lateral eyes. Other ostracodes either have a tripartite median eye in the anterodorsal part of the body, or are blind.

Reproduction

Ostracodes are always divided into separate sexes. Not all of them, however, reproduce sexually. Among Recent fresh-water ostracodes some species are known which reproduce parthenogenetically, as the populations consist only of females which lay fertile eggs. Some fresh-water species reproduce by syngamy (copulation) in warmer areas, whereas in higher latitudes they reproduce parthenogenetically. This phenomenon, known also in other crustaceans, has been called **geographical parthenogenesis.**

Ostracode eggs develop into an initial larval stage, the **nauplius,** which is characteristic for many other crustacean groups. It has three pairs of appendages and possesses a shell. Like other crustacean larvae, ostracode larvae grow by **ecdysis,** molting the old body cover and secreting a new one. During the short ecdysis the ostracode approximately doubles its volume and adds new appendages. Nine growth stages have been observed in most Recent ostracodes, the ninth being the adult animal. Some species, however, have a smaller number of larval stages. The number of growth stages within the same species may vary with climatic conditions, being smaller in warmer waters.

The rate of development is correlated with temperature. Fresh-water ostracodes mature in about one month, the marine ones in several weeks to about three years.

The carapace

The ostracode body is wholly covered by a continuous cuticle secreted by the epidermis. When first formed, the cuticle in all the crustaceans is soft but becomes largely hardened by a complex tanning process called **sclerotization,** the deposition of mineral salts. However, parts of the cuticle, called joints, remain permanently soft so that movement is possible.

The ostracode carapace is an integral part of the cuticle. It develops as a single cuticular fold originating on the head region and completely enveloping the body. The two valves arise through mineralization of its left and right sides and are united in their dorsal part by a narrow strip of soft cuticle called **ligament.** The left and right valves are connected by an adductor muscle which traverses the soft body in its median region. An articulation or joining of the valves, called the **hinge,** is also developed in the dorsal margin of the two valves in many ostracodes (Fig. 4).

Each valve consists of two lamellae. The outer lamella forms the outer surface of the valve and bends at the valve margin into the

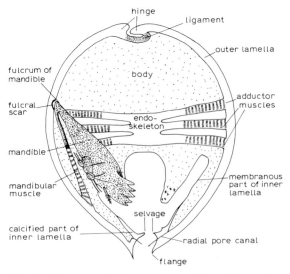

Fig. 4. Diagrammatic transverse section of an ostracode. (After Harding, 1965, slightly modified.)

inner lamella which, in the dorsal part of the body, passes into the body cuticle. Whereas in modern species the outer lamella is almost invariably wholly calcified, the inner lamella is calcified on its peripheral margin only. The space between the outer and inner lamella is called the **valve cavity**. It contains diverse organs and is connected with the main soft body in the dorsal region (Fig. 5). In the exclusively or predominantly Paleozoic orders of ostracodes, an inner calcified lamella has not been definitely recognized. The most complicated marginal structures are characteristic of Cenozoic ostracodes.

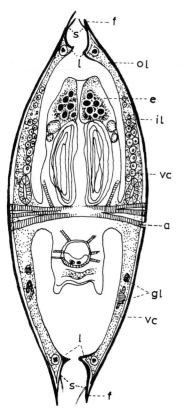

Fig. 5. *Hungarocypris madaraszi* (Örley), a Recent fresh-water cypridid. Longitudinal section. Legend: *a* = adductor scars; *e* = eggs; *f* = flange; *gl* = glands in the valve cavity; *il* = inner lamella; *l* = list; *ol* = outer lamella; *s* = selvage; *vc* = valve cavity. (After Daday in Zalányi, 1929, explanations modified.)

The contact margin of each valve is subdivided into the **cardinal**, also called the **hinge margin**, and the **extracardinal margin** which is composed of the anterior, ventral and posterior margins. The extracardinal margin is com

monly also called the **free margin**. The line along which valves articulate, seen when the carapace is complete, is the **hinge line**.

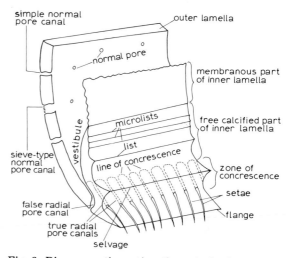

Fig. 6. Diagrammatic section through the free margin of a podocopid ostracode valve.

Different marginal structures, such as spines, denticles or ridges may be developed near the free edge of the valves. Most important of these is the **selvage**. This is a chitinous fringe along the extracardinal margin serving to seal the closed valves. Its base is calcified and can be followed on fossil valves as a continuous crest (Figs. 6, 7).

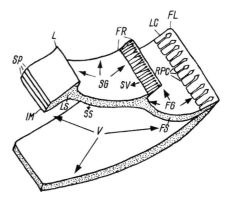

Fig. 7. Diagrammatic section through the marginal zone of *Cypris pubera* O.F. Müller, a Recent freshwater cypridid. Legend: *FG* = flange groove; *FL* = flange; *FR* = fringe; *FS* = flange strip; *IM* = inner margin; *L* = list; *LC* = line of concrescence; *LS* = list strip; *RPC* = radial pore canals; *SG* = selvage groove; *SP* = microlists; *SS* = selvage strip; *SV* = selvage; *V* = vestibule. (After Sylvester-Bradley, 1941; terminology slightly modified.)

The hinge

In some ostracodes, as in the extinct archeocopids or in many myodocopins or cladocopins no hinge structure exists. It is believed that this unhinged state is primitive and that hinges later evolved independently in several ostracode groups.

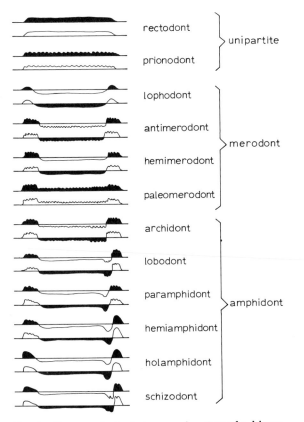

Fig. 8. Some principal types of ostracode hinges.

Hinges may be composed of variously shaped bars, grooves, teeth and sockets effecting the articulation of the two valves and a rather complicated terminology has been developed for the designation of different types. The three main categories of hinges are (Fig. 8):

1. **Unipartite hinges.** This category includes all the hinge types in which no subdivision into terminal and median elements developed.
2. **Merodont hinges.** This category comprises all the hinges which possess terminal teeth in one valve only.
3. **Amphidont hinges** have teeth and sockets in both valves. They arose from merodont hinges by the differentiation of the median element into an anteromedian and a posteromedian part.

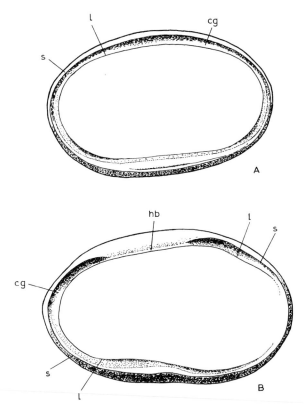

Fig. 9. A. A holosolenic type of valve closure, exemplified by *Cytherella* sp. B. A hemisolenic type of closure displayed by a thlipsurid, *Thlipsurella discreta* (Jones) from the Silurian of Gotland. Legend: *cg* = contact groove; *hb* = hinge bar; *l* = list; *s* = selvage. (After Pokorný, 1957.)

There is one type of valve closure in which the cardinal and extracardinal regions are not morphologically differentiated into a true hinge region. This is a box-like closure (Fig. 9) of some platycopins. The simple edge of one valve fits into the marginal groove of the other valve and has been called **holosolenic**; that is, having a groove all the way around. Although morphologically simple, this is probably a phylogenetically derived hinge-type evolved in parallel ostracode lineages from different ancestors. In a more general way this term is used to designate a valve on which all the elements of the cardinal and extracardinal margins are either raised (positive) or depressed (negative). In **hemisolenic** closure, at least the median part of the hinge is raised in valves with otherwise depressed features of the contact margin and vice versa. The term **virgatodont** has been used for the hinges of Paleozoic forms on which longitudinal grooves and bars prevail (Fig. 10).

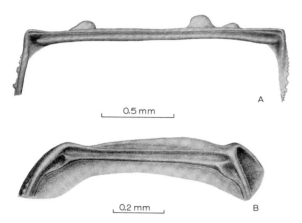

Fig. 10. Virgatodont hinges. A. *Beyrichia (Beyrichia) dactyloscopica* Martinsson, a beyrichiid. Right valve. B. *Clavofabella multidentata* Martinsson, a primitiopsidid. Right valve. Both species from the Silurian of Gotland. (After Pokorný, 1959.)

Muscle scars

On the internal surface of a well-preserved ostracode valve there are small spots of a somewhat different shell structure. They are confined to two areas which, according to their position on the valve, are termed the **central muscle scar field** (Fig. 11) and the **dorsal muscle scar field** (Fig. 12). Each field contains a large number of scars of different origin, the majority of which are attachments of short sinews which connect with muscles.

The central muscle scar field includes three groups of scars:

1. The **adductor muscle scar group** which represents the imprints of closing, adductor muscles.
2. The **mandibular** group which lies in front of the ventral part of the adductor muscle scar group; its scars are not formed by muscles, but by chitinous support rods.
3. The **frontal group**, situated above the mandibular scars. In groups hitherto studied it belongs to a mandibular muscle and to a muscle attaching itself to the chitinous endoskeleton.

Apart from these three groups, a single **fulcral scar** of different character may be seen in some species between the frontal and the adductor group. This is formed by the pivoting motion of the dorsal apex of the mandible against the body wall. The dorsal group of muscle scars is variously developed and includes scars the origin of which has been studied only in a few species.

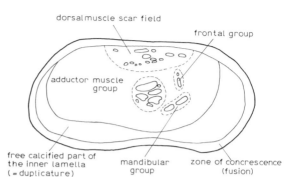

Fig. 11. Arrangement of muscle scars in *Amplocypris recta* (Reuss), a mesohaline cyprydid from the Upper Miocene (Pannonian) of the Vienna Basin. The adductor, mandibular and frontal groups together constitute the central muscle scar field. (After Pokorný, 1954, modified.)

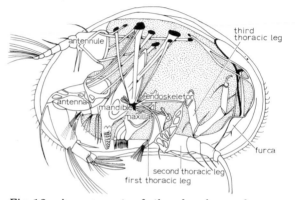

Fig. 12. Arrangement of the dorsal muscle scars (drawn as black areas below the dorsal margin) and their function in *Cypridopsis vidua* (O.F. Müller), a Recent fresh-water cypridacean. (After Smith, 1965.)

The most distinctive group of muscle scars which is also most resistant to fossilization is the adductor group. The basic pattern of adductor scars is characteristic at the familial level and may be sometimes used for generic or specific designations.

Pore canals

Ostracode valves are pierced by diversely shaped holes termed **pore canals** (Fig. 13). According to their position two categories of pore canals may be distinguished: the **normal (lateral)** pore canals and the **radial (marginal)** pore canals. Lateral pore canals traverse the outer lamellae usually at a right angle and display a large variety of shapes. Basically, two types may be distinguished: the simple pore canals (Fig. 13) and the sieve-type pore canals (Fig. 14). The former have a simple

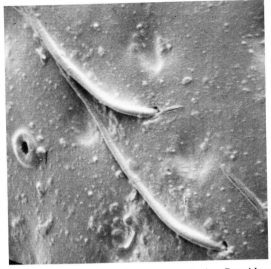

Fig. 13. Simple lateral pore canals in *Cypridopsis vidua* (O.F. Müller), a Recent fresh-water cypridid. Tactile setae pass through two pore canals. × 1400. (SEM photo courtesy of H.J. Oertli.)

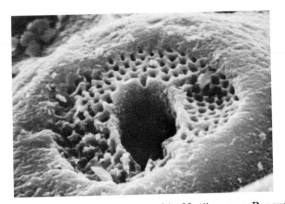

Fig. 14. Sieve-type pore canal in *Mutilus* sp., a Recent hemicytherid from the Mediterranean. × 2625. (SEM photo courtesy of H.J. Oertli.)

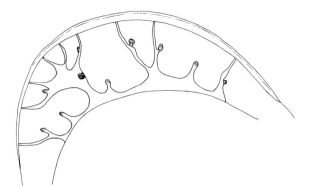

Fig. 15. Radial pore canals along the anterior margin of the right valve in *Leptocythere moravica* Pokorný, a mesohaline leptocytherid from the Upper Miocene of Moravia, Czechoslovakia. The branched pore canals consist of two types of branches. The simple ones reach the outer margin of the valve and correspond to simple true radial pore canals. The sieve-type, open at the outer lateral surface of the valves, correspond to normal pore canals. The branched radial pore canals developed through phylogenetic widening of the zone of concrescence. Their stem represents a common passage of soft tissues to simple pore canals which figure morphologically as branches of a single canal. × 246. (After Pokorný, 1952).

outer opening (pore), while the latter are variously shaped sieve-plates. Many species have more than one of lateral pore canals so that both simple and sieve-type pore canals may occur on the same species (Fig. 15).

Many, but not all, pore canals contain sensory structures, hair-like **setae**. Some setae are long and thick, others are more delicate. A tactile function has been demonstrated for the stronger type; a more sensitive function, such as perception of slight water movements or of sound waves, is presumed for the finer ones. As suggested by G.W. Müller (1894), some sieve-type pore canals with underlying pigment cells may function as photoreceptors.

The pore canals which do not bear setae may represent ducts of some glands, as suggested by Van Morkhoven (1962) or may perform a still unknown function.

The radial (marginal) pore canals originate at the line of concrescence. True radial pore canals run through the plane of concrescence while false radial pore canals open at the external surface of the valves and are homologous with the normal pore canals. Radial pore canals may be either simple or branched.

Valvular vaulting

In many species with long and straight dorsal margins the general vaulting of the valve surface is modified by elevations and depressions which are also reflected on the internal surface of the valve. The elevations are called **lobes**, and the intervening depressions **sulci** (Fig. 16). The sulci are designated from anterior to posterior as S_1, S_2, and S_3; the lobi in the same way as L_1 to L_4. The most persistent sulcus is the S_2, called also the **adductorial** or **median sulcus**, and this is the sulcus developed in unisulcate species. Considerable importance has been assigned to the number and shape of the lobi and sulci for

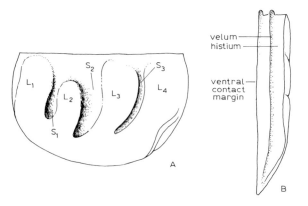

Fig. 16. A. Terminology of lobi and sulci in a beyrichicopid ostracode. B. Adventral structures in a hollinacean (tetradellid) in ventral view.

supraspecific classification of Paleozoic ostracodes. The function of sulci has been ascertained through comparative study of Recent ostracodes, such as the fresh-water genus *Ilyocypris* (Fig. 17) to correspond to muscular attachments inside the valves.

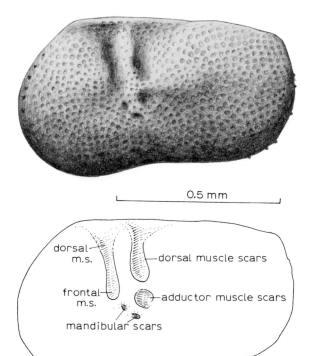

0.5 mm

Fig. 17. *Ilyocypris bradyi* Sars, a fresh-water ilyocypridid from the Lower Pleistocene of Czechoslovakia. The depressions of the valve surface correspond to muscle scar attachment. From the two vertical sulci, the anterior corresponds to S_1, the posterior to S_2 of Paleozoic ostracodes (see Fig. 16).

The eyes are one of the few soft organs which leave traces on the carapace. Their position on the anterodorsal part of the valve may be marked by a hyaline circular area on the wall, by a subglobular elevation, or quite exceptionally, by a telescopic tubular process bearing a subglobular terminal cup. On the internal surface the position of the eye is either not reflected on the surface or is indicated by a depression or a deep eye pit. (Fig. 18).

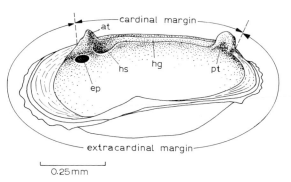

0.25 mm

Fig. 18. *Hemicytheria folliculosa* (Reuss), a hemicytherid from the mesohaline Upper Miocene of Czechoslovakia. Right valve from the inside. Legend: at = anterior hinge tooth; ep = eye pit; hg = hinge groove; hs = hinge socket; pt = posterior hinge tooth. (After Pokorný, 1952, modified.)

Sculpture

Sculpture is defined as only those features of the outer valve surface which are not reflected by the general vaulting of the inner valve surface.

The simplest sculpture on Recent ostracodes consists of an undifferentiated fine reticular network. Through selective thickening or suppression of walls between neighbouring meshes of this primitive type of sculpture, large and diversified meshes, ridges, tubercles, spines and other sculptural features may be derived. To a certain extent the process of sculptural diversification may be followed in the ontogeny of some species.

Orientation of the carapace

For descriptive purposes one recognizes the following views of the carapace (Fig. 19): the lateral or side view normal to the contact (**sagittal**) plane of the valves; the dorsal view — the view on the hinge margin, with line of

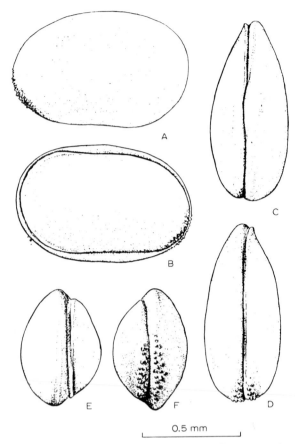

Fig. 19. Descriptive terminology of ostracode carapace exemplified by *Cytherella posterospinosa* Herrig, an Upper Cretaceous cytherellacean. Female carapace. A, side (lateral) view from the right side; B, side view from the left side; C, dorsal view; D, ventral view; E, frontal view; F, posterior view (end view). (After Herrig, 1966.)

sight in the sagittal plane of the carapace; the ventral view — the appearance of the carapace in line of the sagittal plane seen from below; the frontal view — the carapace seen from the anterior end in line of the sagittal plane; and the posterior view — the carapace seen from the posterior end in the sagittal plane.

The term "carapace length" is not uniformly conceived. In straight-backed ostracodes it is defined as the maximum dimension of the carapace in the direction parallel to the hinge line. In specimens with an arched dorsum the length is understood as maximum distance of end points of the carapace measured parallel to the basal line (Fig. 20). The height is measured as the maximum distance perpendicular to the length. The width is the maxi-

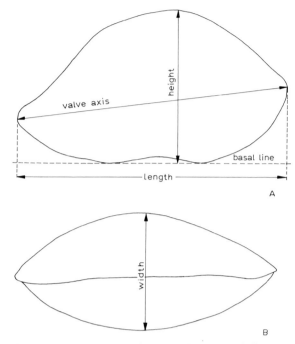

Fig. 20. Measurements of ostracodes, on a bairdiacean. A, right side view; B, dorsal view.

mum distance of the carapace outline perpendicular to the sagittal plane.

One of the most reliable guides to correct orientation is the position of the muscle scars. The adductor muscle scar group of adult ostracodes is, with very few exceptions, situated in front of the valve midlength. Frontal and mandibular scars, if present, definitely mark the position of the anterior end. When muscle scars are effaced by fossilization, their position may be derived from associated features, such as the position of the median sulcus, the lower portion of which corresponds to the position of the adductor scar group or sculptural features which are often arranged concentrically or radially around the central muscle scars.

Ontogeny and morphology of the carapace
Perusal of literature on ostracode taxonomy reveals numerous cases where larval stages were described as new taxa. Most often, such errors result from the facts that: (1) larval ostracode carapaces often differ considerably from those of the adults (Figs. 21, 23); and (2) descriptive work has been done by specialists with inadequate biological background. Some mistakes are due to insufficient material.

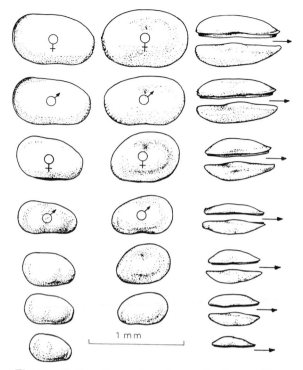

Fig. 21. *Cytherella posterspinosa* Herrig, an Upper Cretaceous cytherellid from the Rügen Island, German Democratic Republic. Ontogenetic development from the fifth to the ninth (adult) instar. Left: left valves; center: right valves; right: both valves in dorsal view. (After Herrig, 1966.)

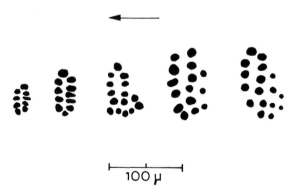

Fig. 22. *Reubenella kramtchanini* (Gramm). A Triassic cytherellacean from the Soviet Far-East. Adductor muscle scars of different growth stages arranged in the order of increasing length of the valves. All are drawn from the outside of left valves. (After Gramm, 1972.)

The most outstanding difference between the larval and adult valves is size. By means of a simple biometric study, even in fossil populations, it is often possible to discriminate the successive larval stages and the adults. Some-

times, however, the fields of different developmental stages are overlapping in a scattergram, due chiefly to ecologically determined size variation of members of successive biological populations which enter into the composition of a fossil sample.

Valves at different ontogenetic stages differ also in their shapes. The length/height ratio is smallest in earliest larval stages, as these have a small number of appendages and consequently a shorter body.

Considerable differences in sculpture between larvae and adults are seen especially in species with highly differentiated adult sculpture (see Fig. 23). Strong ridges or nodes develop gradually during ontogeny. Cases are known, however, where some elements of larval sculpture disappear or are weaker in the adult stage.

Larvae of species with highly developed adult hinges are characterized by more primitive hinge types. For example, the larval hinges of some cytheracean ostracodes do not surpass the merodont stage, although adults possess amphidont hinges. The ontogenetic trend towards a more complex hinge is a general phenomenon linked with a progressively heavier carapace, but rare cases of ontogenetic simplification of the hinge have also been observed.

The zone of concrescence is narrow and traversed by unbranched and usually straight radial pore canals in larval stages. In the adult stage the calcification of the inner lamella becomes broader and sometimes an irregular inner margin develops. The number of radial pore canals usually increases as development progresses. In some species calcification and fusion of the more proximal parts of the inner lamella leads to formation of common passages for soft tissues which enter the pore canals, forming composite marginal pore canals (see Fig. 15). The vestibule of adult specimens may be traversed by calcified septa or pillars.

The pattern of muscle scars may also develop during ontogeny (see Fig. 22). The early larval stages of some genera invariably have a smaller number of adductor muscle scars than the later larval stages and adults. The same is true for the dorsal group of muscle scars. Mandibular and frontal scars may be absent in carapaces of young growth stages.

Sexual dimorphism

There exists no other group of fossil invertebrates whose sexual dimorphism has been so thoroughly studied as the ostracodes. Dimorphism occurs not only in shape and function of the soft body parts, size, shape or sculpture of the carapace, but may manifest itself also in finer structural details, such as the pattern of muscle scars or even in behavior and habitat. The dimorphism of ostracode carapaces may be classified into two categories: (1) the domiciliar dimorphism which affects the size and shape of the carapace; and (2) the extradomiciliar (sculptural) dimorphism in which the proper carapace cavity (**domicilium**) is not affected. Dimorphic features are usually absent in larvae, although a weak sexual dimorphism is sometimes noticeable during the later larval stages.

Determination of the sex of fossil ostracodes is facilitated when dimorphic features can be compared with those of living taxa. A second criterion for sex recognition is the sex ratio. Among living ostracodes both sexes may be equally numerous, but more often the percentage of females is higher than that of males . Similar ratios may be observed in fossil ostracode populations.

Females may be larger, equal in size, or smaller than the males. The two sexes may also differ in lateral outline or length/height ratio (Fig. 23). Strong dimorphism is shown by species with brood-care. In modern ostracodes and their fossil relatives the brood pouches are situated in the posterior part of the carapace so that the female domicilium is characterized by an inflation. This type is called **kloedenellid**, also **cytherellid dimorphism**. In Paleozoic beyrichiomorphs the brood pouches termed **cruminae** (singular: **crumina**) are anterioventral or centroventral in position, and the type of dimorphism is called **cruminal dimorphism** (Fig. 24).

Different types of extradomiciliar dimorphism are abundant and strongly developed in Paleozoic hollinaceans (Fig. 25). An important role in this group is played by the two adventral structures, **velum** and **histium** (Fig. 16), both of which may be dimorphic. The part of the velum or histium which is modified in adult females when compared with adult males has been called the **dolon**. In females of the hollinacean ostracodes the dolo-

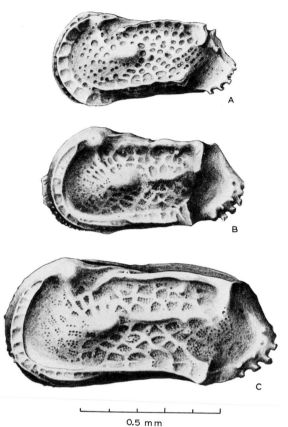

0.5 mm

Fig. 23. Sexual dimorphism and difference in larval and adult sculpture in the trachyleberidid *Rehacythereis* (?) *kodymi* (Pokorný) from the Upper Cretaceous of Czechoslovakia. A, larval stage; B, female; C, male. (After Pokorný, 1967a.)

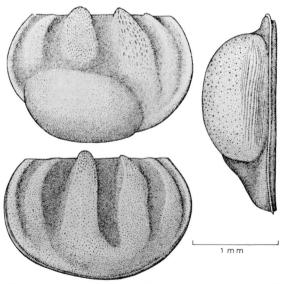

1 mm

Fig. 24. Cruminal dimorphism in beyrichiomorphid ostracodes, shown by *Londinia reticulifera* Martinsson from the Silurian of Sweden. Above and right: heteromorph; below; tecnomorph. (After Martinsson, 1963.)

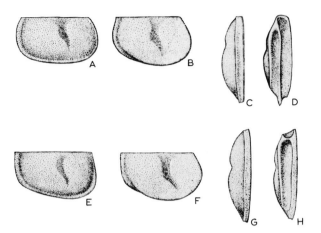

Fig. 25. Antral dimorphism in Ordovician hollinacean (tetradellid) genera. A—D, *Pentagona pentagona* (Jaanusson), a species with biantral dimorphism; A, C, lateral and ventral view of right tecnomorphic valve; B, D, lateral and ventral views of right heteromorphic valve. E—H, *Sigmobolbina variolaris* (Bonnema), a species with supravelar antral dimorphism; E, G, lateral and ventral views of a right tecnomorphic valve; F, H, lateral and ventral views of a right heteromorphic valve. Not to scale. (After Jaanusson, 1966.)

nal portions of adventral structures border concave areas termed **antra** (singular: **antrum**). When situated between the marginal ridge and the velum, the are called **infravelar antra**, when between the velum and histium, **supravelar antra**. Accordingly, the dimorphism is classified by Jaanusson (1966) as infravelar, supravelar or biantral (Fig. 25). The antrum may be simply channel-like, or may be partitioned into separate loculi. In modern ostracodes sculptural dimorphism is generally much less strong and expressed by dimorphic development of ridges, spines, etc.

Because of problems of sex recognition in the large group of predominantly Paleozoic beyrichiocopids, a neutral designation of the two dimorphs has been considered useful. The term **tecnomorph** has been introduced for carapaces of larval stages and those of adults which are essentially similar to larval stages. Those adult carapaces displaying features not present in tecnomorphs have been termed **heteromorphs** and are supposed to be females.

ECOLOGY

Ostracodes probably originated in a marine environment and the largest number of species still inhabits the pelagic and benthic realms of the ocean from the shoreline down to several thousand meters, and from the equator to polar seas. Some species flourish in brackish waters and some are found even in hypersaline environments. Ostracoda have also undergone ecologic radiation in fresh-water environments, from which they are known since the Carboniferous. Some lineages of both fresh-water and marine ostracodes have even invaded terrestrial niches, living in the moist humus of the forests (Harding, 1955), in the aerial part of the fresh-water floating plant accumulations (Danielopol and Vespremeanu, 1964), or, as reported by Schornikov (1969) for the aptly named genus *Terrestricythere*, in the fine, wet gravel on the sea shore of the Kuril Islands.

Nutrition

Ostracodes have evolved a wide variety of nutritional systems including filter-feeding and deposit-feeding. Numerous species feed on marine plants and small living animals such as annelids, turbellarians, nemerteans, or small crustaceans. Some eat detritus from decaying vegetal or animal tissues, while others are **limnivorous**, eating bottom sediments without any selection. Some have their oral apparatus transformed into piercing and sucking organs which are used for the intake of plant juices. Other species of the same family were observed to suck on dead polychaetes, amphipods and other animals. About thirty ostracode genera are known to be commensals, clinging to the appendages or gill cavities of other crustaceans and to the body surface of echinoderms. Some ostracodes interpreted as parasitic were described from the gills and nostrils of fishes (Harding, 1966). Others have glands along the valve margins which secrete a sticky substance to which food adheres. This is then brushed off by mandibular palps and brought to the mouth.

Distribution of marine ostracodes

A comparatively small number of marine ostracodes inhabit the pelagic realm, some living in surficial waters, others distributed through the water column. The greatest number of ostracode species are benthic and their distribution is controlled by a large number of physical, chemical and biological factors.

Salinity

Salinity is the most fundamental factor determining the distribution of ostracodes, as it has a decisive influence on the physiology of the organism. Fresh-water assemblages are taxonomically distinct from marine faunas and few species can thrive in both marine and fresh-water environments. For these species the salinity extremes usually represent marginal conditions of their existence. The near-shore species are often strongly euryhaline, some capable of supporting oligohaline to normal marine conditions, since there are wide oscillations of salinity caused by run-off and rain waters. The brackish waters of lagoons, estuaries, marshes, inland salt pools and lakes are inhabited by a characteristic assemblage of euryhaline and typically brackish species which have several traits in common: with decreasing salinity foraminifera and other marine groups gradually disappear and the dominant position in the microfaunal assemblage is assumed by the ostracodes. The brackish communities are invariably composed of a relatively small number of species. The reduced diversity of organisms may best be explained by the instability of most brackish-water environments. Sometimes a single species may build up a considerable biomass. According to Soviet authors the density of *Cyprideis torosa* (Jones), a typical Recent brackish-water ostracode, may be startling: in the Azov Sea which has been called also the "Ostracode Sea", 14,000 to 30,000 individuals were counted on one square meter of sea bottom. In the Kuban River estuaries the same species attains up to 670,000 individuals per square meter.

More stable brackish environments developed when inland seas were cut off from the open ocean. For example, in the younger Cenozoic of central and southeastern Europe a system of basins collectively called the Paratethys extended north of the Mediterranean Sea (Fig. 26). Before the late Miocene there were free connections between the Paratethys and the Mediterranean. Later the Paratethys

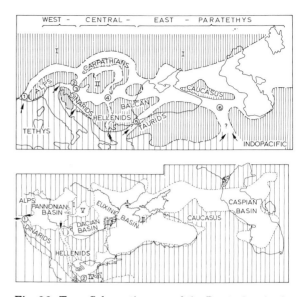

Fig. 26. Top: Schematic map of the Paratethys in the early and middle Miocene. *I*, Bohemian and Podolian massifs; *II*, intermediate mass between the Carpathians and the Dinarids; *1—6*, possible connecting routes with the Mediterranean (Tethys) and Indopacific. (After Seneš et al., 1971.) Bottom: Schematic map of the Paratethys in the late Miocene and Pliocene: *I*, Apuseni Mts.; *II*, Dobrodgea Massif; *III*, Crimea; *IV*, Aegean Continental Mass; *V*, Transsylvanic Depression; *1—2*, possible connections with the Mediterranean. (After Seneš and Marinescu, 1974.)

was isolated into a number of large separate basins, including the Vienna Basin, the Pannonian Basin, the Dacian Basin, the Black Sea and Caspian Basins. Comparatively stable salinity conditions existed in these basins for several hundred thousand years, so that diversified ostracode communities consisting of many endemic species and genera developed.

Temperature

Among ostracodes some species are widely eurythermal, while others are bound to a narrow temperature range. The cold-loving deep-sea species are termed **psychrospheric**; the cold-loving marine shallow-water species, **cryophilic**, and the warm-loving, **thermophilic**.

As is true for many other marine organisms, shallow marine ostracode assemblages of the low latitudes are considerably more taxonomically diverse than those of high latitudes. Temperature alone, however, need not account for diversity differences. Extended evolution under stable conditions, especially under a stable nutrition supply, would allow increased diversification of assemblages.

Highly zonal climates like those of the younger Cenozoic created thermal barriers not only in the horizontal but also in a vertical sense and contributed to the present provinciality of shallow marine ostracode faunas. As no analogous thermal barriers exist in the deep sea, the provinces of this habitat are very broad, much less clear-cut and characterized by distinctive taxa only at the specific level.

Substrate

The nature of the substrate has a pronounced effect on the composition of ostracode communities. Benthic ostracodes inhabit either the bottom or live on marine plants or animals. The sediment-inhabiting species live either at the surface of the sediment, or within the sediment, thus forming part of the infauna. That the composition of the ostracode communities as well as the density of individuals are strongly dependent on the kind of sediment can best be seen when considering populations living in the same area, at the same depth, and under similar salinities and temperatures. Coarse-grained sediments, like clean sands or oolites, support only a small ostracode population, whereas mud-mixed sands and pelitic sediments usually have a larger and much more diversified ostracode fauna. Size and shape of the sedimentary particles as well as the degree of their compaction are factors which control the distribution of the ostracode infauna. The ostracodes are more numerous in the few upper centimeters of the sediment, but may live at least to 15 cm. The size of interstices of the sand is a size-limiting factor for burrowing species, but this size-limitation does not affect mud-burrowing ostracodes. The phytal ostracode community (the plant-dwelling community) is also rich and diversified and distinctive assemblages characterize different marine plants.

Depth

It is difficult to assess the influence of depth, as other decisive factors change in close correlation with depth. With increasing depth, stability of the environment generally increases, whereas the energy level of the environment decreases. Progressive increase in depth is generally accompanied by decreasing grain size of the sediments, decreasing light penetration and vegetation cover. Below the photic zone, the food supply also decreases. Observations suggest that the depth-correlated factors are of greater importance for the ostracode distribution than depth itself.

In high-energy shallow waters both diversity and density of ostracodes are lower than in deeper and more stable offshore environments. Somewhat below the photic zone, however, the ostracode populations become again less diverse and less dense.

According to investigations by Benson (1975a), the true deep-sea ostracode fauna of present-day oceans consists of fewer species than found at many single shallow-water localities. This deep-sea, psychrospheric fauna occurs in marine regions where water temperature does not exceed $10°C$ and is typical in waters where temperature is $4°C$ or cooler.

Pressure may be a physical barrier to the distribution of species which are adapted to specific depth conditions (**stenobathic**). Many ostracodes, however, are adapted to considerable depth range (**eurybathic**).

Food supply

High organic content of the sediment has been considered to be a factor controlling ostracode distribution on the west coast of Florida by Hulings and Puri (1965). The deep-sea ostracode fauna of the Mediterranean Sea seems to be controlled by the amount of nutrients as, according to Puri and others (1969) the abyssal plains closer to land have a higher number of ostracode species than more distant abyssal regions.

PALEOECOLOGY

Paleoecological analysis using ostracodes is based on several methods: (1) actualistic comparison, or comparison to the mode of life of living species; (2) functional morphology of carapace characters and their changes in time and space; (3) population structure of species (mode of reproduction, sex ratio, patterns of distribution); (4) structure of ostracode communities, diversity, dominance, pattern of distribution in space and time; (5)

analysis of accompanying fossil groups; (6) biostratonomic study (ratio valves/carapaces, larvae/adults, general mode of preservation, study of *post-mortem* distribution); and (7) analysis of accompanying sediment.

Actualistic comparison

The actualistic comparison of the mode of life of extinct members of a taxon with its living representatives may perhaps seem a very simple and reliable method for the beginner, but there are definite problems which prevent the drawing of conclusions in a strictly uniformitarian way. Members of phyletic lines often considerably change their mode of life through the course of time. This is clearly exemplified by several psychrospheric ostracode species which developed from shallow-water ancestors. The uniformitarianistic interpretation would lead to grave errors in these cases. On the other hand, the migration in opposite sense, from deep and cold to shallow and warm water seems improbable, as it would require a competitive elimination of phylogenetically younger and well-adapted shallow-water taxa. Consequently it may be presumed that extant shallow, warm-water taxa are descendants from ancestors with a similar habitat. When Sohn (1962) stated that the platycopid genus *Cytherelloidea* lives only in waters above 10°C, we may conclude with a high degree of probability that the fossil representative of this "living fossil" inhabited warm waters. Reliable comparison at the specific level is possible for the youngest geological times only. Thus, the occurrence of cold- versus warm-loving ostracode species can be used to discern the Quaternary history of the Mediterranean Sea and Atlantic Ocean. At the generic level, the poor state of taxonomy often prevents comparison of Recent and fossil forms. This is the case for a large group of species lumped together under the name "*Bairdia*", a genus known since the Devonian. Recent species placed within this broadly conceived genus by the great majority of ostracodologists inhabit both shallow and deep, cold and warm waters. It is only recently that "*Bairdia*" is being split into more natural taxa which may prove useful in paleoecological interpretations. Another problem

with actualistic comparison is that we are living in an interglacial period rather exceptional in geologic history, in which the **thermosphere**, the marine water masses above 10°C, occupies much less of the water column than during the pre-Pleistocene. A considerable correction is therefore needed when evaluating the depths and temperatures of ancient oceans using actualistic data.

Adaptive morphology of the carapace

The carapace of an ostracode is subject to stresses from the outside and pull of the muscles from the inside. From a purely mechanical point of view the ideal stress-resistant form of the carapace would be a spherical shape which would yield the maximum resistance with minimum material output; such a form, however, would be at variance with the needs of most ostracodes. The usually elongate and laterally compressed body needs a totally enveloping carapace suitable for locomotion. The carapace must therefore conform to the body shape and be articulated in its dorsal part.

The nearly globular shape is an exception. It evolved in the largest living ostracode genus, *Gigantocypris* (see Fig. 34B), a form up to 23 mm long, which is extremely well adapted to a pelagic life. A lenticular shape is also rare, found only in some cladocopins (see Figs. 30, 37). Most ostracodes have elongate carapaces, more or less flattened laterally. Two principal shapes may be discerned: the shape with convex ventral margin and the type where at least the part of the ventral margin near the mouth region is flattened or concave.

The conflict between the mechanically ideal shape and the need for a sufficiently long, efficient hinge has been solved in several ways, as demonstrated in Fig. 27.

The mechanical strengthening of the carapace walls has been realized by corrugation of the valve wall in the lobed Paleozoic forms (see Figs. 39, 40, 42, 43B, C, D), by developing a thicker carapace or by developing reticulation or ridges. This last way, used abundantly in modern ostracodes, saves carbonate material when compared with the solution of simply thickening the wall. As underlined by Benson (1974) the study of the carapace design from the standpoint of engineering (Fig. 28) seems to be a promising field of study, greatly facilitated by the inven-

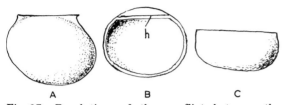

Fig. 27. Resolution of the conflict between the mechanically ideal globular shape and the need of a sufficiently long efficient hinge. A. The hinge forms a tangent to the circular outline of the valves; it is prolongated by addition of terminal triangular "ears" on both the anterodorsal and posterodorsal corners. B. The hinge appears as a cord to the oval or globular outline of the valves; it lies in a depression and the regular vaulting of the valves continues above it. C. The hinge forms a cord to the elliptic outline of the valves, but does not lie in a depression, so that the regular valve vaulting is broken. Many carapaces of this type develop sulci, explained by Triebel (1941) as the effect of muscle tension. By formation of sulci the stress resistance is heightened (corrugation).

tion of the scanning electron microscope.

The influence of the deep-sea habitat upon ostracode carapaces may best be studied when comparing species or genera with their shallow-water relatives. Benson and Sylvester-Bradley (1971) demonstrated that deep-sea forms have thinner walls with a more delicate, but often highly complicated sculpture, simpler hinges and missing eye tubercles; forms with ventrolateral extensions (alate), rare in shallow waters, are common in the deep-sea and the surface/volume ratio of the carapace is elevated (Fig. 29).

Hartmann (1973) summarized the adaptations of species inhabiting the interstitial systems of sands, coral detritus, gravels and *Posidonia* roots. Characteristic is their reduced size which rarely exceeds 0.5 mm and often ranges between 0.1 and 0.3 mm. The size reduction of the entire body is accompanied by reduction of different soft body parts, such as the eyes, genital organs and the number of appendages. Two carapace types are dominant: the lenticular type of the poly-

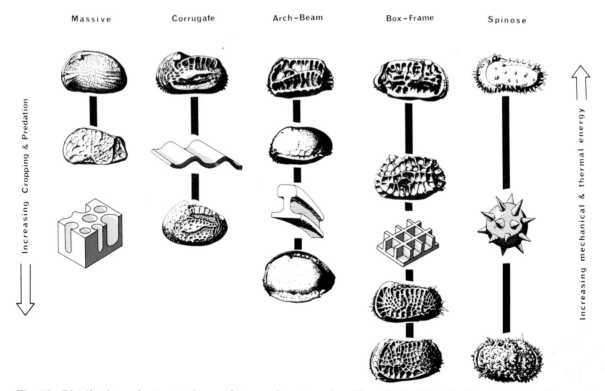

Fig. 28. Distribution of structural morphotypes in ostracodes. Hypothetical distribution of five architectural styles of forms (structural morphotypes) in response to various selective forces. Features used in their definition tend to intergrade between some of the styles and all tend to become more massive in regions of higher mechanical energy. (After Benson, 1975b.)

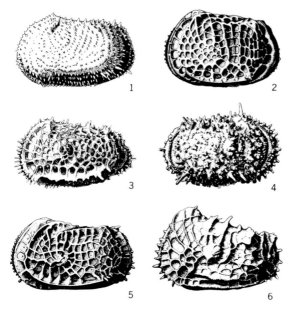

Fig. 29. Typical representatives of Recent deep-sea ostracodes. 1. *Echinocythereis echinata* (Sars). 2. *Poseidonamicus pintoi* Benson. 3. *Agrenocythere spinosa* Benson. 4. *Thalassocythere acanthoderma* (Brady). 5. *Bradleya dictyon* (Brady). 6. *Abyssocythere casca* Benson. (Courtesy R.H. Benson.)

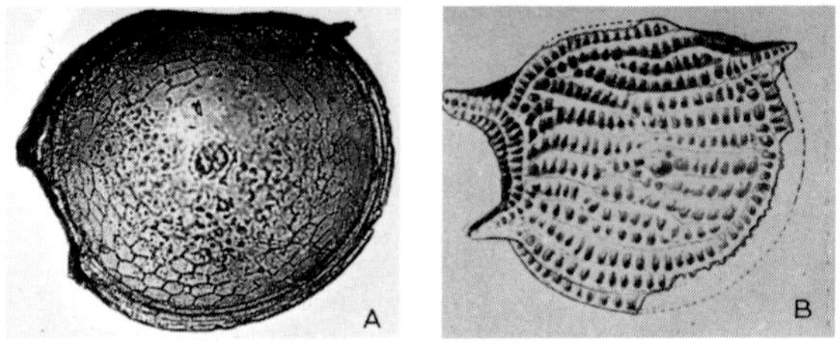

copids, which is connected with their swimming mode of life in interstitial systems of wide lumen; and the elongated shape of the cytheraceans which are often dorsoventrally or laterally flattened.

The carapaces of plant-dwelling ostracodes are mostly thin and smooth while those of burrowing species are also preponderantly smooth but thick-walled.

Carapaces of planktonic species are uncalcified or weakly calcified. Some of them have long spines which function as stability organs. Marine myodocopids have in the anterior part of the carapace a **rostral incisure** (accompanied by an overhanging **rostrum**) which facilitates the movement of the antennae (see Figs. 34, 35). This feature, however, is in no way confined to planktonic members of this group. The anteroventral beak and notch of the fossil fresh-water cyprideids, though morphologically similar to the myodocopid rostrum and incisure, is of a different origin (see Fig. 55E).

On the average, brackish-water ostracodes generally have less pronounced sculptures than the marine ones and excessive sculptural features are almost lacking among them. A

Fig. 30. Thaumatocypridids with different ecology. A. *Danielopolina orghidani* (Danielopol). Left valve. Length 0.52 mm. A recent species from brackish water of a Cuban cave. The only other known species of this genus is abyssal. The cave and abyssal environments are similar in lack of light, climatic stability and scarcity of food. B. *Pokornyopsis feifeli* (Triebel). Left valve. Total length 1.02 mm. An Upper Jurassic shallow-water species with well-calcified carapace, indicating the original habitat of the family. German Federal Republic.

characteristic of brackish waters is the much discussed phenomenon of noding of different species and genera of cytherideids (Howe, 1971; Vesper, 1972).

The above general correlations between habitat and carapace are of statistical nature, occurring in the majority of cases only. Numerous exceptions exist, so that shape and sculpture of the carapace alone cannot usually be used as absolute criteria for evaluating habitat.

That no exact correlation exists between mode of life and carapace shape is perhaps best illustrated by the similarity of some ostracode species with different modes of life.

There does exist a definite correlation between size and environmental factors, but it is sometimes difficult to single out these factors. Temperature is responsible in some cases. Many species, both benthic and planktonic, attain larger individual size in the cold waters of high latitudes. Some marine species decrease in size with decreasing salinity, although this does not hold for euryhaline species. Several studies show that still other factors, such as population density, food supply or degree of oxygenation may also bear on size.

Statistical studies by Elofson (1941) have shown that the mean length of the sand-inhabiting species is the shortest, that of the soft mud-dwelling species the longest. This is due to the size-limiting effect of the interstitial cavities for species burrowing in sands.

In pelitic sediments no such limitation is present.

Population structure

Characteristics of population structure, such as mode of distribution, continuity or patchiness, or the mode of reproduction are also potential paleoecological tools. Pokorný (1964b) and Donze (1970) found intraspecific changes in the sex ratio or even alternation of sexually and asexually reproducing populations within some marine Cretaceous species. As parthenogeny is practically unknown in Recent exclusively marine species, only hypotheses may be offered about the significance of these changes in sex ratio. The fact that parallel changes in the mode of reproduction occurred within several species in the same community suggests clearly environmental causes.

Analysis of ostracode communities

This involves the study of their diversity and density. An analysis of diversity may help in determining the paleobathymetric evolution of a marine basin (Fig. 31). Dominance is roughly inversely related to diversity. Thus, the preponderance of one or a few ostracode species in a community characterizes marginal, unstable environments. Dominance in fossil assemblages, however, may result from selective preservation of thick-shelled species, so that care must be taken when interpreting such data.

Biostratonomic study

This is another valuable approach. A fossil ostracode assemblage rarely represents the original living community (biocenosis) and often contains displaced elements whose origin is difficult to discern. Knowledge of the transport of ostracodes in Recent environments, the state of preservation of the community components, and the mode of their occurrence in sediments can shed light on this problem. Kilenyi (1971) showed that the ratio of weight/surface area of ostracode carapaces governs the transport characteristics of a species. He distinguished two modes of

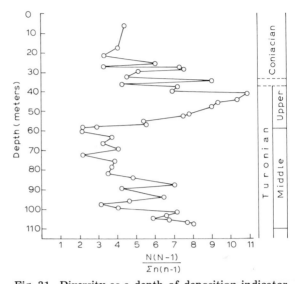

Fig. 31. Diversity as a depth of deposition indicator. An example from a borehole section from the Upper Cretaceous of Bohemia, Czechoslovakia. The interpretation is based on the supposition that the diversity increases gradually from the shoreline towards greater depths and decreases again below the photic zone. The regressive movement of the shallow middle Turonian sea is characterized by gradual decrease of diversity. The following late Turonian transgression is marked by a general trend towards diversity increase. The transgression continued during the early Coniacian, but the diversity shows a generally decreasing trend; this could be because the depth of the basin may have surpassed the value at which the maximum diversity is attained. This explanation is corroborated by the composition of the Coniacian assemblage in which certain small shallow-water trachyleberidids become rare or disappear. The diversity values were computed according to Simpson's formula, where N means the number of specimens, n_1, n_2, to n_z the number of specimens within the species. (After Pokorný, 1971.)

transport in marginal marine environments: limnic to oligohaline species, characterized by light valves, are displaced seawards in suspensions, whereas the more heavily calcified brackish and marine forms are moved landwards within the sediment. Another possible mode of distribution of ostracode carapaces for shorter distances is egestion by fish.

One criterion for autochthony of a fossil assemblage is the degree of sorting of the ostracode remains. The occurrence of numerous larval stages together with adult forms usually indicates a lack of transport. The occurrence of only the larval stages of the larger ostracode species indicates sorting; but the reverse case, occurrence of exclusively adult stages does not necessarily mean sorting by transport. Very often this is merely the result of selective preservation. Even in Recent deposits thin-walled larval stages of some species are not preserved, whereas the thick-walled ones, like those of cytherellids, are abundant.

CLASSIFICATION

The classification of Recent ostracodes is based primarily on the morphology of the soft body, chiefly on appendages, but also on other features, such as shape and position of gonads, or presence or absence of eyes and heart. Since World War II, neontologists, influenced by intensive systematic pale-ontological studies, began to attribute more importance to the features of the carapace as taxonomic criteria, especially at the generic and specific level.

There are no characters which are equally applicable for taxonomic differentiation of ostracodes. Thus, a character may be diag-nostic at the ordinal level for one group of ostracodes, whereas for other groups the same character can be used at lower taxonomic levels only. For instance, the unique mor-phology of muscle scar groups in the leperditi-copids is a good diagnostic criterion for this order, whereas in the myodocopids the muscle scars are so variable that they cannot be reasonably used as a diagnostic criterion for the order as a whole. In certain genera of podocopid ostracodes the adductor muscle scars may be rather invariable, while in others they vary considerably even within the same species. Evolutionary studies of muscle scars, which include consideration of phylogenetic and ontogenetic development as well as individual variation and possible sexual di-morphism, are extremely useful, especially at the superfamily and lower taxonomic levels.

Sexual dimorphism has been used as the main criterion for discrimination of suborders within the beyrichiocopid ostracodes and, in general, is a very useful feature in the tax-onomy of Paleozoic ostracodes.

Features of the marginal zone (extent of calcification, characters of fused areas, course of inner margin, form and shape of radial pore

canals) are sometimes characteristic at the family-group level.

For the definition of a genus the shape of the carapace, basic pattern of gross sculpture, character of the hinge, presence or absence of eye spots, details of muscle scars, the course of the line of concrescence, the shape, posi-tion and number of pore canals, and the structure and width of the marginal zone are among the most frequently used features.

For species recognition, modifications of the carapace shape, sculpture, hinge, number and position of pore canals, development of vestibula, presence or absence of opaque spots in the valve walls, their shapes, as well as other minor features are employed.

We recognize that many diagnostic features evolved repeatedly in different evolutionary lines, for example, the calcified inner lamella, blindness or the amphidont hinge. This so-called **iterative evolution** considerably com-plicates ostracode taxonomy.

The notion stated and underlined already by Darwin that taxa are best defined by the constant co-occurrence of several characters which are not directly related by their func-tions, remains invaluable to present-day taxonomy.

MAJOR MORPHOLOGIC GROUPS

The main purpose of this section is to introduce the reader to the great morphologic variety amongst ostracodes without going into excessive details of taxonomy. Table I sum-marizes the classification of ostracodes. The diagnosis of the ordinal groups are given in the Table, subordinal and suprafamilial levels are briefly described in the text. The more common or typical genera are illustrated.

1. The archeocopids (Fig. 32)

Until recently, this order has been divided into two suborders: the bradoriids and the phosphato-copins. The latter were thought to differ from the former by phosphatized inner lamella. Revision by Kozur (1974) has shown, however, that phosphatiza-tion of bradoriids occurs only in specific types of sediment, for example, the Upper Cambrian "Stink-kalk", and thus is secondary; and only in phospha-tized specimens is an inner lamella found. It extends over most of the inner valve and its preservation speaks for it being elastic during lifetime. Thus, it shows better correspondence to the uncalcified part of the inner lamella of extant ostracodes than to its calcified portion.

There is a wide agreement among recent authors that the archeocopids are ancestral to all post-Cambrian ostracodes.

2. The leperditicopids (Fig. 33)

The leperditicopids are characterized by a combination of several primitive characters: general carapace shape, muscle scars pattern and the simple structure of the free margins. These, together with their large size and thick shells make them an isolated

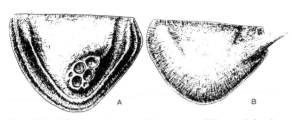

Fig. 32. The archeocopids. A. *Hipponicharion loculatum* Andres. Reconstruction of a right valve. Length 1.48 mm. Middle Cambrian. Öland Island, Sweden. (After Andres, 1969.) B. *Longispina oelandica* Andres. Reconstruction of the left valve. Length 2.48 mm. (After Andres, 1969.)

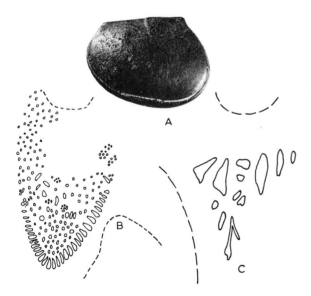

Fig. 33. The leperditicopids. A. *Leperditia* sp., leperditiid. Carapace from the left side. × 3. Upper Silurian, Sweden (Gotland). B. *Moelleritia moelleri* (Schmidt). Scheme of muscle scar pattern in the "chevron" group, corresponding to the frontal scars of other ostracodes. × 31. Middle Devonian, U.S.S.R. (After Abushik, 1958.) C. *Sibiritia ventriangularis* Abushik. Scheme of the muscle scar pattern in the "chevron" group. × 6⌄. Silurian, eastern Siberia, U.S.S.R. (After Abushik, 1958.)

group which represents an evolutionary "dead end". To assign some Cambrian species either to the archeocopids or to the leperditicopids is difficult and suggests the close relation between the two. The latter probably evolved from the former.

After studying leperditicopids in thin section, Levinson (1951) considered it possible that they were "not Ostracoda, as we know them today, but rather an early, specialized branch". Langer (1973) came to a similar conclusion after an electron-microscopic study of the ultrastructure of their carapace which he found to differ from that of other ostracodes.

3. The myodocopids (Figs. 34—38)

The presence of a heart and of compound eyes in some members of this order is a primitive character. Since few myodocopids have strongly calcified valves, their fossil record is inadequate. Most fossil representatives are from the Paleozoic.

A. Myodocopins (Fig. 34) have a chitinous to well-calcified carapace usually bearing a rostral incisure

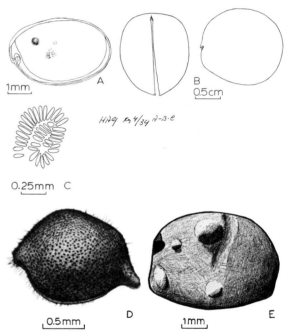

Fig. 34. The myodocopins. A. *Cypridina mediterranea* Costa, a cypridinid. Recent, Gulf of Naples. (After Müller, 1894). B. *Gigantocypris muelleri* Skogsberg, a cypridinid. Femal carapace in ventral and side views. Recent, Antarctic. (After Skogsberg, 1920.) C. *Cyclasterope fascigera* G.S. Brady, a cylindroleberidid. Male. Muscle scar field as seen on the inside of the right valve. Recent, Java. (After Skogsberg, 1920.) D. *Spinacopia sandersi* Kornicker, a sarsiellid. Adult male, seen from the left side. Recent, off Bermuda. (After Kornicker, 1969b.) E. *Cypridella* sp., a cypridinellid. Lower Carboniferous, Belgium. (After Sylvester-Bradley, 1953.)

TABLE I

Suprageneric classification of Ostracoda

1. ORDER ARCHEOCOPIDA

Diagnosis: Carapace equivalved or nearly so, with long, straight dorsal margin and strongly convex ventral margin. Valves flexible, chitinous or slightly calcified or phosphatized, without hinge. Some genera with tubercles presumed to correspond to compound eyes. Surface smooth, wrinkled or with ribs, spines or crests. Inner lamella not hardened by inorganic deposits.
Range: Lower to Upper Cambrian; ?Lower Ordovician.

Classification
Principal families: Bradoriidae, Hipponichariidae.

2. ORDER LEPERDITICOPIDA

Diagnosis: Includes the largest ostracodes, ranging up to 80 mm in length. Carapace heavily calcified, slightly to strongly unequivalved with long straight hinge margin, distinct dorsal angles and convex free margins, fuller and higher posteriorly. Anterodorsal part of the valves in many species with a tubercle, interpreted generally as corresponding to compound eye. Adductor muscle scar group in front of the valve midlength, extraordinarily large, attaining a diameter equalling almost one third of the valve height, with the number of individual scars sometimes exceeding 200. Frontal muscle scar group in front of the adductor muscle scar group and below the eye tubercle, composed of a large number of scars, V-shaped in Silurian and Devonian species, less pronounced in Ordovician ones. Smaller groups of muscle scars, up to 9 or 10 in number, occur above the eye tubercle. Valve surface mostly smooth, sometimes sculptured. Sexual dimorphism unknown. Benthic.
Range: Oldest species assigned with some doubt to leperditicopids are of late Cambrian age: Ordovician to Devonian.

Classification
Principal family: Leperditiidae.

3. ORDER MYODOCOPIDA

Diagnosis: It is impossible to give a detailed diagnosis for this order because of the great morphological variety of its members. Carapace mostly with a convex ventral margin and a peripheral calcification of the inner lamella. Some forms possess a rostral incisure which facilitates the exit and movement of the antennae. Incisure overhung by a rostrum. In living species the exopodite of the second antennae consists of several joints and is adapted for swimming. In some genera sexual dimorphism strongly expressed in the general shape of the carapace. All living representatives marine. Some pelagic, other live on the bottom surface or within the sediment.
Range: Ordovician to Recent.

Classification

A. Suborder Myodocopina
   i. Superfamily Cypridinacea. Principal families: Cypridinidae, Cylindroleberididae, Cypridinellidae, Sarsiellidae, Rutidermatidae.

B. Suborder Halocypriformes
   i. Superfamily Halocypridacea. Family Halocyprididae.
   ii. Superfamily Thaumatocypridacea. Family Thaumatocyprididae.
   iii. Superfamily Entomoconchacea. Family Entomoconchidae.

C. Suborder Cladocopina
      Principal families: Polycopidae, Entomozoidae.

4. ORDER BEYRICHICOPIDA

Diagnosis: Carapace well calcified, in both larval and adult specimens typically with a more or less straight, usually long cardinal margin and convex extracardinal margin. Many genera with pronounced lobation and sulcation. Calcified inner lamella absent. Muscle scars of varying pattern known in a few families only. Cruminal and antral dimorphism confined to this order only. Some forms exhibit sculptural or proportional dimorphism, others are non-dimorphic. Marine.
Range: Known since the Ordovician. Most of them are restricted to the Paleozoic, a few survived into the Triassic. One Cenozoic family, the punciids of New Zealand and Japan, has been attributed to this order.

TABLE I (*continued*)

Classification

A. Suborder Hollinomorpha
   i. Superfamily Hollinacea. Principal families: Ctenonotellidae, Ctenoloculinidae, Hollinellidae, Hollinidae, Tetradellidae, Quadrijugatoridae, Tvaerenellidae, Euprimitiidae.
   ii. Superfamily Eurychilinacea. Principal families: Eurychilinidae, Oepikiidae.
   iii. Superfamily Primitiopsacea. Family Primitiopsididae.

B. Suborder Beyrichiomorpha
   i. Superfamily Beyrichiacea. Family Beyrichiidae.

C. Suborder Binodicopina
   i. Superfamily Drepanellacea. Principal families: Aechminidae, Bolliidae, Drepanellidae.

5. ORDER PODOCOPIDA

Diagnosis: Calcified carapace which never bears a rostral incisure. Larval valves with a more or less straight ventral margin. Adductor muscle scar group consisting of many individual scars in primitive representatives, much reduced in number in advanced taxa. Neither heart nor lateral eyes present. Marine to fresh water.
Range: Ordovician to Recent.

Classification

A. Suborder Metacopina
   i. Superfamily Healdiacea. Principal families: Healdiidae, Bairdiocyprididae, Pachydomellidae, Sigilliidae (=Saipanettidae).
   ii. Superfamily Thlipsuracea. Principal families: Thlipsuridae, Bufinidae, Ropolonellidae, Quasillitidae.

B. Suborder Podocopina
   i. Superfamily Cytheracea. Principal families: Tricorninidae, Berounellidae, Bythocytheridae, Paradoxostomatidae, Glorianellidae, Cytheruridae, Leptocytheridae, Cytherideidae, Loxoconchidae, Xestoleberididae, Krithidae, Progonocytheridae, Protocytheridae, Trachyleberididae, Hemicytheridae, Cytherettidae.
   ii. Superfamily Bairdiacea. Principal families: Bairdiidae, Beecherellidae.
   iii. Superfamily Darwinulacea. Family Darwinulidae.
   iv. Superfamily Cypridacea. Principal families: Cyprididae, Cypridopsidae, Macrocyprididae, Pontocyprididae, Candonidae, Ilyocyprididae, Cyprideidae.

C. Suborder Platycopina
   i. Superfamily Kloedenellacea. Principal families: Kloedenellidae, Glyptopleuridae.
   ii. Superfamily Cytherellacea. Principal families: Cytherellidae, Cavellinidae.

6. UNCERTAIN ORDER

Suborder Kirkbyocopina
Superfamily Kirkbyacea. Principal families: Kirkbyidae, Amphissitidae, Arcyzonidae.

---

and rostrum. Heart and compound eyes (the latter sometimes rudimentary) are present. They range from Silurian to Recent.

Within this suborder are classed the following still living families: The **cypridinids** (Fig. 34A, B) have mostly well-calcified valves and smooth to strongly sculptured surface. Adductor muscle scars are arranged in an inverse V-pattern or in several subhorizontal rows. Oldest, somewhat dubious, representatives are known since the Silurian. The **cylindroleberidids** (Fig. 34C) cannot be reliably distinguished from other myodocopins by carapace features alone. Their adductor muscle scars are numerous and arranged in a more or less clear spiral. No fossil representatives of this family are known. The **sarsiellids** (Fig. 34D) have rounded or broadly elliptical, more or less sculptured, carapace with a caudal process. ?Devonian and Recent. The **rutidermatids** are mostly living forms with carapace similar to that of the sarsiellids.

B. Halocypriformes (Figs. 35, 36). The Cretaceous to Recent halocyprids have a chitinous to slightly calcified carapace with straight dorsal margin, smooth or with fine sculpture. Zone of concrescence is narrow, with few marginal pore canals. The heart is present, the paired eyes are lacking. Adductor muscle scar group consists of rows of narrow elongate scars. They are strictly marine, pelagic forms. The **thaumato-cypridids** include marine pelagic forms with a sub-circular, slightly calcified carapace which bears large spines on its anterior end. The adductor muscle scar group is subcircular, rosette-like, consisting of four closely arranged large scars. Representatives are known from Permian to Recent seas. Whereas the Permian to Jurassic species were shallow-water inhabitants, the Recent representatives are bathyal or abyssal and one species has been found in the brackish water of a submarine cave in Cuba (Fig. 30A, B). Halocypriformes' fossil record is very meager.

Kornicker and Sohn (1974) suggest that the Devonian and Carboniferous **entomoconchaceans** (Fig. 36) are probable ancestors of the thaumato-cyprids. Their hypothesis is based upon discovery of Recent deep-sea myodocopids showing affinities to

both the entomoconchaceans and the thaumato-cypridids. The entomoconchaceans include species without rostrum and rostral incisure. Their carapaces have an opening at the posterior margin which may extend into a short siphon, a feature known also in some myodocopins. Their adductor muscle scar group is suboval and consists of numerous radially arranged, extremely elongated scars. These show a certain similarity to those found in the myodocopin family of the cylindroleberidids and may suggest a common derivation.

C. The cladocopins (Figs. 37, 38) include genera with oval to circular carapaces with short dorsal and hinge margins and toothless hinges. The zone of concrescence is narrow, the calcified part of the inner lamella is relatively broad, rostral incisure is absent or very slightly indicated. Adductor muscle scar group is round, with few large scars. There are no paired eyes, no heart, only one pair of thoracic legs. Marine forms, inhabiting the sediment interstices. Include one family, the **polycopids**. Earliest forms assigned with some doubts to the cladocopids are known since the Devonian. Continuous record from Carboniferous to Recent.

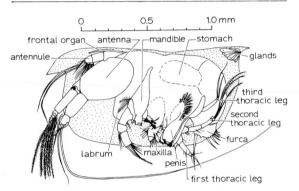

Fig. 35. The halocypriformes. *Euconchoecia chierchiae* Müller, a halocypridid. Internal anatomy. Recent, vicinity of Sapelo Island, Georgia, U.S.A. (After Darby in Kesling and others, 1965.)

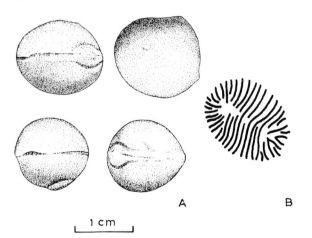

Fig. 36. The entomoconchaceans. A. *Entomoconchus scouleri* McCoy. Lower Carboniferous, Ireland. B. *Entomoconchus* sp. Muscle scar pattern, × 13.6. (Both figures after Sylvester-Bradley, 1953.)

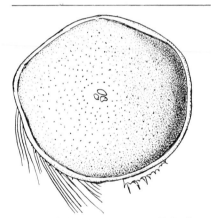

Fig. 37. *Polycope punctata* G.O. Sars, a cladocopin — polycopid. Left valve. Recent, Norway. Length 0.78 mm. (After Sars, 1923.)

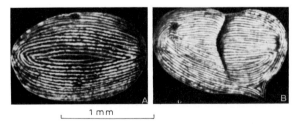

Fig. 38. The entomozoids. A. *Richterina (Volkina) zimmermanni* (Volk). Left valve. Upper Devonian, Ober-Harz, Altenau. (After Rabien, 1954.) B. *Frank-linella multicostata* Rabien. Mold of left valve. Upper Devonian. Federal Republic of Germany. (After Rabien, 1954.)

The extinct family of the **entomozoids** (Fig. 38) is placed within the myodocopids by most authors. According to Kozur (1972a) they are ancestors of polycopids. They are large, with chitinous or slightly calcified valves, ellipsoidal, subquadrate or subtriangular in side view, with straight or curved dorsal margin. No rostrum or incisure developed. Some of them have a long and strong sulcus concave towards the anterior. In other genera only a muscle scar pit is present or neither sulcus nor pits are developed. The surface is smooth or with fine ridges. Ordovician to Permian. The representatives of this family were marine and are interpreted by most workers as planktonic. They attained the peak of their development in the Upper Devonian and lowermost Carboniferous. Many of the species are abundant and widespread and are excellent index fossils.

4. The beyrichicopids (Figs. 39—43)

Being mostly of Paleozoic age, the beyrichicopids are of limited importance in oceanic micropaleontology.

A. *Hollinomorphs* (Figs. 39, 40, 41) are defined chiefly by their antral dimorphism. A few are non-dimorphic. It includes the following superfamilies: The **hollinaceans** (Fig. 39) which are characterized by a subvelar or supravelar dimorphism, either dolonate or locular. Their valves are tri- to nonsulcate. Ordovician to Triassic. The **eurychilinaceans** (Fig. 40) are characterized by a velum consisting of radially arranged hollow tubuli. Uni- or non-sulcate, exceptionally bisulcate. Ordovician to Permian. The **primitiopsaceans** (Fig. 41) have a compact velum (i.e. not consisting of tubuli), which may be dissolved into spines. Vela dimorphic, situated in the posterior part of the carapace, may form a closed pouch in the heteromorphs. Unisulcate and bisulcate, Ordovician to Permian.

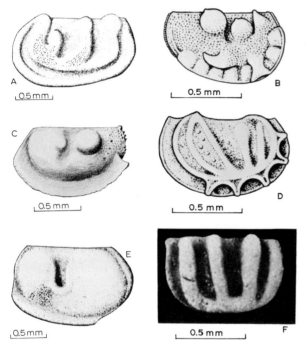

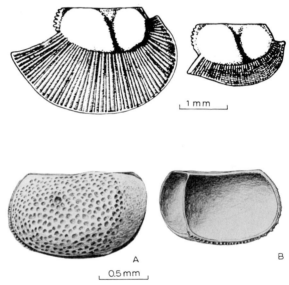

Fig. 39. The hollinaceans. A. *Tallinella dimorpha* Öpik, a primitive ctenonotellid. Middle Ordovician, Baltic area. (After Jaanusson, 1957.) B. *Abditoloculina pulchra* Kesling, a ctenoloculinid. Female, right valve. (After Kesling, 1969.) C. *Hollinella (Hollinella) bassleri* (Knight), a hollinellid. Left valve. Pennsylvanian, Missouri, U.S.A. D. *Tetradella quadrilirata* (Hall and Whitfield), a tetradellid. Female right valve. Upper Ordovician, Ohio, U.S.A. (After Kesling, 1969.) E. *Euprimites effusus* Jaanusson, a tvaerenellid. Reconstruction of left heteromorphic valve. Middle Ordovician, Sweden. (After Jaanusson, 1957.) F. *Quadrijugator permarginatus* (Foerste), a quadrijugatorid. Right valve. Upper Ordovician, Michigan, U.S.A. (After Kesling and Hussey, 1953.)

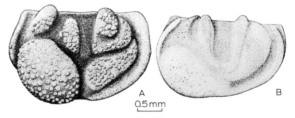

Fig. 42. The beyrichiomorphs. A. *Nodibeyrichia tuberculata* (Kloeden), a beyrichiid. Left female valve. Siluro-Devonian of the Baltic area. (After Martinsson, 1965.) B. *Zygobolba decora* (Billings), a beyrichiid. Reconstruction of female left valve. Silurian, Anticosti Island, North America. (After Martinsson, 1962.)

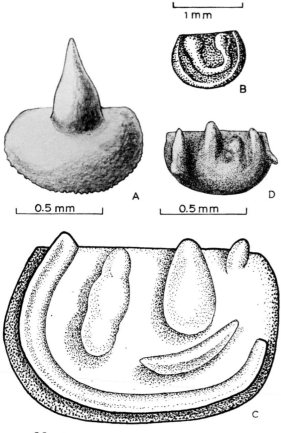

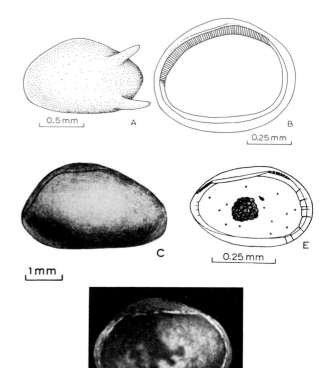

Fig. 43. A—C. The binodicopins — the drepanellaceans. A. *Aechmina bovina* Jones, an aechminid. Right valve. Silurian, Mulde, Gotland. B. *Bollia bicollina* Jones and Hall, a bolliid. Left valve, Silurian, England. (After Bassler and Kellet, 1934.) C. *Drepanella crassinoda* Ulrich, a drepanellid. Left valve. Ordovician, Kentucky. (After Ulrich and Bassler, 1923.) D. *Nodella svinordensis* Zaspelova, a hollinacean. Right valve. Upper Devonian, U.S.S.R. (After Zaspelova, 1952.)

Fig. 44. The healdiaceans. A. *Healdia anterodepressa* Blumenstengel, a healdiid. Left valve. Upper Devonian, German Democratic Republic. (After Blumenstengel, 1965.) B. *Ogmoconcha amalthei* (Quenstedt), a healdiid. Left valve from the inside. Lower Jurassic, Paris Basin, France. C. *Bairdiocypris prantli* Pokorný, a bairdiocypridid. Carapace from the right side. Middle Devonian, Czechoslovakia. (After Pokorný, 1950.) D. *Tubulibairdia antecedens* (Kegel), a pachydomellid. Carapace from the right. Middle Devonian, Federal Republic of Germany. (After Krömmelbein, 1955.) E. *Cardobairdia balcombensis* McKenzie, a sigilliid. Internal view of the left valve. Neogene, Victoria, Australia. (After McKenzie, 1967a.)

B. Beyrichiomorphs (Figs. 24, 42) are characterized by the occurrence of cruminal dimorphism. Histium is absent. Silurian to Lower Carboniferous.

C. Binodicopins (Fig. 43) contain genera with a node in front of the median sulcus and another behind it in the dorsal portion of the carapace or genera which can be derived from this architectural plan. Majority of genera are non-dimorphic. Ordovician to Permian.

5. The podocopids (Figs. 44—57)

This order includes by far the greatest number of post-Paleozoic fossil ostracodes. All extant freshwater species and the large majority of extant marine species belong to it.

A. Metacopins (Figs. 44, 45) were defined originally on fossil material as unit of horizontal classification. Their most outstanding feature is the circular aggregate of many adductor muscle scars.

The **healdiaceans** (Fig. 44) are convex-backed, short-hinged ostracodes with holosolenic type of contact margin. Calcified inner lamella is usually very narrow, but species with a broad calcification and vestibula have also been described. Sexual dimorphism in typical members is unknown. Marine. Ordovician to Recent. Includes the families of the healdiids, the bairdiocypridids, the pachydomellids and the most primitive living podocopins, the sigilliids. The superfamily of the **thlipsuraceans** (Fig. 45) have a convex or straight dorsal margin and hemisolenic type

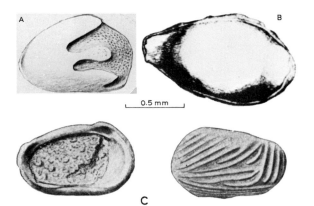

Fig. 45. The thlipsuraceans. A. *Neothlipsura furca* (Roth), a thlipsurid. Left valve, Lower Devonian, Oklahoma, U.S.A. B. *Ropolonellus kettneri* (Pokorný), a ropolonellid. Carapace from the right side. Middle Devonian, Czechoslovakia. × 64. (After Pokorný, 1950.) C. *Svantovites primus* Pokorný, a quasillitid. Left valve from the inside and carapace from the left side. Middle Devonian, Czechoslovakia. × 52. (After Pokorný, 1950.)

Fig. 46. The tricorninids. A. *Nagyella longispina* Kozur. Left valve. Middle Triassic, Hungary. Courtesy of H. Kozur. B. *Bohemina (Bohemina) extrema* Blumenstengel. Left valve. Upper Devonian, German Democratic Republic. (After Blumenstengel, 1965.)

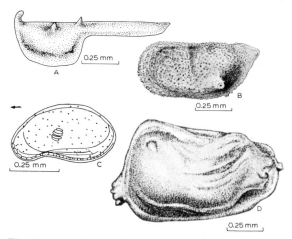

Fig. 47. A. *Berounella tricerata* Blumenstengel, a berounellid. Left valve. Upper Devonian, German Democratic Republic. (After Blumenstengel, 1965.) B. *Bythoceratina (Bythoceratina) umbonata* (Williamson), a bythocytherid. Left valve. Lower Cretaceous. England. C. *Sclerochilus levis* G.W. Müller, a paradoxostomatid. Female, Recent, Bay of Naples, Italy. (After Müller, 1894.) D. *Glorianella vassoevichi* Schneider, a glorianellid. Carapace from the left side. Lower Triassic, Caspian Region, U.S.S.R. (After Schneider, 1960.)

of contact margin. Calcified inner lamella is either not seen or narrow to wide. Marine. Ordovician to Carboniferous. Includes the families of the thlipsurids, the bufinids, the ropolonellids and the quasillitids.

B. Podocopins (Figs. 46—55). Carapace has a midventral incurvature of the valve margin. Duplicature is usually wide, with or without vestibule. Adductor muscle scars are reduced in number. Marine to fresh water. Ordovician to Recent. In the post-Paleozoic this is the most important ostracode suborder. The following principal taxa are included in it:

(i) The cytheraceans (Figs. 46—52). A superfamily of a variety of different shapes and sculptures. Their most characteristic feature is the arrangement of the adductor muscle scars which form a vertical series of four or rarely five scars, sometimes secondarily divided.

The cytheraceans are the most advanced superfamily of the podocopids and in modern marine ostracode faunas they are the ostracode group richest in genera and species. Include the following important families:

The tricorninids (Fig. 46) are thought to be the ancestral family of the cytheraceans. They have a triangular, nearly equivalved carapace with straight dorsal margin, broadly rounded anterior margin and low to acute posterior margin. Valves of nearly all genera have long spines. Marine. Ordovician to Triassic. The berounellids (Fig. 47A) may have been derived from the tricorninids. They are small, subquadrate, lobed and spinose, with straight dorsal margin and characteristic extension of the posterodorsal part of the carapace. Silurian to Triassic. The bythocytherids (Fig. 47B) are the most primitive family of extant cytheraceans. Their carapace has a

straight dorsal margin and usually a caudal process and median sulcus. Marine. Silurian to Recent. The paradoxostomatids (Fig. 47C) have thin-walled, usually smooth carapaces with four or five adductor scars. Because of the fragility of their valves they have a poor fossil record. The oldest known members are from the Lower Cretaceous. The glorianellids (Fig. 47D) are a group of Permian and Triassic ostracodes which are probably ancestral to most of the post-Triassic cytheraceans. They have a rectangular to subreniform carapace, a merodont tripartite hinge, one to two frontal scars and two mandibular scars. They lived chiefly in brackish waters. The cytherids (Fig. 48A) were a basket for many genera the position of which was unclear. They are now practically restricted to the type genus. Quaternary. The cytherurids

(Fig. 48B) are usually small, mostly with a pronounced caudal process, often with ventrolateral extensions. Marine to oligohaline. Upper Triassic to Recent. The **leptocytherids** (Fig. 48C) include genera with low reniform carapaces, merodont hinges, mostly irregularly running line of concrescence, branched marginal pore canals. Mostly marine to brackish waters. Tertiary to Recent. The **cytherideids** (Fig.

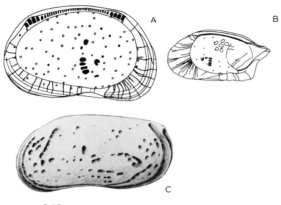

0.25 mm

Fig. 48. A. *Cythere lutea* (O.F. Müller), a cytherid. Scheme of a left female valve seen from the inside. Holocene, The Netherlands. × 60. (After Wagner, 1957.) B. *Semicytherura angulata* (Brady), a cytherurid. Scheme of the right valve from the inside. Holocene, The Netherlands. × 60. (After Wagner, 1957.) C. *Leptocythere pellucida* (Baird), a leptocytherid. Carapace from the left side. Recent, England.

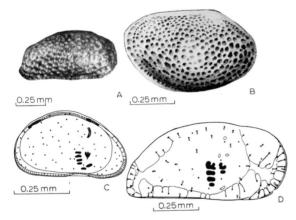

0.25 mm   A   0.25 mm   B

0.25 mm   C

0.25 mm   D

Fig. 49. A. *Cytheridea acuminata* Bosquet, a cytherideid. Right valve, Miocene, Czechoslovakia. B. *Loxoconcha bairdii* G.W. Müller, a loxoconchid. Right valve, Recent, Gulf of Naples, Italy. C. *Xestoleberis aurantia* (Baird), a xestoleberidid. Scheme of the left valve seen from the interior. Note the "xestoleberidid spot" in the anterodorsal region. Holocene, The Netherlands. (After Wagner, 1957.) D. *Krithe undecimradiata* Ruggieri. Male, right valve. Quaternary, Italy. (After Greco and others, 1974.)

49A) include genera with strongly calcified carapaces, with broadly rounded anterior and usually narrowly rounded to acute posterior margins. Inner margin and line of concrescence mostly coincident, sometimes narrow vestibula developed. Marginal pore canals usually densely arranged, simple. Lateral pore canals of sieve-type. Marine to fresh water. Jurassic to Recent. The **loxoconchids** (Figs. 49B) have a characteristic rhomboid shape. Their marginal zone is wide, traversed by a few simple straight pore canals. Lateral pore canals are of sieve type. Marine to oligohaline. Jurassic to Recent. The **xestoleberidids** (Fig. 49C) have a subovate, swollen, smooth, or punctate carapace with arched dorsum. Their most characteristic feature is the "xestoleberidid spot" in the eye region. Most are shallow marine inhabitants. Cretaceous to Recent. The **krithids** (Fig. 49D) include species with elongate to highly arched carapaces. The calcified part of the inner lamella is broad, often with irregularly running inner margin. A pocket-like vestibulum is often developed at the anterior margin. Marine, shallow water to hadal. Upper Cretaceous to Recent. The **progonocytherids** (Fig. 50A, B), as originally defined, comprised genera having an archidont hinge, i.e. quadripartite hinge in which the anteromedian part is more coarsely crenulate than the posteromedian. Radial pore canals are straigth, not abundant. Triassic to Cretaceous. The **protocytherids** (Figs. 50C, D) have pear-shaped to quadrangular, usually, strongly calcified carapaces, and broad

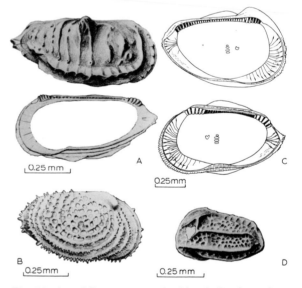

0.25 mm   A   C

0.25mm

B   D

0.25mm   0.25 mm

Fig. 50. A and B, progonocytherids. A. *Lophocythere propinqua* Malz. Right valve from the outside and from the inside. Middle Jurassic, France. B. *Centrocythere denticulata* Mertens. Right valve. Lower Cretaceous, England. C and D, protocytherids. C. *Protocythere triplicata* (Roemer). Left and right female valves from the inside. Lower Cretaceous, German Federal Republic. (After Triebel, 1938.) D. *Pleurocythere impar* Triebel. Left valve, Middle Jurassic.

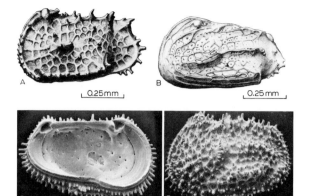

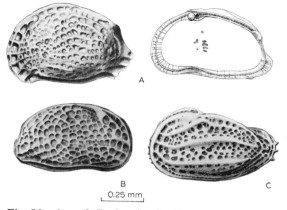

Fig. 51. The trachyleberidids. A. *Oertliella reticulata* (Kafka). Female, carapace from the left side. Upper Cretaceous, Bohemia, Czechoslovakia. (After Pokorný, 1964a.) B. *Mosaeleberis interruptoidea* (Van Veen). Left valve. Upper Cretaceous, Bohemia, Czechoslovakia. C. *Henryhowella asperrima* (Reuss). Pliocene, Italy. Left: interior; right: exterior of left valve. SEM photos. (Courtesy R.H. Benson.)

Fig. 52. A and B, hemicytherids. A. *Pokornyella limbata* (Bosquet). Left: left valve from the outside; right: interior of right valve. Paleogene, Paris Basin, France. × 45. (After Keij, 1957.) B. *Hemicythere villosa* (Sars). Right valve, Skagerrak, Sweden. C. *Cytheretta gracillicosta* (Reuss), a cytherettid. Left valve, Paleogene, German Democratic Republic.

zone of concrescence traversed by long and sinuous simple marginal pore canals. Sculpture is basically with three longitudinal ribs. Jurassic to Recent, with many good index species in the Middle Jurassic through Lower Cretaceous. The **trachyleberidid** (Fig. 51) have mostly strongly calcified carapaces, frequently strongly ornate, amphidont hinges, usually a V-shaped frontal scar, a well-developed zone of concrescence with fairly numerous to abundant marginal pore canals. Marine, shallow water to abyssal. Middle Jurassic to Recent. The trachyleberidids are among those families which have the best fossil record and are most important for the stratigraphy and paleoecology of the Cretaceous and the Cenozoic. The **hemicytherids** (Fig. 52A, B) undoubtedly arose from the trachyleberidids. Instead of a V-shaped frontal muscle scar characteristic for most trachyleberidids, all hemicytherids have two to three frontal scars and some of their adductor scars are often secondarily divided. The oldest genera appeared towards the end of the Cretaceous, but the family belongs to the most characteristic taxa of the shallow-water Cenozoic assemblages. Mostly marine, not uncommon in brackish waters, exceptionally in fresh water. The **cytherettids** (Fig. 52C) are phylogenetically close to the trachyleberidid—hemicytherid group. Their carapaces are oval to cylindrical, their most characteristic feature is the extraordinarily broad zone of concrescence. Inner margin coincident with the line of concrescence. Hinge amphidont, lateral pore canals simple. Upper Cretaceous to Recent.

(ii) The **bairdiaceans** (Fig. 53) have changed very little since early Paleozoic times and are a good example of what is called the "living fossils". A calcified inner lamella is present even in oldest forms. From their central longeval lineage short-lived off-

shoots were repeatedly derived. The superfamily attained its maximum development during the late Paleozoic and the Triassic. A small adaptive radiation occurred in the Cenozoic. They comprise one living and several fossil families:

The **bairdiids** (Fig. 53A, B) include covex-backed ostracodes with typical "bairdioid" shape. Their carapaces are smooth to highly ornate. Hingement short, of ridge-and-groove type. Ordovician to Recent. The **beecherellids** (Fig. 53C) include fairly elongate smooth ostracodes with straight to slightly vaulted dorsum. Spines are present on anterior or anterior

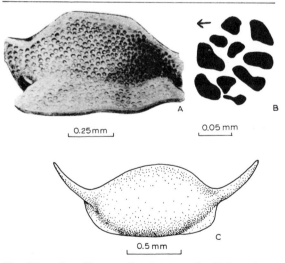

Fig. 53. A. *Havanardia havanensis* Pokorný, a bairdiid. Right valve. B. Adductor muscle scars of the same. Recent, off Havana, Cuba. (After Pokorný, 1968b.) C. *Acanthoscapha volki* Blumenstengel, a beecherellid. Left valve. Upper Devonian, German Democratic Republic. (After Blumenstengel, 1965.)

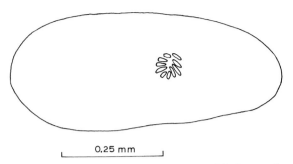

Fig. 54. *Darwinula stevensoni* (Brady and Robertson), a darwinulid. Right valve from the inside. Holocene, Bohemia.

and posterior ends of one or both valves. Ancestral to bairdiids. Ordovician to Triassic.

(iii) The **darwinulaceans** (Fig. 54) have mostly a long-oval smooth carapace, more narrowly rounded in front. Adductor muscle scars form a characteristic rosette. Inner lamella is only peripherally calcified. with a few marginal pore canals and brood room in the posterior part of the carapace. Fresh water to oligohaline. Carboniferous to Recent. Includes a small family, the **darwinulids**.

(iv) The **cypridaceans** (Fig. 55) have a thin carapace, usually with little evolved hinge, smooth or slightly sculptured. Inner lamella is calcified, vestibula are usually present. Mostly fresh water, some live in brackish and marine waters, exceptionally terrestrial. ?Silurian, Upper Paleozoic to Recent.

The **macrocypridids** (Fig. 55A) are the most primitive group of the cypridaceans. They have an elongate carapace with arched dorsum and a narrow to pointed posterior margin with the apex situated near or at the base. Right valve is larger than the left. Hinge is merodont. Adductor muscle scars are numerous, forming an irregular rosette. Zone of concrescence is wide and vestibula deep. Marine, bottom-dwelling. Species with general shape of macrocypridids appear since the early Ordovician. The **pontocypridids** (Fig. 55B) mostly have a triangular carapace. Valves are thin, hinge is adont. Adductor muscle scars are arranged in two subvertical rows, the anterior with three, the posterior with two scars. Marine. Triassic to Recent. The **candonids** (Fig. 55C) have a very variable carapace shape, their valves are thin to thick. Hinge is adont and vestibula deep. The adductor muscle scar group consists of an elongate scar above and two subvertical rows below; the anterior has three, the posterior two scars. The lowermost scars of both rows may be reduced in size or missing. Fresh water to marine. Silurian to Recent. The **cypridids** (Fig. 55D) are a large family comprising several extramarine subfamilies. A general diagnosis is based on the soft parts only. Their carapaces are of very variable shapes and also other features vary widely. Fresh water to slightly brackish. Jurassic to Recent. The **ilyocypridids** (see Fig. 17) have a characteristic subquadrate coarsely punctate carapace

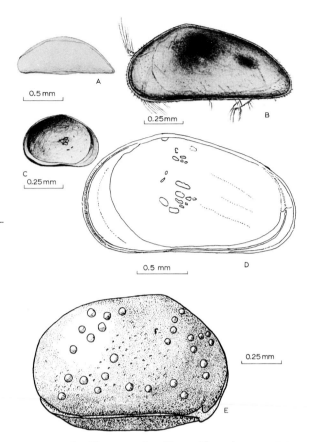

Fig. 55. A. *Macrocypris siliqua* (Jones), a macrocypridid. Carapace from the left side. Upper Cretaceous, England. B. *Pontocypris mytiloides* (Norman). A pontocypridid. Female from the left side. Recent, Norway. (After Sars, 1923.) C. *Cyclocypris ovum* (Jurine), a candonid. Left valve from the inside. Holocene, Bavaria. (Courtesy of A. Absolon.) D. *Eucypris clavata* (Baird), a cypridid. Female, right valve from the inside. Pleistocene, Bohemia, Czechoslovakia. E. *Cypridea granulosa* (Sowerby), a cyprideid. Carapace from the right side. Middle Purbeckian (vicinity of the Jurassic/Cretaceous boundary), England. (After Sylvester-Bradley, 1949.)

with anterior and posterior ends of similar height and two prominent sulci. Nodes may be present. Hinge is adont, calcified part of the inner lamella is narrow. Fresh to slightly brackish water. Jurassic to Recent. The **cyprideids** (Fig. 55E) are characterized by an anteroventral beak. Fresh and slightly brackish water. Jurassic to Paleogene. Members of this family are very useful for intercontinental stratigraphic correlation and for study of continental drift.

C. Platycopins (Figs. 56, 57). The suborder of the platycopins has usually strongly calcified, unequivalved carapaces with inner lamellae uncalcified or calcified only in their most peripheral parts. A contact groove is present along the free margin and often along the hinge margin. Dimorphism is expressed by the swelling of the posterior part of the female carapace. The brood pouch may be set off from the remaining carapace cavity by an internal vertical ridge.

(i) The **kloedenellaceans** (Fig. 56) are characterized by a more or less straight dorsal margin and a convex to concave ventral margin. Valves are quadrilobate to non-lobate. Marine, rarely brackish. Ordovician to Triassic.

(ii) The **cytherellaceans** (Fig. 57) have non-lobate valves with a contact groove along the entire periphery of the larger valve. Adductor muscle scar group is composed of many small scars in Paleozoic and partly also in Triassic representatives; in younger species it is biserial, composed of two curved subvertical closely adjacent rows of scars. Inner lamella is calcified in its outermost parts only. Marginal pore canals, when present, are simple, short, not differing morphologically from the lateral ones. Silurian to Recent. Includes the Silurian to Triassic **cavellinids**, which are ancestral to the only living platycopine family, the **cytherellids**, known since the Triassic.

6. Unknown order

A. **Kirkbyocopins** (Figs. 58, 59). The kirkbyocopins (Fig. 58) have been traditionally included with the beyrichiocopids. Recently, however, some authors have classified them within the podocopids, considering them to be either platycopins or podocopins. They are distinguished through the combination of

Fig. 56. The kloedenellaceans. A. *Poloniella symmetrica* (Hall). Carapace from the right side. Silurian, Maryland, U.S.A. B. *Glyptopleura reniformis* Croneis and Thurman. Carapace in left lateral view. Lower Carboniferous, Illinois, U.S.A. (After Cooper, 1941.)

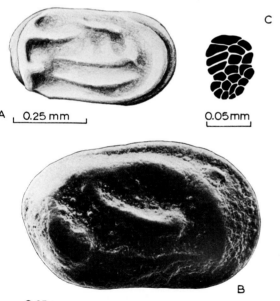

Fig. 57. The cytherellids. A. *Cytherelloidea chapmani* (Jones and Hinde). Right valve. Lower Cretaceous, Sussex, England. B. *Leviella rudis* Kristan-Tollmann. Carapace from the right side. Upper Triassic, Austria. (After Kristan-Tollmann, 1973.) C. *Leviella dichotoma* Kristan-Tollmann. Male, adductor muscle scars of the left valve from the outside. Upper Triassic, Austria. (After Kristan-Tollmann, 1973.)

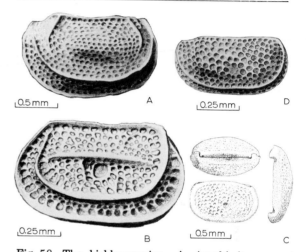

Fig. 58. The kirkbyocopins. A. *Amphissites remesi* Pokorný, an amphissitid. Right valve. Middle Devonian, Czechoslovakia. B. *Amphizona asceta* Kesling and Copeland, an arcyzonid. Left valve. Middle Devonian, New York. C. *Cardiniferella bowsheri* Sohn, a cardiniferellid. Top left: dorsal view; bottom left: carapace from the right side; right: right valve in dorsal view. (After Pokorný, 1958.) D. *Knightina allerismoides* (Knight), a kirkbyid. Right valve, Pennsylvanian, U.S.A.

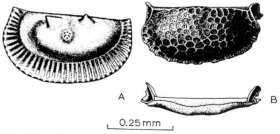

Fig. 59. The punciids. A. *Puncia novozealandica* Hornibrook. Left valve. Recent, New Zealand. B. *Manawa tryphena* Hornibrook. Left valve in lateral and dorsal views. Recent, New Zealand. (Both figures after Hornibrook, 1963.)

several characters: a long straight dorsal margin, a slightly convex ventral margin, an adductor scar pit ("kirkbyan pit") and reticulate surface. Ordovician to Triassic.

The position of the Neogene to Recent punciids (Fig. 59) is still under discussion. Some authors see their affinities to the hollinomorphs (eurychilinaceans), others to the kirkbyaceans.

IMPORTANCE OF OSTRACODES FOR PALEOGEOGRAPHY

Marine ostracodes are not as well suited for interregional and intercontinental stratigraphic correlation as are other groups of planktonic microfossils. Since benthic ostracode species have no planktonic larvae, the shallow, warm-water species cannot easily cross geographic barriers. This limitation, however, qualifies them as excellent paleobiogeographical markers. Ostracodes are also efficient tools for the study of paleobathymetry and paleosalinity. Hence, their study may be extremely helpful for tracing of paleogeographic changes.

Of considerable paleogeographical interest is the history of the cosmopolitan deep-sea fauna in the Mediterranean Province studied by Benson and Sylvester-Bradley (1971). Its characteristic elements are found in Paleocene to Middle Miocene and in Pliocene sediments from different areas of the Mediterranean province. In the late Miocene, the evolution of the normal marine ostracode fauna of the Mediterranean Sea was interrupted, as this sea was cut off from the Atlantic and transformed into a series of lagoons. Some of these dried up, others desalinified and developed a peculiar endemic fauna. This fauna had many elements in common with assemblages from the contemporary Parathethyan basins (see Fig. 26). The Parathethyan basins of late Miocene and Pliocene time were characterized by low salinities and endemic ostracode communities. These endemics are excellent indicators of changing communication between these basins, of their paleosalinities and are most useful as stratigraphic markers in the search for oil and coal within these basins.

At the beginning of the Pliocene, communication between the Mediterranean and Atlantic was re-established in the west, so that Atlantic euhaline species, even the deep-sea psychrospheric fauna, re-invaded the Mediterranean.

As the connection between the Mediterranean Province and the Indo-Pacific region has been interrupted since the middle Miocene, Recent Mediterranean ostracodes are chiefly of Atlantic origin and differ substantially from their Tertiary forerunners. Towards the end of the Tertiary the shallowing of the sill at the Straits of Gibraltar prevented the entrance of cold deep oceanic waters into the Mediterranean, leading to the elimination of the psychrosphere and consequently of the psychrospheric fauna from this area. The present Mediterranean fauna is entirely thermospheric as the temperature even at abyssal depths is near $13°C$.

The present Mediterranean has another aspect of great interest to a biogeographer. With the opening of the Suez Canal more than a hundred years ago, and especially through its reconstruction in the years after World War II, a seaway has been established between two large provinces: the Indo-West Pacific and the Mediterranean. The main stream of immigrants goes from the Red Sea to the Mediterranean and includes shallow-water ostracodes. McKenzie (1973) pointed out that in older collections from the Mediterranean some elements are absent which were observed in recent collections and are thus probably of Red Sea origin. This is a special case of increasing human affect on the composition of naturally established regional faunas, as our seaways and extensive sea traffic tend to introduce many new elements into what were indigenous faunas.

In the Neogene of the Caribbean region Van den Bold (1974) was able to distinguish two faunal provinces, and three subprovinces. Some Caribbean Cenozoic shallow-water ostracodes even reached the remote Galapagos Islands, probably by dispersal on drifting objects, in a warm-water current system. Two of these lineages underwent a spectacular insular evolutionary radiation on the Galapagos (Fig. 60).

A similar colonization by long distance, rather chance dispersal, called sweepstakes routes, has been demonstrated also for the islands of the equatorial Pacific. According to McKenzie (1969), several genera disappear progressively as we go east in the Pacific from New Caledonia to Fiji to Samoa and finally to

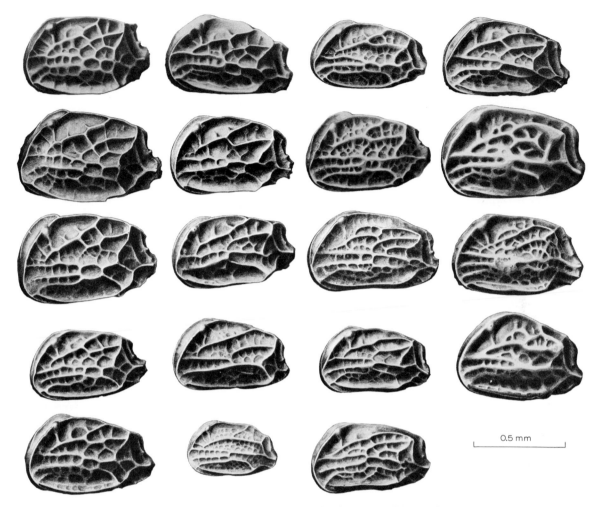

Fig. 60. Insular radiation of the hemicytherid genus *Radimella* Pokorný in the Galapagos Islands. The individuals of this shallow-water genus display a discontinuous variation of sculpture which speaks of the existence of genetically discrete populations. A larger number of these discrete forms occur at the same locality, suggesting their genetic isolation. The distinction of such closely similar forms is possible only by careful morphologic and biometric analyses. (After Pokorný, 1970.)

Bora Bora near Tahiti. Similarly the endemism of the New Zealand fauna has been affected by its position at the end of a sweepstakes route.

Information about paleocirculation patterns has been derived from the study of ostracodes. The importance of the West-Wind drift for the distribution of Austral-Asian ostracodes has been underlined by McKenzie (1973). Because of continental drift, this current was not operative until late Cretaceous time, and in some parts of the area not until the Oligocene. According to McKenzie, the South Equatorial Current which flows from West Africa to the Caribbean via the coasts of Brazil and Venezuela has acted as another sweepstakes route and is responsible for some elements common to the East African and Caribbean provinces.

Ostracodes have been used also as paleoclimatic indicators. During the Quaternary some north European Atlantic species temporarily penetrated the Mediterranean, so that the occurrence of these "northern guests" has been used for the recognition of cold climatic periods (Ruggieri, 1971).

Hazel (1970, 1971) made detailed studies of Recent bottom samples in the region from Nova Scotia to Long Island where many sublittoral, cryophilic ostracode species are not present in the southern parts of the region and several thermophilic (warm-loving) species are absent from its northern part. He applied his results to explain the paleoclimatic conditions of the Neogene Yorktown Formation of Virginia and North Carolina, and was also able to trace the boundary shift of cold and mild-temperate faunal provinces on the Atlantic coast of the U.S.A. during the Pleistocene.

The deep-sea ostracode fauna

It is only since the onset of the Deep Sea Drilling Project that the deep-sea ostracode fauna has been intensively studied. Deep-sea ostracode communities, Recent and past, have some common characteristics:

(1) In each of them the phylogenetically most progressive taxa of that geological time are absent. This is clearly seen in the Recent psychrospheric faunas studied by Benson and Sylvester-Bradley (1971), in which the most characteristic and strongly diversified family of shallow-water ostracodes, the hemicytherids, are absent. Kozur (1972a, b) reported the same phenomenon for the Triassic and Oertli (1972), for Jurassic deep-sea assemblages.

(2) The survival of old lineages which are absent in contemporaneous shallow-water faunas. This has been observed in Triassic deep-sea assemblages (Kozur, 1972b; and Kozur and Mostler, 1972) and in the Jurassic (Oertli, 1972). The phenomenon is also well known in Recent deep-sea faunas. Because of the absence of progressive forms, the deep-sea fauna acquires a conservative character at generic and higher levels.

(3) An important component of deep-sea ostracode assemblages is the extremely long-lived, slowly evolving lineages which also have contemporaneous shallow-water representatives. In Recent deep-sea assemblages this component is represented by the bairdiids and bythocytherids, both known since the early Paleozoic, and by the cytherellids which evolved during the Triassic. The sigilliids (=saipanettids), are the only surviving representatives of the metacopins.

The idea of the constancy of the deep-sea environment which had many proponents

until recent years, is untenable. Paleoclimatologic evidence shows that temperature of deep-sea water has undergone considerable change during geological times. As pointed out by Menzies and others (1973) "the basic tenet of paleoecology is that thermal tolerance is a conservative feature of a species". It is obvious that the temperature at the deep-sea bottom must have a decisive influence on the formation of the deep-sea fauna.

Our knowledge of the history of Campanian to Recent deep-sea ostracodes has been provided primarily by Benson (1971, 1972, 1975a). He has shown that the Recent deep-sea ostracode fauna is widely cosmopolitan at the generic level. The only reasonable explanation of this fact is that these genera derived from warm-water ancestors which originated in the ancient Tethys and became gradually adapted to cooler waters. This is confirmed by findings of fossil species of the Recent psychrospheric genera *Abyssocythere* and *Agrenocythere* in early Tertiary and younger deposits ranging from Trinidad to central and southern Europe.

According to Benson the deep-sea ostracode assemblages of the Campanian through middle Eocene do not show any major alteration and are different from those of early Oligocene and younger ages, which are similar to the Recent psychrospheric fauna. The discontinuity marks the beginning of the psychrosphere. The precise date of this event could not be fixed because of shortage of available samples. His observations further indicated that the evolution of deep-sea ostracodes does not proceed at constant rates.

Times with warmer, deep-water masses were favorable to the penetration of warm and shallow-water lineages into the deep sea. Such conditions ended in the late middle Eocene. In the present-day low latitudes the boundary between the thermosphere and the psychrosphere creates a formidable barrier to virtually all warm-water species. This is one of the reasons for the absence of young, progressive lineages in the deep sea.

ORIGIN AND PHYLOGENETIC TRENDS IN OSTRACODES

Ostracode morphology is unique among the

Crustacea and neither the morphology of Recent species nor their ontogeny give a definite indication of their origin. It has been repeatedly postulated that ostracodes arose from primitive phyllopods (branchipods), a primitive group of crustaceans still living in Recent fresh to hypersaline waters. This idea is untenable in view of the profound differences between the appendages of each group; as shown by Hartmann (1963), the tube-like extremities of a basically bifurcating type found in ostracodes, cannot be derived from the phylogenetically advanced leaf-like extremities of the phyllopods. Furthermore, nothing is known about the soft-body organization of extinct Paleozoic ostracode groups. The presence of the complete carapace, however, indicates that they were probably not radically different from modern species. In view of the small dimensions of most ostracodes it seems reasonable that morphological reduction and simplification have occurred in the formation of this subclass.

Proposed features of the more primitive, ancestral form may be summarized as follows: It had a carapace completely enveloping the body. It is almost certain that the body had a higher number of appendages and showed the original segmentation more clearly. The trend towards reduction of the number of segments is a common one in crustaceans and generally in arthropods. It is seen even in Recent ostracodes where the hind limb pairs have been lost in some families. Moreover, in some Paleozoic ostracodes an elevated number of instars, documented by Spjeldnaes (1951), suggests the presence of more limbs.

Lobation and sulcation were probably not original features, but certainly are very old ones. There is a trend common in Paleozoic lineages of lobate ostracodes towards a reduction of lobation and sulcation. The strengthening of the valves by corrugation, very common in ancient groups, has been replaced in most modern groups by strengthening through reticulation or ridges. From a more or less uniform and minute reticulation highly differentiated sculpture has arisen in several different evolutionary lines.

The carapaces of primitive ostracodes were probably uncalcified or only weakly calcified and continuous along the dorsal margin. Con-

sequently no hinges were developed at all. The trend towards elaboration of hinge structures is thus intimately related to calcification of the valves. Weakly mineralized species, such as some extinct archeocopids, or the still living pelagic myodocopids and halocypriformes, thin-walled marine plant-dwellers or fresh-water forms, usually have none or weakly developed hinges. On the other hand, bottom-dwelling species of shallow, high-energy marine environments are characterized by strongly calcified valves and strong amphidont hinges with high and very strong teeth. These teeth are especially powerful in strongly inequivalved species with a high vaulted dorsum, and with hinge structures shortened during phylogeny. As the anterior of the carapace is heavier in most ostracodes, the anterior hinge element is stronger and phylogenetically more advanced than the posterior. Many cases of parallelism have been discovered in the evolution of hinges. In the podocopids, amphidont hinges evolved several times from the merodont type. As in other cases of parallelism this is due to similar responses of ostracodes to similar environmental conditions. A secondary simplification of the hinges in some lineages has also occurred. In many benthic ostracode lineages a clear trend towards the shortening of the hinge margin occurs and is correlated with increasing complexity of the hinge structure.

The calcification of the inner lamella was originally absent and developed independently in several lineages. Such calcification and the complication of the marginal zone reached their most complicated state in some Cenozoic podocopid ostracodes.

The adductor muscle scars were numerous in ancient ostracodes and their number has been gradually reduced in many ostracode lineages (Fig. 61). The opposite phylogenetic trend, towards the division of muscle scars, has been observed in some podocopid ostracodes, for instance in some Cenozoic hemicytherids. As young larval stages possess a smaller number of adductor scars than adults, it appears that the multiplication of muscle scars during ontogeny reflects merely the ontogenetic stages of the individual. The phylogenetic reduction of the number of adductor scars has been explained by Gramm

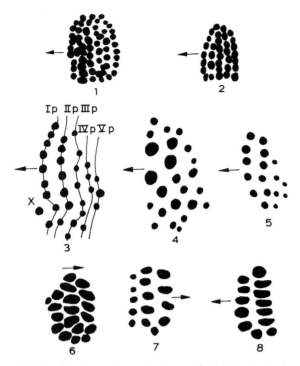

Fig. 61. Phylogenetic reduction of adductor muscle scars in the evolutionary lineage cavellinids — cytherellids (a morphogenetic, not a phylogenetic scheme!): 1. *Cavellina* sp. 2. *C. missouriensis* (Knight). 1—2: molds from the left sides, Upper Carboniferous of U.S.A. 3. *Ussuricavina rakovkensis* Gramm. 4. *Issacharella zharnikovae* (Gramm). 5. *Reubenella kramtchanini* (Gramm). 6. *Reubenella amnekhoroshevi* (Gramm). 7. *Leviella egorovi* (Gramm). 3—7: all specimens seen from the outside. From the Triassic of the Primorye region, U.S.S.R. 8. *Cytherella* sp., right valve from the inside, Recent, Gulf of Aden. (1 and 2 after Scott in Gramm, 1970; 3—8 after Gramm, 1970.) Magnification of 1 and 2 smaller than that of 3—8.

(1973) through persistence of larval conditions.

Structures which are in the process of phylogenetic reduction often show an increased variability. This phenomenon has been shown by Gramm (1967) for the adductor scars of the cavellinid—cytherellid evolutionary series. He also noticed (1969) an **atavistic** (occurring in the forerunners, but absent in the descendents) reappearance of scars in a cytherellid of latest Cretaceous age.

Definite trends may be observed also in the evolution of pore canals. Among the normal pore canals the simple type may be considered as the ancestral one. The sieve type is

derived and evidence shows that it evolved independently in several evolutionary lines.

The primitive marginal pore canals were undoubtedly short, straight and simple. With the phylogenetic broadening of the zone of concrescence, there developed long and sinuous pore canals often inflated in their median parts. The broadening of the zone of concrescence in many species caused the formation of branching radial pore canals. Each of their branches corresponds to an originally independent pore canal and their common stem is nothing but a common duct for soft tissues entering the pore canals.

GEOLOGIC DISTRIBUTION

The stratigraphic ranges of some of the more important genera illustrated in text figures are given in Table II. A general discussion of the geologic distribution of ostracodes follows.

Lower Paleozoic

The ostracodes appeared in the early Cambrian. The bulk of the Cambrian ostracodes belong to the archeocopids, although rare species assigned tentatively to the leperditiids were recorded from the late Cambrian.

Ordovician seas witnessed a great expansion of ostracodes. The number of taxa multiplied and all orders made their appearance at that time. The most characteristic feature of Ordovician ostracode faunas is the common occurrence of many quadrilobate hollinomorph genera. All genera with a histial structure are confined to this system.

During the Silurian the beyrichiaceans developed from an eurychilinacean ancestor. All undisputed members of this group are restricted to the Silurian to Lower Carboniferous and have been successfully used in biostratigraphy. Martinsson (1967), divided the Silurian sequence of Gotland by means of beyrichiaceans into nineteen biostratigraphic zones. The majority of primitiopsaceans lived also in the Silurian.

Among the important components of Devonian faunas are the hollinaceans, thlipsuraceans, healdiaceans, bairdiaceans, kloedenellaceans and cytherellaceans. The

TABLE II

Stratigraphic ranges of some ostracode genera figured in the text

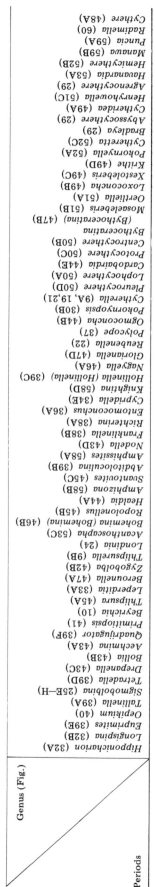

beyrichiaceans are in decline, although many new genera developed. Compared with their Silurian predecessors, they generally have a weakened to effaced lobation and sulcation. Knowledge of Devonian ostracode assemblages has been greatly enhanced by the wide application of the technique of dissolving limestones in weak acids. Assemblages of silicified ostracodes, largely differing from those found in entomozoacean-rich facies, were discovered in many Devonian clayey limestones which originated as a quiet basin facies. These limestones contain a large proportion of highly spinose species. Silicified faunas of similar character are known from comparable Silurian and Lower Carboniferous facies. For the stratigraphy of Upper Devonian and lowermost Carboniferous the entomozoaceans are very important. Rabien (1954 and later) established a modern basis for their taxonomy and stratigraphic use. For their evolutionary trends, stratigraphic distribution and literature the paper by Gründel (1962) may be consulted.

Upper Paleozoic

In Carboniferous and Permian faunas the kirkbyaceans are among the most characteristic elements. Other important components are the kloedenellaceans, cytherellaceans, healdiaceans, hollinaceans, and the bairdiaceans.

Mesozoic

In the Triassic ostracodes constitute one of the most important elements of the microfauna. According to Kozur (1971) there is little difference in the composition of the deep-water assemblages of Triassic and Permian age. The common feature of these assemblages is the bizarre spinosity if many species belonging to the hollinaceans, beecherellids, berounellids, bairdiids, the bythocytherids and the polycopids. According to this author, the deep-sea fauna of most of the Triassic (except the Rhaetian stage) has a Paleozoic character. Shallow-water ostracode assemblages of the early Triassic still contain Paleozoic survivors such as kirkbyaceans which, according to Kozur, disappeared from shallow waters in the upper part of the early Triassic. The shallow-water facies of middle to late Triassic age display a substantial radiation of the bairdiids, especially the strongly sculptured forms. The sculptured cytherellids are also characteristic and important. The late Triassic is a time of intense adaptive radiation of the cytheraceans.

Before the end of the Triassic (at the boundary between the Norian and Rhaetian stages), an extensive reduction of ostracode assemblages occurred both in deep- and shallow-water habitats (Kozur, 1971). In deep-sea assemblages chiefly bythocytherids, smooth bairdiids, polycopids and healdiids survived. The last-named are represented by *Ogmoconcha*, a genus which was also abundant in shallow waters and which, together with closely related genera, often dominated ostracode assemblages during the Rhaetian and lowermost Jurassic. Many shallow-water cytheracean species disappeared at the Norian/Rhaetian boundary.

The early and middle part of the early Jurassic (Liassic) is characterized by ostracode assemblages of low diversity in which the healdiid genus *Ogmoconcha* and the progonocytherid *Procytheridea* and related genera dominate. At the end of the middle Liassic *Ogmoconcha* suddenly disappears along with many other species leaving different species of the *Procytheridea*, to dominate the late part of the early Jurassic. In the middle Jurassic, genera of the cytheracean families of the progonocytherids and trachyleberidids are important; for example, *Progonocythere*, *Oligocythereis*, *Lophocythere*. The deep-water facies so far known contain smooth bairdiids and cytherellids.

In late Jurassic sediments different species of *Lophocythere* serve as useful stratigraphic markers. Cytherellids and several genera of cytherideids are numerous at this time. In many parts of the world the nearly world-wide marine regressions at the end of the Jurassic established large areas of brackish and limnic sedimentation. In brackish and shallow marine water sediments various species of *Macrodentina* are important stratigraphic markers. In slightly brackish to limnic sediments various cyprideids characterized by an anteroventral beak are important. The macrodentinas survived into the early Cretaceous, especially in the Tethys.

Many genera persist from Jurassic into early Cretaceous faunas. For stratigraphic correlation, species of *Protocythere* and allied genera, known since the late Jurassic, are very important. Species of the trachyleberidid genus *Rehacythereis* and related genera appeared in the Cretaceous and were ancestral to a wide variety of stratigraphically important genera; *Neocythere*, *Centrocythere*, *Cythereis* and *Platycythereis* appeared in the early Cretaceous and many genera originated in the late Cretaceous, for instance, *Krithe*, *Brachycythere*, *Curfsina*, *Mauritsina*, *Mosaeleberis*, *Oertliella*, *Phacorhabdotus*, *Pterygocythereis*, *Spinoleberis* and *Trachyleberidea*. Some of these continue into the Cenozoic.

Cenozoic

Because of their strongly calcified and often richly sculptured valves the trachyleberidids as well as families derived from them, the cytherettids and the hemicytherids, are the most important elements of Cenozoic assemblages. *Abyssocythere*, *Actinocythereis*, *Agrenocythere*, *Ambocythere*, *Aurila*, *Bradleya*, *Cnestocythere*, *Cyprideis*, *Cytheretta*, *Cytheridea*, *Henryhowella*, *Loculicytheretta*, *Orionina*, *Pokornyella*, *Trachyleberidea* and *Urocythereis* may be cited as a few examples of Cenozoic cytheraceans.

SUGGESTIONS FOR FURTHER READING

Coryell, H.N., 1963. *Bibliographic Index and Classification of the Mesozoic Ostracoda*. University of Dayton Press, Ohio, 1175 pp. (2 vols.) [Reference work on Mesozoic Ostracoda described prior to 1962. Includes classification outline, diagnoses of supraspecific taxa, selected faunal lists, bibliography, catalogue of species, genera and families.]

Grekoff, N., 1960. *Aperçu sur les Ostracodes fossiles*. Technip, Paris, 97 pp. [Introduction to the study of fossil ostracodes.]

Hartmann, G., 1966, 1967, 1968, and in print. Ostracoda. In: *Dr. H.G. Bronns Klassen und Ordnungen des Tierreichs. Band 5: Arthropoda, I. Abt. Crustacea. 2. Buch, IV. Teil.* Akademische Verlagsgesellschaft Geest und Portig, Leipzig, 568 pp. (3 volumes). [Basic neontological treatise on ostracodes.]

Hartmann, G. and Puri, H.S., 1974. Summary of neontological and paleontological classification of Ostracoda. *Mitt. Hamburg. Zool. Mus. Inst.*, 70: 7—73. [An attempt at classification of ostracode genera and higher taxa based

both on their soft parts and carapace features. Only those families having Recent representatives are included.]

Howe, H.V., 1962. *Ostracod Taxonomy*. Louisiana State University Press, Baton Rouge, La., 366 pp. [Important reference book. Deals with genera and higher taxa, refers to original publications.]

Howe, H.V., 1971. Preliminary list of new ostracod taxa 1961—1971. *Mélanges, Mus. Geosci.*, No. 3: 28 pp.

Howe, H.V., 1972. Corrections and additions to preliminary list of new ostracod taxa 1961—1971. *Mélanges, Mus. Geosci.*, No. 5. [Publication complementing *Ostracod Taxonomy* by the same author.]

Howe, H.V. and Laurencich, L., 1958. *Introduction to the Study of Cretaceous Ostracoda*. Louisiana State University Press, Baton Rouge, La., 536 pp. [Description and figures of all species of Cretaceous ostracodes published up to the end of 1956 as well as the relevant bibliography.]

Moore, R.C. and Pitrat, C.W. (Editors), 1961. *Treatise on Invertebrate Paleontology. Part Q, Arthropoda, 3. Crustacea. Ostracoda*. Geological Society of America and University of Kansas Press, Lawrence, Kansas, 442 pp. [Important reference book on Ostracoda.]

Oertli, H.J., 1963. *Mesozoic Ostracod Faunas of France*. Brill, Leiden, 57 pp. [A survey of the stratigraphical sequence of Mesozoic ostracodes of France.]

Pokorný, V., 1965. *Principles of Zoological Micropalaeontology*, 2. Pergamon Press, London, 465 pp. (Translation of German edition.)

CITED REFERENCES

Abushik, A.F., 1958. In: *Novye rody i vidy ostrakod. Mikrofauna S.S.S.R., Sb.*, 9: 232—287.

Alexander, C.I., 1933. Shell structure of the ostracod genus *Cytheropteron*, and fossil species from the Cretaceous of Texas. *J. Paleontol.*, 7: 181—214.

Andres, D., 1969. Ostracoden aus dem mittleren Kambrium von Öland. *Lethaia*, 2: 165—180.

Bassler, R.S. and Kellet, B., 1934. Bibliographic index of Paleozoic Ostracoda. *Geol. Soc. Am., Spec. Pap.*, No. 1: 500 pp.

Benson, R.H., 1971. A new Cenozoic deep-sea genus, *Abyssocythere* (Crustacea: Ostracoda: Trachyleberididae) with descriptions of five new species. *Smithson. Contrib. Paleobiol.*, 7: 25 pp.

Benson, R.H., 1972. The *Bradleya* problem, with descriptions of two new psychrospheric ostracode species, *Agrenocythere* and *Poseidonamicus* (Ostracoda: Crustacea). *Smithson. Contrib. Paleobiol.*, 12: 138 pp.

Benson, R.H., 1974. The role of ornamentation in the design and function of the ostracode carapace. *Geosci. Man*, 6: 47—51.

Benson, R.H., 1975a. The origin of the psychrosphere as recorded in changes of deep-sea ostracode assemblages. *Lethaia*, 8: 69—83.

Benson, R.H., 1975b. Morphologic stability in Ostracoda. *Bull. Am. Paleontol.*, 65: 13—46.

Benson, R.H. and Sylvester-Bradley, P.C., 1971. Deep-sea ostracodes and the transformation of ocean to sea in the Tethys. In: H.J. Oertli (Editor), *Paléoécologie d'Ostracodes, Pau 1970. Bull. Centre Rech., Pau — SNPA*, 5 (suppl.): 63—91.

Blumenstengel, H., 1965. Zur Taxionomie und Biostratigraphie verkieselter Ostracoden aus dem Thüringer Oberdevon. *Freiberger Forschungsh.*, C 183: 127 pp.

Brady, G.S., 1868. A monograph of the Recent British Ostracoda. *Trans. Linn. Soc. Lond.*, 26: 353—495.

Brady, G.S., 1880. Report on the Ostracoda dredged by H.M.S. Challenger during the years 1873—1876. *Rep. Sci. Res. Voyage H.M.S. Challenger, Zool.* 1, Pt. 3: 184 pp.

Bronstein, Z.S., 1947. Ostracoda presnykh vod. In: *Fauna S.S.S.R., Rakoobraznye II*, 1. Zool. Inst. Akad. Nauk S.S.S.R., N.S., 31: 339 pp.

Cannon, H.G., 1925. On the segmental excretory organs of certain fresh-water ostracods. *Philos. Trans. R. Soc. Lond., Ser. B*, 124: 1—27.

Cannon, H.G., 1926. On the feeding mechanism of a fresh-water ostracod "Pionocypris vidua" (O.F. Müller). J. Linn. Soc. Lond., 36: 325—375.

Cannon, H.G., 1931. On the anatomy of a marine ostracod, Cypridina (Doloria) levis Skogsberg. Discovery Rep. Cambridge, 9: 435—482.

Cannon, H.G., 1933. On the feeding mechanism of certain marine ostracods. Trans. R. Soc. Edinb., 57: 739—764.

Cannon, H.G., 1940. On the anatomy of Gigantocypris mülleri. Discovery Rep. Cambridge, 19: 185—244.

Cooper, C.L., 1941. Chester ostracodes of Illinois. Ill. Geol. Surv. Inv., 77: 101 pp.

Danielopol, D.L., 1972. Sur la présence de Thaumatocypris orghidani n. sp. (Ostracoda — Myodocopida) dans une grotte de Cuba. C. R. Acad. Sci. Paris, 274: 1390—1393.

Danielopol, D.L. and Vespremeanu, E.E., 1964. The presence of ostracods on floating fen soil in Rumania. Fragm. Balc. Mus. Macedonici Sci. Nat., 5: 29—40.

Donze, P., 1970. Influence du milieu sur le mode de reproduction de Cythereis? castellanensis, Ostracode nouveau de Cénomanien—Turonien de la Foux (Basses-Alpes). Bull. Soc. Géol. Fr., Sér. 7, 12: 932—936.

Elofson, O., 1941. Zur Kenntnis der marinen Ostracoden Schwedens mit besonderer Berücksichtigung des Skagerraks. Zool. Bidr. Upps., 19: 215—534.

Gramm, M.N., 1967. Rudimentarnye muskul'nye pyatna u triasovykh Cytherelloidea (Ostracoda). Dokl. Akad. Nauk S.S.S.R., 173: 931—934.

Gramm, M.N., 1969. Otpechatki atavisticheskogo (?) adduktora u iskopaemykh ostrakod. Paleontol. Zh., 1969: 131—133.

Gramm, M.N., 1970. Otpechatki adduktora triasovykh citerellid (Ostracoda), Primorya i nekotorye voprosy teorii filembriogeneza. Paleontol. Zh., 1970: 88—103.

Gramm, M.N., 1972. Marine Triassic ostracodes from South Primorye. On the evolution of the adductors of cavellinids. In: Proc. Int. Geol. Congr., 23rd, Czechoslovakia 1968, pp. 135—148.

Gramm, M.N., 1973. Neotenicheskie yavleniya u iskopaemykh ostrakod. Paleontol. Zh., 1973: 3—12.

Greco, A., Ruggieri, G., Sprovieri, R., 1974. La sezione calabriana di Monasterace (Calabria). Boll. Soc. Geol. Ital., 93: 151—159.

Gründel, J., 1962. Zur Phylogenetik und Taxionomie der Entomozoidae (Ostracoda) unter Ausschluss der Bouciinae. Geologie, 11: 1184—1203.

Harding, J.P., 1955. The evolution of terrestrial habits in an ostracod. In: Symposium on Organic Evolution. Bull. Nat. Hist. Inst. Sci. India, 7: 104—106.

Harding, J.P., 1965. Crustacean cuticle with reference to the ostracode carapace. Pubbl. Stn. Zool. Napoli, 33 (suppl.): 9—31.

Harding, J.P., 1966. Myodocopan ostracods from the gills and nostrils of fishes. In: H. Barnes (Editor), Some Contemporary Studies in Marine Science. Allen and Unwin, London, pp. 369—374.

Hartmann, G., 1963. Zur Phylogenie und Systematik der Ostracoden. Z. Zool. Syst. Evolutionsforsch., 1: 1—154.

Hartmann, G., 1973. Zum gegenwärtigen Stand der Erforschung der Ostracoden interstitieller Systeme. Ann. Spéléol., 28: 417—426.

Hazel, J.E., 1970. Atlantic continental shelf and slope of the United States — Ostracode zoogeography in the Southern Nova Scotian and Northern Virginian faunal provinces. U.S. Geol. Surv., Prof. Pap., 529-E.

Hazel, J.E., 1971. Ostracode biostratigraphy of the Yorktown Formation (upper Miocene and lower Pliocene) of Virginia and North Carolina. U.S. Geol. Surv., Prof. Pap., 704: 13 pp.

Herrig, E., 1966. Ostracoden aus der Weissen Schreibkreide (Unter-Maastricht) der Insel Rügen. Paläontol. Abh., A, 2: 693—1024.

Hornibrook, N. de B., 1963. The New Zealand ostracode family Punciidae. Micropaleontology, 9: 318—320.

Howe, H.V., 1971. Ecology of American torose Cytherideidae. In: H.J. Oertli (Editor), Paléoécologie d'Ostracodes, Pau 1970. Bull. Centre Rech., Pau — SNPA, 5 (suppl.): 349—359.

Hulings, N.C. and Puri, H.S., 1965. The ecology of shallow water ostracods of the west coast of Florida. Pubbl. Stn. Zool. Napoli, 33 (suppl.): 308—344.

Jaanusson, V., 1957. Middle Ordovician ostracodes of central and southern Sweden. Bull. Geol. Inst. Upps., 37: 173—442.

Jaanusson, V., 1966. Ordovician ostracodes with supravelar antra. Publ. Paleontol. Inst. Univ. Uppsala, 66: 1—30.

Keij, A.J., 1957. Eocene and Oligocene Ostracoda of Belgium. Mém. Inst. R. Sci. Nat. Belg., No. 136: 210 pp.

Kesling, R.V., 1969. Copulatory adaptations in ostracods, 3. Adaptations in some extinct ostracods. Contrib. Mus. Paleontol. Univ. Mich., 22: 273—312.

Kesling, R.V. and Hussey, R.C., 1953. A new family and genus of ostracod from the Ordovician Bill's Creek shale of Michigan. Contrib. Mus. Paleontol. Univ. Mich., 11: 77—95.

Kesling, R.V. and others, 1965. Four Reports of Ostracod Investigations. Natl. Sci. Found. Proj. GB-26, University of Michigan, Ann, Arbor, Mich.

Kilenyi, T.I., 1971. Some basic questions in the palaeoecology of ostracods. In: H.J. Oertli (Editor), Paléoécologie d'Ostracodes, Pau 1970. Bull. Centre Rech. PAU — SNPA. 5 (suppl.): 31—44.

Kornicker, L.S., 1969a. Bathyconchoecia deeveyae, a highly ornamented new species of Ostracoda (Halocyprididae) from the Peru—Chile Trench system. Proc. Biol. Soc. Wash., 82: 403—408.

Kornicker, L.S., 1969b. Morphology, ontogeny and intraspecific variation of Spinacopia, a new genus of myodocopid ostracod (Sarsiellidae). Smithson. Contrib. Zool., 8: 50 pp.

Kornicker, L.S. and Sohn, I.G., 1974. Evolution of the Entomoconchacea. In: Int. Symp. Evolution Post-Paleozoic Ostracoda, Hamburg 1974 (abstract).

Kozur, H., 1971. Die Bairdiacea der Trias. Teil II: Skulpturierte Bairdiidae aus mitteltriassischen Tiefschelfablagerungen. Geol. Paläontol. Mitt. Innsbruck, 1: 1—21.

Kozur, H., 1972a. Einige Bemerkungen zur Systematik der Ostracoden und Beschreibung neuer Platycopida aus der Trias Ungarns und der Slowakei. Geol. Paläontol. Mitt. Innsbruck, 2: 1—27.

Kozur, H., 1972b. Die Bedeutung triassischer Ostracoden für stratigraphische und paläoökologische Untersuchungen. Mitt. Ges. Geol. Bergbaustud., 21: 623—660.

Kozur, H., 1974. Die Bedeutung der Bradoriida als Vorläufer der post-kambrischer Ostracoden. Z. Geol. Wiss. Berlin, 2: 823—830.

Kozur, H. and Mostler, H., 1972. Die Bedeutung der Mikrofossilien für stratigraphische paläoökologische und paläogeographische Untersuchungen in der Trias. Mitt. Ges. Geol. Bergbaustud., 21: 341—360.

Kristan-Tollmann, E., 1973. Zur Ausbildung des Schliessmuskelfeldes bei triadischen Cytherellidae (Ostracoda). Neues Jahrb. Geol. Paläontol. Monatsh., 1973: 351—373.

Krömmelbein, K., 1955. Arten der Gattungen Condracypris und Pachydomella im Mittel-Devon. Senckenberg. Lethaea, 36: 295—310.

Langer, W., 1973. Zur Ultrastruktur, Mikromorphologie und Taphonomie des Ostracoda-Carapax. Palaeontographica, Abt. A, 144: 1—54.

Levinson, S.A., 1951. Thin sections of Paleozoic Ostracoda and their bearing on taxonomy and morphology. J. Paleontol., 25: 553—560.

Martinsson, A., 1962. Ostracodes of the family Beyrichiidae from the Silurian of Gotland. Publ. Paleontol. Inst. Univ. Upps., 41: 369 pp.

Martinsson, A., 1963. Kloedenia and related ostracode genera in the Silurian and Devonian of the Baltic area and Britain. Bull. Geol. Inst. Univ. Upps., 42: 62 pp.

Martinsson, A., 1965. The Siluro-Devonian genus Nodibeyrichia and faunally associated Kloedeniines. Geol. Fören. Stockh. Förh., 87: 109—138.

Martinsson, A., 1967. The succession and correlation of ostracode faunas in the Silurian of Gotland. Geol. Fören. Stockh. Förh., 89: 350—386.

McKenzie, K.G., 1967a. Saipanellidae: A new family of podocopid Ostracoda. Crustaceana, 13: 103—113.

McKenzie, K.G., 1967b. The distribution of Caenozoic marine Ostracoda from the Gulf of Mexico to Australasia. In: C.G. Adams and D.V. Ager (Editors), *Aspects of Tethyan Biogeography. Syst. Assoc. Publ. Lond.*, 7: 219—238.

McKenzie, K.G., 1969. Discussion. In: J.W. Neale (Editor), *The Taxonomy, Morphology and Ecology of Recent Ostracoda.* Oliver and Boyd, Edinburgh, pp. 484—485.

McKenzie, K.G., 1973. Cenozoic Ostracoda, In: A. Hallam (Editor), *Atlas of Palaeobiogeography.* Elsevier, Amsterdam, pp. 477—487.

Menzies, R.J., George, R.Y. and Rowe, G.T., 1973. *Abyssal Environment and Ecology of the World Oceans.* John Wiley, New York, N.Y., 488 pp.

Müller, G.W., 1894. *Die Ostracoden des Golfes von Neapel und der angrenzenden Meeresabschnitte. Fauna und Flora des Golfes von Neapel, 21, Monogr. 8.* Friedländer, Berlin, 404 pp.

Neale, J.W. (Editor), 1969. *The Taxonomy, Morphology and Ecology of Recent Ostracoda.* Oliver and Boyd, Edinburgh, 553 pp.

Oertli, H.J. (Editor), 1971. *Colloquium on the Paleoecology of Ostracodes, Pau 1970. Bull. Centre Rech. Pau — SNPA* 5 (suppl.): 953 pp.

Oertli, H.J., 1972. Jurassic Ostracodes of DSDP Leg 11 (sites 100 and 105) — Preliminary account. In: C.D. Hollister, J.I. Ewing and others, *Initial Reports of the Deep Sea Drilling Project, 11.* Government Printing Office, Washington, D.C., pp. 645—657.

Pokorný, V., 1950. The ostracods of the Middle Devonian Red Coral Limestones of Čelechovice. *Sb. Ústřed. Ústav. Geol.,* 17: 513—632.

Pokorný, V., 1952. The ostracods of the so-called Basal Horizon of the *Subglobosa* Beds at Hodonín (Pliocene, Inner Alpine Basin, Czechoslovakia). *Sb. Ústřed. Ústav. Geol.,* 19: 229—396.

Pokorný, V., 1954. *Základy zoologické mikropaleontologie.* Akademia, Prague, 651 pp.

Pokorný, V., 1957. The phylomorphogeny of the hinge in Podocopida (Ostracoda, Crustacea) and its bearing on the taxionomy. *Acta Univ. Carol. Geol.,* 1957: 22 pp.

Pokorný, V., 1958. *Grundzüge der zoologischen Mikropaläontologie, II.* Dtsch. Verl. Wiss., Berlin, 453 pp.

Pokorný, V., 1959. Hinge and free margin structures of some Silurian ostracods. *Acta Univ. Carol. Geol.,* 1959: 321—341.

Pokorný, V., 1964a. *Oertliella* and *Spinicythereis,* new ostracode genera from the Upper Cretaceous. *Věstn. Ústřed. Ústav. Geol.,* 39: 283—284.

Pokorný, V., 1964b. The phylogenetic lines of *Cythereis marssoni* Bonnema, 1941 (Ostracoda, Crustacea) in the Upper Cretaceous of Bohemia, Czechoslovakia. *Acta Univ. Carol. Geol.,* 1964: 255—274.

Pokorný, V., 1967a. New *Cythereis* species (Ostracoda, Crustacea) from the Lower Turonian of Bohemia, Czechoslovakia. *Acta Univ. Carol. Geol.,* 1967: 365—378.

Pokorný, V., 1967b. The genus *Curfsina* (Ostracoda, Crustacea) from the Upper Cretaceous of Bohemia, Czechoslovakia. *Acta Univ. Carol. Geol.,* 1967: 345—364.

Pokorný, V., 1970. The genus *Radimella* Pokorný, 1969 (Ostracoda, Crustacea) in the Galápagos Islands. *Acta Univ. Carol. Geol.,* 1969: 293—334.

Pokorný, V., 1968. *Havanardia* g. nov., a new genus of the Bairdiidae (Ostracoda, Crust.). *Věstn. Ustřed. Ústav. Geol.,* 43: 61—63.

Pokorný, V., 1971. The diversity of fossil ostracode communities as an indicator of palaeogeographic conditions. In: H.J. Oertli (Editor), *Paléoécologie d'Ostracodes, Pau 1970. Bull. Centre Rech. Pau — SNPA,* 5 (suppl.): 45—61.

Puri, H.S. (Editor), 1965. Ostracods as ecological and palaeoecological indicators. *Pubbl. Stn. Zool. Napoli,* 33 (suppl.): 612 pp.

Puri, H.S., Bonaduce, G. and Gervasio, A.M., 1969. Distribution of Ostracoda in the Mediterranean. In: J.W. Neale (Editor), *The Taxonomy, Morphology and Ecology of Recent Ostracoda.* Oliver and Boyd, Edinburgh, pp. 356—411.

Rabien, A., 1954. Zur Taxionomie und Chronologie der oberdevonischen Ostracoden. *Abh. Hess. Landesamt. Bodenforsch.,* 9: 268 pp.

Ruggieri, G., 1971. Ostracoda as cold climate indicators in the Italian Quaternary. In: H.J. Oertli (Editor), *Paléoécologie d'Ostracodes, Pau 1970. Bull. Centre Rech. Pau — SNPA,* 5 (suppl.): 285—293.

Sars, G.O., 1866. Oversight af Norges marine Ostracoder. *Förh. Vidensk. Selsk. Christiania,* 1865: 1—130.

Sars, G.O., 1922—1928. *An Account of the Crustacea of Norway, IX. Ostracoda.* Bergen, 277 pp. (119 pls.).

Schneider, G.F., 1960. Fauna ostrakod nizhnetriasovykh otlozhenii Prikaspiiskoi nizmennosti. In: *Geologia i neftegazonosnost yuga S.S.S.R. Turkmenistan i Zapadnyi Kazakhstan. Tr. Kompleksn. Yuzhnoi Geol. Ekspeditsii,* Akad. Nauk S.S.S.R., 5: 287—303.

Schornikov, E.I., 1969. Novoe semeistvo rakushkovykh ratchkov (Ostracoda) iz supralitorali Kuril'skikh ostrovov. *Zool. Zh.,* 48: 494—498.

Seneš, J. and Marinescu, F., 1974. Cartes paléogéographiques du Néogène de la Paratethys. *Mém. Bur. Rech. Géol. Min.,* 88(2): 767—774.

Seneš, J. and working group, 1971. Korrelation des Miozäns der Zentralen Paratethys (Stand 1970). *Geol. Zb., Geologica Carpathica,* 22(1): 3—9.

Skogsberg, T., 1920. Studies on marine ostracods. Part I (cypridinids, halocyprids and polycopids). *Zool. Bidr. Upps., suppl.-bd.* 1: 784 pp.

Smith, R.N., 1965. Musculature and muscle scars of *Chlamydotheca arcuata* (Sars) and *Cypridopsis vidua* (O.F. Müller) (Ostracoda — Cyprididae). In: R.V. Kesling and colleagues, *Four Reports of Ostracod Investigations.* Natl. Sci. Found. Proj. GB-26, Rep. 3: 40 pp.

Sohn, I.G., 1962. The ostracode genus *Cytherelloidea,* a possible indicator of paleotemperature. *U.S. Geol. Surv. Prof. Pap.,* 450D: 144—147.

Spjeldnaes, N., 1951. Ontogeny of *Beyrichia jonesi* Boll. *J. Paleontol.,* 25: 745—755.

Swain, F.M. (Editor), 1975. Biology and Paleobiology of Ostracoda. A Symposium, University of Delaware, 1972. *Bull. Am. Paleontol.,* 65: 697 pp.

Sylvester-Bradley, P.C., 1941. The shell structure of the Ostracoda and its application to their palaeontologic investigation. *Annu. Mag. Nat. Hist.,* 8: 1—33.

Sylvester-Bradley, P.C., 1949. The ostracod genus *Cypridea* and the zones of the Upper and Middle Purbeckian. *Proc. Geol. Assoc. Engl.,* 60: 125—153.

Sylvester-Bradley, P.C., 1953. The Entomoconchacea; a new superfamily of macroscopic ostracods of Upper Palaeozoic age. *Q. J. Geol. Soc. Lond.,* 108: 127—134.

Triebel, E., 1938. Ostracoden-Untersuchungen, 1. *Protocythere* und *Exophthalmocythere,* zwei neue Ostracoden-Gattungen aus der deutschen Kreide. *Senckenbergiana,* 20: 179—200.

Triebel, E., 1941. Zur Morphologie und Ökologie der fossilen Ostracoden. Mit Beschreibung einiger neuen Gattungen und Arten. *Senckenbergiana.* 23: 294—400.

Ulrich, E.O. and Bassler, R.S., 1923. Ostracoda. Silurian. In: *Systematic Paleontology of Silurian Deposits.* Maryland Geol. Surv., Baltimore, Md., pp. 500—704.

Van den Bold, W.A., 1974. Ostracode associations in the Caribbean Neogene. *Verh. Naturforsch. Ges. Basel,* 84: 214—221.

Van Morkhoven, F.P.C.M., 1962, 1963. *Post-Palaeozoic Ostracoda. Their Morphology, Taxonomy and Economic Use.* Elsevier, Amsterdam, Volume I, General, 204 pp. Volume II, Generic Descriptions, 478 pp.

Vávra, V., 1892. Monografie českých korýšů skořepatých. *Arch. Přír. Výzk. Čech.,* 8: 110 pp.

Vesper, B., 1972. Zum Problem der Buckelbildung bei *Cyprideis torosa* (Jones, 1850) (Crustacea, Ostracoda, Cytheridae). *Mitt. Hamburg. Zool. Mus. Inst.,* 68: 79—94.

Wagner, C.W., 1957. Sur les ostracodes du Quaternaire récent des Pays-Bas et leur utilisation dans l'étude géologique des dépôts holocènes. Thesis, The Hague, 158 pp.

Zalányi, B., 1929. Morpho-systematische Studien über fossile Muschelkrebse. *Geol. Hung., Ser. Paleontol.,* 5: 152 pp.

Zaspelova, V.S., 1952. Ostrakody semeistva Drepanellidae iz otlozhenii verkhnego devona Russkoi platformy. *Tr. Vses. Neft. Nauchno-Issled. Geologorazved. Inst. NS,* 60: 157—216.

5

PTEROPODS

YVONNE HERMAN

INTRODUCTION

Pteropods, also known as sea butterflies, are marine gastropods adapted to pelagic life. Some species possess delicate external calcareous shells while others are devoid of it. Pteropods are widespread in the world oceans and after the animals' death their empty shells, together with the skeletal remains of other calcareous planktonic organisms, settle to the sea floor. In certain regions sediments are composed in large part of the remains of calcareous organisms; these sea-floor deposits are termed calcareous oozes. When pteropods constitute a high percentage of the ooze the deposit is called pteropod ooze.

The aragonitic shell of pteropods is much more susceptible to solution than the calcitic skeletal remains of coccoliths and foraminifers. For this reason the depth range of pteropod oozes is considerably more limited than that of coccolith and foraminiferal oozes. They are preserved between about 700 and 3000 m, but depths differ in the various marine basins and depend in part on bottom-water temperatures, circulation, and rates of sedimentation of biogenic and clastic materials. Pteropods are better preserved in basins having high bottom temperatures, sluggish circulation, and rapid rates of sedimentation such as the Mediterranean and the Red seas.

The existence of pteropods has been known since the seventeenth century; a century later the Swedish naturalist Linnaeus described and illustrated several pteropods in his *Systema Naturae*. However, the systematic position and rank of pteropods was debated for a long time; they were grouped by some naturalists with the cephalopods, while others considered them a separate class within the Phylum Mollusca.

Careful study of their soft parts eventually led several zoologists to recognize the affinities of pteropods with other opistobranch gastropods.

MORPHOLOGY OF SOFT PARTS

The animal is divisible into four regions: (1) **head**; (2) **foot**; (3) **visceral mass**, in which the internal organs are concentrated; and (4) **mantle**, which secretes the calcareous shell.

The head. The head bears a pair of tentacles and eyes and is surrounded by the paired fins (Figs. 1 and 2).

The foot. The most typical structure in pteropods is their fin-shaped foot which surrounds the mouth and is utilized for swimming and gathering food (Fig. 2).

The digestive system. The food of pteropods like that of most plankton feeders is composed of diatoms, dinoflagellates and minute crustaceans; it is gathered by the action of cilia on the wings and foot lobes. The mouth with jaws leads into the **buccal cavity**, or pharynx, which contains the **radula**, a ribbon-like band bearing minute teeth which are added continuously in the radular sac, replacing the worn teeth. The buccal cavity, continued by a long tube, the esophagus, leads into the stomach which is lined with several masticatory plates. The stomach is followed by the intestine which bends and terminates in an anus; the liver is generally located on the right side (Fig. 1).

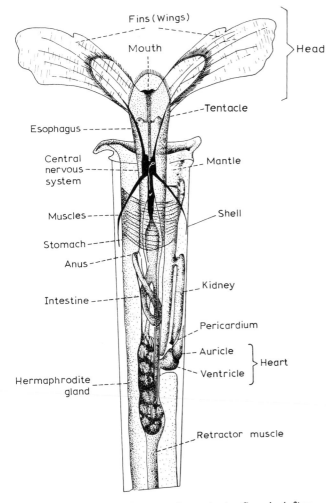

Fig. 1. Morphology of soft parts in *Creseis* (after Tregouboff and Rose, 1957).

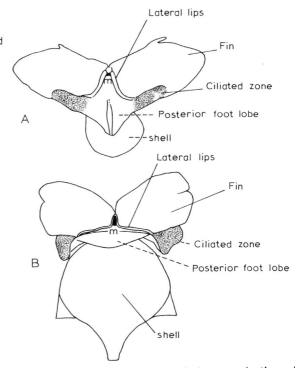

Fig. 2. Diagrammatical view of the organization of Euthecosomata. Ventral view. A. Limacinidae. B. Cavoliniidae. (After Meisenheimer, 1905.) *m* = mouth.

The reproductive system. All the Opistobranchia are hermaphroditic, meaning that each individual has both male and female sexual organs and is capable of producing both ova and spermatozoa. A period of male maturity during which spermatozoa are generated, precedes that of female maturity. Both ova and spermatozoa are formed in one reproductive organ, the **hermaphrodite gland**.

The nervous system. This consists of a number of nerve centers or ganglia, joined by nerve cords and communicating by means of nerves with various parts of the organism. The cerebral ganglia supply nerves to the head and tentacles; the pedal ganglia supply the foot lobes and fins; and the visceral ganglia supply nerves to the mantle, heart, kidney, and the gonads.

The circulatory system. The heart consists of an auricle and a ventricle, enclosed in the pericardial sac. Blood is pumped from the heart through numerous vessels to various parts of the body. The kidney is the main organ for excretion of metabolic waste products; it adjoins both the digestive gland and the pericardium and opens into the mantle cavity.

The mantle. The mantle is an overgrowing sheet of tissue. It appears that special structures in the mantle edge may be responsible for the shell formation or secretion. The space between the mantle and the underlying tissues, known as the mantle cavity, constitutes the respiratory chamber; gills are present in only one genus.

ECOLOGY

Pteropods are exclusively marine and generally live in the open ocean, swimming in the uppermost 500 m; however, some forms are known to live at great depths.

The present-day distributional patterns of pteropods are fairly well known. They are ubiquitous and abundant and about eighty species and subspecies inhabit the world's oceans. Their distribution is controlled by various physical and chemical parameters of the environment, such as temperature, salinity, food, oxygen and water depth.

Temperature

Temperature is the main factor governing the distribution of pteropods. At present, well-defined latitudinal temperature gradients exist; from the cold polar regions, temperatures increase progressively toward the equator. This gradual water temperature change is reflected in pteropod population composition. Thus, the two polar seas are inhabited by one pteropod species, *Limacina helicina* Phipps, whereas the warm, tropical zones are populated by many genera and species (Table I).

Salinity

Marine holoplanktonic invertebrates are cold-blooded and have body fluids isotonic with the surrounding water. For this reason they are limited to the narrow salinity ranges of oceanic water. The average sea-water salinity varies between $35^0/oo$ and $36^0/oo$. As mentioned in the previous paragraph, the warm regions of the three major oceans, Atlantic, Indian and Pacific, support a diversified pteropodal fauna. However, in land-locked warm seas where evaporation exceeds precipitation and runoff from land, salinities are much higher than in the open ocean. In the Red Sea with surface-water salinities greater than $40^0/oo$ but with temperatures similar to those of the oceans, only about 50% of the oceanic species are known to occur (Herman-Rosenberg, 1965). In the Mediterranean where salinities are intermediate between the Red Sea and the open ocean and temperatures are similar to the oceans at comparable latitudes, about 75% of the open-ocean species

TABLE I

Present-day distribution of some pteropods in the oceans

Warm	Cold-temperate	Cold-polar
Cavolinia gibbosa		
C. globulosa		
C. inflexa		
C. longirostris		
C. tridentata		
C. uncinata		
Clio cuspidata		
C. polita		
C. pyramidata convexa	C. pyramidata pyramidata	
Creseis acicula		
C. virgula		
C. conica		
Cuvierina columnella		
Diacria quadridentata		
D. trispinosa		
Hyalocylix striata		
Styliola subula		
Limacina bulimoides		
L. inflata		Limacina helicina
L. lesuerii		
L. trochiformis	Limacina retroversa	
Peraclis spp.		

have been recorded. Their number decreases from the western Mediterranean basins, where conditions are milder, towards the eastern sector of the sea where salinities are higher, suggesting that salinity rather than temperature controls the distribution of certain species. A small number of pteropods have adapted to low salinities. Occasionally, a few hardy forms survive in deltaic or estuarian regions where large volumes of fresh water drain into the sea and lower salinities considerably. Pteropods apparently cannot survive in the Black Sea where salinities are much lower than those of open ocean waters.

MAJOR MORPHOLOGICAL GROUPS

The systematic classification of pteropods is based both on shell characteristics and on the organization of the soft parts. The information available is taken from various sources and may be summarized in the form of a key. Only the calcareous shell-bearing forms will be described since they are the forms preserved in sediments.

KEY

Class GASTROPODA
   Subclass OPISTOBRANCHIATA
      Order GYMNOSOMATA
      Order THECOSOMATA
         Suborder PSEUDOTHECOSOMATA
            Family PERACLIDIDAE
            Family CYMBULIDAE
         Suborder EUTHEOCOSOMATA
            Family LIMACINIDAE
            Family CAVOLINIIDAE

Suborder Pseudothecosomata

Shell is not always present, in most genera it is replaced by a cartilaginous pseudoconcha. Fins fused into a continuous swimming plate, ventral to the mouth. Tentacles equal in size.

Family Peraclididae

This is the only shell-bearing family in the suborder. Aragonitic shell spirally coiled, sinistral, possessing a twisted prolongation of the columella. Shell generally sculptured with a network of delicate reticulation. Sever-

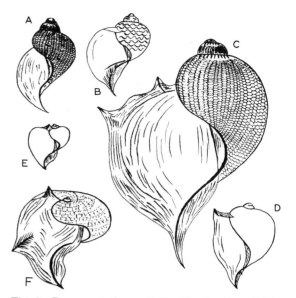

Fig. 3. Representatives of the Family Paraclididae. A. *Peraclis reticulata* (d'Orbigny), apertural view. × 20. B. *Peraclis apicifulva* Meisenheimer, apertural view. × 20. C. *Peraclis bispinosa* Pelseneer, apertural view. × 8.5. D. *Peraclis moluccensis* Tesch, apertural view. × 8.5. E. *Peraclis depressa* Meisenheimer, apertural view. × 10. F. *Peraclis triacantha* (Fischer), apertural view. × 8.5. (After Tesch, 1946.)

al species are known, they live at depths greater than 100 m and are seldom preserved in sediments (Fig. 3).

Suborder Euthecosomata

Shell present, aragonitic. Fins separated, dorsal to the mouth. Tentacles unequal in size.

Fig. 4. *Limacina bulimoides* (d'Orbigny). A. Apertural view. × 45. B. Oblique apertural view. × 58.

Fig. 5. A, B. *Limacina trochiformis* (d'Orbigny). A. Spiral view. B. Apertural view. × 30. C. *L. inflata* (d'Orbigny), spiral view. × 30.

Fig. 6. *Limacina retroversa* (Fleming). Apertural view, with partly broken shell at lower left. × 65.

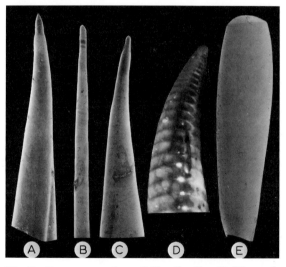

Fig. 7. Some examples of Family Cavoliniidae. A. *Styliola subula* Quoi and Gaimard. × 77. B. *Creseis acicula* (Rang). × 77 C. *C. clava* (Rang). × 20. D. *Hyalocylix striata* (Rang). × 12. E. *Cuvierina columnella* (Rang). × 7.

Family Limacinidae
Shell spirally coiled, sinistral, operculum present; mantle-cavity dorsal (Figs. 2A, 4—6). Genus *Limacina* has a spiral shell with whorls gradually increasing towards aperture (Figs. 4—6).

Family Cavoliniidae
Shell uncoiled, ventral and dorsal sides distinct, operculum absent; mantle-cavity ventral (Figs. 1, 2B, 7, 8, 9, 10). Genus *Styliola* has a conical shell, a circular aperture and a longitudinal ridge over the dorsal side (Fig. 7A). Genus *Hyalocylix* has a shell that is conical and slightly curved dorsally, with distinct transverse striation; the aperture is oval (Fig. 7D). The shell of genus *Cuvierina* is bottle-shaped, bulging in the middle, and the aperture is bean-shaped (Fig. 7E). Genus *Creseis* has an uncoiled shell, circular in transverse section and no operculum (Fig. 7B,

C). Genus *Diacria* has also an uncoiled shell with a thickened, dorsal aperture (Fig. 8C, D). Genus *Cavolinia* shows an uncoiled shell and a variable aperture. The dorsal aperture is not thickened (Figs. 8E—I, 9). Genus *Clio* has a

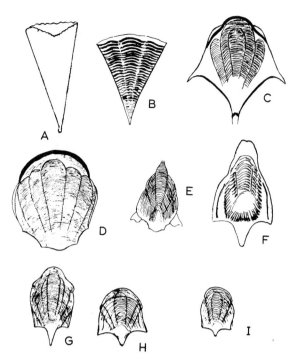

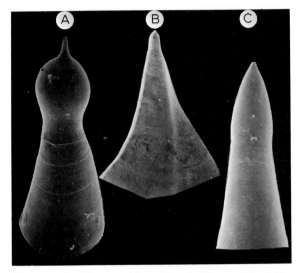

Fig. 10. A. *Clio cuspidata* (Bosc). Upper 1/4 of the shell. × 42. B. *C. pyramidata* Linné, forma *convexa* (Boas). × 9. C. *C. pyramidata* forma *pyramidata* Linné, upper 1/4 of the shell. × 60.

Fig. 8. A. *Clio polita* (Craven), dorsal view. × 2.1. B. *Clio chaptali* (Souleyet), dorsal view. × 2.1. C. *Diacria trispinosa* (Lesueur), forma *major*, dorsal view. × 3.5. D. *Diacria quadridentata* (Lesueur) forma *costata*, dorsal view. × 10. E. *Cavolinia longirostris* (Lesueur), dorsal view. × 4.2. F. *Cavolinia tridentata* Forskal, dorsal view. × 2.1. G. *Cavolinia gibbosa* (Rang), dorsal view. × 2.8. H. *Cavolinia uncinata* (Rang), dorsal view. × 2.1. I. *Cavolinia globulosa* (Rang), dorsal view. × 2.1. (After Tesch, 1946.)

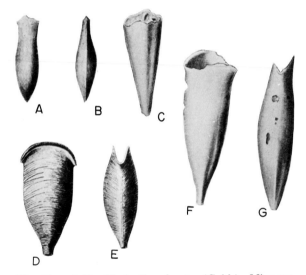

Fig. 11. A, B. *Vaginella clavata* (Gabb), Miocene, California. A. Ventral view. B. Lateral view. × 6.3. C. *Vaginella bicostata* (Gabb), Miocene, Costa Rica. × 10. D, E. *Vaginella chipolana* Dall, Miocene, Florida, × 10. F, G. *Vaginella floridana* Collins, Miocene, Florida. × 8.5. (After Collins, 1934.)

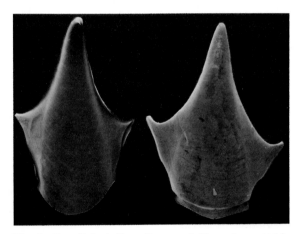

Fig. 9. *Cavolinia inflexa* (Lesueur). Two views. × 10.

three-sided shell which has dorsal ribs and well-defined lateral borders. The cross-section of the shell is triangular with maximum breadth at the opening (Figs. 8A, B, 10A—C). Genus *Vaginella* is similar to *Cuvierina*. The shell is vase-shaped, laterally compressed, and lateral keels are generally present. Aperture is elliptical (Fig. 11).

EVOLUTIONARY TRENDS

Phylogenetic relationships among pteropods have been a matter of debate and the precise

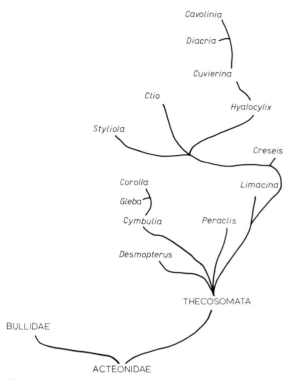

Fig. 12. Suggested relationships amongst pteropod taxa.

ancestry and interrelationships within this group are still unknown. It is probable that pteropods evolved from bottom-dwelling littoral gastropods and then adapted to a pelagic mode of life. Fig. 12 shows the probable relationship between the major groups and their ancestry.

The coiled *Limacina* and *Peraclis* are thought to be primitive genera, whereas the uncoiled Cavoliniidae are more advanced forms, the coiling thought to have been lost in the course of evolution.

Changes that have occurred within each group may have been adaptive to changing environments and/or adaptation to cope better with the existing environment. For example, the shell of pteropods is lighter than that of bottom-dwelling gastropods; the reduction of the external shell to an internal pseudoconcha or its absence in pseudothecosomes and gymnosomes as well as the change from a muscular foot into a wing-shaped swimming plate are adaptive characters to pelagic life.

A number of species belonging to the genera *Limacina*, *Peraclis* and *Clio* live at great depths. These bathypelagic species have

several characters in common: the shape of the wings, as well as the digestive and nervous systems. These forms belonging to different genera bear common traits as a result of adaptation to common life conditions. At first bathypelagic pteropods were thought to be primitive representatives of the group, but now it is accepted that all deep-living species are highly specialized.

FOSSIL RECORD AND BIOSTRATIGRAPHY

Turning to the fossil record, we find Cambrian beds with abundant remains of *Hyolites* and *Orthoteca*, Silurian deposits with *Tentaculites*, and Devonian sediments with *Styliolina* described from various localities. The opinion of earlier paleontologists that these fossils are true pteropods has now been abandoned.

Two living families with known fossil representatives are the Limacinidae, ranging from Eocene to Recent, and the Cavoliniidae. *Clio*, which belongs to the Cavoliniidae, has been found in Upper Cretaceous rocks. *Vaginella* (Fig. 11), one of the earliest pteropods, resembling the living *Cuvierina*, ranges from Upper Cretaceous through Miocene. Although many fossil pteropods have been described, their identification and geologic range is still disputed. The rare preservation of pteropods in the geologic record, particularly in pre-Pleistocene sediments, is mainly due to their thin and fragile aragonitic tests which are more susceptible to solution than are those of other marine calcitic microfossils. Therefore stratigraphic divisions as well as correlations over widely separated geographic regions have not been attempted. On the other hand, the usefulness of pteropods in local correlations is well established, particularly in the Mediterranean and Red Sea basins where Quaternary deep-sea sediments are composed of pteropodal, calcareous oozes.

PALEOECOLOGY

The Pleistocene was marked by repeated world-wide temperature fluctuations; cold episodes (glacials) were interrupted by mild periods (interglacials) with climates similar to or warmer than those of today. During glacials, air and sea temperatures dropped and continental ice sheets grew as water was extracted from oceans and precipitated on land covering large portions of continental areas on both hemispheres. The repeated waxing of glaciers during cold periods brought about world-wide lowering of sea level by more than 100 m, whereas waning of glaciers during interglacials resulted in sea-level rise.

As mentioned in a previous section the distributional patterns of living pteropods indicate that many species have a limited tolerance to changes in temperature and salinity. Accordingly, variations in faunal composition in consecutive sediment layers should reflect changes in climatic and hydrologic conditions at the time of their burial. In addition, other variables determine the composition of faunal remains in sediments. Important among these are variations in production rates, redistribution by currents and burrowing animals, accumulation rates of detrital sediments and solution of calcareous tests.

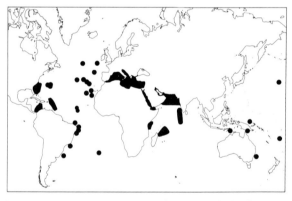

Fig. 13. Known occurrence of pteropods in Recent marine sediments.

Recent studies of pteropodal distribution in time and space indicate their usefulness in paleoecology, paleohydrology and paleobathymetry (Fig. 13). By studying Pleistocene deep sea cores from the Mediterranean and Red seas, Herman (1971) observed that the replacement of one pteropodal population by another with time coincides generally with changes in planktonic foraminiferal populations which have proven useful indices of past

climates. As compared with the major oceans, Pleistocene climatic changes in low-latitude inland seas were relatively marked due to their size and proximity to land. Furthermore, water exchange with the open ocean during lowered sea levels was much more restricted than it is today. The Red Sea which is separated from the Indian Ocean by a narrow, shallow sill was almost completely cut off from the ocean during glacial periods. Water temperatures were lower and salinities higher than today. As a result, environmental conditions must have been rigorous as compared with those prevailing today, and many planktonic organisms including pteropods were not able to survive in the Red Sea. The few species that adapted to the environment include epiplanktonic *Creseis* and *Limacina trochiformis* (d'Orbigny), forms that are known to tolerate wide ranges of salinities.

The Mediterranean water temperature and salinity fluctuations during the Pleistocene were of a different kind than the oscillations in the tropical Red Sea. Lower temperatures brought into the Mediterranean subpolar, North Atlantic and Norwegian Sea immigrants. The composition of preserved assemblages suggests that during glacials, salinities were close to those prevailing today. However, during the warming trends, surface-water salinities decreased as large volumes of glacial meltwater and fresh water derived from the adjacent Black Sea, the Nile, Rhone and Po rivers discharged into the sea. The temporary low-salinity intervals are recognized in sediments by the dominance of planktonic species that are known to be able to tolerate wide ranges of salinities. They include the foraminifers *Globoquadrina dutertrei*, *Globigerina quinqueloba* and the pteropod *Creseis*.

SUGGESTIONS FOR FURTHER READING

Bonnevie, K., 1913. Pteropoda. *Rep. Sci. Res. "Michael Sars" North Atlantic Deep Sea Exped., 1910*, 3: 1—69. [Discussion of anatomy and present-day distribution of pteropods.]
Hida, T.S., 1957. Chaetognaths and pteropods as biological indicators in the North Pacific. *U.S. Fish. Wildl. Serv., Spec. Sci. Rep. Fish.*, 215: 1—13. [Present-day distribution in the North Pacific.]

Selected references on fossil pteropods

Avnimelech, M., 1945. Revision of fossil Pteropoda from Southern Anatolia, Syria and Palestine. *J. Paleontol.*, 19: 637—647.

Collins, R.L., 1934. A monograph of the American Tertiary pteropod mollusks. *Johns Hopkins Univ., Stud. Geol.*, 11: 137—234.

Curry, D., 1965. The English Palaeogene pteropods. *Proc. Malacol. Soc. Lond.*, 36: 357—371.

Dollfus, G. and Ramon, G., 1886. Liste de ptéropodes du terrain tertiaire parisien. *Mém. Soc. Malacol. Belg.*, 20: 36—44.

Jung, P., 1971. Fossil molluscs from Carriacou, West Indies. *Bull. Am. Paleontol.*, 61: 147—262.

Kittl, E., 1886. Ueber die miocenen Pteropoden von Oesterreich—Ungarn. *Ann. K.K. Naturhist. Hofmus.*, 1: 47—74.

CITED REFERENCES

Herman, Y., 1971. Vertical and horizontal distribution of pteropods in Quaternary sequences. In: B.M. Funnell and W.R. Riedel (Editors), *Micropaleontology of Oceans*. Cambridge University Press, Cambridge, pp. 463—486.

Herman, Y. and Rosenberg, P.E., 1969. Pteropods as bathymetric indicators. *Mar. Geol.*, 7: 169—173.

Herman-Rosenberg, Y., 1965. Etudes des sédiments quaternaires de la Mer Rouge. *Ann. Inst. Océanogr.* (Masson et Cie.), 42: 343—415.

Meisenheimer, J., 1905. Pteropoda. *Wiss. Ergebn. Dtsch. Tiefsee-Exped. "Valdivia" 1898—1899*, 9: 1—314.

Tesch, J.J., 1946. *The Thecosomatous Pteropods, I. The Atlantic*. Carlsberg Foundation, Copenhagen, Dana Rep. No. 28.

Tregouboff, G. and Rose, M., 1957. *Manuel de planctonologie Mediterranéene*. C.N.R.S., Paris.

6

CALPIONELLIDS

JÜRGEN REMANE

INTRODUCTION

Calpionellids are pelagic protozoans possessing an axially symmetrical calcareous test (Fig. 1). They are typically found in deep-sea deposits of the late Jurassic and, less frequently, the early Cretaceous. There are no Recent calpionellids, hence their systematic position is uncertain. Together with other simple microfossils, some organic and some calcareous, calpionellids have been assigned by most

workers to the organic-walled tintinnids which has resulted in much confusion concerning taxonomy.

Calpionellids vs. tintinnids

Recent tintinnids possess an organic test, the **lorica**. This lorica is generally bell-shaped, although in some cases it may be tubular with two openings. The surface may be smooth or variously sculptured (Fig. 2). Some species bear surficially agglutinated materials, such as minute quartz grains, coccoliths or diatom frustules.

The living cell is attached to the posterior part of the test by its aboral extremity so that nearly all of its surface is separated from the lorica wall (Fig. 3). The most prominent feature of the living cell is the powerful locomotory apparatus, made up of a row of **membranelles** situated along the outer edge of the **peristome** (Fig. 3).

Calpionellids were mostly thought to be fossil tintinnids because of the overall resemblance of their tests with the loricae of certain tintinnids. They are, however, distinguished from tintinnids by their calcareous test, and it must be emphasized that mineralized tests are unknown not only in Recent tintinnids but also in ciliates as a whole. Moreover, in all cases where mineralized tests occur in Recent Protozoa, they are in direct contact with the living cell over most of its surface, quite unlike the organic lorica of tintinnids. For these reasons, none of the calcareous forms, including calpionellids, can be assigned to the order Tintinnida nor to the class Ciliatea. There are only a few fossil species with an organic lorica which may be fossil tintinnids (Fig. 4). In the case of some forms reported

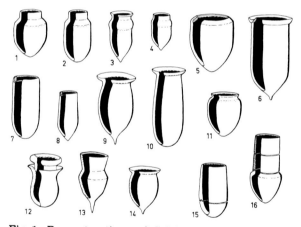

Fig. 1. Reconstructions of Calpionellidea, with parts of the loricae cut away. × 140. (After Remane, 1971.) Family Calpionellidae: *1, Calpionella alpina* Lorenz; *2, Calpionella elliptica* Cadisch; *3, Crassicollaria intermedia* (Durand Delga); *4, Crassicollaria parvula* Remane; *5, Calpionellites darderi* (Colom); *6, Remaniella cadischiana* (Colom); *7, Calpionellopsis simplex* (Colom); *8, Calpionellopsis oblonga* (Cadisch); *9, Tintinnopsella carpathica* (Murgeanu and Filipescu); *10, Tintinopsella longa* (Colom); *11, Lorenziella hungarica* Knauer and Nagy; *12, Chitinoidella bermudezi* (Furrazola-Bermúdez); *13, Chitinoidella cristobalensis* (Furrazola-Bermúdez); *14, Chitinoidella boneti* Doben. Family Colomiellidae: *15, Colomiella recta* Bonet; *16, Colomiella mexicana* Bonet.

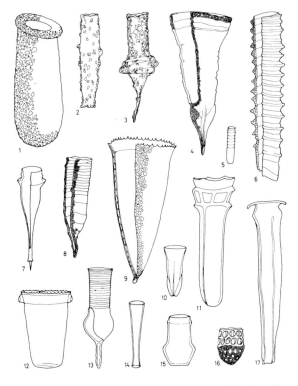

Fig. 2. Loricae of some Recent tintinnids showing differences from calpionellids mainly by the presence of an external sculpture or of two openings (*2, 6, 14*). × 120. (From Remane, 1971, after Kofoid and Campbell, 1929.)

1, Tintinnidium neapolitanum; 2, Leprotintinnus pellucidus; 3, Tintinnopsis prowazeki; 4, Favella helgolandica; 5, Metacyclis annaluta; 6, Climacocyclis elongata; 7, Xystonellopsis inaequalis; 8, Coxliella helix; 9, Cyttarocyclis magna; 10, Amphorella calida; 11, Stelidiella stelidium; 12, Cymatocyclis situla; 13, Codonellopsis pacifica; 14, Tintinnus macilentus; 15, Undellopsis entzi; 16, Dictyocysta magna; 17, Daturella ora.

from middle Jurassic phosphate nodules in northern Germany (Rüst, 1885) and three genera from the Albian—Cenomanian of the western interior of America (Eicher, 1965) their attribution to tintinnids seems justified (Fig. 4).

HISTORY OF CALPIONELLID STUDY

The first calpionellid species (*Calpionella alpina* Lorenz, Fig. 1, *1*) was discovered from the Swiss Alps and ascribed to the foraminifera because of its superficial resemblance to members of the Lagenidae. Colom (1934) for the first time interpreted the five known

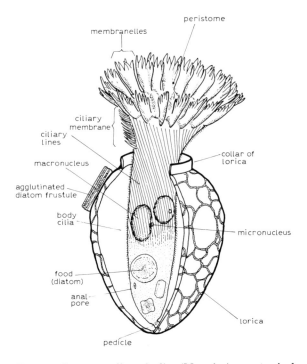

Fig. 3. *Stenosemella nivalis* (Meunier), a typical Recent tintinnid. Part of the lorica cut away in order to show soft parts; note also the powerful locomotory apparatus. × 450. (Simplified after Campbell, 1954.)

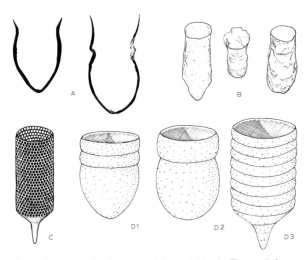

Fig. 4. Possible fossil tintinnids. × 185. A. Two of the forms from the Silurian of North Africa figured by Chennaux (1967) (thin sections). B. Organic loricae found in Triassic palynomorph assemblages by Visscher (1970). C. The most typical of the forms described by Rüst (1885) in the Middle Jurassic of northern Germany. D. Albian to Cenomanian forms from the western interior of the United States, after Eicher (1965). D1, *Dicloeopella borealis* Eicher; D2, *Codonella bojiga* Eicher; D3, *Coxliella coloradoensis* Eicher.

species as fossil tintinnids and nearly all workers have since accepted this interpretation. Colom believed first in a replacement of organic loricae by secondary calcite. Deflandre (1936) maintained that only quartz agglutinating forms could have been preserved in this manner, a view also shared by Colom. But the frequent replacement of crystallized silica by calcite is mineralogically highly improbable, and especially in the case of calpionellids which are often found together with well-preserved arenaceous foraminifera. Recently, Tappan and Loeblich (1968) suggested that most calpionellids correspond to Recent tintinnids with an organic lorica completely covered by agglutinated coccoliths. During diagenesis the coccoliths would have recrystallized and formed a secondary thickening of the agglutinated cover.

Bonet (1956), on the other hand, had already pointed out that the wide geographical distribution and general abundance of calcitic calpionellids make it much more plausible that the calcite is primary. In addition, loricae are never replaced by other materials, such as pyrite, and disappear from the stratigraphic record despite the persistence of favorable pelagic facies. These facts demonstrate that the theory of a diagenetic calcification was based neither on mineralogical considerations, nor on observed facts.

Recent studies of well-preserved calpionellids under the scanning electron microscope (Aubry and others, 1976) have revealed a wall structure of spirally arranged calcite prisms normal to the surface of the lorica, supporting the primary calcitic nature of calpionellid loricae and the idea that calpionellids are not fossil tintinnids (Fig. 5), thus we have to class them among Protozoa *incertae sedis*.

BIOLOGY AND PALEOECOLOGY

Although calpionellids are extinct organisms, their paleoecology can be reconstructed with a high degree of confidence. Calpionellids occur typically in very fine grained (micritic) limestones or marly limestones which are very poor in benthic fossils but rich in micro- (radiolarians) and nannoplankton (coccoliths, nannoconids). This clearly indicates that calpionellids were also marine planktonic protozoans, which is confirmed by their resemblance with Recent tintinnids. The

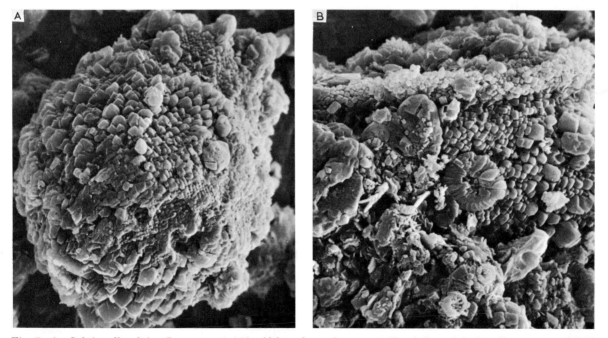

Fig. 5. A. *Calpionella alpina* Lorenz. ×1,440. Although partly recrystallized the original wall structure of helicoidally arranged radial crystals is still clearly recognizable. B. *Tintinnopsella* sp. × 1,800. Partial view showing wall structure and agglutinated (?) coccolith. (Photos courtesy of Dr. Bignot, Paris.)

presence of a large oral opening makes it very probable that calpionellids possessed some sort of locomotory organelle similar, but nevertheless unlike the membranelles of tintinnids. In other words, calpionellids show ecological analogies with Recent tintinnids but their systematic position is quite different.

Finally, it should be mentioned that calpionellid limestones are very representative of uppermost Jurassic and Lower Cretaceous pelagic facies of the Tethyan province. These sediments often contain thousands of individuals per cubic centimeter.

Preparation for microscopy

Under favorable circumstances, Calpionellidae may be isolated from marly sediments (Fig. 6), but routine stratigraphic work depends entirely on thin-sections where loricae are encountered in random sections of unknown orientation (see Fig. 10). The morphology of the collar is still recognizable under these conditions, but the proportions of the loricae, important in specific determination, are difficult to estimate.

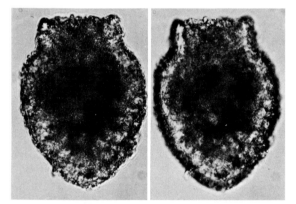

Fig. 6. *Calpionellia alpina* Lorenz. × 770. Two views of the same specimen at different optical focus.

MAJOR GROUPS OF CALCAREOUS FORMS

Problems of delimitation of calpionellids

As stated earlier, extinct calcareous microfossils cannot be assigned to tintinnids. The question remains then, how closely such forms may be related to calpionellids.

Since there is evidence that calpionellids, typically possessing hyaline loricae, evolved directly from the mid-Tithonian genus *Chitinoidella* (see Figs. 8, 10A) with a microgranular test, older hyaline forms (occurring since the Silurian) can be excluded from calpionellids with certainty. Younger genera are, of course, not as easy to place. However, the Tertiary forms described by Tappan and Loeblich (1968) under the generic names *Remanellina*, *Tytthocoris*, *Yvonniellina*, and *Pseudarcella* are so different in size, morphology, and habitat that a relationship with calpionellids is highly improbable.

This means that calpionellids are restricted to uppermost Jurassic and Lower Cretaceous. From this time interval, 26 genera containing 91 species of calcareous "fossil tintinnids" have been described to date. Theoretically, any of them could be calpionellids, but many are not.

One important group which can be eliminated from the calpionellids are the so-called "aberrant tintinnids". These are large forms with "loricae" often more than 2 mm long; the maximum length for calpionellids is about 150 μm. Some sections show several stacked trumpet-shaped tubes with two openings (Fig. 7). They also differ from calpionellids by their sparitic wall structure with an irregular fabric of large calcite crystals, resulting from recrystallization. These "aberrant tintinnids" occur widely in uppermost Jurassic and lowermost Cretaceous back-reef facies, unlike calpionellids which are pelagic.

Other forms resembling calpionellids occur sometimes with them in pelagic limestones. For example, the genus *Patelloides* was created for abnormally flat forms which are, in reality, oblique sections of ostracode carapaces. They occur together with calpionellids but may precede them in time and thus lead to erroneous age assignments. After such considerations only ten genera containing about thirty species remain which can be firmly established as calpionellids. It seems best to group calpionellids in one superfamily (Calpionellidea), placed among Protozoa *incertae sedis*.

Calpionellidea may then be defined as organisms possessing a calcitic lorica of axial symmetry. The oral opening is always large

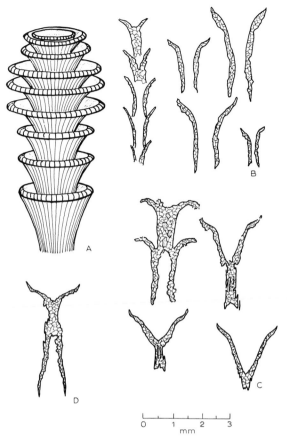

Fig. 7. *Campbelliella striata* (Carozzi), a calcareous alga interpreted as an aberrant tintinnid. (A: reconstruction after Bernier, 1974; B—D: variously oriented sections, most of isolated whorls, after Radoičić, 1969.) B. Axial sections. C. Longitudinal sections parallel to the median axis. D. Slightly oblique tangential sections. Most of the differences are only apparent and due to angle of intersection.

and surrounded by an oral collar of variable form, whose internal structure may be different from that of the lorica wall (see Fig. 1). In thin-section, the lorica appears generally transparent (hyaline) or sometimes microgranular and brownish opaque.

According to its original definition, the superfamily comprises two families: Calpionellidae and Colomiellidae. The main difference between the two families is that in Colomiellidae the collar is always much longer than the lorica itself and may be made up of several pieces separated by fine sutures.

Both families, above all the Calpionellidae, contain typically planktonic forms, abundant in fine-grained pelagic limestones and marls.

Until now no significant biogeographical differentiation could be observed in Calpionellidae, which are nevertheless very important for biostratigraphy.

Family Colomiellidae

This family comprises only one genus, *Colomiella* (type species: *C. mexicana* Bonet) which may be characterized as follows: lorica very short, aborally rounded, with a long, cylindrical collar separated by a suture from the lorica. The collar may be constituted of several successive rings. Lorica and collar are hyaline and calcitic, but with different angles of extinction under crossed nicols (Fig. 1, *15, 16*). Some transverse sections show a black cross at a 45° angle, which is not seen in the Calpionellidae. Total length: 90—140 μm, width 40—75 μm. Range: Upper Aptian to Albian.

At present the Colomiellidae are known only from Mexico, Cuba, Tunisia, and the Aquitaine Basin (SW France).

Family Calpionellidae

This family may be defined as follows: Calcitic lorica with axial symmetry, with only one oral opening which is wide and surrounded by an oral collar that is usually formed by a prolongation of the lateral walls but may also be an independent element of different structure or crystallographic orientation. In some genera, both types of collar occur together. In the earliest representatives (*Chitinoidella*) the microgranular wall appears brownish opaque in transmitted, and white, in reflected light. Nearly all other Calpionellidae possess a purely hyaline wall of radially fibrous calcite; transverse sections show a black cross under crossed nicols. Transitional forms (*Praetintinnopsella*, Fig. 8, *2*; Fig. 10B) possess an outer microgranular and an inner hyaline layer. Range: Middle Tithonian to "Middle" Valanginian; later occurrences not confirmed by ammonite stratigraphy.

Calpionellidae are more widespread and cover the whole Tethyan Province, from eastern Mexico, Texas, Venezuela and Cuba in the west to Oman, Iran and the Kiogar nappes of Tibet in the east. They have also been found

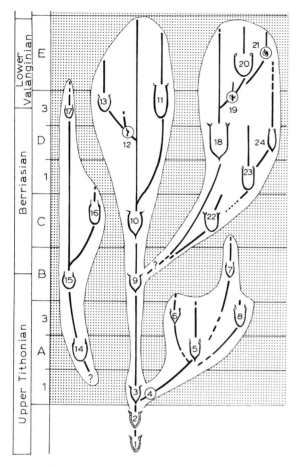

Fig. 8. Phylogenetic relationships between important species of Calpionellidae. *A* to *E* are the calpionellid zones and subzones distinguished in the Western Alps (see Table I). All species are drawn to the same scale, except collars in *12, 19* and *21*, which are enlarged. (Modified from Remane, 1971).
1, Chitinoidella boneti Doben; *2, Praetintinnopsella andrusovi* Borza; *3*, small variety of *Tintinnopsella carpathica* (Murgeanu and Filipescu); *4, Tintinnopsella remanei* Borza; *5, Crassicollaria intermedia* (Durand Delga); *6, Crassicollaria brevis* Remane; *7, Crassicollaria parvula* Remane; *8, Crassicollaria massutiniana* (Colom); *9, 10, Tintinnopsella carpathica* (Murgeanu and Filipescu) showing the increase of lorica dimensions in the beginning of the Berriasian; *11, Tintinnopsella longa* (Colom); *12*, left collar of *Lorenziella plicata* Remane; *13, Lorenziella hungarica* Knauer and Nagy; *14, 15, Calpionella alpina* Lorenz showing the evolution from large, more elongated forms (*14*) to the spherical variety of the uppermost Tithonian/basal Berriasian; *16, Calpionella elliptica* Cadisch; *17*, small, atypical *Calpionella alpina* Lorenz of the Upper Berriasian; *18, Remaniella cadischiana* (Colom); *19*, right collar of *Remaniella murgeanui* (Pop); *20, Calpionellites darderi* (Colom); *21*, right collar of *Calpionellites coronata* Trejo; *22, Remaniella ferasini* (Catalano); *23, Calpionellopsis simplex* (Colom); *24, Calpionellopsis oblonga* (Cadisch).

in DSDP Leg XI, sites 100 and 105 in the western Atlantic and in offshore wells on the Scotian Shelf (Canada) and the Grand Banks, southeast of Newfoundland.

The phylogenetic relationships of the principal genera are shown in Fig. 8 and need not be discussed here in more detail. Among the figured genera, the following are stratigraphically important and will be briefly characterized:

(1) *Chitinoidella*; type species: *C. boneti* Doben (Fig. 1, *12—14*; Fig. 8, *1*; Fig. 10A)
  Range: Middle Tithonian to basal Upper Tithonian.
  Length: 45—75 μm. Width 30—45 μm.
Distinctive character: the microgranular wall. Morphology of the collar extremely variable and very complex in some species. Ten species have been described up to the present day, but most of them are not very well known and exact ranges are not clear. The genus as a whole is, however, stratigraphically very important. There is a continuous transition from *C. boneti* Doben to a small variety of *Tintinnopsella carpathica* (Murgeanu and Filipescu) through *Praetintinnopsella andrusovi* Borza.

(2) *Tintinnopsella*; type species: *T. carpathica* (Murgeanu and Filipescu) (Fig. 1, *9—10*; Fig. 8, *3, 9—11*; Fig. 10J, K)
  Range: Upper Tithonian to "Middle" Valanginian.
  Length: 50—150 μm. Width: 40—70 μm.
Distinctive character: the very large oral opening surrounded by a rather simple collar, corresponding to a right-angle outward deflection of the lateral wall. A small variety of *T. carpathica* (Murgeanu and Filipescu) is abundant for a very short time span in the earliest hyaline faunas, but then the genus almost vanishes until Lower Berriasian. Cretaceous forms are larger and more diversified; in this sense *Tintinnopsella* is a typically Cretaceous genus.

(3) *Crassicollaria*; type species: *C. brevis* Remane (Fig. 1, *3, 4*; Fig. 8, *5—8*; Fig. 10C—F)
  Range: Upper Tithonian to Lower Berriasian.
  Length: 65—110 μm. Width 35—60 μm.
Most species of *Crassicollaria* were formerly assigned to *Calpionella elliptica* Cadisch because of their elliptical outline in longitudinal sections. The collar is, however, different, although variable in detail. The most constant feature is a more or less marked convexity just below the collar; the transition from this widening to the collar is always gradual. There is a great proliferation of forms in the lower part of the Upper Tithonian, but only one small species (*C. parvula* Remane) continues to the Berriasian.

(4) *Calpionella*; type species: *C. alpina* Lorenz (Fig. 1, *1, 2*; Fig. 8, *14—17*; Fig. 10G—I)
  Range: Upper Tithonian to uppermost Berriasian.
  Length: 45—105 μm. Width: 35—75 μm.

Distinctive character: cylindrical collar which is distinctly narrower than the lorica. The genus attains its maximum frequency around the Tithonian/Berriasian boundary. Uppermost Tithonian faunas may be almost exclusively constituted of *C. alpina* Lorenz.

(5) *Calpionellopsis*; type species: *C. oblonga* (Cadisch) (Fig. 1, *7, 8*; Fig. 8, *23, 24*; Fig. 9; Fig. 10L—O)
Range: Middle to Upper Berriasian, locally until Lower Valanginian.
Length: 90—120 μm. Width 40—60 μm.

Distinctive characters: extinction of the collar under crossed nicols and its morphology. Morphologically the collar is very inconspicuous, it forms a simple ring surrounding the oral opening which appears in thin section as a straight prolongation of the lateral walls (Fig. 9). Under crossed nicols, extinction appears when the lateral walls are in a 45° position. In the southern Mediterranean Province, *C. oblonga* (Cadisch) is sometimes a predominant element of Upper Berriasian calpionellid faunas.

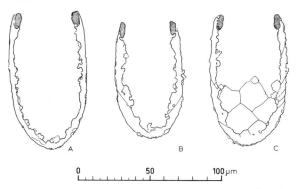

Fig. 9. Longitudinal sections of *Calpionellopsis simplex* (Colom), showing morphological details of collar (stippled) which is distinguished from the lateral walls mainly by its angle of extinction between crossed nicols. Note also recrystallization of loricae. (After Remane, 1965.)

(6) *Calpionellites*; type species: *C. darderi* (Colom) (Fig. 1, *5*; Fig. 8, *20, 21*; Fig. 10P)
Range: Lower Valanginian.
Length: 80—140 μm. Width: 65—90 μm.

Distinctive characters: with respect to the terminal portion of the lateral walls, the main part of the collar is parallel internally. In thin section the whole appears as an inward directed terminal bifurcation of the lateral walls. The extinction is the same as in *Calpionellopsis*. In *C. darderi* (Colom), the collar forms a simple conical ring, but in two recently discovered species (*C. coronata* Trejo and *C. caravacaensis* Allemann), it continues beyond the rim of the oral opening with a crescentic outward deflection. *C. darderi* (Colom), although not very frequent, is a characteristic element of Lower Valanginian calpionellid faunas.

CALPIONELLID BIOSTRATIGRAPHY AND ZONATION

Since Colom (1948) first published a range chart of eighteen calpionellid species, zonations have been suggested for many regions. Much of the stratigraphic data, however, were of little utility due to taxonomic ambiguities. The consensus of the calpionellid workers at the Second Planktonic Conference in Rome in 1970 was the first step towards the establishment of zonation schemes of wider applicability (e.g., see Allemann and others, 1971).

Our present knowledge of calpionellid biostratigraphy and correlation with ammonite or nannoplankton zones is summarized in Table I and some of the biostratigraphically important species are figured in Fig. 10.

Strata with *Chitinoidella* (age: middle Tithonian to early late Tithonian) are known to precede the *Crassicollaria* Zone from Cuba to the Carpathians. A *Chitinoidella* Zone is, therefore, in general use, but its lower boundary has never been formally defined and the stratigraphic distribution of different species is imperfectly known.

The *Crassicollaria* Zone covers the lower part of Upper Tithonian. Its lower boundary is defined by the first occurrence of a small variety of *Tintinnopsella carpathica* (Murgeanu and Filipescu). The characteristic element is the genus *Crassicollaria* which rapidly becomes dominant. *Calpionella alpina* Lorenz may be frequent in the uppermost part but is never predominant.

The *Calpionella* Zone extends from uppermost Tithonian to Lower Berriasian. The lower boundary is defined by a sudden increase in the abundance of *Calpionella alpina* Lorenz, corresponding to a change from large, somewhat elongated forms to a smaller spherical variety (see Fig. 8; Fig. 10G, H). *C. alpina* is largely dominant in the lower part of the zone, where the diversity of calpionellid faunas is greatly reduced. *Calpionella elliptica* Cadisch appears only in the upper half of the zone, just above the base of the Berriasian. In the uppermost part, *Tintinnopsella carpathica* may be the most abundant species.

The *Calpinellopsis* Zone ranges from Middle Berriasian to basal Valanginian. The lower boundary is defined by the first occur-

TABLE I

Correlation between calpionellid, ammonite, and calcareous nannoplankton zones

Periods/stages				Ammonite zones	Calpionellid zones	W.A.	Nannoplankton zones
LOWER CRETACEOUS	Valanginian	Upper		verrucosum	Calpionellites	E	Calcicalathina oblongata
		Lower		campylotoxus			
				roubaudi			
				pertransiens	Calpionellopsis	D	Cretarhabdus crenulatus
	Berria-sian			boissieri			
				occitanica	Calpionella	C	Nannoconus colomi
				grandis		B	— ? —
UPPER JURASSIC	Tithonian	Upper		jacobi			
				"Durangites"	Crassicollaria	A	Conusphaera mexicana
				microcantha			
		Lower/Middle		ponti	Chitinoidella	Chit.	
				fallauxi	— ? —	— ? —	

W.A.: calpionellid zones of the Western Alps, after Remane. Ammonite zones are at present mostly tentative: after Thieuloy (1974) for Valanginian; Le Hégarat (1971) for Berriasian; and Enay and Geyssant (1975) for Tithonian. Nannoplankton zones after Thierstein (1975). For faunal associations of calpionellid zones see Fig. 8.

Fig. 10. Important calpionellid species, in thin-section. ca. × 420.
A. *Chitinoidella boneti* Doben. Basal to middle Upper Tithonian. B. *Praetintinnopsella andrusovi* Borza. Wall structure predominantly hyaline, whereas the collar is still entirely microgranular. Basal Upper Tithonian. C. *Crassicollaria intermedia* (Durand Delga). Lower part of Upper Tithonian. D. *Crassicollaria brevis* Remane. Lower part of Upper Tithonian. E. *Crassicollaria massutiniana* (Colom). Lower part of Upper Tithonian. F. *Crassicollaria parvula* Remane. Upper Tithonian—Lower Berriasian. G. *Calpionella alpina* Lorenz. Upper Tithonian—Berriasian. Large form typical for the lower part of the Upper Tithonian. H. *Calpionella alpina* Lorenz. Upper Tithonian—Berriasian. Smaller, spherical variety typical for the Jurassic—Cretaceous boundary beds. I. *Calpionella elliptica* Cadisch. Lower Berriasian. J. *Tintinnopsella carpathica* (Murgeanu and Filipescu). Upper Tithonian—Lower Valanginian. Large form, typical of Cretaceous (exceptional section showing the caudal appendage along its whole length). K. *Tintinnopsella carpathica* (Murgeanu and Filipescu). Upper Tithonian—Lower Valanginian. L. *Calpionellopsis simplex* (Colom). Middle—Upper Berriasian. M. *Calpionellopsis simplex* (Colom). Middle—Upper Berriasian. N. *Calpionellopsis oblonga* (Cadisch). Upper Berriasian. Axial section with badly preserved collar. O. *Calpionellopsis oblonga* (Cadisch). Upper Berriasian. Oblique section with well-preserved collar. P. *Calpionellites darderi* (Colom). Lower Valanginian.

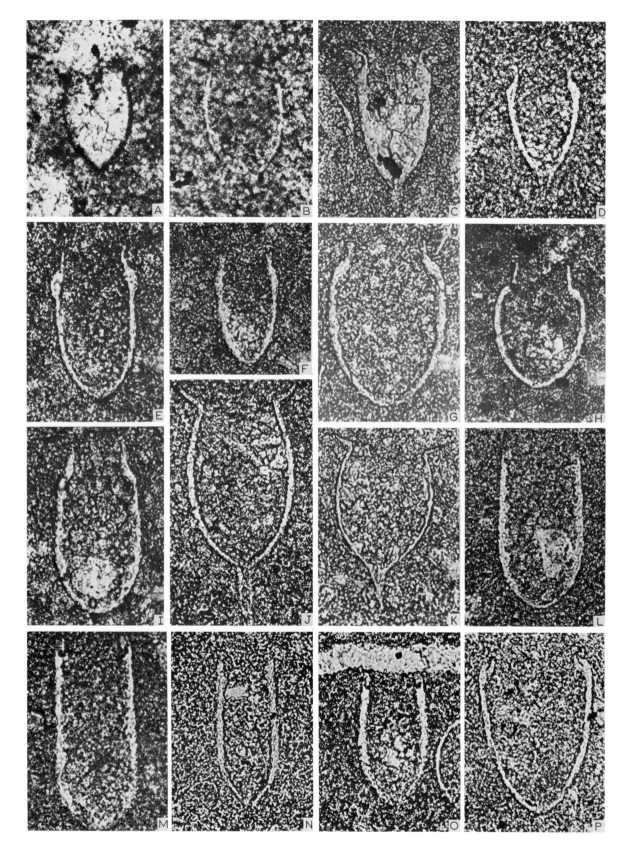

rence of *Calpionellopsis simplex* (Colom) which never becomes abundant. *T. carpathica* is always frequent and often predominant. *Calpionellopsis oblonga* (Cadisch) is frequent in the middle to upper part of the zone, but in southeastern France the genus *Calpionellopsis* does not range to the upper boundary of the zone.

The *Calpionellites* Zone covers the Lower, perhaps also part of the Upper Valanginian. The lower boundary is defined by the first occurrence of *Calpionellites darderi* (Colom), the upper boundary by the extinction of Calpionellidae. *T. carpathica* is predominant through the entire zone.

More detail on faunal associations can be obtained from Fig. 8 which also shows that finer subdivisions are possible on a regional scale. One great advantage of calpionellid stratigraphy is that in most pelagic limestones, sufficient material is available to determine statistically significant first occurrences. Moreover, three of the five zonal boundaries are based on phylogenetic changes which then cannot be confused with ecologic events.

SUGGESTIONS FOR FURTHER READING

Borza, K., 1969. *Die Mikrofazies und Mikrofossilien des Oberjuras und der Unterkreide der Klippenzone der Westkarpaten*. Slovak Academy of Science, Bratislava, 301 pp. [Calpionellidae are dealt with in great detail and descriptions of all stratigraphically important species are given with many illustrations.]

Colom, G., 1948. Fossil tintinnids: loricated infusoria of the order of the Oligotricha. *J. Paleontol.*, 22: 233—266. [This is a classic monograph of the Calpionellidae, in which all genera and species known at that time are described and figured. Ecology, systematics, and history of study are discussed in great detail.]

Remane, J., 1971. Les Calpionelles, Protozoaires planctoniques des mers mésogéennes de l'époque secondaire. *Ann. Guébhard*, 47: 4—25 (Neuchâtel. [Arguments against including calpionellids in Tintinnida and critical evaluation of forms erroneously considered as Calpionellidae are given. Phylogeny, stratigraphy and the problem of definition of species in thin sections are also discussed.]

Tappan, H. and Loeblich Jr., A.R., 1968. Lorica composition

of modern and fossil Tintinnida (ciliate Protozoa): systematics, geologic distribution, and some new Tertiary taxa. *J. Paleontol.*, 42: 1378—1394. [Reevaluation of arguments in favor of including calpionellids in the order Tintinnida, with a thorough discussion of the biology of living tintinnids. Calpionellids and also most non-calpionellid genera interpreted as fossil tintinnids are discussed, but only Tertiary forms are described at specific level.]

CITED REFERENCES[1]

Allemann, F., Catalano, R., Farès, F. and Remane, J., 1971. Standard calpionellid zonation (Upper Tithonian—Valanginian) of the western Mediterranean Province. *Proc. Second Planktonic Conf., Roma 1970*, II: 1337—1340 (Ediz. Tecnoscienza, Rome).

Aubry, M.-P., Bignot, G., Bismuth, H. and Remane, 1976. Premiers résultats de l'observation au M.E.B. de la lorica des Calpionelles et de quelques microfossiles qui leurs sont associés. *Rev. Micropaléontol.*, 18: 127—133.

Bernier, P., 1974. *Campbelliella striata* (Carozzi): Algue Dasylcladacée? Une nouvelle interprétation de l'"Organisme C" Favre et Richard, 1927. *Geobios*, 7: 155—175.

Chennaux, G., 1967. Tintinnoïdes et microorganismes incertae sedis du Siluro-Devonien saharien. *Publ. Serv. Géol. Algérie, N.S., Bull.* No. 35: 93—99.

Enay, R. and Geyssant, J., 1975. Faunes d'ammonites du Tithonique des chaînes bétiques (Espagne méridionale). *Colloque Limite Jurassique—Crétacé, Lyon—Neuchâtel 1973. Mém. Bur. Rech. Géol. Min.*, 86: 39—55.

Le Hégarat, G., 1971. Le Berriasian du Sud-Est de la France. *Doc. Lab. Géol. Fac. Sci. Lyon*, No. 43: 576 pp.

Radoičić, R., 1969. Aberantna grana fossilnih Tintinnina (podred Tintinnina). (La branche aberrante des Tintinnines fossiles (sous-ordre Tintinnina.) *Paleontol. Jugosl.*, 9: 71 pp.

Remane, J., 1969. Nouvelles données sur la position taxonomique des Calpionelliidae Bonet (1956) et leurs rapports avec les Tintinnina actuels et les autres groupes de "Tintinnoïdiens" fossiles. *Proc. First Planktonic Conf., Genève 1967*, II: 574—587 (Brill, Leiden).

Thierstein, H.R., 1975. Calcareous nannoplankton biostratigraphy at the Jurassic—Cretaceous boundary. *Colloque Limite Jurassique—Crétacé, Lyon—Neuchâtel 1973. Mém. Bur. Rech. Géol. Min.*, No. 86: 84—94.

Thieuloy, J.P., 1974. The occurrence and distribution of boreal ammonites from the Neocomian of Southeast France (Tethyan Province). In: R. Casey and P.F. Rawson (Editors), *The Boreal Lower Cretaceous. Geol. J., Spec. Iss.*, No. 5: 289—302.

Visscher, H., 1970. On the occurrence of chitinoid loricas of Tintinnida in an early Triassic palynological assemblage from Kingscourt, Ireland. *Geol. Surv. Irel., Bull.*, 1: 61—64.

---

[1] For further references see also Tappan and Loeblich (1968).

7

CALCAREOUS ALGAE

JOHN L. WRAY

INTRODUCTION

If we examine the sand-sized fraction of carbonate sediments accumulating today in tropical and subtropical marine shelf environments, we often find that skeletal remains of calcium-carbonate-depositing benthic red and green algae make up a high proportion of these deposits (Fig. 1). And so it is that similar kinds of ancient calcareous algae, in addition to laminated structures caused by algae, are important microfossils in marine and nonmarine carbonate rocks of nearly all ages, and include some of the oldest fossils known. Calcareous algae are important in micropaleontology as records of ancient life, and they can be used in the interpretation of paleoenvironments and for age determinations of strata. Algae have been quantitatively significant producers of carbonate sediment and influential in sedimentological processes, such as the construction of reef frameworks and in the trapping and binding of fine-grained sedimentary particles.

Algae represent a large and diversified assemblage of aquatic photosynthetic plants, varying from minute plankton, a few microns in size, to huge marine benthic plants, tens of meters in length. The calcareous algae comprise an artificial group that cuts across both taxonomic and disciplinary boundaries. By that we mean algae with calcified hard parts are found among several major taxa, along with noncalcified species, and in geology calcareous algae are treated by paleontologists as biological entities and considered by petrographers as rock constituents and sedimentary structures.

Investigations of calcareous algae traditionally have included only the benthic forms, and have excluded planktonic calcareous algae, the coccolithophores. This division remains today because of radically different analytical procedures, as well as interpretative ends. Whereas the purpose of most studies of benthic calcareous algae are in connection with lithological and paleoenvironmental interpretations of carbonate rock facies, the principal application of coccolithophores (see Chapter 3 on nannoplankton) is in biostratigraphic zonation, notwithstanding the volumetrically significant "rock-building" contribution of coccoliths to Cretaceous and Cenozoic chalks.

HISTORY OF STUDY

Recent calcareous algae have been studied systematically since the 1700's, when they were believed to belong to the animal kingdom, but relatively few fossil forms were

Fig. 1. Recent carbonate sand, south Florida reef tract. Principal constituents: calcareous green alga *Halimeda* (*H*), coralline red algae (*C*), foraminifera, and molluscan fragments. Transmitted light, thin section of plastic impregnated sediment.

described prior to the late 1800's. Between 1900 and 1925 a moderate amount of interest was generated in the subject and several published works from this period serve as a sound basis for our present knowledge (e.g., Lemoine, 1911; Pia, 1920). Unfortunately, fossil algae have frequently been misidentified and misinterpreted throughout the history of their study, partly because of a general lack of understanding of analogous living forms. All too often this group has served as a "wastebasket" for various unidentifiable biotic and nonbiotic constituents "presumed" to be algae, which has further complicated their taxonomy.

The tremendous growth of information during the past twenty years in the field of carbonate sedimentology, together with an understanding of the role algae have played in the development of these sediments, have helped clarify both the systematics and geological interpretations of calcareous algae, and thus accentuated the usefulness of this group of microfossils. Johnson (1961) presented the most comprehensive English-language treatment of the taxonomy of fossil calcareous algae. He also prepared numerous catalogs and review articles on the subject, in addition to three bibliographies (Johnson, 1943, 1957, 1967). Other relatively modern general works on fossil calcareous algae include Maslov (1956), Maslov and others (1963), and Němejc (1959). Ginsburg and others (1971) recently summarized the geological aspects of calcareous algae.

METHODS

Fossil calcareous algae are studied chiefly in thin sections of lithified rocks with the light microscope at magnifications of about 10 to 150 times (Kummel and Raup, 1965). Microscopic examination in transmitted light reveals both external shape and internal skeletal arrangements. Most benthic calcareous algae are macroscopic plants ranging from several millimeters to tens of centimeters in overall size, yet they are usually treated as microfossils. This is because they disaggregate into small segments on the death of the plant, break up into fine particles in the depositional environment, or require the microscope to

discern small-scale morphological features. Calcified reproductive bodies of the non-marine charophytes (see Fig. 20) are generally separated from the rock matrix by washing and studied as free forms in reflected light.

The scanning electron microscope (SEM) has been used to examine surface features and internal structures of Recent calcareous red and green algae. This technique provides insight into minute structural elements of some skeletal calcareous algae (Fig. 2), but has not yet contributed significantly to a better understanding of their taxonomy. The SEM would seem to be of limited value in studying fossil calcareous algae in thoroughly indurated sediments; also, diagenetic alteration of the carbonate skeletons can obliterate details of some features, especially in those forms originally composed of aragonite.

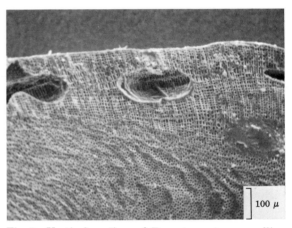

Fig. 2. Vertical section of Recent crustose coralline alga *Lithophyllum* showing reproductive organs (conceptacles) embedded in calcified cellular tissue of perithallium; top of specimen is growth surface. Scanning electron micrograph.

BIOLOGY OF CALCAREOUS ALGAE

General aspects

Algae are aquatic, autotrophic, nonvascular plants. That is to say, they live in water or moist environments, manufacture their own food, and do not have vascular tissues characteristic of higher plants. The plant body of an alga, called a thallus, is not differentiated into true roots, stems, and leaves; in addition, nearly all cells within a plant can carry on photosynthesis. Like other autotrophic plants, algae contain chlorophyll *a*, require oxygen for respiration, and produce oxygen during photosynthesis. Beyond these common attributes the algae constitute a heterogeneous group with extreme variations in size, morphology, cellular organization, biochemistry, and reproduction. In fact, the algae display a far greater diversity of form and metabolism than exists among the higher plants. The algae include both of the fundamentally different patterns of cellular organization, the primitive **procaryotes** and the specialized **eucaryotes**. Blue-green algae are procaryotic organisms (as are bacteria and viruses) in which the DNA is not separated from the cytoplasm by a membrane. In contrast, all other algae (and higher forms of life) have eucaryotic cells characterized by chromosomal nuclei enclosed in a membrane.

Recall then, that benthic calcareous algae cut completely across taxonomic boundaries and constitute a highly artificial group — yet thoroughly practical association — based solely on the presence of biochemically precipitated or mechanically accumulated deposits of calcium carbonate. Thus, it is virtually impossible to summarize all aspects of their biology for purposes of this discussion. It is more important to be aware of the wide range of biological specialization represented by the calcareous algae, and that each group must be considered separately to understand their skeletal morphology and criteria used in their classification.

Calcification

It is estimated that between 5 and 10% of the approximately 10,000 living species of marine benthic algae are calcified to any extent, and these belong mainly to a few families of red and green algae (Dawson, 1966). Most nonmarine calcareous algae belong to the Charophyta, although a few nonmarine calcified blue-green algae (Cyanophyta) are known.

The mechanisms for calcium-carbonate precipitation, and the function of skeletal carbonate in algae, are incompletely understood, partly because only a few physiological and biochemical studies of these processes

have been undertaken. Lewin (1962) and Arnott and Pautard (1970) have summarized the state of knowledge on the subject. Calcium-carbonate precipitation in some algae is a cell surface phenomenon resulting from carbon-dioxide extraction from water during photosynthesis. Other kinds of calcification are the result of intracellular metabolic processes and the selective absorption or assimilation of specific carbonate salts.

The location of skeletal calcium-carbonate deposits in algae varies, and this has an influence on the quality of morphologic features preserved in the fossil record. Skeletal calcium carbonate may occur: (1) within the cell; (2) associated directly with the cell wall; (3) outside the cell; and (4) as surficial deposits on the plant. Skeletal carbonates in living calcareous algae are either calcite or aragonite, never mixtures of the two. All representatives of the Corallinaceae (red algae) deposit calcium carbonate in the form of calcite, but with varying amounts of magnesium in the crystal lattice, whereas calcified members of the families Dasycladaceae and Codiaceae (green algae) deposit aragonite. Calcified reproductive organs of the Charophyta are calcite. The skeletal mineralogy is a major factor in the fossil preservation of morphologic features, because different mineral compositions show varying susceptibility to dissolution and alteration during diagenesis.

Classification

Current usage defines calcareous algae as: (1) those kinds of algae that have the ability to precipitate calcium carbonate within, between, and upon their tissues; or (2) distinctively laminated calcareous structures produced by the mechanical accumulation of fine carbonate sediment on organic films generated by colonies of noncalcareous filamentous and coccoid algae. Thus, in dealing with fossil calcareous algae we can distinguish two major categories: (1) preserved skeletal remains representing direct or indirect evidence of algal tissues; and (2) nonskeletal biosedimentary structures, generally called stromatolites, caused by algae.

Most skeletal calcareous algae include microfossils that can be treated adequately by existing biological classifications. Phyla of living algae are classified on the basis of pigments, nature of food reserves, chloroplast ultrastructure, and kind of flagellation. Many algae show obvious color differences which provide an approximate guide to a primary classification. Details of vegetative structures and reproductive processes are essential for the classification of living algae into classes, orders, families, genera, and species. The presence of skeletal carbonate *per se* is not a classification criterion. Fossil skeletal calcareous algae are classified according to morphologic characteristics utilized in the classifica-

TABLE I

A classification of fossil skeletal calcareous algae

	Approximate number principal genera	Some common genera
Phylum CYANOPHYTA (blue-green algae) Class CYANOPHYCEAE	5*	*Girvanella*

Phylum RHODOPHYTA (red algae)
    Class RHODOPHYCEAE
        Order CRYPTONEMIALES		

Family SOLENOPORACEAE	2	*Parachaetetes* *Solenopora*

Family CORALLINACEAE
        Subfamily MELOBESIEAE
            (crustose corallines)
	15	*Archaeolithothamnium*
		Lithothamnium
		Lithophyllum
		Lithoporella
		Melobesia

Subfamily CORALLINEAE (articulated corallines)	5	*Corallina* *Jania*
Family UNCERTAIN (ancestral corallines)	5	*Archaeolithophyllum* *Cuneiphycus*

Phylum CHLOROPHYTA (green algae)
    Class CHLOROPHYCEAE
        Order SIPHONALES		

Family CODIACEAE	15	*Ortonella* *Penicillus* *Halimeda*
Order DASYCLADALES		
Family DASYCLADACEAE	30	*Acetabularia* *Cymopolia* *Dasycladus* *Mizzia* *Neomeris*

Phylum CHAROPHYTA
    Class CHAROPHYCEAE
        Order CHARALES
	15	*Chara*
		Trochiliscus

*Some genera classified as blue-green algae may belong to other algal phyla or are organisms of uncertain affinities.

tion schemes of analogous living forms. A classification of fossil skeletal calcareous algae and some common genera are given in Table I.

Morphological details of vegetative structures and reproductive organs are absent or poorly preserved in many fossil calcareous algae, and this accounts for some of the uncertainties in their classification. Needless to say, fossil forms with morphologically similar extant representatives have a better chance of being properly classified than ancient algae that lack modern descendants or morphological counterparts. Similarly, skeletal calcareous structures representing only a portion of the plant may be difficult to identify, whereas classification would be relatively simple if associated noncalcareous vegetative structures were also preserved.

Knowledge of fossil calcareous algae, although admittedly incomplete and fragmentary, is still being accumulated and major discoveries have been made within recent years. Skeletal calcareous algae provide the best preserved record of ancient benthic marine algae, yet we know they constituted only a small percentage of the total flora at any one time, and not a very representative one. While some ancient algae have living representatives, many others do not, and there are major gaps in the record and in our understanding of the principal courses of algal evolution.

Nonskeletal calcareous algae or stromatolites (see Fig. 22) have been interpreted in various ways and classified on both biological and nonbiological bases. The practice of applying a binary nomenclature to fossil stromatolites was used extensively by early workers who interpreted ancient stromatolites to be direct remains of organisms. The first paleontological description dates from 1883 when James Hall described *Cryptozoon proliferum* from the Upper Cambrian of New York. Most workers now agree that stromatolites cannot be classified on a biological basis, but are suitable to a grouping based on geometry, including both internal and external physical attributes (Hofmann, 1969). However, stromatolite classifications are currently a matter of debate, and no single scheme has been generally accepted.

SKELETAL CALCAREOUS ALGAE

Blue-green algae (Cyanophyta)

Blue-green algae are some of the most common marine and nonmarine algae today and have been since Precambrian time, yet only a few taxa have produced biochemically precipitated skeletal remains. All plants are individually microscopic, consisting of very simple **unicells** or **filaments** that occur within a **mucilaginous sheath**. Many of these forms have been influential in trapping and binding fine-grained sediment to build laminated sedimentary structures or stromatolites. Recent nonmarine skeletal blue-greens are known, but apparently no living species secrete skeletal calcium carbonate in normal marine waters. Most fossil skeletal blue-green algae comprise polygenetic groups of filamentous forms that are difficult to differentiate into real taxa because of their exceedingly simple skeletal morphology. The common fossil genus *Girvanella* (Fig. 3) consists of unsegmented tubes 10 to 50 μm in diameter and ranges from early Cambrian to late Cretaceous. Filamentous skeletal blue-green algae encrust other objects and have developed nodular and columnar growth forms.

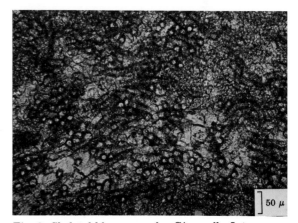

50 μ

Fig. 3. Skeletal blue-green alga *Girvanella*. Interwoven network of unsegmented tubular filaments. Upper Cambrian, Texas.

Red algae (Rhodophyta)

Most calcareous red algae are characterized by distinctive internal cellular tissue. Substantial amounts of calcium carbonate precipitated within and between cell walls preserve details of the tissue and reproductive organs (see Fig. 2), although the degree of calcification varies between taxa. Three principal groups of calcareous red algae are recognized in the fossil record: (1) Corallinaceae; (2) Solenoporaceae; and (3) an informal group called "ancestral corallines".

Corallinaceae

The family Corallinaceae appeared in the Jurassic and includes most of the present-day carbonate secreting red algae. The cellular tissue in representatives of the Corallinaceae, or corallines, usually can be differentiated into two kinds of cellular arrangements, a basal or central (depending on growth form) **hypothallium**, and an upper or marginal **perithallium** (Fig. 4). Spore-producing bodies occur in the **perithallium**, either as isolated sporangia (Fig. 5) or collected into a cavity called a conceptacle (see Figs. 2, 4).

Historically, two groups of coralline algae have been distinguished on the basis of

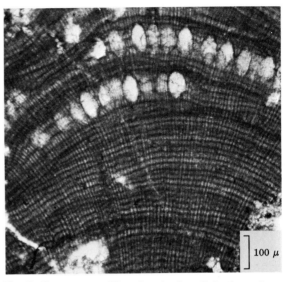

Fig. 5. Crustose coralline alga *Archaeolithothamnium*. Section of perithallium containing rows of individual sporangia. Eocene, Guam.

growth form: (1) those characterized by an encrusting or **crustose** habit; and (2) the erect **articulated-segmented** forms. Recently, the family Corallinaceae was subdivided into seven subfamilies based on morphological criteria recognized in living forms (Adey and Johansen, 1972; Adey and MacIntyre, 1973). However, it is not certain whether all of these subgroups can be distinguished in fossil material. Thus, the more traditional breakdown into two subfamilies, the Melobesieae (crustose corallines) and the Corallineae (articulated corallines), on the basis of gross external morphology remains a practical classification scheme in dealing with imper-

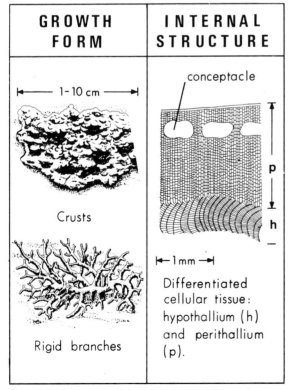

Fig. 4. Typical growth forms and internal morphology of crustose coralline algae (subfamily Melobesieae).

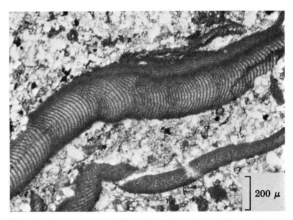

Fig. 6. Crustose coralline alga showing differentiation of cellular tissue. Eocene, Guam.

fectly preserved fossil material. Taxa are distinguished chiefly on the basis of internal morphology, namely, the size and arrangement of cellular tissue (Fig. 6); character of cell walls, especially the presence or absence of secondary pit-connections between cells; and the kind of reproductive organs.

The crustose coralline algae have developed a variety of growth forms, including laminated crusts, nodules, highly irregular masses, and rigid branching habits (see Figs. 4, 6, 7). They range in size from a few millimeters to several centimeters. Most are firmly attached to the substrate, although a few forms develop free on the sea bottom. The articulated coralline algae are erect, segmented plants about 5 to 20 cm high (Figs. 8, 9). Crustose coralline algae are important in binding and cementing coral and other skeletal material together to construct modern reefs, but they also disintegrate into clastic sedimentary particles. Delicate branching forms and articulated-segmented types generally disaggregate into sand-sized fragments in the depositional environment.

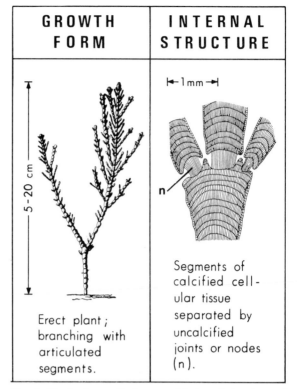

Fig. 8. Typical growth form and internal structure of articulated coralline red algae (subfamily Corallineae).

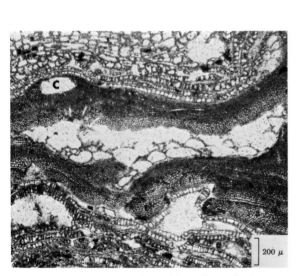

Fig. 7. Multiple crusts of coralline algae. At bottom of photograph: *Lithoporella* thalli composed of single layers of large cells. Middle part of photograph: *Lithothamnium* showing differentiation of tissue into a basal portion with curving rows of cells (hypothallium) and an upper portion with a regular network of cells (perithallium) containing a conceptacle (*C*). Top part of photograph: *Lithoporella* encrusting *Lithothamnium* subsequently encrusted by foraminifer *Gypsina*. Pleistocene, Gulf of Mexico.

Solenoporaceae

The Solenoporaceae comprise an extinct family of calcareous red algae ranging from the late Cambrian to the Paleocene. They occur mainly as encrusting, rounded, nodular masses a few millimeters to a few centimeters in size. Internally these algae have relatively simple cellular tissue that is not differentiated into more than one kind of cellular arrangement (Fig. 10). Reproductive organs are unknown in this extinct group; thus, the Solenoporaceae are subdivided into taxa on the basis of the size and shape of the cells. An important generic characteristic is the degree of development of horizontal layers or cross partitions within vertical threads of cells.

Ancestral corallines

Several genera of calcareous algae with internal structure similar to that of modern Corallinaceae appeared in the late Paleozoic, and were particularly common in the Pennsylvanian period. These algae, known informally as "ancestral corallines", display both

Fig. 9. Articulated-segmented coralline alga *Corallina* showing arcuate rows of cellular tissue (hypothallium) within segments. Miocene, Saipan.

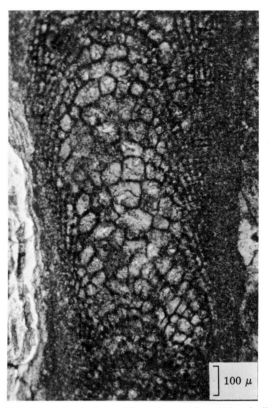

Fig. 11. Ancestral coralline alga *Archaeolithophyllum* showing arrangement of cells in hypothallium (central portion) and outer perithallium. Pennsylvanian, Missouri.

Fig. 10. Solenoporacean red alga *Parachaetetes*. Vertical section of nodular growth form showing cellular tissue. Upper Devonian, Western Australia.

encrusting and erect growth habits, and are volumetrically important contributors to limestone development. *Archaeolithophyllum* (Fig. 11), ranging from late Mississippian to middle Permian, is one of the more common and distinctive taxa belonging to this extinct group. This crustose alga, possessing **conceptacles** (cavity-containing sporangia) and differentiated cellular tissue, is remarkably similar to the modern coralline genus *Lithophyllum*.

Green algae (Chlorophyta)

Carbonate-secreting habits have been developed by two families of marine green algae, the Codiaceae and the Dasycladaceae. Skeletal carbonates in the green algae are exceedingly variable; they include rather thoroughly calcified **thalli**, fragile skeletal elements composed of aragonite needles, and surficial deposits of calcium carbonate representing molds of the thallus.

The microfossils of both codiaceans and dasycladaceans consist mainly of whole or broken segments, rather than entire plants. The shape of individual segments, in addition to the internal organization of filaments and pores, are the bases for classifying individual taxa. Erect, stalked codiaceans with bushy crowns, such as *Penicillus*, frequently disaggregate completely into micron-sized aragonite crystals which cannot be identified taxonomically. This is of interest sedimentologically because the process is considered to be the origin of much of the carbonate mud and silt accumulating in Recent tropical and subtropical shallow marine environments (Stockman and others, 1967). Although the amount of calcium carbonate produced per plant is small, they are so numerous and grow so rapidly that they produce volumetrically significant amounts of sediment. Similar kinds of calcareous green algae were probably the source of fine-grained sediment in ancient shallow-shelf carbonate environments. Both codiaceans and dasycladaceans have a long geologic record, extending from the Cambrian.

Codiaceae

Fossil calcareous Codiaceae exhibit two distinct growth habits: (1) crustose or nodular forms; and (2) erect plants, commonly consisting of segmented branches (Fig. 12). Nodular forms of the Codiaceae occur most often in the Paleozoic and genera are distinguished on the basis of the character of internal branching tubular filaments. *Ortonella* and *Garwoodia* are two common Paleozoic genera.

Most Mesozoic and Cenozoic calcareous codiaceans are erect plants (Fig. 13) several centimeters high, generally segmented and branching, and possess an internal structure composed of interwoven tubular filaments (Fig. 14). The shape and segmentation of the overall plant and the internal organization and branching of the filaments differentiate individual taxa. The extant genus *Halimeda*, which appeared first in the Cretaceous, is an important representative of the erect calcareous codiaceans. Skeletal carbonate in

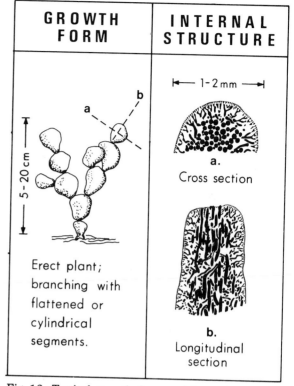

Fig. 12. Typical growth form and internal structure of erect calcareous Codiaceae (green algae).

Halimeda and related genera consists of minute elongate aragonite crystals. Calcification is more complete in older portions of the plant compared to younger parts, and outer regions of the thallus are more thoroughly calcified than the inner portion. These differences in degree of calcification are observed in both living and fossil specimens.

Dasycladaceae

Calcareous Dasycladaceae are erect, segmented, branching plants usually several centimeters high. Most species are characterized by a large central stem surrounded by tufts or whorls of smaller radiating branches; thus, dasycladacean skeletal remains are mainly molds of segments preserving arrangements of these features. Most fossils appear as hollow perforated cylinders or spheres (Figs. 15, 16, 17), while others are perforated discs or blade-like objects (Fig. 18). **Sporangia** (reproductive organs) are developed adjacent to the stem or branches and their outline may be preserved by calcification.

This group of calcareous green algae is

Fig. 13. Underwater photograph of living calcareous codiacean green algae, *Halimeda* and *Penicillus*. Individual plants are 10 to 15 cm high. Shallow lagoonal environment, Eleuthera Island, Bahamas.

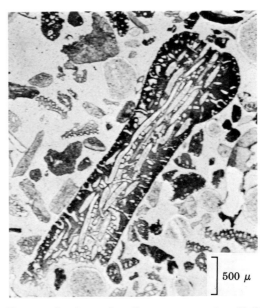

Fig. 14. Section of codiacean green alga *Halimeda* showing internal tubular filaments. Recent, Florida.

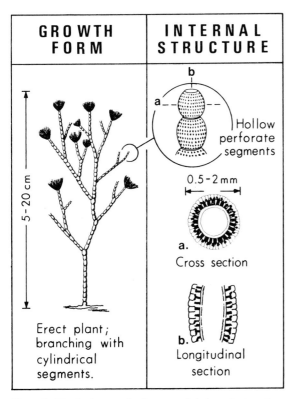

Fig. 15. Typical growth form and internal structure of calcareous Dasycladaceae (green algae).

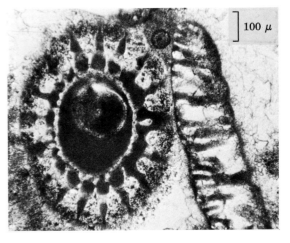

Fig. 16. Transverse and longitudinal sections of dasycladacean alga *Cymopolia*. Paleogene, North Africa.

classified on the basis of several morphological features: the overall shape of the plant, kind of segmentation, the arrangement and number of branches, and the shape and location of sporangia. Because of the large number of diagnostic criteria used in taxonomic differentiations, numerous genera and species of dasycladaceans have been recognized in the fossil record.

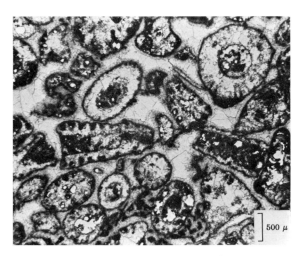

Fig. 17. Limestone composed entirely of dasycladacean segments. Paleogene, Guatemala.

Fig. 18. Fragment of dasycladacean alga *Acicularia*. Paleogene, North Africa.

Charophyta

The Charophyta are physiologically and biochemically similar to the Chlorophyta, but differ markedly in their morphology, and therefore are usually classified as a separate phylum. The charophyte plant attains a size of several centimeters to as much as a meter, and as a whole resembles the higher plants in its general habit and external differentiation of parts (Fig. 19). Most charophytes calcify only the reproductive organs, principally the **gyrogonite**, sometimes called **oogonium**, which is borne on a short branch or stem. Characteristically, oogonia are elliptical in shape, usually 0.5 to 1 mm in dimension, and are composed of spirally arranged tubes which appear as ridges on the exterior (Fig. 20). Modern charophytes occupy fresh- and brackish-water environments, in contrast to the marine habitats of most other benthic calcareous algae. However, oogonia may be transported by streams and deposited in nearshore marine sediments. Fossil species (Fig. 21) are known from the late Silurian.

STROMATOLITES

Algal stromatolites are exceedingly variable in size, ranging from a few millimeters (forms that are the concern of the micropaleontol-

Fig. 19. Sketch of charophyte plant which commonly attains a size of a few tens of centimeters.

Fig. 20. Calcified reproductive body (gyrogonite of) fresh-water Charophyta.

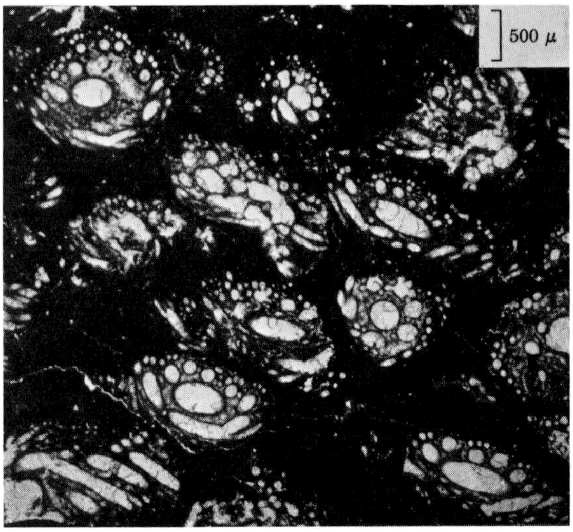

Fig. 21. Sections of charophyte gyrogonites (oogonia) in a fresh-water limestone. Jurassic, Colorado.

ogist) to large-scale biosedimentary structures. Microscopic blue-green algae are the essential elements, because they generate organic films that mechanically agglutinate fine-grained carbonate sediments and thereby create distinctive algal-laminated calcareous structures; however, the algae themselves normally are not calcified and the sedimentary record preserves only the laminations caused by the algae.

Stromatolites include a variety of external shapes, ranging from forms with flat-lying laminations to domical and columnar structures, some with branching or digitate habits. Commonly, forms with flat-lying laminae are called "algal-laminated sediments", while the

Fig. 22. Sketch of growing columnar stromatolites with appreciable vertical relief. Commonly these biosedimentary structures range in size from a few centimeters to several tens of centimeters. Forms similar to these are developing now in Shark Bay, Western Australia.

term stromatolite is applied to forms with pronounced vertical relief (Fig. 22). **Oncolites** are concentrically layered, algal-laminated bodies. Most stromatolites are measured in centimeters, with a single lamina about 0.5 to 1 mm thick, while others are minute and occur within the dimensions of a standard thin section. Stromatolites and oncolites may contain identifiable skeletal calcareous algae, such as *Girvanella*.

Investigations of Recent algal stromatolites during the past decade or so, such as those in Shark Bay, Western Australia (e.g., Logan, 1961; Davies, 1970; Logan and others, 1974) and elsewhere (e.g., Monty, 1967; Gebelein, 1969), have increased our understanding of the origin of these structures and shown the complex interrelationships of biological and physical environmental processes. Empirical classifications (Hofmann, 1969, 1973) based on geometric properties seem to be the most useful way of grouping these kinds of nonskeletal calcareous algae.

ECOLOGY AND PALEOECOLOGY

Major groups of Recent calcareous algae have preferred habits and habitats, and these associations can be used to understand the distribution of similar kinds of algae in ancient environments. The principal groups of calcareous algae and their usual environments are:

Group	Environment
Blue-green algae	marine/nonmarine; muddy substrate
Red algae	marine; reef and rocky substrates
Green algae	marine; sandy and muddy substrates
Charophytes	fresh and brackish water

The distribution pattern of modern calcareous algae provides a useful perspective in the interpretation of algal-rich biofacies in the Mesozoic and Paleozoic. Fig. 23 illustrates the distribution of Cenozoic calcareous algae in a model of a marine carbonate shelf and adjacent environments. Charophytes occur in nonmarine waters, and their calcified remains may make a high proportion of some lake sediments. Blue-green algae have produced stromatolites in a variety of environments, but most often in shallow marginal waters of

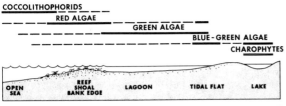

Fig. 23. Generalized environmental distribution of Cenozoic calcareous algae.

Group	Province
Blue-green algae (including stromatolites)	tropical to polar
Crustose coralline algae	tropical to polar
Articulated coralline algae	tropical to cool-temperate
Codiacean green algae	tropical to subtropical
Dasycladacean green algae	tropical to warm-temperate

marine basins and saline lakes (for example, the Eocene Green River Formation: Surdam and Wray, 1976). Deep-water marine stromatolites are known and numerous freshwater forms have been described. Skeletal blue-green algae occur generally in shallow marine and nonmarine deposits.

Skeletal calcareous green algae, both codiaceans and dasycladaceans, although ranging to depths of several tens of meters, are most abundant in relatively shallow, protected lagoonal environments, where they contribute substantial amounts of calcareous material to the sediments. Crustose coralline red algae have a depth range from sea level down to at least 230 m, and are dominant constituents in reef and rocky-shore environments. Both the crustose and articulated coralline groups thrive in exposed, high-energy environments. Calcareous planktonic algae (coccolithophores) are significant sediment contributors in the deep ocean basins, but they also occur in continental shelf and near-shore deposits (see Chapter 3).

Biogeography

Numerous physical, chemical, and biological factors influence the distribution of individual genera and species of benthic calcareous algae. Important physical factors include light, temperature, substrate, and water movement. Because algae are photosynthetic plants, light intensity and its spectral composition (wave length) are significant factors in controlling their growth and depth distribution. Temperature is critical in the latitudinal distribution of major taxa and species. The biogeography of the major groups of marine benthic calcareous algae follows:

Depth and latitudinal distribution are broad at the phylum or division level (red algae, green algae, etc.); however, at lower taxonomic levels, especially those of genus and species, there is evidence that depth and biogeographic limits are relatively narrow. Unfortunately, the environmental distribution limits of most living calcareous algae have not been established. As a result, there is a limited amount of data with which to interpret the paleoenvironments of ancient algae (Riding, 1975).

Adey (1966, 1970) has provided some of the only quantitative ecological data on the distribution of living skeletal marine algae. In studies of boreal—subarctic crustose coralline species in the North Atlantic, Adey (1970) determined that the principal factors affecting their distribution are temperature, partly a function of latitude, and light, which is mainly a function of depth. Fig. 24 illustrates schematically the depth distribution of crustose coralline species in the northwestern North Atlantic (Adey, 1966).

Studies of the distribution of fossil calcar-

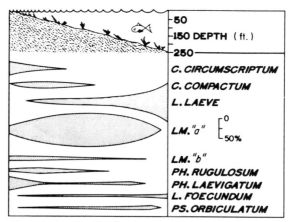

Fig. 24. Depth distribution and relative abundance of crustose coralline algae species in the northwestern North Atlantic (data from Adey, 1966).

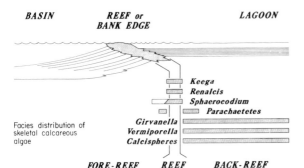

Facies distribution of skeletal calcareous algae

Fig. 25. Environmental distribution of principal taxa of skeletal calcareous algae in Upper Devonian reef complexes. *Keega* is a problematical ancestral coralline alga; *Parachaetetes* is a solenoporacean; *Renalcis*, *Sphaerocodium* and *Girvanella* are blue-green algae; *Vermiporella* is a dasycladacean; and calcispheres are believed to be reproductive bodies (gametangia) of unknown dasycladaceans. (From Wray, 1972.)

eous algae in Devonian carbonate reef complexes (Wray, 1972) indicate that individual taxa are restricted to particular depositional facies (Fig. 25). In this example, all of the algae are extinct and the relationship of most taxa to living groups is unclear; consequently, their paleoecology cannot be inferred from analogous living groups. Yet this empirical distribution pattern of Devonian calcareous algae within a facies complex does yield a comprehensible paleoenvironmental picture. Thus, calcareous algae can provide an important complement to other benthic organisms in determining biofacies in ancient carbonate shelf environments, despite the fact that the environmental limits of many fossil forms, at least at the lower taxonomic levels, have not been determined.

GEOLOGIC DISTRIBUTION

Major groups of benthic calcareous algae and many minor taxa have long geologic ranges. As a result, calcareous algae have limited value for age determinations and in biostratigraphy, although a few forms provide useful marker horizons in some parts of the section. Some algae, notably the primitive blue-greens, are characterized by extreme evolutionary conservatism, and living species are almost indistinguishable from species that lived millions of years ago. Also, in contrast to planktonic organisms, benthic calcareous

algae were often restricted to narrow paleoecologic niches, and are not useful for widespread biostratigraphic correlations. Wray (1971) outlined the geologic distribution of calcareous algae considered to be important in ancient and modern reef development.

Stromatolites range from the Archaean; the oldest ones known (2.6 billion years) occur in southern Rhodesia. The late Precambrian of the Soviet Union has been subdivided into four zones based on stromatolite assemblages, and this chronology has been applied in other regions (Cloud and Semikhatov, 1969); however, many discrepancies in ranges and methodology have been noted, and a worldwide Proterozoic stromatolite zonation is open to criticism. Stromatolites of various ages have been used in the physical correlation of beds within sedimentary basins. The apparent decline in abundance of stromatolites in the Phanerozoic has been explained by the expansion of grazing and burrowing animals that destroyed algal laminae (Garrett, 1970).

The time distribution, relative abundance, and suggested evolution of the principal groups of marine skeletal calcareous algae are illustrated in Fig. 26. The time ranges of important genera are summarized in Fig. 27. From these one can recognize major trends in the occurrence of calcareous algae in time, and distinguish assemblages characteristic of particular time intervals.

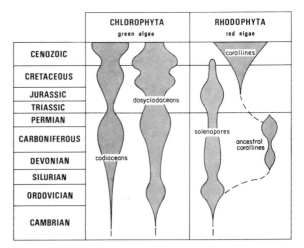

Fig. 26. Geologic distribution and inferred evolution of major groups of marine skeletal calcareous algae. Abundance and diversity is suggested by width of patterns.

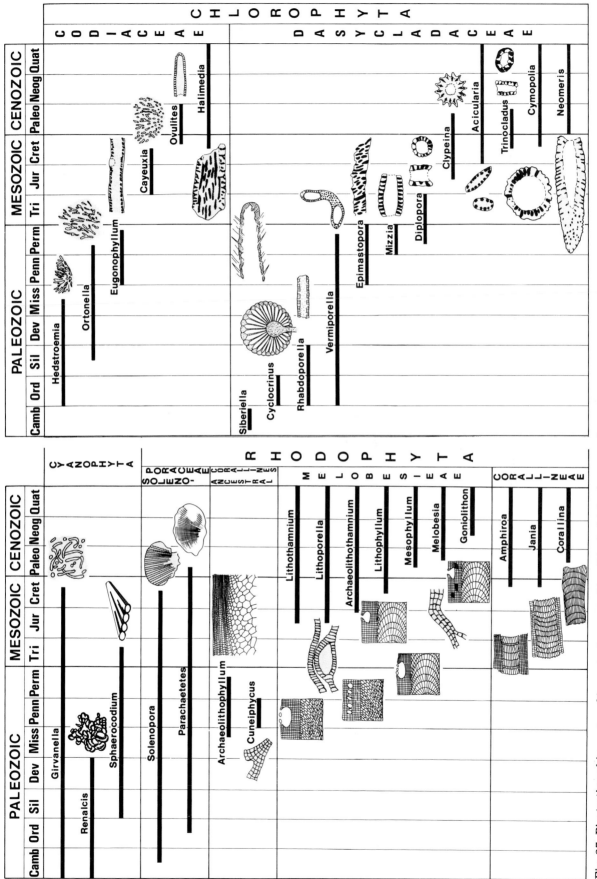

Fig. 27. Biostratigraphic ranges of important genera of benthic skeletal calcareous algae.

The early Paleozoic is characterized chiefly by taxa belonging to the Dasycladaceae and Solenoporaceae which were especially abundant and diverse during the Ordovician. Codiaceans became progressively more common in the late Paleozoic and reached a peak during the Mississippian and Pennsylvanian. Dasycladaceans show a second Paleozoic period of abundance in the Permian. Taxa assigned to the "ancestral corallines" range from the Devonian into the Permian and were a distinctive element in the Pennsylvanian. Skeletal blue-green algae and various problematical algae were common at various times throughout the Paleozoic.

Early Mesozoic rocks contain relatively few Codiaceae and Dasycladaceae; on the other hand, the Solenoporaceae are common and reached the height of their development in the Jurassic. Codiaceans and dasycladaceans again became abundant and widespread in the Cretaceous. Skeletal blue-greens are only occasional constituents in Mesozoic rocks and are rare in the Cenozoic.

The Corallinaceae evolved rapidly into many genera and species in the late Mesozoic and Cenozoic. The corallines comprise the most diverse, abundant, and widespread group of skeletal calcareous algae in modern seas. Solenoporaceae were locally important until they became extinct at the end of the Paleocene. Calcareous dasycladaceans appear to be more common in the early Cenozoic than the codiaceans, whereas the reciprocal seems to be the case in the late Cenozoic.

These occurrence patterns through time provide some understanding of the evolution of major groups of calcareous algae, but the evolutionary history of lesser taxa is largely speculative. The two families of calcareous green algae, Codiaceae and Dasycladaceae, presumably evolved from a common ancestor in early Cambrian or Precambrian time. The earliest known representatives of calcareous red algae, the Solenoporaceae, probably gave rise to the "ancestral corallines" during the middle Paleozoic. The ancestry of the Corallinaceae is not understood, but its members may have descended from the Solenoporaceae, the "ancestral corallines", or other ancient red algae.

SUGGESTIONS FOR FURTHER READING

Dawson, E.Y., 1966. *Marine Botany: An Introduction*. Holt, Rinehart and Winston, New York, N.Y., 371 pp. [Textbook overview of physiology, ecology and geographic distribution of marine plants; emphasis on benthic algae. Includes a brief chapter on calcareous algae.]

Fritch, F.E., 1935—1945. *The Structure and Reproduction of the Algae*. Cambridge University Press, London, Vol. 1 (1935), 791 pp.; Vol. 2 (1945), 939 pp. [Massive two-volume compilation of world's literature on morphology and reproduction of algae. Still a useful and inclusive reference work on aspects of these subjects; obviously deficient in modern advances in physiology and classification.]

Ginsburg, R., Rezak, R. and Wray, J.L., 1971. *Geology of Calcareous Algae: Notes for a Short Course*. University of Miami, School of Marine and Atmospheric Science, Miami, Fla., 64 pp. [Up-to-date summary (illustrated outline format) of various aspects of calcareous algae. Emphasis on their use in environmental interpretation of carbonate sediments.]

Johnson, J.H., 1961. *Limestone-Building Algae and Algal Limestones*. Colorado School of Mines, Golden, Colo., 297 pp. [Descriptions, illustrations, geologic ranges and geographic distribution of important genera of fossil and recent calcareous algae. Also chapters on classification, chemical composition and ecology.]

Walter, M.R. (Editor), 1976. *Stromatolites*. Elsevier, Amsterdam, 790 pp. [Compilation of 43 studies on all aspects of stromatolites; methodology, systematics, biology, biostratigraphy, and their use in paleoenvironmental and basin analysis.]

Wray, J.L., 1977. *Calcareous Algae*. Elsevier, Amsterdam, 185 pp. [Comprehensive illustrated text-reference covering major groups of marine and non-marine calcareous algae. Descriptions, classification, and stratigraphic and environmental distribution; also chapters on paleoecology, sediment-producing algae and algal facies in time.]

CITED REFERENCES

Adey, W.H., 1966. The distribution of saxicolous crustose corallines in the northwestern North Atlantic. *J. Phycol.*, 2: 49—54.

Adey, W.H., 1970. The effects of light and temperature on growth rates in boreal—subarctic crustose corallines. *J. Phycol.*, 6: 269—276.

Adey, W.H. and Johansen, H.W., 1972. Morphology and taxonomy of Corallinaceae with special reference to *Clathromorphum*, *Mesophyllum* and *Neopolyporolithon* gen. nov. (Rhodophyceae, Cryptonemiales). *Phycologia*, 11: 159—180.

Adey, W.H. and Macintyre, I.G., 1973. Crustose coralline algae: a re-evaluation in the geological sciences. *Geol. Soc. Am. Bull.*, 84: 883—904.

Arnott, H.J. and Pautard, F.G.E., 1970. Calcification in plants. In: H. Schrader (Editor), *Biological Calcification: Cellular and Molecular Aspects*. Appleton-Century-Crofts, New York, N.Y., pp. 375—446.

Cloud, P.E. and Semikhatov, M.A., 1969. Proterozoic stromatolite zonation; *Am. J. Sci.*, 267: 1017—1061.

Davies, G.R., 1970. Algal-laminated sediments, Western Australia. In: B.W. Logan, G.R. Davies, J.F. Read and D.E. Cebulski, *Carbonate Sedimentation and Environments, Shark Bay, Western Australia*. *Am. Assoc. Pet. Geol. Mem.*, 13: 169—205.

Garrett, P., 1970. Phanerozoic stromatolites: noncompetitive ecologic restriction by grazing and burrowing animals. *Science*, 169: 171—173.

Gebelein, C.D., 1969. Distribution, morphology, and accretion rate of Recent subtidal algal stromatolites, Bermuda. *J. Sediment. Petrol.*, 39: 49—69.

Hofmann, H.J., 1969. Attributes of stromatolites. *Can. Geol. Surv. Pap.* 69-39: 58 pp.

Hofmann, H.J., 1973. Stromatolites: characteristics and utility. *Earth-Sci. Rev.*, 9: 339—373.

Johnson, J.H., 1943. Geologic importance of calcareous algae with annotated bibliography. *Colo. School Mines Q.*, 38(1): 102 pp.

Johnson, J.H., 1957. Bibliography of fossil algae: 1942—1955. *Colo. School Mines Q.*, 52(2): 92 pp.

Johnson, J.H., 1967. Bibliography of fossil algae, algal limestones, and the geological work of algae, 1956—1965. *Colo. School Mines Q.*, 62(4): 148 pp.

Kummel, B. and Raup, D., 1965. *Handbook of Paleontological Techniques*. W.H. Freeman and Co., San Francisco, Calif., 852 pp.

Lemoine, M., 1911. Structure anatomiques des Melobesiées. Application à la classification. *Inst. Océanogr. Monaco Ann.*, 2(1): 215 pp.

Lewin, J.C., 1962. Calcification. In: R.A. Lewin (Editor), *Physiology and Biochemistry of Algae*. Academic Press, New York, N.Y., pp. 457—465.

Logan, B.W., 1961. *Cryptozoon* and associated stromatolites from the Recent, Shark Bay, Western Australia. *J. Geol.*, 69: 517—533.

Logan, B.W., Hoffman, P. and Gebelein, C.D., 1974. Algal mats, cryptalgal fabrics, and structures. In: B.W. Logan, J.F. Read, G.M. Hagan, P. Hoffman, R.G. Brown, P.J. Woods and C.D. Gebelein, *Evolution and Diagenesis of Quaternary Carbonate Sequences, Shark Bay, Western Australia. Am. Assoc. Pet. Geol. Mem.*, 22: 358 pp.

Maslov, V.P., 1956. Iskopaemye isvestkovye vodorosli S.S.S.R. (Fossil calcareous algae of the U.S.S.R.). *Akad. Nauk S.S.S.R., Tr. Inst. Geol. Nauk*, 160: 301 pp.

Maslov, V.P. et al., 1963. Vodorosli (Algae). In: Yu.A. Orlov, *Osnovy paleontologiǐ. Moskva Akad. Nauk S.S.S.R.*, 14: 19—312.

Monty, C.L.V., 1967. Distribution and structure of Recent stromatolitic algal mats, eastern Andros Island, Bahamas. *Soc. Geol. Belg. Ann.*, 90: 55—100.

Němejc, F., 1959. Paleobotanika. *Praha, Nakladalestvi Česk. Akad. Věd*, 1: 402 pp.

Pia, J., 1920. Die Siphoneae verticillatae vom Karbon bis zur Kreide. *Zool.-Bot. Ges. Wien Abh.*, 11(2).

Riding, R., 1975. *Girvanella* and other algae as depth indicators. *Lethaia*, 8: 173—179.

Stockman, K.W., Ginsburg, R.N. and Shinn, E.A., 1967. The production of lime mud by algae in south Florida. *J. Sediment. Petrol.*, 37: 633—648.

Surdam, R.C. and Wray, J.L., 1976. Lacustrine stromatolites, Eocene Green River Formation, Wyoming. In: M.R. Walter (Editor), *Stromatolites*. Elsevier, Amsterdam, pp. 535—541.

Wray, J.L., 1971. Algae in reefs through time. *North Am. Paleontol. Conv., Chicago, 1969, Proc.*, J: 1358—1373.

Wray, J.L., 1972. Environmental distribution of calcareous algae in Upper Devonian reef complexes. *Geol. Rundsch.*, 61: 578—584.

8

BRYOZOA

KRISTER BROOD

INTRODUCTION

The bryozoans are a little known but common group of organisms which occur abundantly as fossils and in modern seas. The phylum Bryozoa comprises approximately 4,000 living and 15,000 fossil species. They live in colonies containing several microscopic individuals but colonies may range in size from a few mm to about 10 cm in diameter (Fig. 1). Colonies may appear bush-like, fungiform or encrusting, forming carpet-like covers on stones, shells or other hard substrates (Fig. 2). It is this carpet-like appearance that first earned them the epithet "moss-animals", a name which is still commonly used in the literature. The bryozoans are sessile animals, living chiefly in the marine environment. In the ocean, Bryozoa occur from intertidal to abyssal depths, but a large majority of the species inhabit relatively shallow waters of the continental shelf.

The majority of bryozoans have a calcified skeleton that is easily preserved, so that they are found abundantly as fossils. They range from Ordovician to Recent. In spite of their abundance in post-Cambrian sediments and their rapid evolution in the Paleozoic and Mesozoic, Bryozoa have been mostly ignored in micropaleontology as index fossils. They can be useful fossils since an individual in a bryozoan colony is usually about 1 mm in

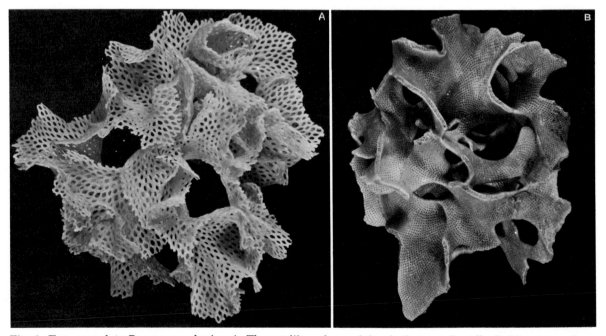

Fig. 1. Two complete Bryozoan colonies. A. The net-like colony of the cheilostome *Sertella*. B. Bilaminar colony of a membraniporid cheilostome. Ca. × 1.5.

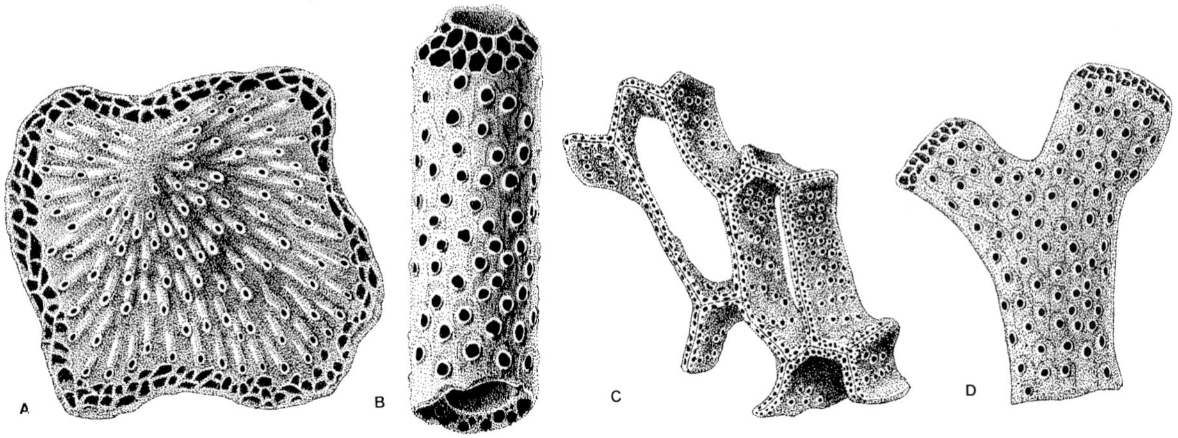

Fig. 2. Diagrams of four different colony growth forms occurring in the cyclostome genus *Diastopora*. A. Wholly encrusting form, prevalent in turbulent waters. B. Erect, hollow stem, typical for calm waters. C. Retiform colony typical for turbulent waters. D. Erect, bifoliate form typical for calm waters.

length and can often be identified in cores and well cuttings. In addition, bryozoan skeletons are often highly adapted to environmental conditions, thus making them invaluable for the interpretation of ancient environments.

Reference to bryozoan species appears in the literature as early as the sixteenth century. However, it was during the nineteenth century that bryozoan studies were initiated in various European countries, especially England where a large number of specialists and amateurs collected and studied fossil and living Bryozoa. Strong impetus to bryozoan research came at the turn of the century when many workers in North America undertook to describe and catalogue American assemblages.

In recent years research in Bryozoa has been directed more towards the biogeographic and paleoecological aspects of the group. As the interest in the study on Bryozoa increases, it is expected that they will assume their proper place with other fossil groups as useful stratigraphic and paleoecological tools.

BIOLOGY OF BRYOZOA

The soft parts of the individual, the zooid, are called the **polypide**; whereas the hard parts, the skeletal structure, are referred to as the **zooecium** (Fig. 3). One might picture a bryozoan polypide as being enclosed in a sort of carbonate envelope or case with one opening called the zooecial aperture or

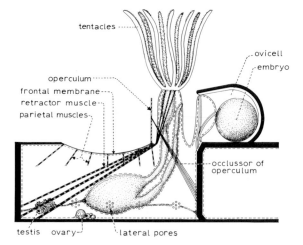

Fig. 3. Diagram of the internal structure of a cheilostomatous bryozoan (simplified after Ryland, 1970).

orifice, located at the distal end of the zooecium. Its shape varies from circular to highly complex in outline. The aperture may be flush with the surface of the colony or surrounded by a raised rim, the **peristome**, or partly covered by a calcareous hood-like projection, the **lunarium**, or apertural spines. The aperture may be covered with a movable, chitinoidal lid, the **operculum**. As in Brachiopoda the food-catching organ is termed the **lophophore**. This structure bears a tentacular crown of eight to over a hundred tentacles, which are arranged in a circle around the mouth. The tentacles bear cilia which produce a water current to sweep food particles towards the mouth. The food consists of microorganisms, bacteria and organic detritus. A mouth at the end of the lophophore leads into a complete alimentary canal, which includes pharynx, oesophagus, a u-shaped stomach, and a slender intestine leading to the anus, located close to the lophophore.

That no special respiratory organs are developed is considered as an adaptation to small size, but respiration does take place through the body wall, especially through the tentacles. Communication between the individual zooids takes place through interzooidal pores, through the zooidal walls, or by the zooids being united with each other by soft tissues at the surface of the zoarium.

Reproduction and life cycle

Bryozoan zooids contain small gonads within the zooidal cavity. The individual colonies are generally hermaphroditic, containing differentiated male and female zooids, or hermaphroditic zooids. Rarely, a colony may be completely male or female and have zooids of only one sex.

The two major classes, Stenolaemata and Gymnolaemata behave differently during reproduction. In the Stenolaemata the zooids are strictly differentiated with regard to sex. During reproduction the embryos are brooded in a large **ovicell**, which is actually a specialized female zooid. This large ovicell contains many embryos that have been produced by fission of a single embryo. In the Gymnolaemata, which commonly have hermaphroditic zooids, the fertile egg is generally placed in a small ovicell located near the distal end of the mother zooid (Fig. 4) where it is housed during its early stages. The ovicell in these forms only contains a single embryo. In some gymnolaematous forms the egg or embryo may be liberated directly into the water. These latter forms have a long larval stage termed the **cyphonautes**.

When the larva has left the ovicell, it settles to the substrate generally after only a few hours of existence and undergoes metamorphosis into the primary zooid, the **ancestrula**. During metamorphosis all the larval organs are absorbed and replaced by adult organs. The ancestrula may then produce a whole new colony by asexual budding of daughter zooids, from which new zooids are produced and so on. In the Gymnolaemata the new zooids are produced in a single distal bud which will continue budding in the same

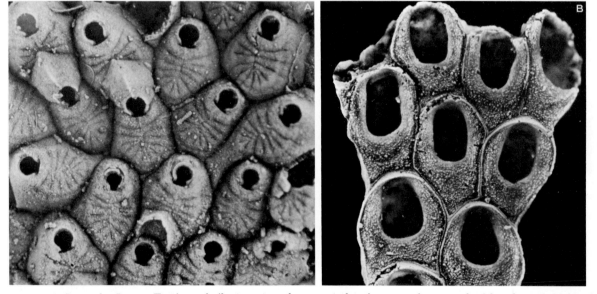

Fig. 4. A. Part of a Lower Tertiary cheilostomatous bryozoan showing zooecia covered with a frontal shield of highly modified avicularia (the transversely oriented rib-like structures) and several ovicells (the helmet-like vesicles on the zooecia in the upper left). B. A Lower Tertiary cheilostome with uncalcified frontal membrane. × 18.

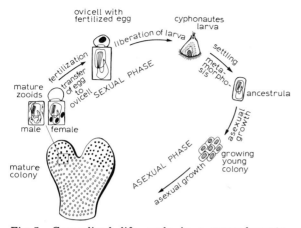

Fig. 5. Generalized life cycle in a gymnolaemate bryozoan.

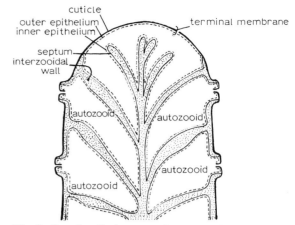

Fig. 6. Growing tip in a stenolaemate colony.

direction, forming a longitudinal series of zooids. Branching is accomplished by a simple division of the end of the bud. This reproductive cycle is illustrated in Fig. 5. At certain periods during the life of the zooid the polypide degenerates and a new one is formed. This process may happen several times during the life of the zooid and is considered a response to unfavourable environmental conditions. It occurs most often during winter, but may also be affected by the organism at other times for excretion of waste materials (though this last mentioned function is not the primary reason for this process).

Growth

In the Stenolaemata the skeleton is produced in the growing zone by the growth of longitudinal septa which form from the basal part of the mother zooid and upwards towards the terminal end, so that a new zooid is separated from the distal growing zone (Fig. 6). As new septa are formed new zooids are produced. Several septa can be formed simultaneously, producing a considerable widening of the growing zone of the colony, allowing it to grow rapidly.

In the Gymnolaemata, new zooecia grow from protuberances produced from existing zooecia. A series of zooecia are produced by growth of transverse septa within the protuberance. The series may divide according to a specific pattern. In this manner, rows of joined identical zooecia radiating from the ancestrula are formed.

Polymorphism

A distinctive feature of zooids of many groups of Bryozoa is polymorphism. Besides the fully developed, feeding individual, or **autozooid**, and the earlier described ovicells, several other types occur. In particular the colonies of most Gymnolaemata are polymorphic and contain modified zooids such as the **avicularium** and the **vibracularium** (see Fig. 7). The avicularium is smaller than the autozooid but like the autozooids contains an operculum. This operculum is often modified to resemble a tiny, movable jaw (Fig. 7), and when an avicularium sits on a stalk it may look like a little bird head

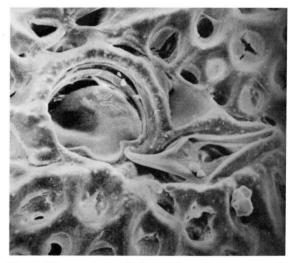

Fig. 7. The aperture of the modern cheilostomatous bryozoan *Schizoporella* showing the operculum and the opening for the ascus (the excavation at the lower part of the aperture). To the right an avicularium with the mandible open. × 100.

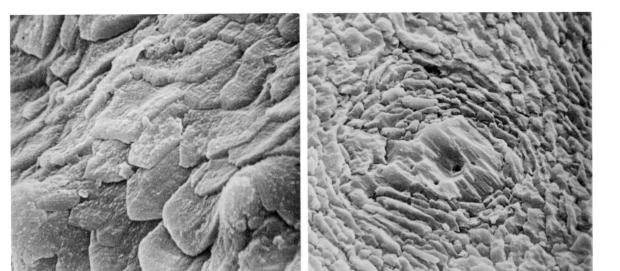

Fig. 8. A. Interior surface of a modern cyclostome showing the tabular crystals of the skeleton. B. Thin section of Silurian cryptostome showing the central rod of an acanthopore and the surrounding calcareous laminae. Scanning electron micrographs, × 5,500.

attached to the colony. The function of avicularia is not known, but apparently these little jaws are capable of a picking motion and may be protective and/or food-capturing devices. The vibracularium is an avicularium whose operculum is modified into a long, slender bristle. The vibracularia produce sweeping movements which move particles, such as larvae, searching for a settling place. Zooids that become modified for a supporting function are called **kenozooids**, and are characterized by thickened zooidal walls.

Skeletal structure

Most bryozoans have skeletons composed of calcium carbonate, although some Gymnolaemata lack a mineralized skeleton. Bryozoan skeletons in different orders demonstrate characteristic patterns of calcite crystal orientation and structure.

In the Stenolaemata the skeleton is composed of calcite only. The crystals may appear as a granular layer without texture, or laminar layers with lath-like crystals (Figs. 8, 9).

Among the Gymnolaemata, however, the primitive members of the order possess a skeleton of calcite only, whereas the more specialized cheilostomes may have a secondary thickening of aragonite (Fig. 10).

Fig. 9. Diagram of wall structure types in stenolaemate Bryozoa. A. Wall structure in a generalized cyclostome. B. Wall structure in a typical trepostome. C. Wall structure in a fenestellid cryptostome.

— cuticle   laminated calcite
------ epithelium   granular calcite

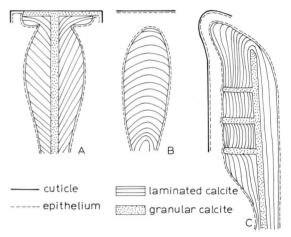

Fig. 10. Generalized diagram of the wall structure in an ascophorate cheilostome.

Major divisions of Bryozoa

The major orders of the two marine bryozoan classes, Stenolaemata and Gymnolaemata, are summarized in Table I. The third class, Phylactolaemata, is exclusively fresh-water. This Table should be used together with the description and geological distribution of taxa at the end of the chapter.

ECOLOGY AND PALEOECOLOGY

Distribution in rocks and sediments

Fossil Bryozoa are most abundant in calcareous rocks such as limestones, calcareous shales and shelly marls. Generally Bryozoa are rare in black shales, dolomites and quartzose clastic rocks. In sedimentary rocks bryozoans

TABLE I

Major divisions of the phylum Bryozoa

Class STENOLAEMATA	zooecia cylindrical with calcified bodywall; new zooecia produced in a common bud by division of septa; marine; ovicell large; Ordovician—Recent; approximately 550 genera
Order CYCLOSTOMATA	aperture circular; zooecia generally with pseudopores and interzooidal pores; with large ovicells; developed polymorphism; chiefly post-Paleozoic; Ordovician—Recent; approximately 250 genera Common genera: *Tubulipora, Diastopora, Pustulopora, Ceriopora, Idmidronea, Hornera, Crisia*
Order CYSTOPORATA	aperture of autozooecia with lunaria; many cystopores between the autozooecia; zooecia interconnected by pores; Ordovician—Permian; approximately 50 genera Common genera: *Ceramoporella, Ceramopora, Fistulipora*
Order TREPOSTOMATA	colonies ramose or massive; autozooecia long and tubular; aperture of autozooecia circular or polygonal; zooecial wall thickens distally; mesopores numerous; diaphragms and acanthopores common; Ordovician—Permian; approximately 100 genera Common genera: *Batostomella, Mesotrypa, Monticulipora, Hallopora*
Order CRYPTOSTOMATA	colonies reticulate fronds, bifoliate or ramose, circular stems; zooecial tubes circular but shorter than in Trepostomata; diaphragms rare; mesopores generally absent; acanthopores common in many forms; large ovicells in some groups; Ordovician—Permian (Triassic?); approximately 130 genera Common genera: *Fenestella, Polypora, Nematopora, Helopora, Ptilodictya, Rhombopora, Archimedes, Phylloporina*
Class GYMNOLAEMATA	zooecia cylindrical or box-like; polymorphism well developed; new zooecia formed in branching longitudinal series; marine; Ordovician—Recent; approximately 650 genera
Order CTENOSTOMATA	zooids cylindrical or flat; walls not calcified; avicularia and ovicells lacking; Ordovician—Recent; approximately 40 genera
Order CHEILOSTOMATA	zooecia boxlike; with calcified walls; aperture of frontal side closed by a hinged operculum; specialized zooids as avicularia common; ovicells small; Jurassic—Recent; approximately 600 genera Common genera: *Onychocella, Coscinopleura, Smittipora, Lunulites, Membranipora, Cellaria, Mucronella, Tubucellaria, Schizoporella, Metrarabdotus*
Class PHYLACTOLAEMATA	zooids cylindrical with a horseshoe-shaped lophophore; bodywall noncalcareous with incorporated muscles; with dormant buds; exclusively fresh-water; (Cretaceous?)—Recent; approximately 12 genera

Fig. 11. Greater than 1 mm fraction of a Danian bryozoan limestone from Denmark showing numerous erect, twig-like cheilostomatous Bryozoa.

occur mostly as fragments since the fragile colonies of many species disintegrate before they are deposited (Fig. 11).

Fossil Paleozoic Bryozoa are generally associated with sessile benthic organisms such as solitary corals, articulate brachiopods, and echinoderms. Post-Paleozoic Bryozoa are mostly found together with mollusks, sponges and octocorals.

Bryozoans are important constituents of both Recent and fossil sediments. In the Recent they occur mainly in sediments of the continental shelves and around coral reefs. In ancient sediments similar facies types contain rich bryozoan faunas. Paleozoic shallow-water sediments are commonly characterized by great abundance of bryozoan detritus, especially limestones associated with organic reefs. In the Upper Paleozoic they are the most significant group in organic reefs. Bryozoans also contributed extensively to organic banks and biostromes in Mesozoic and Cenozoic rocks, such as the bryozoan mounds in the Danian of Scandinavia (Cheetham, 1971).

Calcareous sands consisting largely of Bryozoa occur today over 250,000 square kilometers off the coast of southern Australia, where they are mixed with reworked Pleistocene and Tertiary fossils (Wass and others, 1970). The extant Bryozoa are reported to have lived here at depths of 60—120 fathoms and temperatures of 7—16°C. Fossil bryozoan faunas, which may represent analogous environmental conditions, are contained in the rich Eocene to Miocene calcarenites of the Eucla and Murray Basins in southern Australia and also in the Tertiary of Europe (Buge, 1957).

Factors controlling distribution

Since Bryozoa are sessile, their distributions are controlled by such factors as the nature of the substrate, water turbulence, rate of sedimentation, salinity, temperature and water currents. The bryozoans with their comparatively heavy skeletons generally grow on hard substrates, such as invertebrate shells, stones and large algae. They are therefore most common on stable bottoms. A very few forms have adapted to living on soft bottom sediments. Many encrusting forms live in shallow water, where they form a covering layer on algal fronds. A few Bryozoa may attach themselves to living shell-bearing animals such as crustaceans and gastropods.

The factor that chiefly determines the upper bathymetric limit of Bryozoa is water turbulence. Most of the erect, branching forms cannot, as a rule, survive turbulent near-shore waters where they would be rapidly destroyed. Concentrations of massive and encrusting colonies of Bryozoa, therefore, suggest relatively turbulent environmental conditions, while concentrations of erect, delicately built forms indicate fairly calm water (Brood, 1972). Erect forms may, however, become more flexible by transforming their stems into articulated segments (Fig. 12) or they may lack calcareous parts altogether, thus enabling their stems to withstand considerable agitation. Articulated stems have evolved independently in several groups of Bryozoa, and in fossil forms provide a valuable key to one parameter in an ancient environment.

Some Bryozoa may modify the entire shape of the colony to suit the local environment. For instance, many species have an encrusting mode of life when they live in turbulent waters, while in more sheltered areas they grow erect (Harmelin, 1974). However, the majority of the Bryozoa have fixed

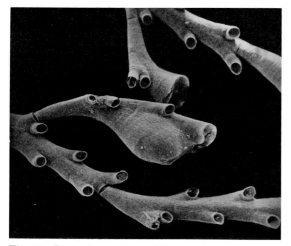

Fig. 12. Part of a colony of the modern cyclostome *Crisia* showing several articulated segments and a large ovicell in the centre. × 20.

growth forms and occur in and depict a specific environment.

Rates of sedimentation are as a rule too low to effect the distribution of the Bryozoa, except in the case of river deltas. Such regions are dominated by forms with large vibraculae, which sweep across the surface of the colony to remove sediment particles.

Generally, the Bryozoa are fairly steno-haline and live only in normal sea water with a salinity of roughly $35^0/_{00}$. Fluctuations in salinity are often the cause of a noticeable decrease in the bryozoan fauna.

Only a few Bryozoa occur in the deep sea. These forms are characterized by tiny and slender colonies. Like other fossils, Bryozoa can be transported into the deep sea from shallower areas by turbidity currents.

GEOLOGICAL OCCURRENCE AND EVOLUTION

The stratigraphic ranges of the major Bryozoa groups is shown in Fig. 13 and Table II. The first fossil bryozoans are found in the Lower Ordovician. A poorly preserved fossil, called *Archaeotrypa*, which was assigned to the Bryozoa, has been reported from the upper Cambrian, but this find is open to question. Because there are large gaps in our knowledge of the bryozoan faunas from many periods, their evolutionary pattern cannot be reconstructed at present. However, the four orders belonging to Stenolaemata all became established at approximately the same

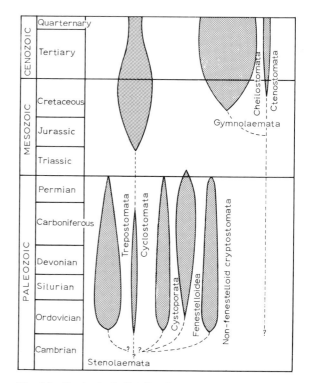

Fig. 13. Geological distribution and suggested evolution of marine Bryozoa. Width represents relative abundance of the group.

Fig. 14. Frontal and back side of *Fenestella penta-gonalis* Hennig. Wenlockian (Middle Silurian), Sweden. × 20.

time in the Middle Ordovician. The dominant bryozoan group in the Lower Paleozoic was the trepostomes, the "stony bryozoans", which evolved rapidly in the Ordovician. The trepostomes generally have robust branching zoaria, but may have large massive or globular colonies. Trepostomatous genera which are common in the Lower Paleozoic are *Hallopora*, *Batostomella* (Ordovician), *Mesotrypa* (Ordovician—Silurian) and *Monticulipora* (Ordovician).

TABLE II

Stratigraphic ranges of important bryozoan genera

Age \ Genus	Batostomella Ulrich	Mesotrypa Ulrich	Monticulipora d'Orbigny	Hallopora Bassler	Ceramopora Hall	Fistulipora McCoy	Phylloporina Ulrich	Fenestella Lonsdale	Polypora McCoy	Nematopora Ulrich	Helopora Hall	Ptilodictya Lonsdale	Rhombopora Meek	Archimedes Owen	Diastopora Lamoroux	Pustulopora Blainville	Ceriopora Hagenow	Idmidronea Canu and Bassler	Tubulipora Lamarck	Hornera Lamoroux	Crisia Lamoroux	Onychocella Jullien	Coscinopleura Marsson	Smittipora Jullien	Lunulites Lamarck	Membranipora Blainville	Cellaria Ellis and Solander	Mucronella Levinsen	Tubucellaria d'Orbigny	Schizoporella Hincks	Metrarabdotus Canu
Quaternary															▮	▮		▮	▮		▮	▮	▮	▮	▮	▮	▮	▮	▮	▮	▮
Tertiary																															
Cretaceous																															
Jurassic																															
Triassic																															
Permian																															
Carboniferous																															
Devonian																															
Silurian																															
Ordovician																															
Cambrian																															

The Cryptostomata which later became the most important group in the Paleozoic evolved rapidly during the Ordovician. Typical Lower Paleozoic genera are the fenestellid relatives *Phylloporina* (Ordovician), *Polypora* (Ordovician–Permian) and *Fenestella* itself (Fig. 14). Also certain cystoporate genera such as *Ceramopora* were common in the Ordovician.

In the Silurian the pattern of the bryozoan fauna changes slowly so that the trepostomes gradually decline in importance and the fauna becomes dominated by the cryptosomes with slender branching colonies and graceful lace-like types such as *Helopora* (Fig. 15) with its articulated stems, *Nematopora* (Fig. 16) and the bifoliate fan-shaped ptilodictyids.

In the Upper Paleozoic these trends become more pronounced. In Carboniferous

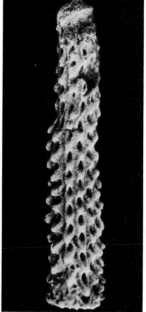

Fig. 15. Colony segment of *Helopora lindstroemi* Ulrich, an articulated cryptostome Bryozoa. The genus can be recognized by its cylindrical articulated stems with large acanthopores (the spines on the exterior side). × 20. Wenlockian (Middle Silurian), Europe.

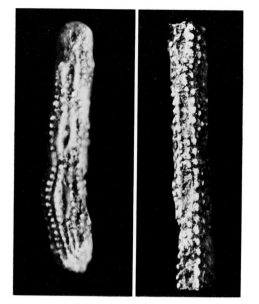

Fig. 16. Two specimens of *Nematopora visbyensis* Brood showing numerous acanthopores and elongated apertures of autozooecia. × 20. Ordovician to Silurian.

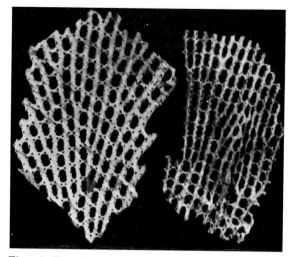

Fig. 18. Two fenestellid species. × 3.5. Visean (Lower Carboniferous), Ireland.

Fig. 17. A cystoporate colony belonging to *Ceramoporella lindstroemi* (Hennig). The Cystoporata can be recognized by the horse-shoe-shaped proximal parts of their apertures. (For generic and specific differentiation thin sections are necessary.) × 7. Wenlockian (Middle Silurian), Sweden.

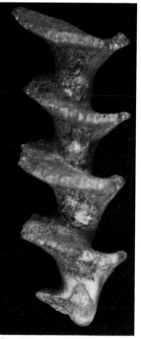

Fig. 19. The screw-like colony of an *Archimedes* species. This genus, a fenestellid, can be identified by its characteristic colony shape. × 5. Restricted to Carboniferous.

and Permian strata the bryozoan fauna is composed almost entirely of lace-like fenestellids and delicate branching types (Fig. 18). Common cryptostomatous genera are the peculiar *Archimedes* (Fig. 19) with its solid, screw-like colony and *Rhombopora* (Devonian—Permian). The Cystoporata is represented by the genus *Fistulipora* (Silurian—Permian). At the end of the Era all three large Paleozoic bryozoan orders disappeared and the cyclostomes were the only group which survived the Permo-Triassic mass extinctions into the Mesozoic.

Bryozoans are rare in the Triassic. However, the Cyclostomata evolved rapidly in the Jurassic and reached the peak of their evolution in the Cretaceous. Common Mesozoic genera are *Diastopora*, *Pustulopora* (Fig. 20), *Ceriopora* (Fig. 21), and *Idmidronea* (Fig. 22).

In the Cenozoic the cyclostomes slowly decreased in number, possibly due to compe-

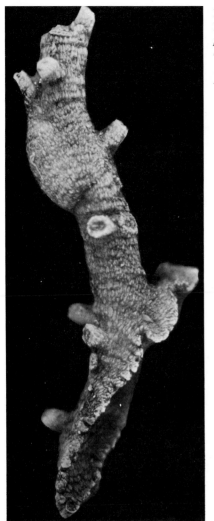

Fig. 20. *Pustulopora virgula* Hagenow, a typical *Pustulopora* species. The elongated capsule on the upper left part is an ovicell. The genus *Pustulopora* is characterized by thin, ramose, cylindrical stems with the apertures of the autozooecia opening irregularly all around the stems. × 25. Maestrichtian (Upper Cretaceous), Europe.

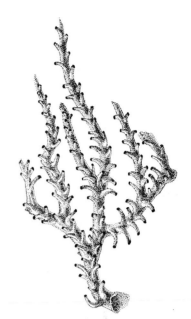

Fig. 22. Diagram of the cyclostome *Idmidronea*. *Idmidronea* is characterized by transverse rows of zooids opening on one side of the stems. × 25. Jurassic to Recent.

Fig. 23. *Theonoa disticha* (Hagenow), a cyclostomatous Bryozoa. The genus *Theonoa* is characterized by the arrangement of the autozooecia in stellate groups. The genus is common in the Cretaceous. × 18. Maestrichtian (Upper Cretaceous), Europe.

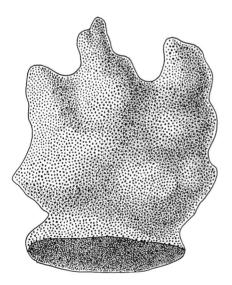

Fig. 21. Diagram of a colony of the cyclostome *Ceriopora*. *Ceriopora* is characterized by robust colonies with the apertures of the autozooids opening at right angles to the surface. The structure of the zooidal walls separates it from the trepostomes which have colonies of the same general shape. × 1.5. Jurassic to Tertiary.

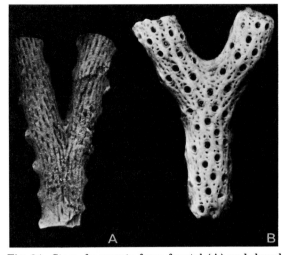

Fig. 24. Stem fragments from frontal (A) and dorsal (B) side of *Hornera striata* Edwards. The genus *Hornera* is typical for the Tertiary deposits and can be recognized by its heavily calcified stems with longitudinal branching ribs which enclose numerous small pores. The apertures of the autozooecia open on one side of the stem only. × 15. Miocene, Italy.

Fig. 26. *Lunulites saltholmiensis* Berthelsen. The genus *Lunulites* is characterized by having a cupuliform, disc-like colony and zooecia arranged in radial rows with interspaced avicularia. × 10. Danian (Lower Paleocene), Denmark.

Fig. 25. *Coscinopleura angusta* Berthelsen. The genus *Coscinopleura* is characterized by the porous avicularia located at the lateral sides of the stem. × 20. Danian (Lower Paleocene) of Denmark.

Fig. 27. *Mucronella hians* Hennig. The genus *Mucronella* is characterized by its clover-leaf-shaped aperture, its almost inperforate zooecium and its small spines around the peristome. × 20. Danian (Lower Paleocene), Denmark.

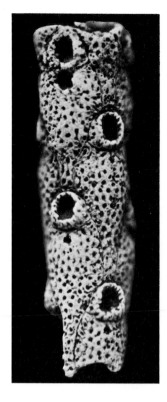

Fig. 28. A *Tubucellaria* species. The genus *Tubucellaria* is recognized by its articulated, cylindrical stems with tubular zooecia and porous zooecial front pierced by an ascopore. × 24. Pliocene of Italy.

←

Fig. 29. *Metrarabdotus helveticum* Roger and Buge. The genus *Metrarabdotus* is recognized by its bifoliate colonies with elongated autozooecia with a row of pores at their margin and an avicularium at each side of the aperture. × 20. Miocene, Italy.

→

tition with the newly evolved cheilostomes, and in present-day seas they are numerically small. Typical Tertiary genera are *Hornera* (Fig. 24) and *Crisia* (see Fig. 12).

The Ctenostomata are found first in the Upper Ordovician. As this group bears no calcified parts, their fossil record is based on shallow impressions and excavations in shell-bearing organisms.

The cheilostomes evolved from the ctenostomes during the late Jurassic. It was a successful group which developed rapidly. Genera which appear in the Cretaceous are *Onychocella*, *Coscinopleura* (Fig. 25) and *Lunulites* (Fig. 26). The Cheilostomata have dominated the bryozoan fauna since the beginning of the Tertiary. Tertiary genera are *Cellaria, Smittipora, Membranipora, Mucronella* (Fig. 27), *Tubucellaria* (Fig. 28), *Schizoporella* and *Metrarabdotus* (Fig. 29).

SUGGESTIONS FOR FURTHER READING

Bassler, R.S., 1953. In: R.C. Moore (Editor), *Treatise on Invertebrate Paleontology. Part G, Bryozoa.* Geol. Soc. Am., New York, N.Y., and Kansas Univ. Press, Lawrence, Kansas, pp. 1—253. [Contains short descriptions of all the genera known till that time and techniques of sample preparation for Bryozoa.]

Brien, P., 1960. In: P. Grasse (Editor), *Traité de Zoologie*, 5(2): pp. 1054—1335. [General descriptions of the phylum in French.]
Hyman, L., 1959. *The Invertebrates. 5, Smaller Coelomate Groups. Ectoprocta.* McGraw-Hill, New York, N.Y., 619 pp. [General descriptions of the phylum.]
Ryland, J.S., 1970. *Bryozoans.* Hutchinson, London, 175 pp. [An excellent general introduction to the phylum.]
Silén, L., 1966. On the fertilization problems in the gymnolaematous Bryozoa. *Ophelia*, 3: 113—140. [Description of the fertilization process in some Bryozoa.]
Tavener-Smith, R. and Williams, A., 1972. The secretion and structure of the skeleton of living and fossil Bryozoa. *Philos. Trans. R. Soc. Lond.*, 264: 97—159. [Descriptions of the mineralized skeleton and its secretion in Bryozoa.]

CITED REFERENCES

Buge, E., 1957. Les Bryozoaires du Néogene de l'Ouest de la France et leur signification stratigraphique et paléobiologique. *Mem. Mus. Natl. Hist. Nat., Sér. C*, VI: 1—436.
Brood, K., 1972. Cyclostomatous Bryozoa from the Upper Cretaceous and Danian in Scandinavia. *Stockh. Contrib. Geol.*, XXVI: 1—464.
Cheetham, A., 1971. Functional morphology and biofacies distribution of cheilostome Bryozoa in the Danian Stage (Paleocene) of southern Scandinavia. *Smithson. Contrib. Paleobiol.*, 6: 1—85.
Harmelin, J., 1974. *Les Bryozoaires Cyclostomes de Mediterranée. Ecologie et Systematique.* Thesis, University of Aix—Marseille, U.E.R. des Sciences de la Mer et de l'Environment, vol. II.
Wass, R.E., Conolly, J.R. and MacIntyre, R.J., 1970. Bryozoan carbonate sand continuous along southern Australia. *Mar. Geol.*, 9: 63—73.

SILICEOUS MICROFOSSILS

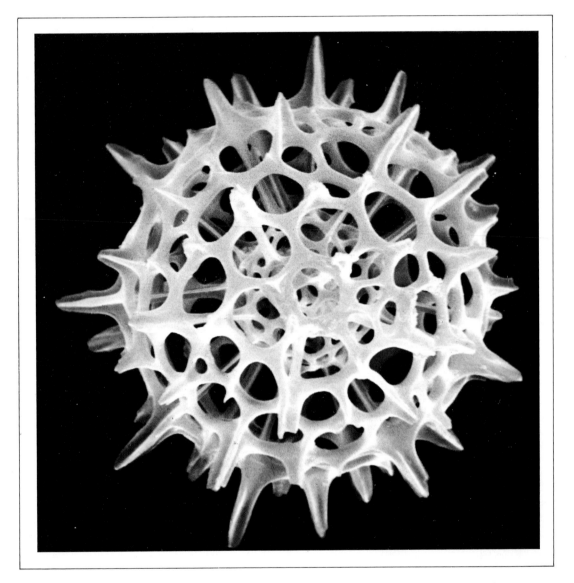

Spumellarian radiolarian showing concentric spherical
shells connected by radial spirals.

9

RADIOLARIA

STANLEY A. KLING

INTRODUCTION

Geologic history reveals few fossil groups with as complete a record as the radiolarians. Ranging throughout Phanerozoic time, these protozoans were apparently as diverse and widespread in the Paleozoic as they are now. And they have left behind a detailed evolutionary record that makes them potentially one of the most important marine microfossil groups.

Radiolarians are distinguished from other protists by division of the cell contents into an inner zone separated by a membrane from the remaining outer zone. Characteristics of the membrane are in turn used to separate the major subgroups of radiolarians. However, basic differences in the composition and geometry of the skeleton reflect these cytological relationships, so that we are able to discover natural relationships among these organisms from studies of their hard parts.

Geologists are concerned almost exclusively with one major radiolarian group, the polycystines, which include the radially symmetrical spumellarians and the helmet-shaped nassellarians. These are common both in the plankton and in sediments. One other important group, the tripyleans, are generally larger forms that are also common in the plankton but rarely preserved in sediments. Fig. 1 illustrates typical examples of these major types of radiolarians.

Skeletons of living radiolarians are composed of amorphous (opaline) silica and this skeletal mineralogy is common through the Tertiary. Although pre-Tertiary fossils are often replaced by other minerals, some siliceous collections are always to be found and we believe that the siliceous composition has

persisted throughout radiolarian history.

The acantharians, a group closely related to radiolarians, bear skeletons composed of celestite (strontium sulfate) which are never preserved in sediments. Although they are no longer included among radiolarians, we should be able to distinguish them in plankton samples. They are constructed on a highly symmetrical pattern of twenty spines which is illustrated along with some typical natural forms in Fig. 2.

Heliozoans, a group of fresh-water protozoans with radial symmetry, resemble radiolarians but also do not produce a preservable skeleton. They may be closely related to radiolarians, but detailed comparisons have not been made and their affinities remain obscure.

HISTORY OF RADIOLARIAN STUDY

The year 1834 marks the first published description of Radiolaria by F.V.F. Meyen, who included two radiolarian species among his description of "several polyps and other lower animals". C.G. Ehrenberg was the first to make extensive investigations of radiolarians. He published a series of papers between 1838 and 1875, and included a number of radiolarians in his *Mikrogeologie* (1854–1856). His descriptions of Eocene (and possibly Oligocene) forms from Barbados were responsible for establishing this as a classic radiolarian locality.

In the mid-nineteenth century, a number of leading biologists became interested in radiolarians. Among them were Johannes Müller and Richard Hertwig, who studied living specimens mainly from the Mediterranean Sea. Müller coined the name Radiolaria

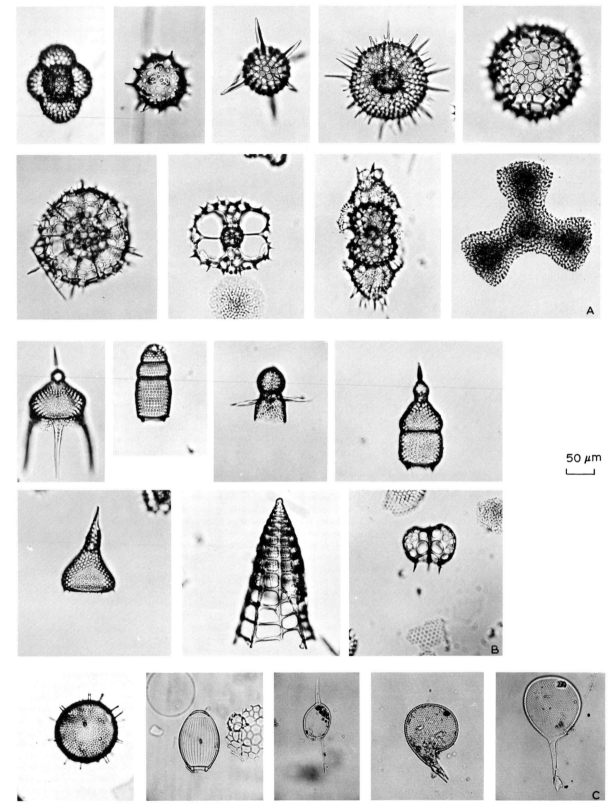

Fig. 1. Main Radiolarian groups. A, B. Polycystines from Quaternary deep-sea sediments comprising spumellarians (A) and nassellarians (B). C. Some tripyleans found in sediments. Reproduced from Stadum and Ling (1969) by permission of the authors and the American Museum of Natural History, Micropaleontology Press.

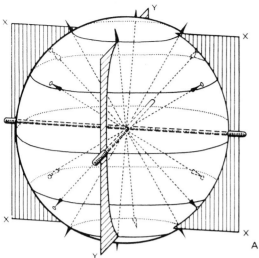

and was the first to include acantharians in his studies. Hertwig, a noted biologist and teacher, was the first to firmly establish the unicellular (acellular) nature of radiolarians, and in his synthesis of 1879, *Der Organismus der Radiolarien*, divided the major radiolarian groups on the basis of the morphology of the central capsular membrane.

Ernst Haeckel was a contemporary of Ehrenberg, Müller, and Hertwig, and is the most famous name among early students of radiolarians. He, too, began his studies with living material from the Mediterranean Sea and published an extensive monograph in

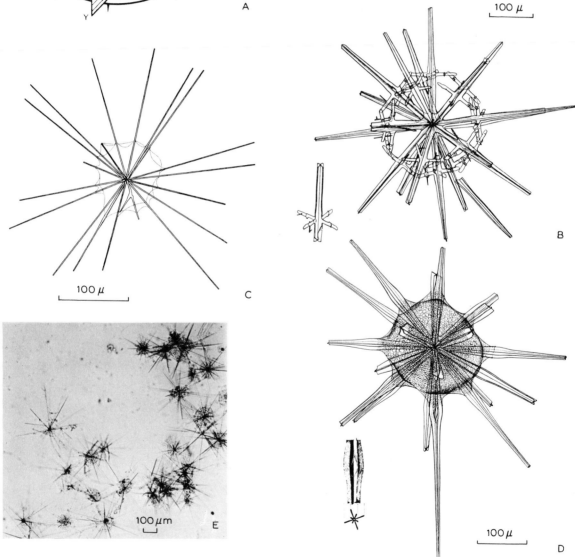

Fig. 2. Acantharians. A. Diagram of the basic spine geometry. B—E. Drawings of representative living specimens and photograph of forms in a plankton bloom. (Courtesy of E.G. Merinfeld.)

1862. His monograph of 1887, based mainly on collections of HMS *Challenger*, presented an exhaustive classification based on skeletal morphology which has been used, with occasional minor modification, by essentially all workers until very recently. Although Haeckel wished to "combine the phylogenetic aims of a natural system with the essentially artificial divisions of a practical classification", the practical won out over the natural and we had a system which obscured the value of radiolarians in biostratigraphy for nearly a century.

The oceanographic expeditions of the late nineteenth and early twentieth centuries initiated an explosion in the study of radiolarians as well as other pelagic organisms. The resulting works on radiolarians in plankton samples provided a wealth of new systematic and distributional data and some important observations of cellular structure.

During most of the twentieth century until fairly recently, radiolarians have received little concentrated attention. Hollande and Enjumet have made important recent contributions to radiolarian biology and this work is being continued by Cachon and Cachon. Working with material from the Mediterranean, their studies have been concerned principally with the cytology and mode of skeletal secretion based on observations of both living and preserved specimens. The resulting elucidation of the pseudopodial system has provided a valuable new contribution to radiolarian taxonomy.

W.R. Riedel, in the early 1950's, began extensive studies of the sequences of radiolarians from deep-sea oozes as well as from numerous localities in tropical areas on land. By careful study of natural relationships based on comparative morphology, he recognized that radiolarians evolved during the Tertiary at rates comparable to other groups. Continuation of these studies is leading to a more natural taxonomy, and a complete revision of Haeckel's system now in progress. Synthesis of these extensive investigations has resulted in a biostratigraphic zonation that has placed radiolarians among the groups important for dating marine sediments.

Following the pioneering work of Khabakov and Lipman, Russian scientists have also undertaken full-scale studies of radiolarians resulting in very important recent contributions, perhaps best exemplified by the extensive work of M.G. Petrushevskaya. Working on material from plankton samples and Recent sediments, she has made detailed comparisons of skeletal homologies in many groups. These have helped to clarify the natural relationships among taxa, providing valuable information upon which to found a new radiolarian system.

BIOLOGY

Organization at the cellular level

The majority of radiolarians occur as single individual cells, but some spumellarians are colonial. The colonies consist of numerous individuals, each of which commonly bears its own skeleton, loosely associated in gelatinous masses which may reach centimeter dimensions.

Only a few radiolarian species have been studied by cytologists, so features common to species which have been relatively thoroughly and recently investigated will be taken for the present as typical of the group. Fig. 3 shows a schematic cross-section of the major structures in the soft anatomy of radiolarians in relation to a hypothetical spherical skeleton.

A central zone of the cell, the **central capsule**, is enclosed in a membrane which separates the inner endoplasm from the outer ectoplasm. The shape of the central capsule varies considerably from species to species. Although it is often spherical, it may be quite irregular, taking on complex lobate shapes in some cases. Its shape is frequently related to the structure of the skeleton; for example, with lobes protruding through pores in internal structures.

Contained in the central capsule is the nucleus or nuclei, as multiple nuclei are not uncommon in radiolarians. In some cases, these multiple nuclei represent arrested stages of multiplication in the reproductive cycle. Nuclei of radiolarians are also noted for the presence of large numbers of chromosomes, counts as high as 1,500 having been reported.

The endoplasm contains vacuoles, lipid droplets of varying composition, albumenoid spherules with concretions or cubic crystals,

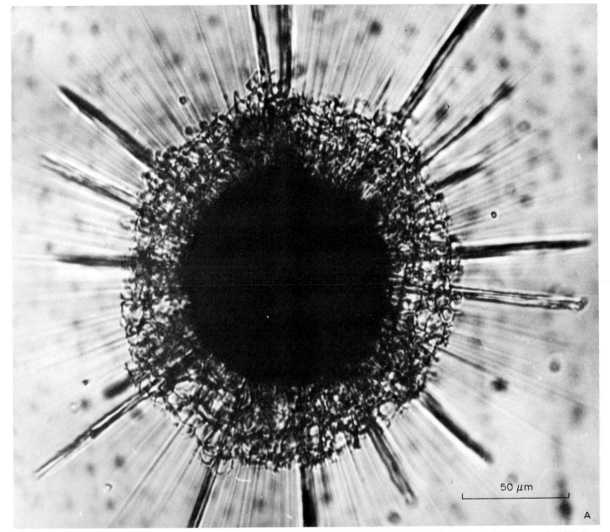

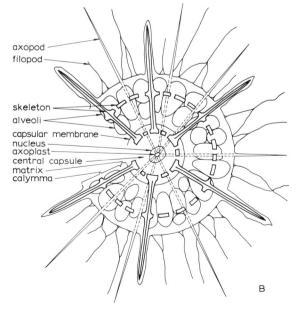

axopod
filopod

skeleton
alveoli
capsular membrane
nucleus
axoplast
central capsule
matrix
calymma

Fig. 3. The living radiolarian. A. Photograph of living specimen. Dark central area is central capsule, dark radial extensions are skeletal spines, light ones are axopodia. (Courtesy of Patricia C. Adshead.) B. Idealized diagram of major features of radiolarian soft anatomy, not all of which may be expected in a given specimen or species.

and crystals presumed to be proteins. The functions of reproduction, anabolism (biochemical synthesis), and catabolism (energy production) are the chief responsibility of the endoplasm.

The central capsular membrane is usually pigmented and easily visible in living and preserved material. In living specimens the colors are diverse — red, yellow, brown, and rarely blue violet, and green — and bright. These brilliant colors seldom persist in preserved specimens which are more likely to be yellow or brown. The colors are characteristic of some groups, and nearly always the same in a species. The membrane is firm but elastic, and its chemical composition is unknown. Generally, it is considered to be "chitinous" or "pseudo-chitinous". The membrane is single in the polycystines but double in the tripyleans. Piercing the membrane are tiny holes whose number and distribution are characteristic of taxa at various levels. These serve basically to permit exchange between endo- and ectoplasm, and as exits from the endoplasm for stiff pseudopodia.

The bulk of the ectoplasm is occupied by the so-called **alveoli**, barely visible ellipsoidal structures occurring in a concentric, frothy mass, the **calymma**, toward the outer perimeter of the organism. Although their exact composition is unknown, they have been described as mucoid in nature, and Brandt suggested that their contents are saturated with carbon dioxide. They are generally believed to function as hydrostatic regulators.

A relatively narrow ectoplasmic zone, the **matrix**, separates the calymma from the central capsule. It is not clearly distinguished structurally, and is generally thought to be the site where nutrient material is assimilated into the cytoplasm.

Partially digested food material can often be seen in the ectoplasm, and in tripyleans, a special structure comprising a cluster of yellow to brown to green pigmented spherules — the **phaeodium** (hence, the alternate name for tripyleans, the phaeodarians) is usually thought to be a collection of waste materials, though its exact nature and function are unknown. Diatom frustules and other remains of probable nutrient objects have been observed in the phaeodium. Symbiotic algal cells (zooxanthellae) are also frequently included in the ectoplasm.

Radiolarians bear two principal kinds of pseudopodia — **axopodia** and **filopodia**. The long, straight axopodia extend radially and are stiffened by characteristic axial filaments which extend through the ectoplasm and the capsular membrane to the interior of the endoplasm. They are inserted on a special structure termed an **axoplast**. The development of axopods and the axoplastic complex is of fundamental importance in radiolarian taxonomy. The thin, delicate **filopodia** originate as simple extensions of the peripheral cytoplasm. They are approximately radial but may anastomose, particularly toward their bases. They are said to gather into clusters during the capture of prey. In some groups the filopodia are woven into a complex reticulate network. Simple, undifferentiated pseudopodia engulf peripheral extensions of the skeleton (see Cachon and Cachon, 1971—1972).

Reproduction

Simple cell division has been observed in a number of forms, but the role of a supposed sexual stage has yet to be confirmed by observation of the complete life cycle of living specimens. At present, the reproductive process can only be sketched from fragmentary reports and by analogy with closely related groups whose biology is better known.

When radiolarians have been kept alive in culture for a few days, simple binary fission has usually been observed. Budding and multiple fission have been reported at various times, but these modes are apparently less well documented. It seems likely that repeated reports of multiple nuclei in radiolarians may be related to these kinds of multiple division as well as to the formation of swarmers.

Early observers of living radiolarians noted their frequent association with numerous tiny flagellated cells, similar to cells that were known to serve a reproductive function in other protistan groups. These so-called spores, or swarmers, are simple cells with a more or less oval shape, pointed at one end, with flagella attached to the blunt end. Two different kinds of these cells, differing mainly in size, were observed. This heterogeneity of reproductive cells is commonly associated in other protistan groups with the production of gametes, which subsequently unite in a sexual reproductive stage. Two different kinds of cells differing mainly in size (thus "anisoswarmers") were reported, but it was subsequently found that one group belongs to algal symbionts. Hollande and Enjumet described the formation of true radiolarian swarmers in some detail, but were unable to observe their ultimate role in the reproductive cycle.

Skeletal dimorphism observed in tripyleans (Kling, 1971) and polycystines may possibly result from alternating generations in the reproductive cycle as it does in foraminifera. Thus, it seems likely that radiolarian swarmers function much as similar cells do in foraminifera, and that radiolarians also undergo a complex alternation of sexual and asexual reproductive phases.

The life-spans of radiolarians have not been measured directly, but estimates based on comparisons between standing crops in the plankton and rates of flux to nearby sediments off southern California range from days to weeks (Casey and others, 1971) to months.

Nutrition

Radiolarians apparently feed on various kinds of planktonic organisms including microflagellates and other protozoans, diatoms, and possibly forms as large and active as copepods. There is, however, little information on specific food preferences or feeding habits.

Most statements regarding nutrition seem to be based on objects observed in the cytoplasm of preserved specimens which could be introduced artificially during sampling or storage. Radiolarians maintained in cultures were fed local phytoplankton by Adshead (1967).

Symbiotic algae also contribute to radiolarian nutrition. Some workers have noted that radiolarians containing algal symbionts are able to exist for relatively long periods without apparent external nutrition as long as light is available to sustain the algae.

The skeleton

Only a very few species of radiolarians lack hard parts. The radiolarian skeleton is encased in the soft cytoplasm. The bulk of the test usually lies in the ectoplasm, with pseudopodia extending outward to encase any protruding spines. Thus, the hard parts are never in direct contact with sea water, so that the skeleton of living organisms is never subject to dissolution in the aqueous environment. Innermost portions of the test may lie within the endoplasm in the central capsule.

Skeletons of the major fossil radiolarian group, the polycystines, are composed exclusively of amorphous silica ($SiO_2 \cdot nH_2O$). In living and well-preserved fossil material the siliceous substance is clear, transparent and isotropic with a glassy appearance in transmitted light. Internal structures of well-preserved material can be observed by focussing at various levels while viewing through the outer layers.

The skeleton is typically constructed of a network of siliceous elements, of which there are two fundamental types. Elongate elements connected at both ends to other elements are known as **bars**, whereas elongate elements attached at one end only are **spines**.

There are groups of polycystines whose skeleton consists solely of a simple association of spines known as a **spicule**. The majority of species, though, have a more complex skeleton in which a definite enclosing wall can be recognized (see Fig. 4).

Wall structure
The basic kinds of radiolarian wall structure are also illustrated in Fig. 4. The **latticed wall** consists of a network of bars forming closely spaced pores. The basic shape of the pores is usually hexagonal, but deposition of silica inside pores produces rounded outlines. Shapes of pores and their distribution are generally consistent in a species and can often be used for taxonomic purposes. The **spongy wall** is an intricate interlacing of relatively thin bars in a thick, usually irregular, three-dimensional network. Clearly defined pore patterns cannot be recognized. A third wall structure, the **perforate plate wall**, is a solid, uniformly thin wall penetrated by pores that are relatively widely spaced.

Skeletal shapes
The two major polycystine groups, the Nassellaria and the Spumellaria, are characterized by different skeletal shapes. Spheres are the commonest shape among spumellarians and radial spines usually extend from the surfaces of the spheres. A skeleton often consists of two or more nested spheres, which are, with very few exceptions, concentric and are connected by radial bars. The main, external shell is referred to as a **cortical shell** and the inner ones as **medullary shells**. Although the innermost spherical shell may be extremely small, it is characteristic of radiolarians that radial elements never actually meet at the center. Paleozoic spumellarians

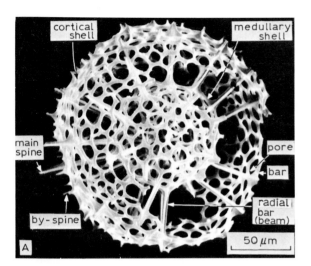

typically bear an internal spicule of converging radial bars, and similar structures are present in some living forms but these structures also are characteristically excentric. Other spumellarian shells are ellipsoidal (one axis elongated), discoidal (one axis shortened), coiled, or based on a series of concentric bands. These, too, may display multiple concentric shells and radial spines.

Nassellarians are characterized by axial symmetry although various modifications of this fundamental pattern can be seen in the multitude of individual forms. The walls of nassellarian tests are usually latticed, although spongy structure appears in some species, and in some forms the wall is a perforate plate.

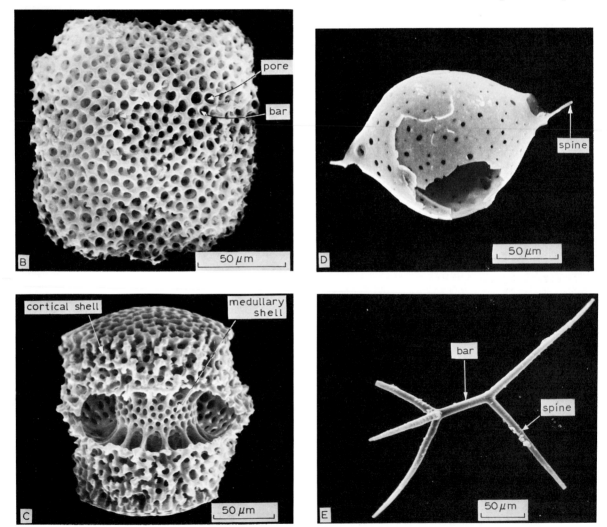

Fig. 4. Terminology of basic radiolarian skeletal elements and illustration of wall types. A. Latticed wall. B. Spongy wall. C. Spongy shell with latticed medullary shell. D. Perforate plate wall. E. Spicule.

Skeletal development

The manner and timing of skeletal secretion in radiolarians have been discussed repeatedly throughout the history of radiolarian study, but are still poorly known. Hollande and Enjumet concluded that the configuration of the endoplasm must control the secretion of the skeleton. However, they found no evidence of an unsilicified layer or template preceding the hard skeletal secretion, so that they believe that silica is precipitated directly at certain surfaces or junctions of surfaces.

It is likely that skeletal secretion is controlled by fundamental laws of fluid and interfacial physics. D'Arcy Thompson, in fact, compared radiolarian skeletal geometry to a number of physical models that could account for both micro- and macro-structural features of some species. He drew attention to the

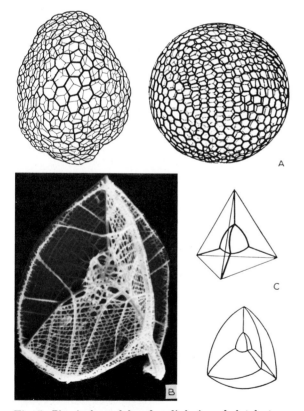

Fig. 5. Physical models of radiolarian skeletal structure. Drawings reproduced from D'Arcy W. Thompson, *On Growth and Form*, Copyright 1942, by permission of Cambridge University Press. A. Natural hexagonal mesh and hexagonal radiolarian lattice shell. B. Scanning electron micrograph of the radiolarian *Callimitra agnesae*. C. Tetrahedral models obtained by dipping wire frames in soap solution.

basic hexagonal pattern of the wall lattice and likened the structure of radiolarian species with simple hexagonal meshwork to the production of a very similar hexagonal meshwork "in certain cell nuclei... by adsorption or partial solidification of interstitial matter in a close-packed system of alveoli". He further showed that a spherical tetrahedron produced by the interaction of soap bubbles in confined space is the basic shape of some nassellarian skeletons (Fig. 5).

Ontogenetic growth of skeletons has rarely been observed in living specimens and has seldom been systematically studied. However, series of closely related forms in preserved material suggest the presence in some species of juvenile, intermediate and adult stages. Failure, in general, to discover juvenile forms has been biased by the fact that both plankton nets and sediment sieves commonly have openings, usually 62 μm, that pass the smaller individuals.

Evidence of growth is easiest to see in the nassellarians. Fig. 6 shows some examples of forms consisting of a single chamber. They are clearly related to multi-chambered forms from the same sample and have identical initial chambers. Growth stages are not easily determinable in the majority of spumellarians which possibly secrete their spherical shells more rapidly. Hollande and Enjumet reported that the skeleton was secreted very rapidly in the spumellarians they studied, and that

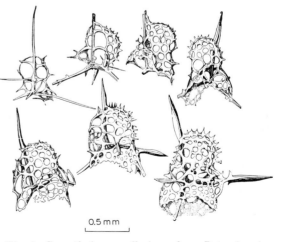

0.5 mm

Fig. 6. Growth in nassellarians, from Petrushevskaya (1962). Reproduced by permission of the author and the Copyright Agency of the U.S.S.R.

initially thin elements thicken secondarily. They observed that forms with multiple concentric shells develop centrifugally. Both nassellarians and spumellarians include rare specimens with what appears to be extra or secondary skeletal material, suggesting growth (Fig. 7).

ECOLOGY

We can make only a sketchy outline of radiolarian ecology because of the lack of comprehensive studies both of living specimens in the laboratory and on plankton samples from different depths in the oceans.

Radiolarians are exclusively marine and are found in all oceans. As far as we know, all species are planktonic, but there is some evidence that radiolarians have the ability to attach themselves to other objects. Individuals in captivity explore the surfaces of laboratory containers by means of pseudopodia with no apparent difficulty, and a specialized pseudopod, the axoflagellum, is similar to an attach-

ment structure in heliozoans. However, no definitely benthic radiolarian are known.

Radiolarians are characteristically open-ocean organisms and distinctive coastal forms are not generally recognized.

They apparently live at all depths in the oceans and are restricted to discrete depth zones as indicated by changing species. The bulk of species and individuals are found in the upper few hundred meters with a rapid decline in abundance in deeper waters (Fig. 8), although deep opening-and-closing net tows do contain specimens with soft parts, showing that they actually can live at great depths. Radiolarians have been recovered from deep oceanic trenches such as the Kurile—Kamchatka Trench in the northwestern Pacific, and some species may be restricted to these great depths (Reschetnjak, 1955).

Symbiotic algae

Radiolarians are among the marine organisms that play host to symbiotic dinoflagel-

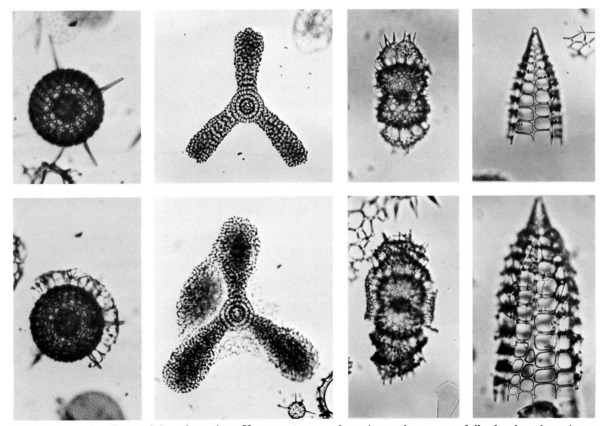

Fig. 7. "Secondary" growth in polycystines. Upper row: normal specimens; lower row: fully developed specimens showing additional delicate skeletal lattice work.

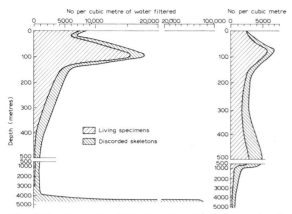

No per cubic metre of water filtered

Fig. 8. Profiles of radiolarian abundance versus depth at two stations in the equatorial Pacific (from Petrushevskaya, 1971c). Reproduced from B.M. Funnell and W.R. Riedel, *The Micropaleontology of Oceans*, Copyright 1971, by permission of the author and Cambridge University Press.

lates known collectively as **zooxanthellae**.

Although the role of symbionts in the life of radiolarians is not known, the radiolarian undoubtedly receives some nutrition from the photosynthetic activity of the symbionts, and in return, production of carbon dioxide in respiration is utilized in the metabolism of the symbionts.

Planktonic adaptations

With our meager knowledge of live radiolarians, we are inclined to conclude that their intricate shapes are an adaptation to the planktonic mode of life. However, if most of the skeleton is enclosed in a mass of cytoplasm, skeletal adaptation to floating may not be very important. Nevertheless, radiating spines and the numerous stiff axopods would inhibit movement through the water. Oil droplets in the cytoplasm have also been cited as enhancing buoyancy. The ability of radiolarians to attach to surfaces in laboratory cultures suggests that they might attach to larger floating objects in the natural environment.

Radiolarians are also noted for a unique apparatus, the alveolar complex, that is thought to be an adaptation for planktonic existence. Brandt (1895) concluded that the alveoli contain carbon-dioxide-saturated water and suggested that radiolarian buoyancy is controlled primarily by adjustments in volume of the alveoli via fluctuations in carbon-dioxide content. This would support earlier observations that radiolarians respond to changes in their environment (e.g., thermal or mechanical agitation) by vertical movement. Another suggestion has been that radiolarians sink during reproduction as a consequence of loss of the alveoli. The intriguing possibility has been mentioned by Pokorný (1963) that regulation of carbon-dioxide balance in the alveoli by diurnal fluctuations in the photosynthetic activity of the zooxanthellae may cause vertical migrations of radiolarians of as much as 200 to 350 m. Daily fluctuations of radiolarian abundances have also been attributed to water movements.

Biogeography

Relatively few studies of radiolarians have treated material from plankton samples and the distribution of hardly any species has been covered in detail sufficient to describe its geographic limits, or to assess its paleoecologic significance. Considerably more attention has been given to surface sediment distributions, which are usually compared to oceanographic variables in overlying waters.

Results of plankton studies by Petrushevskaya (1967, 1971b—d) in the Antarctic, Indian, and Pacific Oceans indicate that the Antarctic Convergence is a major biogeographic boundary separating distinctive Antarctic and Subantarctic species assemblages, and that both of these are distinguishable from equatorial assemblages. Included in the assemblages are species endemic to physically defined areas of the oceans. Renz (1973), studying tropical forms in the Pacific, recognized distinctions between radiolarians inhabiting the equatorial region and those in the adjacent central regions to the north and south. Thus, the major distribution patterns of radiolarians in the oceans, as those of other kinds of plankton, seem to be related to water masses. Circulation is important to maintenance of these water masses and is reflected in radiolarian distributions (as well as those of other plankton) by appearance of equatorial forms displaced poleward in western boundary currents (as the Kuroshio off Japan) or polar forms displaced equatorward in eastern boundary currents (as the California Current off the western United States).

Vertical distributions, critical to understanding radiolarian biogeography, are related to subsurface water masses. Certain tripyleans in the North Pacific occupy the Intermediate water which submerges southward from the subarctic region beneath the surface water masses (Kling, 1976). These radiolarians have a broad geographic distribution representing, however, a narrow range of environmental variables quite different from those of overlying and underlying waters.

Many radiolarians exhibit bipolar distribution patterns with the same or closely related species occupying both polar regions. Some such species could, of course, maintain continuity across the equator in subsurface water masses. Similarly, corresponding water masses in different oceans contain similar species. Equatorial radiolarians are commonly found in all oceans, and, for example, Pacific and Atlantic subarctic assemblages have many species in common.

Radiolarian distributions in surface sediments have been plotted on an ocean-wide scale in the Pacific by Kruglikova (1969), Nigrini (1968, 1970), Casey (1971) and Petrushevskaya (1971c); in the Atlantic Ocean by Goll and Bjørklund (1971, 1974); and in the Indian Ocean by Nigrini (1967) and Petrushevskaya (1971d). Results of these studies, although biased by poor preservation in certain areas such as the central regions where productivity is low, generally confirm the fundamental relationship between distribution patterns and surface water properties.

PREPARATION TECHNIQUES

The basic technique for preparing radiolarians begins with heating the sample in water with an appropriate disaggregating agent (e.g., tetrasodium pyrophosphate). Hydrogen peroxide is usually added to oxidize organic matter, and the bubbling action is useful in physical disaggregation as well. Eventually the material is acidized (hydrochloric or acetic) to eliminate calcareous material. Shales that do not disaggregate in water may respond to saturation of the dried sample, while warm, with kerosene, followed by addition of water after drainage of the kerosene. Treatment in a gentle ultrasonic bath is often helpful in removing coatings and infillings of clay particles. The final concentrate, after sieving (usually with 62-μm openings) is pipetted onto glass microscope slides and cover glasses mounted with Canada Balsam or a medium of similar refractive index. Specimens can be etched from cherts and similar hard siliceous rocks with dilute hydrofluoric acid. Carbonate-replaced specimens can sometimes be differentially etched with dilute acids. (For details see Riedel and Sanfilippo, 1976a.)

MAJOR MORPHOLOGICAL GROUPS

Polycystine radiolarians are classified principally on the basis of skeletal shape and symmetry into two major groups — the spumellarians and the nassellarians. Riedel (1971), Petrushevskaya (1971a) and Petrushevskaya and Koslova (1972) have proposed revised classifications. The groups outlined here (Table I) follow closely those suggested by Riedel. Though not all radiolarians can be accounted for in this still incomplete scheme, most common living and fossil forms can be classified.

TABLE I

Outline of radiolarian classification

Phylum PROTOZOA
    Class ACTINOPODA Calkins
        Subclass RADIOLARIA Müller
            Superorder TRIYPLEA Hertwig
            Superorder POLYCYSTINA Ehrenberg, emend. Riedel
                Order SPUMELLARIA Ehrenberg
                    Family ENTACTINIIDAE Riedel
                    Family OROSPHAERIDAE Haeckel
                    Family COLLOSPHAERIDAE Müller
                    Family ACTINOMMIDAE Haeckel, emend. Riedel
                    Family PHACODISCIDAE Haeckel
                    Family COCCODISCIDAE Haeckel
                    Family SPONGODISCIDAE Haeckel, emend. Riedel
                    Family HAGIASTRIDAE Pessagno
                    Family PSEUDOAULOPHACIDAE Riedel, emend. Pessagno
                    Family PYLONIIDAE Haeckel
                    Family THOLONIIDAE Haeckel
                    Family LITHELIIDAE Haeckel

                Order NASSELLARIA Ehrenberg
                    Family PLAGONIIDAE Haeckel, emend. Riedel
                    Family ACANTHODESMIIDAE Haeckel
                    Family THEOPERIDAE Haeckel, emend. Riedel
                    Family CARPOCANIIDAE Haeckel, emend. Riedel
                    Family PTEROCORYTHIDAE Haeckel, emend. Riedel
                    Family AMPHIPYNDACIDAE Riedel
                    Family ARTOSTROBIIDAE Riedel
                    Family CANNOBOTRYIDAE Haeckel, emend. Riedel
                    Family ROTAFORMIDAE Pessagno
Radiolaria incertae sedis
                    Family ALBAILLELLIDAE Deflandre
                    Family PALAEOSCENIDIIDAE Riedel

Fig. 9. Fundamental nassellarian skeletal elements; A and B, idealized drawings; C—F, SEM photographs. A. The basic elements and their terminology, with letter symbols used to indicate homologies in subsequent figures. B. Interconnections of basic elements to form D-shaped sagittal ring and collar pores. C, D. Basic elements in simple natural forms. E, F. Basic elements in a more complex form. View of broad conical form from base toward apex. Note that lateral and dorsal spines continue from collar pore region into surrounding lattice.

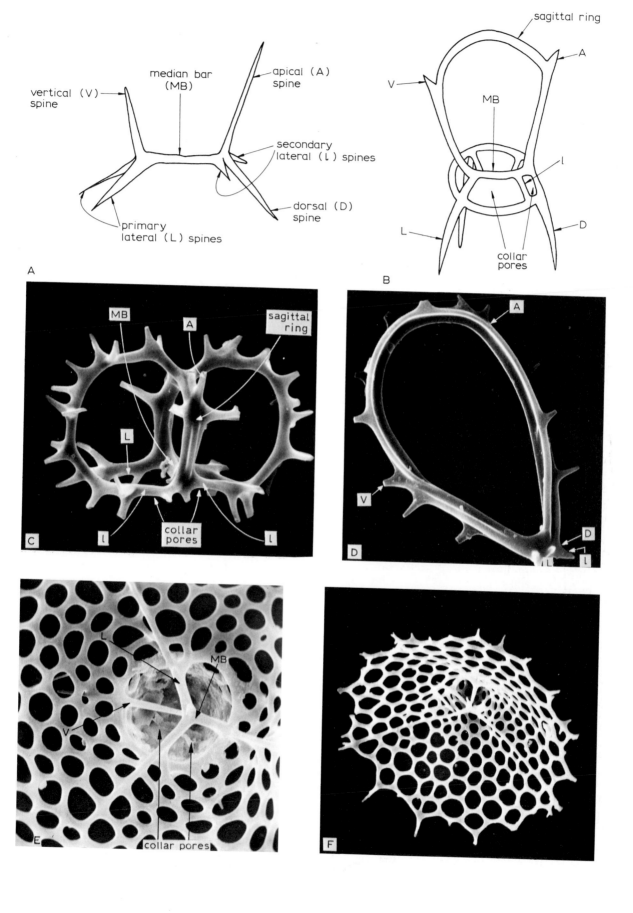

Basis for subdivision

Families are distinguished in the spumellarians on the basis of overall test shape and wall structure, and in the nassellarians on the basis of homologies in basic structures (Fig. 9). Further subdivision depends generally on structures peculiar to individual family-level taxa with little consistency between them. For example, characters such as pore size, shape and arrangement are commonly significant at the species level, but may characterize even family-level groups.

Note: In the descriptions the following abbreviations are used: S = shape; WS = wall structure; DF = distinctive features; R = remarks; GR = geologic range.

Spumellarians

The spumellarians are distinguished by radial symmetry, and several groups can be readily separated by departure from a strict spherical shape, as well as by wall structure. Thus, the ellipsoidal and discoidal groups involve the lengthening or shortening of one axis, and these groups are further subdivided on the basis of structural peculiarities.

Distinctive internal structures in the initial growth stages of several spumellarian groups support close relationships among various discoidal and ellipsoidal groups as suggested by Hollande and Enjumet on the basis of cytological similarities. This leaves a host of spherical forms with widely variable structure, and the systematics of these spumellarians remains one of the most perplexing problems in radiolarian taxonomy.

Entactiniids (Fig. 10)
S: Spherical to ellipsoidal.
WS: Latticed.
DF: Simple, eccentric, internal spicule connected to outer shell by radial bars.
R: Superficially similar internal structures occur in some living actinommids.
GR: Ordovician to Carboniferous.

Orosphaerids (Fig. 11)
S: Spherical to cup-shaped.
WS: Coarse, polygonal, latticed.
DF: Large size (1—2 mm), irregular pore size and shape.
R: Often the only forms preserved in deep sea brown clays (see Friend and Riedel, 1967). May be rare or broken in normal preparations, requiring special separation from coarser size fraction. Fragments of some forms can be identified to genus level for age estimate.
GR: Eocene to Recent.

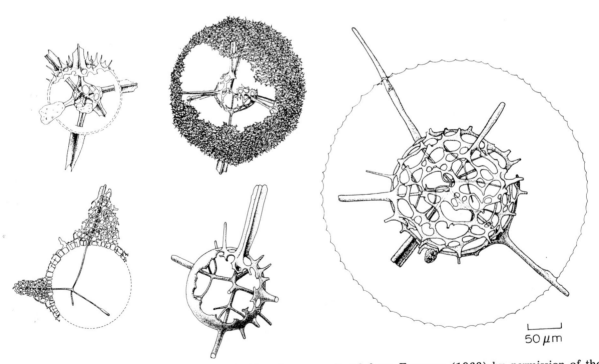

50 μm

Fig. 10. Entactiniids. Note internal spicule. Drawings reproduced from Foreman (1963) by permission of the author and the American Museum of Natural History, Micropaleontology Press.

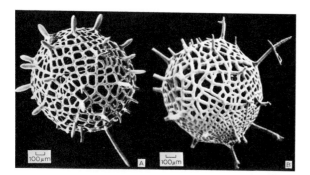

Fig. 11. Orosphaerids. Note coarse, angular lattice.

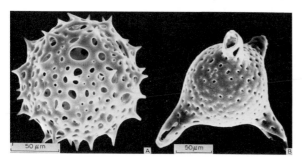

Fig. 12. Collosphaerids.

Collosphaerids (Fig. 12)
S: Spherical to ellipsoidal.
WS: Perforate plate.
DF: Wall structure, thin wall, generally small.
R: Shells often irregular, may bear tubular projections. Colonial forms, colonies reach dimensions of several centimeters, colonies not preserved in sediments.
GR: Lower Miocene to Recent.

Actinommids (Fig. 13)
S: Spherical to ellipsoidal, not discoidal.
WS: Latticed.
DF: Regular latticed wall, little departure from spherical shape, single or multiple shells.
R: A large, polyphyletic family suggested by Riedel for generally spherical forms whose relationships have not yet been determined; thus, subject to future revision and subdivision. Two subfamilies (see below) can be separated at this time.
GR: Paleozoic (?), Triassic to Recent.

Saturnalins (Fig. 14)
S: Spherical.
WS: Latticed.
DF: Outer ring connected to spherical latticed or spongy shell by two or more spines, rarely joins spherical shell directly.
R: Subfamily of actinommids. Forms with spiny outer ring restricted to Mesozoic.
GR: Triassic to Recent.

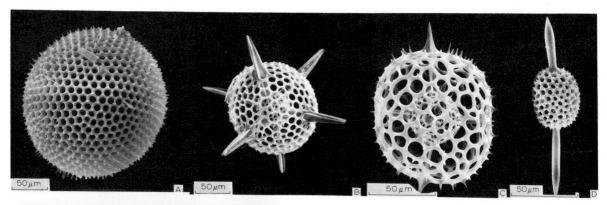

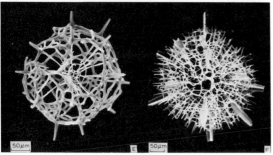

Fig. 13. Actinommids.

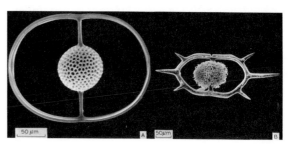

Fig. 14. Saturnalins.

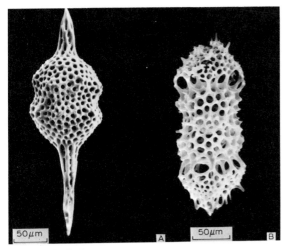

Fig. 15. Artiscins.

Fig. 16. Phacodiscids.

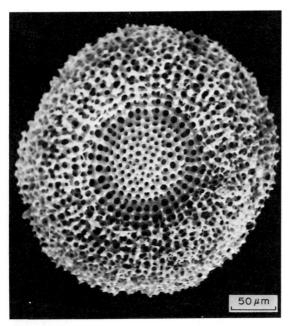

Fig. 17. Coccodiscid.

Artiscins (Fig. 15)
S: Ellipsoidal.
WS: Latticed, some forms with spongy accessory structures.
DF: Equatorial constriction.
R: Subfamily of actinommids, underwent rapid evolution in Miocene; similar forms, some elliptical with solid spines, others discoidal, are excluded.
GR: Oligocene to Recent.

Phacodiscids (Fig. 16)
S: Discoidal, biconvex to flat, rarely triangular.
WS: Latticed outer shell.
DF: Plain, latticed outer shell separates these from other discoidal families.
GR: Mesozoic (possibly Paleozoic) to Recent.

Coccodiscids (Fig. 17)
S: Discoidal, lenticular.
WS: Latticed.
DF: Latticed shell surrounded by chambered girdles or by chambered or spongy arms.
GR: Mesozoic to Oligocene; doubtful reports from Recent.

Spongodiscids (Fig. 18)
S: Discoidal.
WS: Spongy.
DF: Spongy wall of central disc.
R: A polyphyletic family including some with outer porous plate and some with radiating arms or marginal spines.
GR: Devonian to Recent.

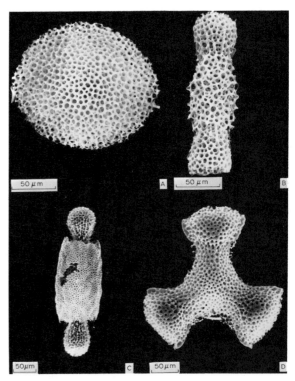

Fig. 18. Spongodiscids.

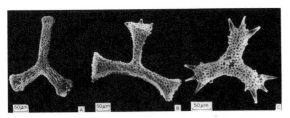

Fig. 19. Hagiastrids. Reproduced from Pessagno (1971) by permission of the author and the Paleontological Research Institution.

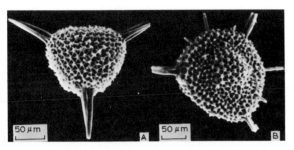

Fig. 20. Pseudoaulophacids. Reproduced from Pessagno (1972) by permission of the author and the Paleontological Research Institution.

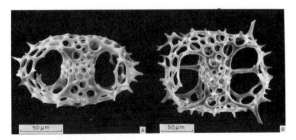

Fig. 21. Pyloniids.

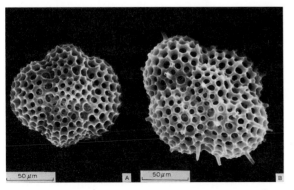

Fig. 22. Tholoniids.

Hagiastrids (Fig. 19)
S: Discoidal, flat.
WS: Spongy, rectangular.
DF: Two, three or four radial arms of regular, rectangular spongy meshwork.
R: Radial arms usually more conspicuous than central area.
GR: Mesozoic.

Pseudoaulophacids (Fig. 20)
S: Discoidal, lenticular to triangular.
WS: Spongy, triangular.
DF: Regular, equilaterial triangular meshwork in concentric layers.
R: Usually bear a few prominent marginal spines.
GR: Mesozoic.

Pyloniids (Fig. 21)
S: Ellipsoidal.
WS: Latticed.
DF: Successively larger, latticed elliptical girdles in three perpendicular planes.
GR: Eocene to Recent; common only from Miocene to Recent.

Tholoniids (Fig. 22)
S: Ellipsoidal.
WS: Latticed.
DF: Cortical shell divided into dome-shaped segments separated by annular constrictions or furrows.
R: Initial chamber of basic pylonid structure; rare forms, few species.
GR: Pliocene to Recent.

Litheliids (Fig. 23)
S: Ellipsoidal (rarely spherical) to lenticular.
WS: Latticed.
DF: Internal structure coiled.
R: Initial chamber of basic pyloniid structure.
GR: Carboniferous to Recent.

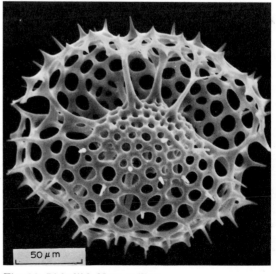

Fig. 23. Litheliid. Note coiling.

Nassellarians

Nassellarians are subdivided on the basis of homologies in a basic skeletal element (see Fig. 9). A prominent apical spine is often termed apical horn, while prominent dorsal and lateral spines may be termed feet. The median bar, apical, dorsal and primary lateral spines, or their homologues are almost invariably recognizable by their relative sizes and angular relationships. This simple spicule itself is the basis for one family, while another is based on a D-shaped ring (Fig. 9D), either isolated or as a prominent sagittal ring (Fig. 9C), resulting from an arched connection between the apical and vertical spines. In other forms, the basic spines are joined together in a latticed chamber called the cephalis, whose size, shape and structure characterize families. In multi-chambered forms, the first two post-cephalic segments are termed **thorax** and **abdomen**.

Plagoniids (Fig. 24)
S: Simple nassellarian spicule or single latticed chamber (cephalis).
DF: Basic spicule without post-cephalic chambers.
R: Wide variety of forms developed from accessory spines and branches, including latticed chamber surrounding spicule; probably a polyphyletic group subject to future subdivision.
GR: Cretaceous to Recent.

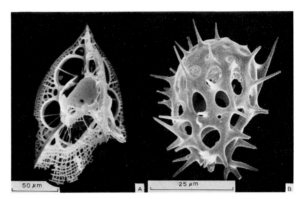

Fig. 24. Plagoniids.

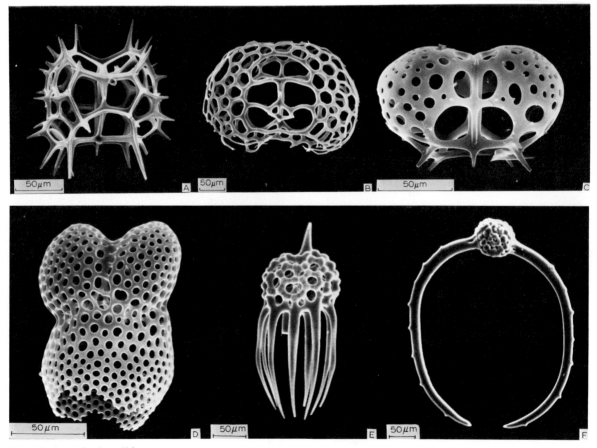

Fig. 25. Acanthodesmiids.

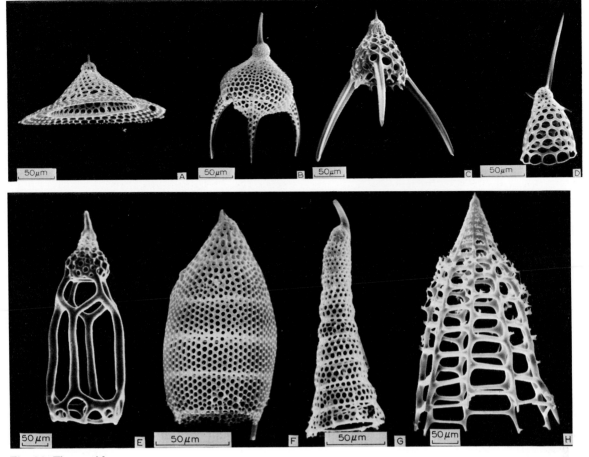

Fig. 26. Theoperids.

Acanthodesmiids (Fig. 25)
S: D-shaped ring or latticed, bilobed chamber with D-shaped sagittal ring.
DF: D-shaped ring always conspicuous externally.
R: Forms range from simple rings to latticed chambers consisting of lobes developed on either side of the D-ring; family has been revised by Goll (1968, 1969) under name Trissocyclidae.
GR: Cenozoic.

Theoperids (Fig. 26)
S: Small spherical cephalis and one or more postcephalic chambers.
DF: Cephalis usually poreless or sparsely perforate.
R: Cephalis contains reduced internal spicule homologous with that of plagoniids; a large, probably polyphyletic group containing majority of ordinary cap- or helmet-shaped nassellarians.
GR: Triassic to Recent.

Carpocaniids (Fig. 27)
S: Small cephalis merging with thorax.
DF: Cephalis nearly indistinguishable from thorax, often reduced to a few bars that are homologous with spicule in other groups.
GR: Eocene to Recent.

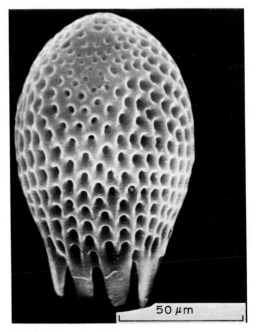

Fig. 27. Carpocaniid.

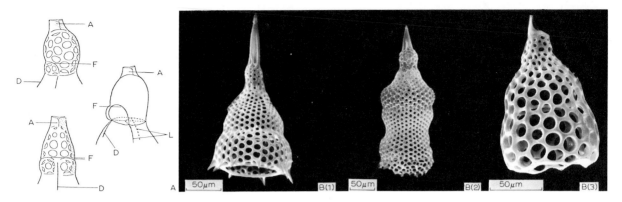

Fig. 28. Pterocorythids. A. Drawings of lobed cephalis showing homologies with fundamental elements shown in Fig. 9. *F*, lateral furrow; see Fig. 9 for other abbreviations. After Riedel (1957); reproduced by permission of the author and the Swedish Natural Science Research Council. B. Photographs of actual specimens.

Pterocorythids (Fig. 28)
S: Large elongate cephalis and one or more post-cephalic chambers.
DF: Paired lateral cephalic lobes separated by furrows directed obliquely downward from apical spine.
GR: Eocene to Recent.

Amphipyndacids (Fig. 29)
S: Elongate cephalis and several post-cephalic chambers.
DF: Cephalis divided into two vertical segments by transverse internal ledge.
GR: Cretaceous to lower Tertiary.

Artostrobiids (Fig. 30)
S: Spherical cephalis and one or more post-cephalic chambers.
DF: A lateral cephalic tubule homologous with vertical spine; pores usually in transverse rows.
GR: Cretaceous to Recent.

Cannobotryids (Fig. 31)
S: Multilobed cephalis, thorax and rarely an abdomen.
DF: Cephalic lobes unpaired and asymmetrical; one lobe homologous with cephalis of theoperids.
GR: Eocene to Recent.

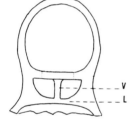

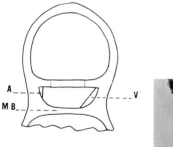

Fig. 29. Amphipyndacids. A. Idealized drawing of cephalic structure. Abbreviations, see Fig. 9. Reproduced from Foreman (1966) by permission of the author and the American Museum of Natural History, Micropaleontology Press. B. Photograph of specimen.

Fig. 30. Artostrobiids. Note cephalic tubule and aligned pores.

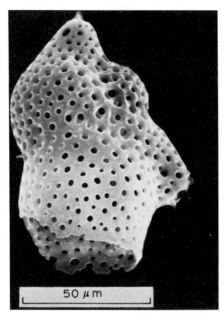

Fig. 31. Cannobotryid.

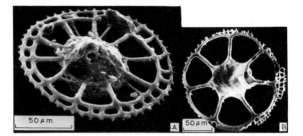

Fig. 32. Rotaformids. Reproduced from Pessagno (1970) by permission of the author and the Paleontological Research Institution.

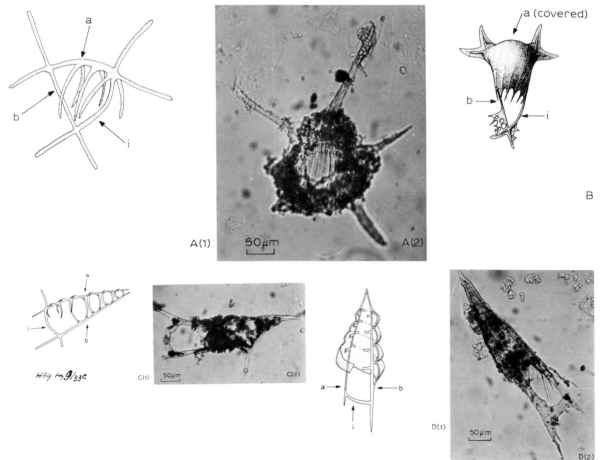

Fig. 33. Albaillellids. Drawings A, C, D from Holdsworth (1969); drawing B from Foreman (1963); reproduced by permission of the authors and the American museum of Natural History, Micropaleontology Press. Photographs courtesy B.K. Holdsworth. Arrows indicate basic skeletal elements as designated by Holdsworth: *a*, "a-spine"; *b*, "b-spine"; *i*, "intersector". A. *Ceratoikiscum* (drawing not to scale). B. *Holoekiscus*. C, D. *Albaillella* (drawings not to scale).

Rotatormids (Fig. 32)
S: Lenticular.
DF: Flattened central area enclosing basic nassellarian spicule is connected by radial bars to outer ring.
R: Outer ring departs from usual axial symmetry of nassellarians, but internal spicule is homologous with that in other families.
GR: Cretaceous.

Uncertain affinity

Albaillellids (Fig. 33)
S: Triangular spicule to cone-shaped chamber.
DF: Basic open triangular element, often enclosed by lamella or rib-like spines resulting in bilaterally symmetrical cone.
GR: Silurian to Carboniferous.

Palaeoscenids (Fig. 34)
S: Simple spicule.
DF: Set of four diverging basal spines connected proximally by flat lamellae and topped by two to four apical spines.
GR: Devonian to Carboniferous.

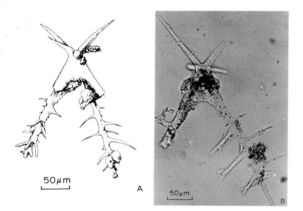

Fig. 34. Palaeoscenids. Drawing reproduced from Foreman (1963) by permission of the author and the American Museum of Natural History, Micropaleontology Press.

EVOLUTION AND GEOLOGIC HISTORY

Spumellarians are the most ancient radiolarians, occurring in all assemblages from the early Paleozoic to the present. The spherical forms have the longest history and are probably the most conservative group. Paleozoic forms are distinguished by a distinctive internal spicule but appear outwardly very similar to more recent forms. On the other hand, spumellarians as a whole, have undergone great proliferation since the Paleozoic, particularly in the Cenozoic.

Nassellarians first appear in the Mesozoic and steadily diversify in the Cenozoic. A trend toward morphologic simplicity with time is suggested in the nassellarians. Although we tend to think of the simple spicular skeletons as primitive, these forms, in fact, appeared late in the fossil record and must have developed from the more complex forms which contain an internal homologue of the simple spicule.

A trend toward lighter tests during the Cenozoic is also apparent. We note, for example, that Eocene assemblages contain generally more heavily silicified forms than upper Tertiary or Quaternary forms. Harper and Knoll (1975) suggested that this trend may be due to selection pressure applied through reduction of silica in the oceans by increasing evolutionary success of diatoms through the Cenozoic.

Radiolarian evolution during the Cenozoic is illustrated by various lineages outlined in the Appendix. Riedel and Sanfilippo (1977b) have noted examples among radiolarians of two somewhat unusual kinds of evolution which they describe as (1) rectilinear, monophyletic change, and (2) coexisting and contrary trends for morphologic change. In the first case, monophyletic lineages can be described in rather simple terms. Thus, for example the *Cannartus—Ommatartus* lineage (see Appendix) shows a reduction in size of spongy polar columns while polar caps become more prominent. Throughout the lineages, there is a tendency for any sample to contain, along with the dominant morphology for any species, a few specimens of both ancestor and descendent forms. Furthermore, many extinct lineages tend to end in very unusual ("bizarre") forms, as for example, in the *Dorcadospyris* and *Podocyrtis* lineages. These traits suggested to Riedel and Sanfilippo an "inertial" kind of evolution, in which trends are "established within the organism, and not influenced (except perhaps at the initiation of the lineage, and certainly at its termination) by the forces of natural selection". This, they pointed out, is unlike Darwinian natural selection, but perhaps a reasonable hypothesis for these fairly simple organisms living in relatively stable tropical environments.

The second evolutionary tendency exemplified by radiolarians involves complex changes in dominance of several coexisting, related morphologies. Thus, for example, in the late Cenozoic development of the genus *Spongaster*, circular, bipolar and polygonal forms are almost invariably present. However, the time-series can be broken up into segments during which first the circular, then the bipolar, then the polygonal forms dominated the assemblages.

PALEOECOLOGY AND PALEO-OCEANOGRAPHY

Principles of deposition of modern radiolarian-bearing sediments

Radiolarians play an important role in the silica cycle in the oceans, a subject which has been reviewed in some detail recently by Calvert (1974), and by Heath (1974). Radiolarian skeletons become buried in the sea-floor sediments to which they sometimes make an important contribution. Mud with 20 to 30% biogenous silica dominated by radiolarians, called **radiolarian ooze**, covers certain areas of the ocean bottom. Mobilization and eventual redeposition of silica after burial is generally believed to result in radiolarian cherts. Accumulation of radiolarian-rich sediments depends on complex, interrelated processes affecting their productivity in the water, their transfer to the ocean bottom, preservation in sediments and dilution by other organic and inorganic components. Heath suggested that not more than about 10% of the siliceous skeletons produced in the water column become permanently added to the geologic record.

Silica is soluble in sea water, so part of the radiolarian assemblage in the water column is dissolved before it reaches the bottom. In laboratory experiments, T.C. Johnson (1974) has shown significant dissolution of radiolarians in a matter of hours to weeks. The rate of dissolution is more rapid in the silica-poor waters of the upper 1000 m in the Pacific, and does not increase at great depths as it does for calcareous forms (Berger, 1968). Dissolution is selective with respect to various taxonomic groups. The rule of thumb that skeletons with thinner elements dissolve more rapidly is consistent with the results of Johnson's experiment.

Further dissolution of radiolarian skeletons takes place at the sediment surface and, after burial, in the sediment column with much of the silica being returned to the water. Factors controlling these processes are poorly known, but saturation levels of bottom and interstitial waters appear to be of prime importance. Siliceous fossils are generally better preserved where rates of sediment accumulation, particularly of organic constituents, are higher. This presumably has the combined effect of rapidly removing skeletons from corrosive bottom waters and buffering the interstitial waters by partial dissolution of the accumulating mass of biogenic silica. Likewise, dissolution of inorganic silicates would have a buffering effect, while uptake of interstitial silica by clays or incorporation of silica in other authigenic silicates would adversely affect skeletal preservation. Nevertheless, radiolarians are preserved in sediments bearing interstitial waters undersaturated with respect to amorphous silica. Surface coatings or complexing with cations (e.g., Mg, Al) have been suggested as protective media.

The role of igneous activity in the origin of siliceous sediments has been a matter of continuing debate, principally because of the frequent association of volcanic with radiolarian-rich rocks. Volcanism is thought to contribute variously, either to direct precipitation of amorphous silica on the sea floor or to an increase in production of siliceous organisms. Heath (1974) calculated that volcanic contributions of silica to the present oceans are minor, and no unequivocal evidence of direct precipitation of amorphous silica on the deep-sea floor is known. Although enrichments of dissolved silica and associated plankton blooms in the vicinity of active volcanoes have occasionally been reported, these would be of limited spatial and temporal extent. In any case, their influence on productivity could hardly compete with the masses of plankton related to oceanographic cycling of nutrients in the vast equatorial and polar regions. Volcanism may contribute indirectly to skeletal preservation by addition of silica to the interstitial environment either by direct fluid emanations or alteration of volcanic glass. But this buffering action would again be restricted to relatively small areas.

Radiolaria accumulate in abundance in equatorial sediments where productivity is high in the water column above. However, the productivity of other organisms is also high, so that radiolarians are often masked by large quantities of foraminifera and calcareous nannoplankton, except in areas, notably in the northern tropical Pacific Ocean, where the sea floor lies below the carbonate compensation level and essentially all calcareous skeletal material is dissolved before or soon after it reaches the bottom. Here, siliceous oozes (dominated largely by radiolarians in the equatorial Pacific, by diatoms in most other places) accumulate (Fig. 35).

Radiolarians are also relatively rich under the high-latitude productivity belts — particularly around Antarctica and in the North Pacific. Here calcareous skeletons are generally lacking, and radiolarians are generally subordinate to diatoms.

In the central parts of the oceans, productivity is low, rates of accumulation of terrigenous sedimentary components is also low, and depths are generally below the carbonate compensation level. Low dilution, then, might be expected to override the low productivity of radiolarians, but the lengthy exposure at the sediment—water interface results in relatively complete dissolution, so it is in these regions that the barren pelagic brown clays accumulate.

A recurring theme in this discussion has been the reciprocal preservational characteristics of calcareous and siliceous microfossils. Such reciprocity causes a striking difference in radiolarian occurrences between the sedi-

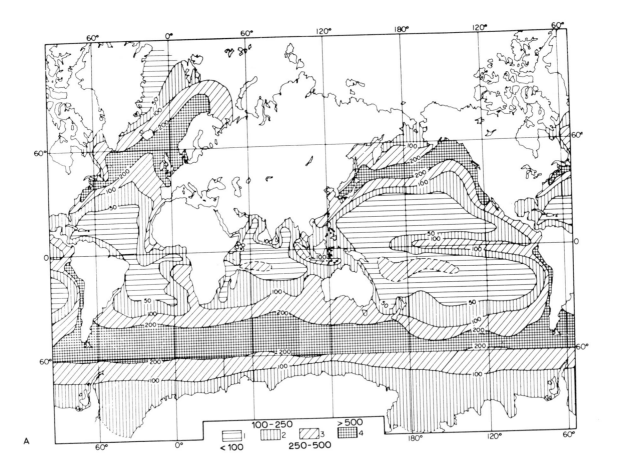

Fig. 35. Relationship between plankton productivity and silica in sea-floor sediments. After Lisitzin (1971); reproduced from B.M. Funnell and W.R. Riedel (Editors), *The Micropaleontology of Oceans*, copyright 1971, by permission of the author and the Cambridge University Press. A. Annual production of silica in the plankton in

ments of the Atlantic Ocean on the one hand, and the Pacific and Indian Oceans on the other. Radiolarians are widely distributed and often abundant in sediments of the Indo-Pacific Ocean, where bottom waters relatively rich in silica and poor in carbonate favor retention of siliceous skeletons. In the Atlantic, radiolarians are produced in abundance in the water, but are relatively scarce in younger Tertiary sediments because bottom waters are relatively deficient in silica and chemical conditions favor preservation of carbonate.

Radiolarians are generally very rare or absent in continental margin sediments where they are diluted by large influxes of terrigenous material, which may also provide a chemical sink for silica. On the continental shelf off the western United States, for example, radiolarians appear near the surface but disappear rapidly at depth in the sediments. They may, on the other hand, be abundant in relatively shallow basins not far from shore where chemical conditions are particularly favorable for preservation and terrigenous dilution is relatively low. For example, in the present-day Santa Barbara Basin, laminated anaerobic sediments rich in radiolarians (and other microfossils) are accumulating in water depths of about 500 m. Similar occurrences are noted in fiords along the coasts of western North America and Norway.

Application to ancient sediments

Interpretation of radiolarian cherts
Radiolarian cherts are prominent during

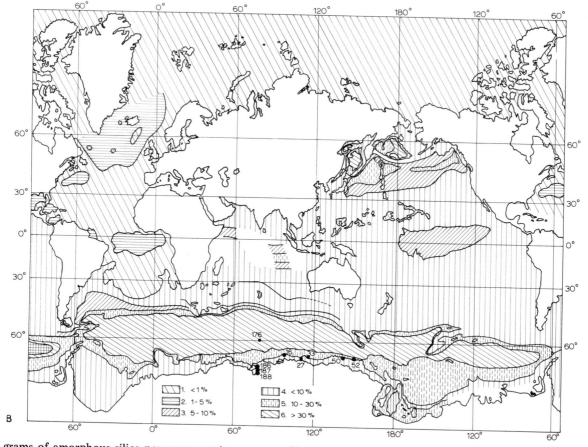

grams of amorphous silica per square meter per year. Similar patterns are shown by plots of phytoplankton and zooplankton standing crops and nutrient chemicals other than silica. B. Amorphous silica (primarily skeletons of diatoms and radiolarians) in surface sediments in percent of the dry sediment.

certain times in earth history with the best known episode being that during Jurassic time in the Tethyan region. The classic Alpine radiolarites were deposited at this time, as well as formations in California and other places. Following Steinmann's (1905, 1925) pioneering comparison of the Alpine radiolarites with abyssal pelagic sediments, recent years have seen a revival of interest in them in light of their apparent role in early stages of sea-floor spreading. Garrison (1974) reports that most radiolarian cherts share with deep-sea deposits a very slow rate of sedimentation (Table II). Typical Alpine sections measuring tens of meters, represent some thirty to forty million years, comparing well with typical pelagic sedimentation rates of one to a few meters per million years.

In cases where radiolarian cherts overlie pelagic limestones, bathymetric interpretation in terms of fluctuation of the carbonate compensation level is appropriate. Thus, a situation like the present can be envisioned, where calcareous sediments, deposited on new sea floor near relatively shallow spreading centers, are overlain by siliceous sediments where ridge flanks subside below the compensation depth. Approximate depths can be interpreted for the siliceous to calcareous facies change based on present-day average compensation levels, as Bosellini and Winterer (1975) have done, but compensation levels in the past may have changed in response to the influence of climatic fluctuations on subsurface water-mass distributions.

In other cases, radiolarian cherts directly overlie oceanic basalts. Often the radiolarian rocks are intimately interlaced in pillow basalts, clearly indicating deposition in a basin remote from terrigenous influx and below the carbonate compensation level. This seems to violate the model of new ocean floor created at shallow spreading centers and thus acquiring an initial cover of carbonates. However, the flexibility of the depth of the compensation level and the fact that the calcareous plankton (planktonic forams, calcareous nannoplankton) first appear in the Jurassic, provide plenty of room for alternative interpretations. Garrison (1974) has suggested useful working examples of non-spreading sea floor such as sea mounts and diapiric structures, while Bosellini and Winterer (1975) analyzed the effect on the compensation level of the Mesozoic shift from a dominantly neritic to pelagic biogenous carbonate budget.

Radiolarian cherts are sometimes associated with turbidites, and in some cases the cherts themselves retain sedimentological evidence of turbidity current origin. These too are subject to varying interpretation depending on supporting evidence. Biogenic debris itself may be redistributed on the flanks of ridges as suggested by Nisbet and Price (1974). On the other hand, association with coarse terrigenous turbidites may indicate juxtaposition to continental sources. One setting in which this may take place is in trenches at the base of tectonically active continental slopes. Such a setting is suggested by inclusion of radiolarites

TABLE II

Characteristics of ancient and modern radiolarian deposits

	Terrigenous component	Average rate of sedimentation	Associated microfossils
Radiolarian ooze	low	1 m/m.y.	diatoms foraminifera calcareous nannoplankton
Radiolarites	low	1 m/m.y.	foraminifera calcareous nannoplankton diatoms ?
Santa Barbara Basin	moderate	1000 m/m.y.	diatoms foraminifera calcareous nannoplankton
Monterey Formation	moderate	1000 m/m.y.	diatoms foraminifera calcareous nannoplankton

in complexes of structurally highly contorted and fractured rock masses termed mélanges, and thought to represent ancient zones of subduction of oceanic against continental crust. This kind of depositional environment has been suggested for Franciscan radiolarites in California by Chipping (1971).

From this brief discussion we see that some radiolarites may be interpreted as early deposits in moderately to very deep basins. On the other hand, it is imprudent to presume a deep-water origin for all deposits rich in radiolarians. Radiolarians and other typically pelagic microfossils can occur in quite shallow-water sediments. For example, the Solnhofen Limestone of southern Germany is replete with radiolarians and calcareous nannofossils, but with macrofossil and sedimentological evidence indicating a shallow-water depositional site (Stürmer, 1963; Barthel, 1970).

The Monterey Formation — another case
Although the Miocene Monterey Formation of California is known principally for its abundance of marine diatoms, it is also rich in radiolarians as well as foraminifera and calcareous nannoplankton. These rocks are different from the typical Alpine radiolarites. Hard chert or chert-like layers are often well developed but are associated with conspicuous amounts of soft, diatom- and radiolarian-rich rocks, the purest of which are termed diatomites. The diatomites are typically rhythmically bedded with the finest laminae consisting of couplets of light and dark layers. These couplets probably represent seasonal fluctuations in supply of organic and/or inorganic sedimentary components and, thus, would be annual accretions properly regarded as marine varves. Also striking is the fact that the rate of deposition, allowing for consolidation to the present fraction of a millimeter, approaches one millimeter per year. This is orders of magnitude greater than any deep-sea sediments (see Table II). This and paleogeographic considerations suggest that land was not far removed from the depositional site.

Bramlette, in his monograph on the Monterey Formation (1946), concluded that the evidence favors deposition in shallow to moderately deep basins in areas of high organic production and relatively low rates of terrigenous sedimentation; that is, low relative to normal continental shelf deposits. Recent investigations of sediments in California borderland basins (Soutar, 1971) suggest a good analogue for the Monterey Formation. The sediments in many of these basins are laminated, with annual increments of light and dark layers of about one millimeter, and are rich in the same kinds of microfossils as the Monterey. They accumulate at depths of around 500 m where the oxygen minimum impinges on the sea floor and burrowing invertebrate animals are unable to live. Although a detailed comparison has yet to be made, this is a sedimentary environment with many traits in common with the Monterey Formation, and an example of a radiolarian-rich deposit accumulating rapidly in a basin of only modest depth, relatively near-shore.

Paleoclimatology

Because of their world-wide distribution and great diversity, radiolarians are an important paleoclimatic tool, particularly in sediments lacking calcareous fossils. Recent years have seen an accelerating use of radiolarians in paleoecologic studies of such carbonate-poor ocean floors as the North Pacific and Antarctic.

In 1970, Nigrini showed similarities between distribution patterns of groups of radiolarian species commonly associated in the sediments and the patterns of present-day surface water masses. She derived from these groups an index related to temperature, one of the defining characteristics of the water masses. Changing compositions of radiolarian assemblages with depth in a core could then be presented as relative temperature fluctuations. Johnson and Knoll (1974) applied Nigrini's index to two Pleistocene cores from the equatorial Pacific. Their curves showed fluctuations during the past 300,000 years which were closely correlated with each other and with calcium-carbonate fluctuations. Using a similar approach, but employing different statistical techniques, studies by Moore (1973) in the North Pacific showed quantitative relationships between relative abundances of radiolarian species and average temperatures of near-surface waters. The resulting radiolarian-temperature equations were used to estimate absolute paleotemperatures in cores through portions of the late Pleistocene dated by radiometric methods. Thus, Moore was able to suggest a warming trend over the last 24,000 years from temperatures about 4°C cooler than the present average off Oregon. Sachs (1973) reported temperature fluctuations ranging from present values to 10°C cooler over the last 200,000 years in the central subarctic Pacific.

Biological productivity and sea-floor spreading

The association of belts of sediments rich in siliceous microfossils with zones of high biological productivity in the overlying waters provides a valuable tool for broad-scale paleo-oceanographic reconstructions of the current systems responsible for these fertile areas. Although calcareous microfossils are similarly affected by productivity, their preservation is adversely affected by dissolution at depth.

Riedel and Funnell (1964) charted microfossil distributions for the Tertiary of the Pacific Ocean, and their maps suggest that the broad patterns of currents and water masses were generally in the same positions and with the same orientations during most of the Tertiary as at present. In detail, however, they noted that two occurrences in the Pliocene suggest higher productivity, that the northern boundary of the Equatorial Current System seems to have been about 5° farther north along 139°W during the Miocene with a similar northward extension during the Oligocene, and that abundant Eocene radiolarians are found both farther south and north of their present equatorial zone of abundance. Thus, it was suggested that the Equatorial Current System narrowed in the late Tertiary, possibly indicating stronger circulation related to cooling during that period.

Subsequent to 1964, however, it became evident that accumulation of these sediments on a spreading sea floor would cause offset of their positions with time. Thus, some of the anomalies mentioned above can be explained in terms of progressive northwestward migration of the Pacific equatorial ocean floor west of the East Pacific Rise (Berger and Winterer, 1974).

BIOSTRATIGRAPHY

Historically, radiolarians have had a bad reputation as biostratigraphic tools. But as we have seen, this has resulted from a lack of concentrated study leading to a natural classification based on phylogenetic relationships. Rapid progress is now being made, however, and radiolarians are recognized as one of the most important groups for long-range age correlations. A detailed zonation has been formulated for the Cenozoic, and recent studies promise comparable success with older faunas.

Paleozoic

Radiolarians have been reported from rocks as old as Precambrian although there is doubt as to the authenticity of either the age of such collections or their identification as radiolarians. Reports of Cambrian radiolaria are similarly open to question, but well-preserved collections from rocks assigned with confidence to the Ordovician are now known (Fortey and Holdsworth, 1972; Dunham and Murphy, 1976). Radiolarians are known from all the remaining Paleozoic periods.

Less is known about Paleozoic radiolarian faunas than those of younger eras, but recent studies by Deflandre, by Foreman and by Holdsworth (see references), have provided accurate descriptions and illustrations of unique Paleozoic forms. Typical Paleozoic radiolarians are spumellarians with internal spicules (entactiniids, Foreman, 1963) and forms such as albaillelids and palaeoscenids (Deflandre, 1960; Holdsworth, 1969, 1971).

Continuing studies of Paleozoic Radiolaria promise further expansion of our knowledge of the systematics and biostratigraphy of this important segment of radiolarian history. Particularly useful will be study of sequences that can be accurately dated on the basis of known stratigraphically important fossil groups such as graptolites and conodonts.

Mesozoic

Triassic radiolarians are scarcely known at all, but a number of Jurassic species have been reported from moderately well-preserved material whose ages are reliable. Results of detailed, up-to-date studies of well-preserved, accurately dated Jurassic assemblages are now beginning to appear.

A number of well-preserved Cretaceous collections have been studied recently, and it now is possible to make some subdivisions of this period on the basis of radiolarians. Pessagno (1976) has proposed a zonation for Upper Cretaceous radiolarians of California that promises to be of great value in an area where other microfossils are rare or lacking; Riedel and Sanfilippo (1974) combine information from land-based samples with those from Deep Sea Drilling Project Leg 26 to propose a coarse zonation for the entire Cretaceous. Although it is not yet possible to present a Cretaceous radiolarian biostratigraphy that has been widely tested, a number of distinctive taxa can be briefly described and illustrated. These occur widely in Cretaceous assemblages and should serve to distinguish Cretaceous assemblages from those of other ages (see Appendix).

Important descriptions of Mesozoic assemblages by Dumitrică (1970), Foreman (1968), and Pessagno (1976) will aid in identification of new collections, and additional contributions by various authors appear in the Initial Reports of the Deep Sea Drilling Project.

Cenozoic

Radiolarians in Tertiary and Quaternary deposits are by far the most thoroughly studied, and we can now make accurate age determinations of unknown assemblages over most of this geologic interval. The studies contributing to our detailed Cenozoic radiolarian biostratigraphy have been carried out largely on deep-sea sediment samples in which calcareous microfossils also occur. This has allowed for comparisons between radiolarian sequences and those of better known and also widely occurring planktonic groups such as the foraminifera and calcareous nannoplankton.

A detailed zonation for the Eocene through Quaternary interval in tropical regions was first proposed by Riedel and Sanfilippo during Leg 4 of the Deep Sea Drilling Project and has since been tested and amplified during many subsequent DSDP legs. (See Riedel and Sanfilippo, 1977b, for a summa-

ry.) Absolute age estimates for these zones have been derived from the paleomagnetic time-scale by Theyer and Hammond (1974a, b).

Hays (1965) proposed a zonation for Antarctic deep-sea sediments of Pliocene–Pleistocene age. This was subsequently used in studies of paleomagnetism in ocean-floor sediments and compared to absolute ages derived from the radiometric–paleomagnetic time scale (Hays and Opdyke, 1967). Another zonation was based on North Pacific sediment cores of late Pliocene–Pleistocene age (Hays, 1970), and again coordinated with radiometric–paleomagnetic dates. The Pleistocene of the eastern tropical Pacific has been divided by Nigrini (1971) into four zones which may be applicable in other areas. Johnson and Knoll (1975) estimated absolute ages for these zones as well as other Quaternary radiolarian datum levels.

It is beyond the scope of this book to present a detailed description of existing zonations. Rather, some of the most important species and evolutionary lineages are briefly described in the Appendix and illustrated with known ranges of species. In doing this, the intention has been to select the groups most commonly encountered and most easily recognized. This should enable a student to estimate the age of a tropical Cenozoic radiolarian assemblage (see below).

APPENDIX AND RANGE CHART

The Appendix presents abridged descriptions of selected stratigraphically important forms. Forms 1 through 68 are from the Cenozoic. They are listed in chronological order and illustrated in the accompanying range chart. Forms 69 through 71 are characteristic Mesozoic forms.

On the range chart, approximate known age ranges are indicated by the height of the illustration or by a vertical line either extending from or beside the illustration. Diagonal arrows indicate probable evolutionary development.

Although reference to photographs in the chart will normally suffice for identification, the notes accompanying the charts summarize the diagnostic features of each illustrated species with brief comments on their evolution.

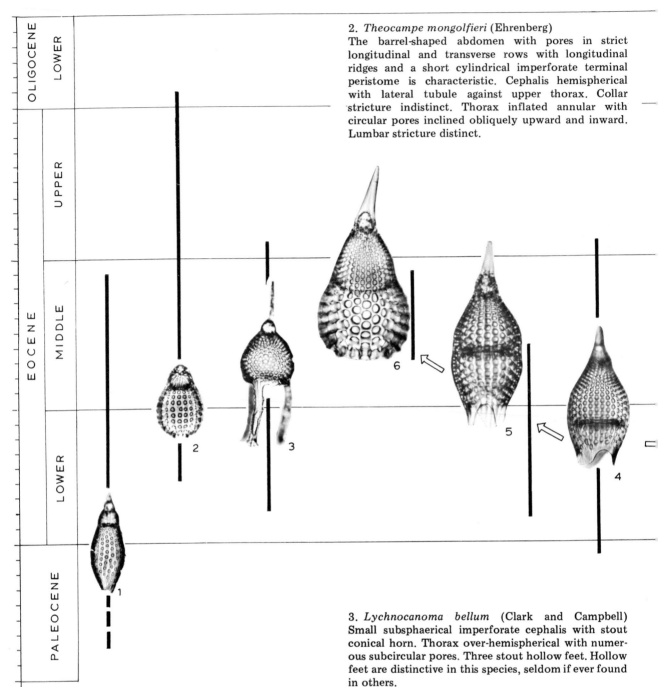

2. *Theocampe mongolfieri* (Ehrenberg)
The barrel-shaped abdomen with pores in strict longitudinal and transverse rows with longitudinal ridges and a short cylindrical imperforate terminal peristome is characteristic. Cephalis hemispherical with lateral tubule against upper thorax. Collar stricture indistinct. Thorax inflated annular with circular pores inclined obliquely upward and inward. Lumbar stricture distinct.

1. *Phormocyrtis striata* Brandt
Characterized by overall fusiform shape. Cephalis hemispherical, with small pores and bladed horn. Collar stricture indistinct. Thorax hemispherical with irregularly arranged circular pores. Lumbar stricture indistinct. Abdomen fusiform with greatest width near middle, with circular pores in longitudinal rows usually separated by ridges. Termination ragged, toothed, or rarely closed. Paleocene and early Eocene specimens with abdomen usually triangular in cross-section.

3. *Lychnocanoma bellum* (Clark and Campbell)
Small subsphaerical imperforate cephalis with stout conical horn. Thorax over-hemispherical with numerous subcircular pores. Three stout hollow feet. Hollow feet are distinctive in this species, seldom if ever found in others.

Podocyrtis papalis — *P. ampla* lineage
This lineage consists of three broadly defined species.

4. *Podocyrtis papalis* Ehrenberg
Diagnostic features are the inflated conical thorax passing into inverted truncate-conical abdomen without external expression of the lumbar stricture, the longitudinal rows of pores separated by ribs, and the pored part of the abdomen shorter than the thorax and below that a poreless part with three large, shovel-shaped feet.

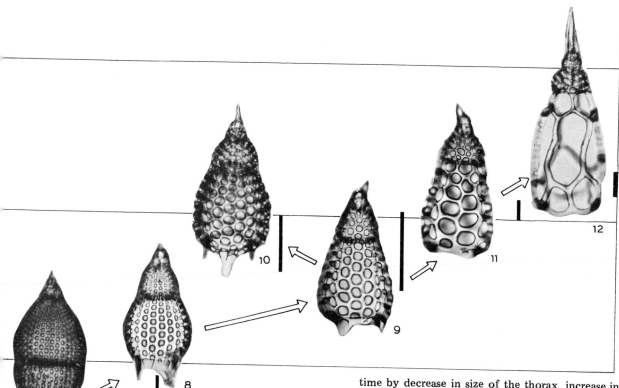

time by decrease in size of the thorax, increase in size of the abdomen and its pores, and a diminution and ultimate loss of the feet.

7. *Podocyrtis aphorma* Riedel and Sanfilippo
Distinguished from *P. papalis* only by less regular abdominal pores and the presence of a slight lumbar stricture.

8. *Podocyrtis sinuosa* Ehrenberg (?)
Distinguished from *P. aphorma* by the larger abdomen and from *P. trachodes* by the smoother surface. With time, the abdomen increases in size while the thorax decreases.

9. *Podocyrtis mitra* Ehrenberg
Distinguished from *P. sinuosa* and *P. trachodes* in having its abdomen widest near the distal end, rather than medially.

10. *Podocyrtis trachodes* Riedel and Sanfilippo
Distinguished by the rough surface of its thorax and (especially) abdomen.

11. *Podocyrtis chalara* Riedel and Sanfilippo
Distinguished from *P. mitra* by larger abdominal pores and in generally lacking feet.

12. *Podocyrtis goetheana* (Haeckel)
Distinguished from *P. chalara* by even larger abdominal pores, a medial transverse row of which are distinctly elongate.

5. *Podocyrtis diamesa* Riedel and Sanfilippo
A form intermediate between *P. papalis* and *P. ampla*, differing from *P. papalis* by its larger size and the presence of a distinct lumbar stricture, and from *P. ampla* in its general spindle-shaped rather than conical form. Thorax and abdomen usually of approximately same length. Pores separated by ridges in early specimens, larger and lacking intervening ridges in later specimens. Three feet shovel-shaped, irregular in some specimens with very restricted apertures.

6. *Podocyrtis ampla* Ehrenberg
Characterized by conical overall shape and abdomen terminating in narrow thickened rim bearing three small shovel-shaped or spathulate feet. Collar and lumbar strictures not pronounced. Abdominal pores larger than thoracic, usually in longitudinal rows without intervening ridges.

Podocyrtis aphorma — *P. goetheana* lineage
P. aphorma evolved from *P. papalis* and then developed through three intermediate forms to *P. goetheana*. *P. trachodes* is apparently a related side branch. The succession is characterized with

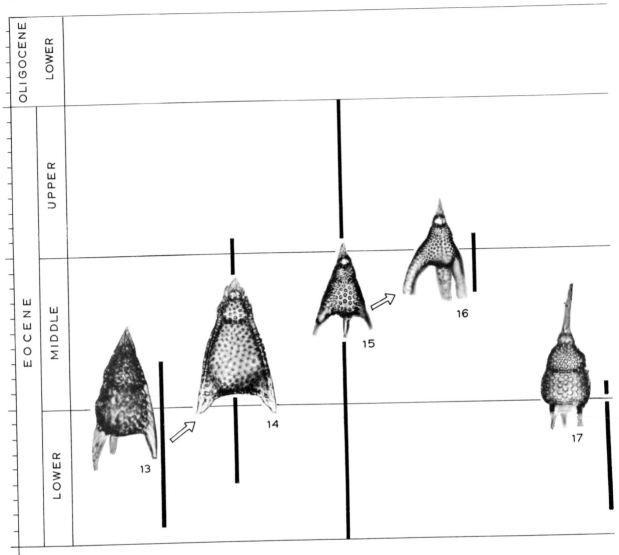

Lithochytris archaea — L. vespertilio lineage
Two broadly defined species constitute this lineage.
These forms resemble superficially the later species
of *Sethochytris* (e.g., *S. triconiscus*) from which
they differ in having two rather than one post-
cephalic segments.

13. *Lithochytris archaea* Riedel and Sanfilippo
Inflated conical thorax with thick wall and circular
pores, distinct lumbar stricture, and abdomen joined
to three divergent feet through most of their length,
with circular pores and truncate distally.

14. *Lithochytris vespertilio* Ehrenberg
Distinguished from *L. archaea* in having shorter free
distal feet, and in having the part of the feet from the
lumbar stricture to the end of the abdomen reduced
to ribs within the abdominal wall.

Sethochytris babylonis — S. triconiscus lineage
This lineage begins with a broadly defined species
group and develops through transitional forms

(not illustrated here; see Riedel and Sanfilippo,
1970) to a terminal species.

15. *Sethochytris babylonis* (Clark and Campbell)
    group
This group includes a wide variety of forms with
small spherical cephalis, pyriform to tetrahedral
thorax with a very restricted mouth, and robust
cylindrical to conical feet and horn.

16. *Sethochytris triconiscus* Haeckel
Distinguished from *S. babylonis* group by having
pyriform thorax drawn out distally into three divergent cylindrical porous tubes which are open terminally.

Thyrsocyrtis hirsuta — T. tetracantha lineage
Three species constitute this lineage.

17. *Thyrsocyrtis hirsuta* (Krasheninnikov)
This species has a campanulate, somewhat inflated,
thorax separated by a distinct lumbar stricture from a

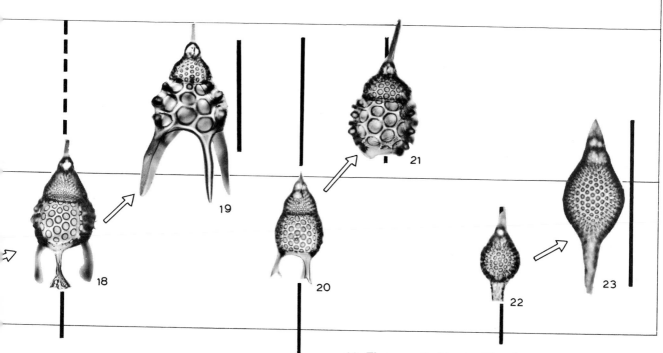

subcylindrical or slightly inflated abdomen of approximately the same length and breadth as the thorax. Three divergent subcylindrical feet arise smoothly from the poreless rim of the abdomen. A form with larger abdominal pores, *T. hirsuta tensa* Foreman 1973, probably gave rise to *T. triacantha*.

18. *Thyrsocyrtis triacantha* (Ehrenberg)
Distinguished from *T. hirsuta* by an abdomen that is much broader than the generally conical thorax, and that has pores distinctly larger than those of the thorax.

19. *Thyrsocyrtis tetracantha* (Ehrenberg)
Typical specimens of this species are distinguished from *T. triacantha* by lack of a differentiated terminal abdominal rim and by strongly divergent feet extending smoothly from the abdominal lattice and which may number up to five or six.

Thyrsocyrtis rhizodon — *T. bromia* lineage
The first of two species in this lineage may have arisen from *T. hirsuta*, but the relationship is still not clear.

20. *Thyrsocyrtis rhizodon* Ehrenberg
The collar stricture and usually lumbar stricture are distinct. Thorax campanulate; abdomen barrel-shaped, wider and usually longer than thorax, and with pores usually two to three times as large as those of the thorax. Surface of thorax and abdomen rough. Three short feet arise smoothly from poreless abdominal rim and are terminally truncate or bluntly pointed, sometimes with an outwardly directed ridge or indistinct thorn, or occasionally more complicatedly branched.

21. *Thyrsocyrtis bromia* Ehrenberg
Characterized by the large, subcircular abdominal pores, often thorny surface, and very short feet (sometimes absent).

Eusyringium lagena(?) — *E. fistuligerum* lineage
Two species are included in this lineage.

22. *Eusyringium lagena* (Ehrenberg) (?)
Characterized by pyriform thorax with thick wall and very constricted mouth. Some late specimens have an abdomen in the form of a narrow tube with very thin wall and irregular pores which is commonly elongated.

23. *Eusyringium fistuligerum* (Ehrenberg)
Distinguished from *E. lagena* by further development of the abdomen such that it arises gradually from the thorax and is distinctly more robust than in *E. lagena*. In some specimens a stricture separates off a narrow proximal part of the thorax. Thorax may bear three inconspicuous wings proximally.

Calocyclas hispida — C. turris lineage
Two species constitute this lineage.

24. *Calocyclas hispida* (Ehrenberg)
Characterized by the numerous spathulate feet arising from the pyriform thorax.

25. *Calocyclas turris* (Ehrenberg)
Apparently developed from *C. hispida* by union of the feet through lateral branches to form an abdomen.

Lithocyclia ocellus — Ommatartus tetrathalamus lineage
This is a long lineage which begins with species of the coccodiscid family, the last of which gives rise to the earliest members of the artiscin subfamily by reduction of the three spongy columns to two and rotation of the axis of radial symmetry through 90°. It includes at least eleven or twelve species in a direct lineage ranging from Eocene to the present, and a side branch including two species.

26. *Lithocyclia ocellus* Ehrenberg group
This, as yet undivided group, is characterized by a phacoid (latticed) cortical shell which is surrounded by a continuous spongy zone which, in late specimens especially, is usually concentrically zoned, the innermost zone commonly being the widest and most distinct. Spines of variable number (sometimes none), bladed, acute, originate within the spongy zone, or at the periphery of the cortical shell, or occasionally at the outer medullary shell.

27. *Lithocyclia aristotelis* (Ehrenberg) group
Distinguished from *L. ocellus* group by division of spongy zone into separate spongy arms. Included are forms with three or four (or more?) arms, with or without terminal spines and with or without a patagium (laminar meshwork connecting arms).

28. *Lithocyclia angusta* (Riedel)
In this species, the spongy outer zone is reduced to three narrow spongy arms.

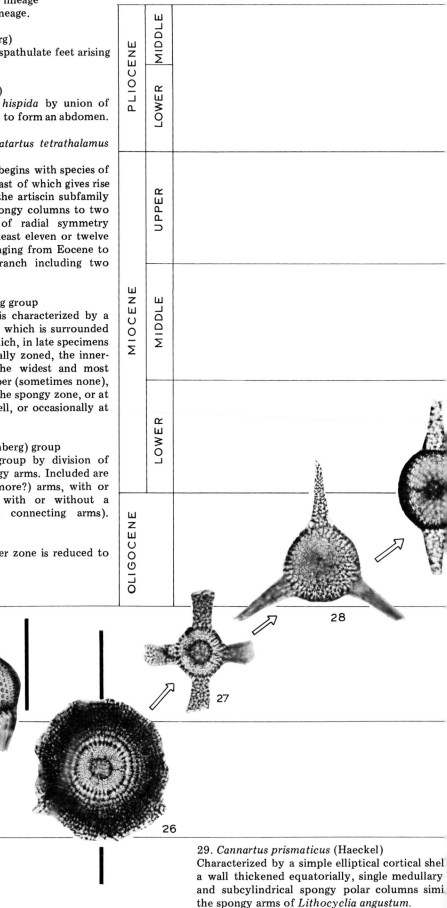

29. *Cannartus prismaticus* (Haeckel)
Characterized by a simple elliptical cortical shell a wall thickened equatorially, single medullary and subcylindrical spongy polar columns simi the spongy arms of *Lithocyclia angustum.*

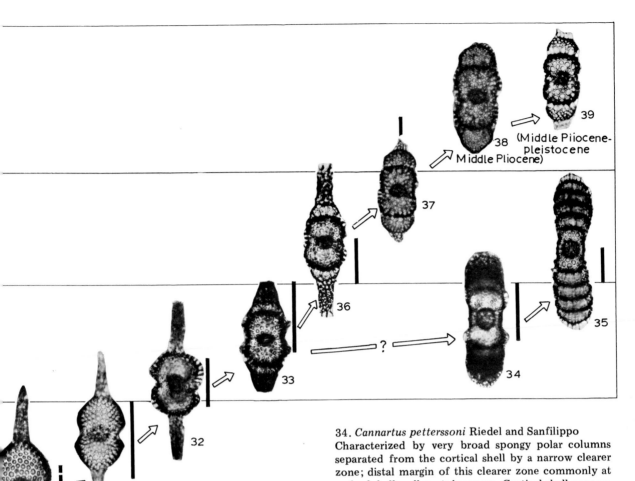

38 (Middle Pliocene-pleistocene)

Middle Pliocene

37

36

35

34

?

33

32

31

30

30. *Cannartus tubarius* (Haeckel)
This and all later artiscins usually have a double medullary shell. Distinguished from *C. prismaticus* by an equatorial constriction at which the shell wall is puckered to form coarse folds. May arise from *C. prismaticus* or more directly from *Lithocyclia angustum*.

31. *Cannartus violina* Haeckel
Thickened folds of cortical shell occur in two broad zones on either side of the equatorial constriction rather than within the constriction as in *C. tubarius*.

32. *Cannartus mammiferus* (Haeckel)
In this species the thickened folds of the wall in the broader parts of the cortical shell are developed into pronounced obtuse mound-like protuberances.

33. *Cannartus laticonus* Riedel
Distinguished by broad conical spongy polar columns. Protuberances of shell wall are similar to those of *C. mammiferus*, and in some specimens tend to be arranged in two girdles around each broadened half of the shell.

34. *Cannartus petterssoni* Riedel and Sanfilippo
Characterized by very broad spongy polar columns separated from the cortical shell by a narrow clearer zone; distal margin of this clearer zone commonly at end of shell wall protuberances. Cortical shell approximately cylindrical (sometimes bulged at equator) with protuberances surrounding each end. May have developed from *C. laticonus*.

35. *Ommatartus hughesi* (Campbell and Clark)
Cortical shell is similar to *Cannartus petterssoni*, but the spongy columns are replaced by series of distinct polar caps, subequal, of approximately the same width as the cortical shell. In early specimens polar caps tend to merge distally into spongy structure. Spongy polar caps of some specimens of *Cannartus petterssoni* are divided into narrow parallel zones, apparently indicating the transition to this species.

36. *Ommatartus antepenultimus* Riedel and Sanfilippo
Distinguished from *Cannartus laticonus* by distinct polar caps between the cortical shell and the conical spongy polar columns. In transitional forms of *C. laticonus*, clear zones between the cortical shell and spongy columns are parallel to the cortical shell wall.

37. *Ommatartus penultimus* (Riedel)
Polar caps are larger (approaching the size of half of the cortical twin-shell) and the spongy columns smaller than in *O. antepenultimus*.

38. *Ommatartus avitus* (Riedel)
Characterized by a tuberculate (knobby) cortical shell, well-developed polar caps and no spongy polar columns.

39. *Ommatartus tetrathalamus* (Haeckel)
Similar to *O. avitus* but with a smooth cortical shell. May have double polar caps.

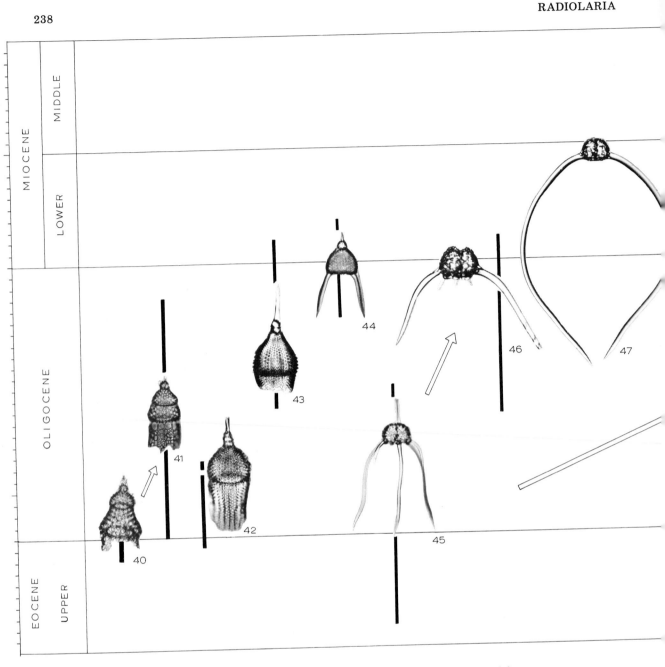

Artophormis barbadensis — A. gracilis lineage
Two species comprise this lineage.

40. *Artophormis barbadensis* (Ehrenberg)

The spherical cephalis bears a loosely spongy (occasionally latticed-bladed) apical horn. Thorax inflated campanulate separated from longer truncate-conical abdomen by distinct lumbar stricture. First three segments with thorny surface. Distinctive fourth segment is short, formed of very irregular latticework of which some elements are longitudinal ribs which commonly extend a short distance as free terminal spines.

41. *Artophormis gracilis* Riedel

Distinguished from *A. barbadensis* by inflated thorax that with the cephalis and abdomen forms a conical portion of the shell, and by a longer, more developed fourth segment. Apical spine may be simple conical, multiple, or latticed.

42. *Theocyrtis tuberosa* Riedel

Characterized by tuberose surface of subhemispherical thorax with pores tending to longitudinal alignment, separated by indistinct lumbar stricture from subcylindrical abdomen with longitudinal ridges (plicae) separated by two to four longitudinal rows of pores.

43. *Theocyrtis annosa* (Riedel)

Characterized by pronounced longitudinal ridges (plicae) on inflated-campanulate thorax separated by three to five longitudinal rows of pores. Abdomen is subcylindrical; thoracic plicae usually extend into abdomen, but less distinctly.

44. *Lychnocanoma elongata* (Vinassa)

Characterized by the two heavy, three-bladed curved feet attached to an inflated hemisperical thick-walled thorax. In some specimens, a small amount of meshwork is developed between the feet. [= *L. bipes* (Riedel) 1959.]

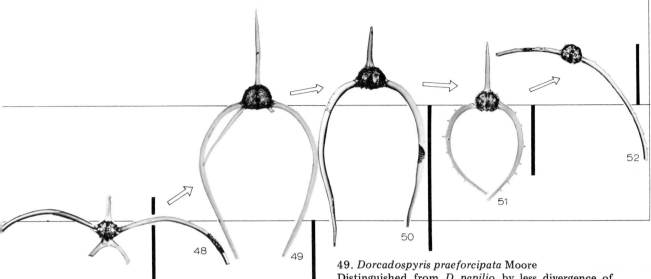

Tristylospyris triceros — Dorcadospyris simplex lineage
This appears to be one branch (three species) of the long, and as yet incompletely understood, development of the genus *Dorcadospyris*.

45. *Tristylospyris triceros* Ehrenberg)

The shell is nut-shaped, tuberculate, thick-walled with an indistinct sagittal stricture characteristic of all members of this genus. Characterized by three robust primary feet, circular in section, curved, convex outward. Secondary feet, developed in some specimens, may range from one to seven (usually three) in number. Rarely a small apical horn is present.

46. *Dorcadospyris ateuchus* (Ehrenberg)

Distinguished by two primary feet, robust, circular in section, usually tending to be straight. Secondary feet, when present, are one to four in number. In some specimens, a small amount of irregular lamellar meshwork is developed in place of the secondary feet. Often with small apical horn.

47. *Dorcadospyris simplex* (Riedel)

Distinguished by widely divergent and subsequently semicircularly curved feet. Short, weak apical horn in some specimens. Outer sides of feet of some specimens bear short, inconspicuous, conical spinules.

Dorcadospyris papilio — D. alata lineage
Tristylospyris triceros apparently gave rise through a series of intermediate forms (Moore, 1971) to *D. papilio* which was succeeded by the remaining four species in this lineage.

48. *Dorcadospyris papilio* (Riedel)

The two primary feet diverge initially at 180° or more, then curve semicircularly to converge terminally. Secondary feet three to eight in number (often four), lamellar or subcylindrical, varying in form and disposition. Most specimens have a stout conical apical horn.

49. *Dorcadospyris praeforcipata* Moore

Distinguished from *D. papilio* by less divergence of the primary feet, and from *D. forcipata* by possession of secondary feet. Primary feet may converge or recurve slightly terminally.

50. *Dorcadospyris forcipata* (Haeckel)

Distinguished from *D. praeforcipata* by the lack of secondary feet.

51. *Dorcadospyris dentata* Haeckel

Distinguished from *D. forcipata* by the presence on the convex side of each foot of four to ten conspicuous conical spines.

52. *Dorcadospyris alata* Riedel

The lattice shell is relatively small and lacks an apical horn. Distinguished by widely divergent feet (180° or more proximally), subsequently curved downward to a greater or lesser extent, and bearing (in most specimens) short conical spinules on the convex sides.

The genus *Calocycletta*
Species of this genus are conspicuous in the late Oligocene and early Miocene. Three of the commonest and most easily recognized forms are illustrated here, but see Moore (1972) for recognition of related forms and discussions of their evolutionary relationships.

53. *Calocycletta robusta* Moore

Characterized by an inflated campanulate thorax; tapering, subcylindrical abdomen which terminates in short, acute triangular teeth (when margin is preserved).

54. *Calocycletta virginis* (Haeckel)

Distinguished from *C. robusta* by eleven to sixteen lamellar, usually truncate, parallel terminal feet, broader than the spaces between them.

55. *Calocycletta costata* (Riedel)

Distinguished from other species of *Calocycletta* by longitudinal ridges separating rows of pores on the thorax, and often the abdomen. Feet are similar to those in *C. virginis*.

Stichocorys delmontensis — *S. peregrina* lineage
The broadly defined *S. delmontensis* gives rise to *S. peregrina* and possibly also to *S. wolffii*.

56. *Stichocorys delmontensis* Campbell and Clark
A multisegmented form with the upper three segments generally more robustly constructed than subsequent segments and forming a distinctive upper conical section of the shell. Fourth and subsequent segments form a cylindrical lower section. Shell broadest near middle of the third segment.

57. *Stichocorys wolffii* Haeckel
Distinguished from *S. delmontensis* by a distinctly clear-appearing thorax resulting from secondary closing of pores by siliceous lamellae, and by longitudinal alignment of pores with intervening longitudinal ridges on the first one or two post-abdominal segments.

58. *Stichocorys peregrina* (Riedel)
Characterized by upper four segments generally more robustly constructed than lower ones, by a truncate conical rather than inflated third segment. Shell broadest near base of third and middle of fourth segment.

Spongaster berminghami — *S. tetras* lineage
This lineage of three species had its origin in a circular spongodiscid, but the relationships are not yet worked out.

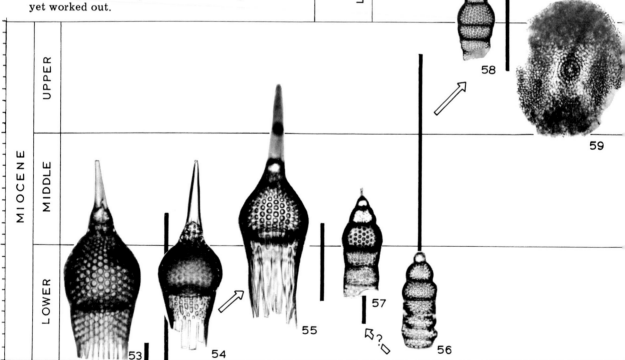

59. *Spongaster berminghami* (Campbell and Clarke)
Spongy disc, elliptical in outline; with thickening at center, at two opposing radii, and in two crescentic zones on either side of the thickened diameter.

60. *Spongaster pentas* Riedel and Sanfilippo
Spongy disc usually pentagonal, occasionally hexagonal. Rays from center to marginal angles generally not markedly denser (but usually slightly thicker) than the spongy structure between them. Central area more dense or thicker, with concentric structure.

61. *Spongaster tetras* Ehrenberg
Distinguished from *S. pentas* by reduction of the number of rays from five or six to four and the resulting square or rectangular outline.

62. *Pterocanium prismatium* Riedel
Large thorax tending to be straight-sided distally and obtusely pyramidal proximally. Ribs extending through the thorax continue as short terminal feet.

63. *Pterocanium praetextum* (Ehrenberg)
Characterized by basically hemispherical thorax modified by swellings between the three thoracic ribs (resembling a "three-cornered biretta") which are extended as three long straight, almost parallel three-bladed feet. Feet are proximally fenestrated and some meshwork (abdomen?) is commonly developed between the feet.

Theocorythium vetulum − *T. trachelium* lineage
These closely similar species apparently constitute a lineage.

64. *Theocorythium vetulum* Nigrini
Characterized by a cupola-shaped thorax separated by a pronounced lumbar stricture from an inflated conical abdomen with (usually) a distal row of three-bladed subterminal feet, a slight terminal constriction to a poreless rim which may or may not bear a row of triangular terminal teeth.

65. *Theocorythium trachelium* (Ehrenberg)
Distinguished from *T. vetulum* by a cylindrical rather than inflated conical abdomen. Two subspecies have been recognized by Nigrini (1967), one of which (*Theocorythium trachelium dianae*) has a distally somewhat inflated abdomen, more similar to that of *T. vetulum*.

Lamprocyrtis heteroporos − *L. nigriniae* lineage
This lineage begins approximately at the Miocene—Pliocene boundary with a form displaying a sharp difference in pore size between thorax and abdomen, which probably evolved from a form lacking such a contrast. With time the stricture between thorax and abdomen disappears, producing a two-segmented form which eventually loses the contrast in pore size in a form that still lives. These forms were first described from high latitudes but they appear to be helpful for subdividing equatorial Pliocene—Pleistocene sequences as well.

66. *Lamprocyrtis heteroporos* (Hays)
A three-segmented form characterized by an abrupt change, by a factor of two or more, in pore size across the stricture (indentation and internal ledge) between thorax and abdomen.

67. *Lamprocyrtis neoheteroporos* Kling
Resembles *L. heteroporos* except that no internal trace of the stricture between thorax and abdomen remains. The number of segments is thus reduced to two.

68. *Lamprocyrtis nigriniae* (Caulet)
In this two-segmented form the abrupt linear contrast in pore size is lost. Pore size gradually increases distally. Exceptional occurrences of sudden increase are not repeated at the same level in contiguous vertical rows of pores. Known also as *L. haysi* Kling, which is considered synonymous with *Conarachnium nigriniae* Caulet.

69. *Dictyomitra multicostata* group (Fig. 36)

These widely occurring forms probably include several species that are generally known under this name. They are long conical nassellarians with numerous segments and distinct longitudinal ribs. They are known only from the Cretaceous.

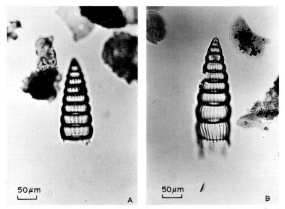

Fig. 36. *Dictyomitra multicostata* group. Somewhat greater morphological variation than is shown here, particularly in the number and distribution of pores, may be allowed.

70. Pseudoaulophacids

These forms with a distinctive triangular surface structure are described on page 219 (see Fig. 20). They are known from Cretaceous strata with a doubtful report of one occurrence in the Eocene.

71. Spiny ringed saturnalins

Members of this subfamily with numerous spines on the distinctive outer ring appear to be restricted to the Cretaceous (see Fig. 14). Often the rings or fragments of rings alone are found, but they can be recognized by remains of the centripetal spines at the ends of which can be seen the attachment points of the former latticed or spongy shell.

72. Hagiastrids

Members of this family, described on page 219, are common in many Mesozoic assemblages (see Fig. 19).

SUGGESTIONS FOR FURTHER READING

Campbell, A.S., 1954. Subclass Radiolaria. In: R.C. Moore (Editor), *Treatise on Invertebrate Paleontology. Protista 3*. Geol. Soc. Am., New York, N.Y., and Kansas Univ. Press, Lawrence, Kansas, pp. D11—D163. [A convenient version of Haeckel's system with some later additions and minor revisions. Should be used with caution because of errors and the unreliability of age ranges compiled from older literature.]

Foreman, H.P. and Riedel, W.R., 1972. *Catalogue of Polycystine Radiolaria. Series 1 (1834—1900), Vol. 1 (Meyer, 1834—Bury, 1862), Parts 1 and 2*. Special Publication, American Museum of Natural History, New York, N.Y. [This catalogue contains original descriptions of and subsequent references to genus- and species-level taxa covered in chronological order from earliest works. Subsequent volumes are in preparation.]

Grunau, H.R., 1965. Radiolarian cherts and associated rocks in space and time. *Eclogae Geol. Helv.*, 58(1): 157—208. [Summarizes geographic and stratigraphic distribution of radiolarian rocks throughout the world. Discusses their (deep-water) origin mainly in terms of sedimentological—geochemical evidence and association with oceanic basalts.]

Haeckel, E., 1887. Report on the Radiolaria collected by H.M.S. *Challenger* during the years 1873—76. *Rep. Voyage Challenger, Zool.* 18: clxxxvii + 1803 pp. (140 pls., 1 map). [The fundamental monograph on radiolarian taxonomy. An indispensable reference work, but should be used with awareness of the artificiality of the classification, and the inadvertent inclusion of reworkerd older specimens in what he took to be Quaternary samples. Includes extensive review of biology and distribution.]

Hollande, A. and Enjumet, M., 1960. Cytologie, évolution et systématique des Sphaeroidés (Radiolaires). *Arch. Mus. Nat. Hist. Nat., Ser. 7*, 7: 7—134. [A key reference on radiolarian biology. Stresses cytology of the spherical spumellarians but includes observations on some nassellarians as well. Proposes revised classification of the former group based on soft anatomy.]

Petrushevskaya, M.G., 1971a. On the natural system of polycystine Radiolaria (Class Sarcodina). In: *Proc. Second Planktonic Conf., Roma, 1970* Edizioni Tecnoscienza, Rome, pp. 981—992. [This summary of a revised radiolarian classification will serve as an introduction, in English, to extensive work on radiolarian systematics, morphology, and biogeography.]

Riedel, W.R. and Holm, E.A., 1957. Radiolaria. In: J. Hedgpeth (Editor), *Treatise on Marine Ecology and Paleoecology*, I. *Geol. Soc. Am., Mem.* 67: pp. 1069—1072. [This annotated bibliography on radiolarian ecology is a good introduction to the monographs resulting from the great oceanographic expeditions of the late nineteenth—early twentieth centuries.]

Riedel, W.R. and Sanfilippo, A., 1977a. Cenozoic Radiolaria. In: A.T.S. Ramsey (Editor), *Oceanic Micropaleontology*. Academic Press, New York, N.Y., in press. [These two papers by Riedel and Sanfilippo summarize numerous studies on radiolarian systematics and biostratigraphy, the second reviewing radiolarian knowledge applicable to paleontology.]

Riedel, W.R. and Sanfilippo, A., 1977b. Stratigraphy and evolution of tropical Cenozoic radiolarians. In: W.R. Riedel and T. Saito (Editors), *Marine Plankton and Sediments. Proc. Third Planktonic Conf., Kiel, 1974*. Micropaleontology Press, in press.

Strelkov, A.A., Khabakov, A.V. and Lipman, R.Kh., 1959. Podclass Radiolaria ili lucheviki. In: Yu.A Orlov (Editor), *Osnovy Paleontologii*. Akad. Nauk S.S.S.R., Moskow, vol. V, pp. 369—482. English Translation: *Fundamentals of Paleontology*. Israel Program for Scientific Translations, Jerusalem, 728 pp. [The English translation is a valuable summary of extensive Russian radiolarian investigations which are largely unknown to western readers. Although the Haeckelian system is followed, the treatment of all aspects of radiolarian study is quite comprehensive.]

CITED REFERENCES

Adshead, P.C., 1967. Collection and laboratory maintenance of planktonic foraminifera. *Micropaleontology*, 13(1): 32—40 (pls. 1—2).

Barthel, K.W., 1970. On the deposition of the Solnhofen lithographic limestone (Lower Tithonian, Bavaria, Germany). *Neues Jahrb. Geol. Paläontol., Abh.*, 135: 1—18 (2 text-figs, 4 pls.).

Berger, W.H., 1968. Radiolarian skeletons: solution at depths. *Science*, 159: 1237—1238.

Berger, W.H. and Winterer, E.L., 1974. Plate stratigraphy and the fluctuating carbonate line. In: K.J. Hsü and H.C. Jenkyns (Editors), *Pelagic Sediments on Land and Under* the Sea. Int. Assoc. Sedimentol., Spec. Publ. No. 1. Blackwell, Oxford, pp. 11—48 (26 text-figs).

Bosellini, A. and Winterer, E.L., 1975. Pelagic limestone and radiolarite of the Tethyan Mesozoic: a genetic model. Geology, 3(5): 279—282 (2 text-figs).

Bramlette, M.N., 1946. The Monterey formation of California and the origin of its siliceous rocks. U.S. Geol. Surv., Prof. Pap., 212: 57 pp.

Brandt, K., 1895. Biologische und faunistische Untersuchungen an Radiolarien und anderen pelagischen Thieren; 1. Untersuchungen über den hydrostatischen Apparat von Thalassicollen und coloniebildenden Radiolarien. Zool. Jahrb., Syst., 9: 27—74.

Cachon, J. and Cachon, M., 1971. The axopodial system of Radiolaria Nassellaria. Origin, organisation and relation with the other cell organelles. General considerations on the macromolecular organisation of the stereoplasm of actinopods. Arch. Protistenkd., 113: 80—97 (9 text-figs., 9 pls.) [in French with English summary].

Cachon, J. and Cachon, M., 1972a. The axopodial system of Radiolaria Sphaeroidae. I. Centroaxoplastidae. Arch. Protistenkd., 114: 51—64 (7 text-figs, 9 pls.) [in French with English summary].

Cachon, J. and Cachon, M., 1972b. The axopodial system in Radiolaria Sphaeroidae. II. The Periaxoplastidae. III. The Cryptaxoplastidae (Anaxoplastidae). IV. The fusules and the rheoplasmic system. Arch. Protistenkd., 114: 291—307 (5 text-figs, 8 pls.) [in French with English summary].

Calvert, S.E., 1974. Deposition and diagenesis of silica in marine sediments. In: K.J. Hsü and H.C. Jenkyns (Editors), Pelagic Sediments On Land and Under the Sea. Int. Assoc. Sedimentol., Spec. Publ. No. 1. Blackwell, Oxford, pp. 273—299 (7 text-figs.).

Casey, R.E., 1971. Radiolarians as indicators of past and present water masses. In: B.M. Funnell and W.R. Riedel (Editors), The Micropaleontology of Oceans. Cambridge University Press, London, pp. 331—341.

Casey, R.E., Partridge, T.M. and Sloan, J.R., 1971. Radiolarian life spans, mortality rates, and seasonality gained from Recent sediment and plankton samples. In: Proc. Second Planktonic Conf., Roma, 1971. Edizioni Tecnoscienza, Rome, pp. 159—165.

Chipping, D.H., 1971. Paleoenvironmental significance of chert in the Franciscan Formation of western California. Geol. Soc. Am., Bull., 82(6): 1707—1711 (1 text-fig.).

Deflandre, G., 1960. A propos du développement des recherches sur les radiolaires fossiles. Rev. Micropaléontol., 2(4): 212—218 (pl. 1).

Dumitrică, P., 1970. Cryptocephalic and cryptothoracic Nassellaria in some Mesozoic deposits of Romania. Rev. Roum. Géol., Géophys., Géogr., Sér. Géol., 14(1): 45—124 (27 pls.).

Dunham, J.B. and Murphy, M.A., 1975. An occurrence of well-preserved Radiolaria from the upper Ordovician (Caradocian) of central Nevada. J. Paleontol., 50(5): 882—887 (2 text-figs., 1 pl.).

Foreman, H.P., 1963. Upper Devonian Radiolaria from the Huron member of the Ohio shale. Micropaleontology, 9(3): 267—304 (pls. 1—9).

Foreman, H.P., 1966. Two Cretaceous radiolarian genera. Micropaleontology, 12(3): 355—359 (11 text-figs.).

Foreman, H.P., 1968. Upper Maestrichtian Radiolaria of California. Palaeontol. Assoc., London, Spec. Pap., no. 3: pp. 1—82 (pls. 1—8, text-fig. 1).

Fortey, R.A. and Holdsworth, B.K., 1972. The oldest known well-preserved Radiolaria. Bull. Soc. Paleontol. Ital., 10(1): 35—41 (pls. 10, 11, text-fig. 1).

Friend, J.K. and Riedel, W.R., 1967. Cenozoic orosphaerid radiolarians from tropical Pacific sediments. Micropaleontology, 13: 217—232.

Garrison, R.E., 1974. Radiolarian cherts, pelagic limestones, and igneous rocks in eugeosynclinal sediments. In: K.J. Hsü and H.C. Jenkyns (Editors), Pelagic Sediments On Land and Under the Sea. Int. Assoc. Sedimentol., Spec. Publ. No. 1. Blackwell, Oxford, pp. 367—396 (6 text-figs.).

Garrison, R.E. and Fischer, A.G., 1969. Deep-water limestones and radiolarites of the Alpine Jurassic. In: G.M. Friedman (Editor), Depositional Environments in Carbonate Rocks, a Symposium. Soc. Econ. Paleontol. Mineral., Spec. Publ., No. 14: pp. 20—56.

Goll, R.M., 1968, 1969. Classification and phylogeny of Cenozoic Trissocyclidae (Radiolaria) in the Pacific and Caribbean basins; Part I, J. Paleontol., 42(6): 1409—1432; Part II, J. Paleontol., 43(2): 322—339.

Goll, R.M. and Bjørklund, K.R., 1971. Radiolaria in surface sediments of the North Atlantic Ocean. Micropaleontology, 17(4): 434—454 (8 text-figs.).

Goll, R.M. and Bjørklund, K.R., 1974. Radiolaria in surface sediments of the South Atlantic. Micropaleontology, 20(1) 38—75 (16 text-figs.).

Harper, H.E. and Knoll, A.H., 1975. Silica, diatoms and Cenozoic radiolarian evolution. Geology, 3(4): 175—177 (1 text-fig.).

Hays, J.D., 1965. Radiolaria and late Tertiary and Quaternary history of Antarctic seas: Biology of the Antarctic Sea II. Antarct. Res., Ser. 5, pp. 125—184.

Hays, J.D., 1970. Stratigraphy and evolutionary trends of Radiolaria in North Pacific deep-sea sediments. In: J.D. Hays (Editor), Geological Investigations of the North Pacific. Geol. Soc. Am., Mem. 126: 185—218 (pl. 1).

Hays, J.D. and Opdyke, N.D., 1967. Antarctic Radiolaria, magnetic reversals, and climatic change. Science, 158: 1001—1011.

Heath, G.R., 1974. Dissolved silica and deep-sea sediments. In: W.W. Hay (Editor), Studies in Paleo-Oceanography. Soc. Econ. Paleontol. Mineral., Spec. Publ., No. 20: 77—93 (11 text-figs.).

Holdsworth, B.K., 1969. The relationship between the genus Albaillella Deflandre and the ceratoikiscid Radiolaria. Micropaleontology, 15(2): 230—236 (pl. 1, text-fig. 1).

Holdsworth, B.K., 1971. The ceratoikiscid nature of the radiolarian Lapidopiscum piviteaui Deflandre. Micropaleontology, 17(2): 244—248 (pl. 1, text-figs. 1—2).

Johnson, D.A. and Knoll, A.H., 1974. Radiolaria as paleoclimatic indicators: Pleistocene climatic fluctuations in the equatorial Pacific Ocean. Quaternary Res., 4: 206—216 (3 text-figs., 2 pls.).

Johnson, D.A. and Knoll, A.H., 1975. Absolute ages of Quaternary radiolarian datum levels in the equatorial Pacific. Quaternary Res., 5: 99—110 (2 text-figs., 1 pl.).

Johnson, T.C., 1974. The dissolution of siliceous microfossils in surface sediments of the eastern tropical Pacific. Deep-Sea Res., 21: 851—864 (6 text-figs., 2 pls.).

Kling, S.A., 1971. Dimorphism in Radiolaria. In: Proc. Second Planktonic Conf., Roma, 1970. Edizioni Tecnoscienza, Rome, pp. 663—672 (5 pls.).

Kling, S.A., 1976. Relation of radiolarian distributions and subsurface hydrography in the North Pacific. Deep-sea Res., 23: 1043—1058 (8 figs.).

Kruglikova, S.B., 1969. Radiolyarii v poverkhmostnom sloe osadkov severnoi polovuny Tikhogo Okeana. (Radiolaria in the surface layer of sediments of the northern half of the Pacific Ocean.) In: Tikhy Okean. Mikroflora i mikrofauna v osadkakh Tikhogo Okeana. Nauka, Moscow, pp. 48—72.

Lisitzin, A.P., 1971. Distribution of siliceous microfossils in suspension and in bottom sediments. In: B.M. Funnell and W.R. Riedel (Editors), The Micropaleontology of Oceans. Cambridge University Press, London, pp. 173—195.

Moore, T.C., 1971. Radiolaria. In: Initial Reports Deep Sea Drilling Project, Leg 8, pp. 727—775.

Moore Jr., T.C., 1972. Mid-Tertiary evolution of the radiolarian genus Calocycletta. Micropaleontology, 18(2): 144—152 (pls. 1, 2).

Moore Jr., T.C., 1973. Late Pleistocene—Holocene oceanographic changes in the northeastern Pacific. Quaternary Res., 3: 99—109 (6 text-figs.).

Nigrini, C., 1967. Radiolaria in pelagic sediments from the Indian and Atlantic Oceans. Scripps Inst. Oceanogr. Bull., 11: 125 pp.

Nigrini, C., 1968. Radiolaria from eastern tropical Pacific sediments. Micropaleontology, 14(1): 51—53 (16 text-figs., 1 pl.).

Nigrini, C., 1970. Radiolarian assemblages in the North Pacific and their application to a study of Quaternary sediments in core V20-130. In: J.D. Hays (Editor), *Geological Investigations of the North Pacific. Geol. Soc. Am., Mem.* 126: 139—183 (4 pls.).

Nigrini, C., 1971. Radiolarian zones in the Quaternary of the equatorial Pacific Ocean. In: B.M. Funnell and W.R. Riedel (Editors), *The Micropaleontology of Oceans.* Cambridge University Press, London, pp. 443—461.

Nisbet, E.G. and Price, I., 1974. Siliceous turbidites: bedded cherts as redeposited, ocean ridge-derived sediments. In: K.J. Hsü and H.C. Jenkyns (Editors), *Pelagic Sediments On Land and Under the Sea.* Int. Assoc. Sedimentol., Spec. Publ. No. 1. Blackwell, Oxford, pp. 351—366.

Pessagno Jr., E.A., 1970. The Rotaformidae, a new family of Upper Cretaceous Nassellaria (Radiolaria) from the Great Valley sequence, California Coast Ranges. *Bull. Am. Paleontol.*, 58(257): 1—32 (9 pls.).

Pessagno Jr., E.A., 1971. Jurassic and Cretaceous Hagiastridae from the Blake-Bahama Basin (Site 5A, JOIDES Leg I) and the Great Valley sequence, California Coast Ranges. *Bull. Am. Paleontol.*, 60(264): 1—83 (pls. 1—19).

Pessagno Jr., E.A., 1972. Cretaceous Radiolaria. Part I: The Phaseliformidae, a new family, and other Spongodiscacea from the Upper Cretaceous portion of the Great Valley sequence. Part II: Pseudoaulophacidae Riedel from the Cretaceous of California and the Blake-Bahama Basin (JOIDES Leg I). *Bull. Am. Paleontol.*, 61(270): 267—328 (pls. 22—31).

Pessagno Jr., E.A., 1976. Radiolarian zonation and stratigraphy of the upper Cretaceous portion of the Great Valley sequence. *Micropaleontology, Spec. Pap.*, No. 1.

Petrushevskaya, M.G., 1962. Znachenie rosta skeleta radiolyarii dlya sistematiki. (The importance of skeletal growth in radiolarians for their systematics.) *Zool. Zh.*, 41(3): 331—341.

Petrushevskaya, M.G., 1967. Radiolyarii otryadov Spumellaria Nassellaria antarkticheskoi oblasti. (Antarctic spumelline and nasselline radiolarians.) In: E.N. Povlovskii (Editor), *Issledovaniya Fauni Morei, 3(12).* Zool. Inst., Akad. Nauk. S.S.S.R., Leningrad, 186 pp. (102 text-figs.).

Petrushevskaya, M.G., 1971b. Radiolyarii Nassellaria v planktone Mirovogo Okeana. (Nassellarian Radiolaria in the plankton of the world ocean.) In: B.E. Bykhovskii (Editor), *Issledovaniya. Fauny Morei, 9(17).* Zool. Inst., Akad. Nauk. S.S.S.R., Leningrad, 295 pp. (146 text-figs.).

Petrushevskaya, M.G., 1971c. Spumellarian and nassellarian Radiolaria in the plankton and bottom sediments of the Central Pacific. In: B.M. Funnell and W.R. Riedel (Editors), *The Micropaleontology of Oceans.* Cambridge University Press, London, pp. 309—317 (6 text-figs.).

Petrushevskaya, M.G., 1971d. Radiolaria in the plankton and Recent sediments from the Indian Ocean and Antarctic. In: B.M. Funnell and W.R. Riedel (Editors), *The Micropaleontology of Oceans.* Cambridge University Press, London, pp. 319—329 (6 text-figs.).

Petrushevskaya, M.G. and Koslova, G.E., 1972. Radiolaria: Leg 14, Deep Sea Drilling Project. In: D.E. Hayes, A.C. Pimm and others, *Initial Reports of the Deep Sea Drilling Project, Leg 14:* 495—648 (41 pls.).

Pokorný, V., 1963. *Principles of Zoological Micropaleontology.* Pergamon Press, London, 652 pp.

Renz, G.W., 1976. The distribution and ecology of Radiolaria in the Central Pacific plankton and surface sediments. *Scripps Inst. Oceanogr., Bull.*, 22: 1—267 (17 text-figs., 8 pls.).

Reschetnjak, V.V., 1955. Vertikalnoe raspedelenie radiolyariy Kurilo-Kamchatskoy Vpadiny. (Vertical distribution of radiolarians in the Kurile-Kamchatka Trench.) *Tr. Zool. Inst., Akad. Nauk S.S.S.R.*, 21: 94—101 (1 pl.).

Riedel, W.R., 1957. Radiolaria: a preliminary stratigraphy. *Rep. Swedish Deep-Sea Exped.*, 6(3): 59—96 (pls. 1—4).

Riedel, W.R., 1971. Systematic classification of polycystine Radiolaria. In: B.M. Funnell and W.R. Riedel (Editors), *The Micropaleontology of Oceans.* Cambridge University Press, London, pp. 649—661.

Riedel, W.R. and Funnell, B.M., 1964. Tertiary sediment cores and microfossils from the Pacific Ocean floor. *Q.J. Geol. Soc. Lond.*, 120: 305—368 (pls. 14—32).

Riedel, W.R. and Sanfilippo, A., 1971. Radiolaria from the southern Indian Ocean, DSDP Leg 26. In: T.A. Davies, B.P. Luyendyk and others, *Initial Reports of Deep Sea Drilling Project, Leg 26:* 771—813 (15 pls.).

Sachs, H.M., 1973. Late Pleistocene history of the North Pacific; evidence from a quantitative study of Radiolaria in Core V21-173. *Quaternary Res.*. 3(1): 89—93.

Soutar, A., 1971. Micropaleontology of anaerobic sediments and the California Current. In: B.M. Funnell and W.R. Riedel (Editors), *The Micropaleontology of Oceans.* Cambridge University Press, London, pp. 223—230 (1 text-fig., 5 pls.).

Stadum, C.J. and Ling, H.-Y., 1969. Tripylean Radiolaria in deep-sea sediments of the Norwegian Sea. *Micropaleontology*, 15(4): 481—489 (1 pl.).

Steinmann, G., 1905. Geologische Beobachtungen in den Alpen. II. Die Schardt'sche Überfaltungstheorie und die geologische Bedeutung der Tiefseeabsätze und der ophiolitischen Massengesteine. *Ber. Naturforsch. Ges. Freiburg*, 16: 18—67.

Steinmann, G., 1925. Gibt es fossile Tiefseeablagerungen von erdgeschichtlicher Bedeutung? *Geol. Rundsch.*, 16: 435—468.

Stürmer, W., 1963. Mikrofossilen in den Mörnscheimer Schichten. *Geol. Bl. N.-O. Bayern*, 13(1): 11—13 (1 pl.).

Theyer, F. and Hammond, S.R., 1974a. Paleomagnetic polarity sequence and radiolarian zones, Brunhes to Polarity Epoch 20. *Earth Planet. Sci. Lett.*, 22: 307—319 (7 text-figs.).

Theyer, F. and Hammond, S.R., 1974b. Cenozoic magnetic time scale in deep-sea cores: completion of the Neogene. *Geology*, 2(10): 487—492)4 text-figs.).

Thompson, D'Arcy W., 1942. *On Growth and Form. A New Edition.* Cambridge University Press, London, 1116 pp.

10

MARINE DIATOMS

LLOYD H. BURCKLE

INTRODUCTION

The marine diatoms are an interesting, useful and relatively uncrowded field for study. Future work in micropaleontology will stress the solution to correlation and paleoecologic problems in high latitudes — regions which contain few, if any, calcareous microfossils but which are exceedingly rich in diatomaceous remains. In spite of this, even in low latitudes the study of fossil diatoms lags far behind that on other major microfossil groups.

Diatoms are useful in biostratigraphy and paleoecology. Their biostratigraphic application has been aided in large part by the use of deep-sea sediment cores, particularly from high latitudes and by the integration of diatom data with paleomagnetic stratigraphy and with data from other microfossil groups. Diatoms can be used as indicators of such environmental parameters as water chemistry, paleosalinity, paleodepth, paleotemperature and paleonutrient concentrations. Fossil diatoms are important as indicators of paleocurrents and particularly bottom currents. Since many diatoms are built to remain in suspension they can easily be transported laterally, in some cases for several thousands of kilometers. In such cases, these displaced diatoms can further be used as indicators of the historical stability of the transporting currents.

Diatoms are photosynthetic, single-celled algae that inhabit many aquatic and subaquatic environments. Some species have adopted a quasi-colonial life style but are capable of living singly. Diatoms may be free-floating (planktonic) or attached to some foreign surface (sessile). They, along with the coccolitho-phorids, make up the major part of the phytoplankton of the sea and, as such, constitute an important basic link in the food chain. Diatoms are important constituents of many environments, including open ocean, the littoral zone, as well as various fresh-water environments.

HISTORY OF STUDY

Fossil diatoms were first studied extensively in the latter half of the nineteenth century. Although these works have little stratigraphic value they did contribute to our understanding of diatom taxonomy as well as ferret out and describe numerous classical diatom collecting localities. Modern work, stressing the stratigraphic applicability of diatoms, began in the middle 1920's with publications on the extremely diatomaceous California sections. These sections, which range in age from Cretaceous to Pliocene, continue to be the object of considerable study. Workers in other countries, particularly the Soviet Union and Japan, also began serious study of diatomaceous sections. Such studies were guided by two definitive works: *The Schmidt Diatom Atlas* (beginning in 1875) and the publications of Hustedt (beginning in 1927) which refined taxonomy and added notes on the ecology and distribution of selected diatom species.

The advent of extensive deep-sea coring programs was accompanied by a renewed interest in diatom biostratigraphy, again with the Russians taking the lead. This interest continues to the present day and most workers in stratigraphic micropaleontology now draw some or all of their material from deep-sea sediments.

BIOLOGY

Organization at the cellular level

The diatom cell contains a nucleus which is usually displaced to one side and is embedded in a protoplasmic mass. This nucleus is invariably small and may assume a number of shapes. Smaller bodies, **nucleoli**, are present within the nucleus. Surrounding the nucleus are photosensitive bodies, the **chromatophores**, which vary in number, size, shape and position within the cell according to species. Much of the color imparted to the ocean by masses of diatoms is due to these chromatophores.

During certain times of the year, diatoms produce a fatty substance which is stored in the cell in the form of several rounded globules. Although the exact purpose of these globules is unknown, it is thought to be a reserve food supply and may play a role in permitting some diatoms to survive through the winter season.

Skeletal construction

Diatoms secrete an external shell, the **frustule**, that is often compared to a pill box in which one half of the box, the **valve**, fits over the other half and encloses the protoplasmic mass (Fig. 1). Frustules are composed of opaline silica and the larger of the two valves is called the **epitheca** while the other is called the **hypotheca**. In the fossil state most frustules are separated, so this distinction is not especially important. The two valves are connected by a usually thin circular band called a **girdle** — a structure frequently seen in the fossil record, but which has no stratigraphic value. The valve **mantle** is at the junction of the valve margin and the girdle. For microscopic examination diatom valves are usually cemented onto a glass slide with the valve face up, and this is referred to as the valve view which is especially desirable since most fossil diatom taxonomy is based upon the valve view.

Diatom symmetry can usually be related to three axes: the **apical axis**, which is parallel to the long dimensions of the valve; the **transapical axis**, which is at right angles to the apical axis; and the **pervalvar axis**, which runs through the center of the two valves (Fig. 2). We usually do not find a complete frustule in the fossil state and thus, fossil diatoms are typically described with reference to the first two axes.

Cell wall

According to Hendey (1964), the cell wall

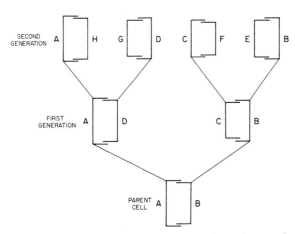

Fig. 1. Girdle views of diatom valves through several reproductive phases. Note the progressive decrease in size of some forms.

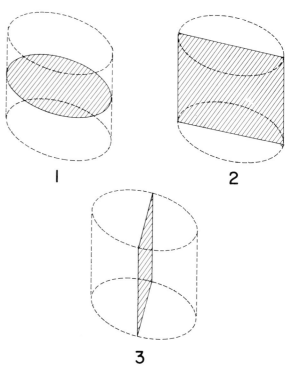

Fig. 2. Diatom symmetry. Oblique view showing both the valve and girdle views. 1. Valvar plane. 2. Apical plane. 3. Transapical plane.

may be of two types. A single laminar wall is composed of a single layer of silica. Although the thickness of the wall is generally uniform, there may be local thickenings. A more complex wall, the **locular wall**, consists of a double silica wall separated by vertical silica "slats".

Valve structure

A brief note on morphology should be introduced here as further discussion will consider diatoms in two categories: **centric** forms and **pennate** forms. Centric forms may be circular, triangular or oblong, but their major distinguishing feature is that the surface markings radiate from a central area. Pennate forms have one long axis and two short axes with the surface markings at right angles to the long axis (see Fig. 3).

As mentioned above, the valve view contains structures significant in taxonomy. Most centric diatoms and many pennate forms have valves covered with a honeycomb structure, termed **areolae**. This structure consists of vertical-sided chambers within the valve wall (Fig. 4). Their geometry and spacing are rather conservative features and can be used to differentiate species. Frequently a finely perforated plate, termed the **sieve plate**, may cover the areolae.

Some genera of centric diatoms possess scattered small pores in the valve view. Such pores perforate the cell wall and, in the pennate forms, may form parallel lines that have the appearance of striations under the light microscope.

Areas of clear structureless silica, termed the **hyaline areas**, characterize some diatom species. Certain species of *Coscinodiscus*, for example, have clear hyaline central areas, while the genera *Asteromphalus* and *Asterolampra* are characterized by large hyaline central areas (see Fig. 4). In the pennate diatoms

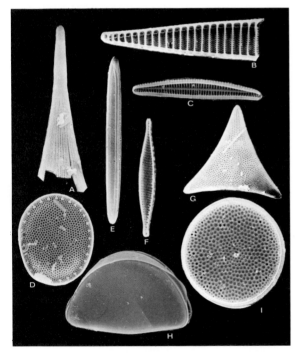

Fig. 3. Representative genera demonstrating basic patterns in centric and pennate diatoms. A. *Rhizosolenia bergonii* H. Peragallo, a pennate form. B. *Nitzschia* sp. a pennate form. C. *Pseudoeunotia doliolus* (Wallich) Grunow, a pennate form. D. *Roperia tesselata* var. *ovata* Heiden (Mann), a centric form. E. *Nitzschia marina* Grunow, a pennate form. F. *Nitzschia bicapitata* Cleve, a pennate form. G. *Triceratium cinnamomeum* Greville, a centric form. H. *Hemidiscus cuneiformis* Wallich, a centric form. I. *Coscinodiscus nodulifer* A. Schmidt, a centric form.

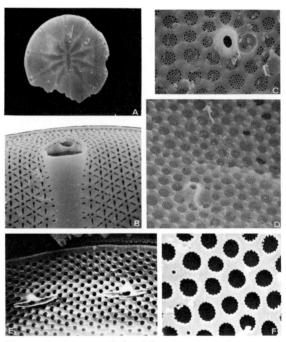

Fig. 4. A. *Asteromphalus hiltonianus* showing central hyaline area and hyaline ray. B. Detail of *Asteromphalus* showing hyaline ray. C, D. Detail of areolae with finely perforated sieve plate. The central structure is a pore. E. Detail of labiate process around periphery of centric valve. F. Detail of areolae. Variability in the size of the "dentition" around the areolae is probably due to dissolution.

such hyaline regions run down the center of the valve parallel to the apical axis and are called the **pseudoraphe**. Some pennate genera have a V-shaped slit, **raphe**, in place of the pseudoraphe.

Additional structures, although seemingly insignificant, are important even at the generic level. A small hyaline area, **pseudonodule**, near the periphery of centric diatoms, for example, identifies the genus *Actinocyclus* while two or more pores near the central area characterize the genus *Thalassiosira*, and a circlet of spines identifies the genus *Stephanopyxis*. It is essential, therefore, to emphasize that a careful scrutiny of the diatom valve is necessary for accurate specific identification.

Reproduction

Diatoms reproduce by simple cell division. Just prior to division, the epitheca and hypotheca move slightly away from each other and separation of the two valves takes place (see Fig. 1). New valves are formed on the exposed protoplasm from the central area outwards and always within the confines of the parent valve. Thus, the original two valves both become epitheca in the new individuals. One can project, therefore, a steady decrease in valve diameter with each succeeding generation. Not only do the valves become smaller, but they also change in geometry as well as in the character of their surface markings.

In some reproductive phases, the diatom will subdivide to a very small size and to a low level of vitality before dying. In others, cell division will cease at some level and a special cell, called an **auxospore**, will be formed. The purpose of this auxospore, although the mechanics are not well understood in most species, is to return the diatom species to its original size from which the process starts anew.

Just as important to species survival as the reproductive cycle is the ability of many diatoms to survive from one growing season to another. Some species will persist through the winter months, their life processes greatly reduced. Most, however, form a heavy resting spore which falls to the bottom and is revived when conditions favorable for growth return.

Nutrition

The nutrient content of the water is extremely important to diatom growth and reproduction. Three nutrients are considered essential for almost all diatom species. These are phosphorus, nitrate and silica. The Russian worker T.V. Belayeva has reported that the giant marine diatom *Ethmodiscus rex* (Rattray) Hendey lives best in waters of low phosphate concentration, but this preference is considered atypical for diatom species. Experience, both in the laboratory and in the field, has shown that when any of these three basic nutrients is missing, diatom growth and reproduction ceases. Some diatoms are so sensitive that even a slight reduction in the supply of these dissolved nutrients will inhibit their reproduction.

Given these facts it is relatively easy to point out areas of high diatom productivity. Any area of upwelling in the ocean will bring a constant supply of these nutrients to the surface and, thus, cause productivity blooms. Coastal regions which receive high concentrations of nutrients from both run-off from land and rain will support large diatom populations.

In addition to the nutrient supply many species have special requirements. A number, for example, are known to require cobalamin (vitamin B_{12}) as well as thiamin (vitamin B_1). Sulphur, iron and manganese are also considered essential for many diatoms. In addition, various diatom species are known to respond to trace elements in the water. Thus, a number of nutritional variables may be operating to promote or inhibit the presence of specific diatom species.

ECOLOGY

Habitats

Within both fresh-water and marine environments diatoms can be found occupying a great number of niches. On land they are found in soils and occasionally on wetted rocks and plants. In streams, lakes and ponds they are found attached to rocks and plants as well as in bottom muds.

In the marine realm, two broad diatom

habitats are recognized. Benthic diatoms live in littoral environments either attached (**sessile**) or capable of some movement along the bottom (**vagile**). They are of considerable importance in that they serve as the major food for many shallow-water marine feeders. In addition, certain species produce enough cementing mucus to cement sediment grains together and thus inhibit submarine erosion. Other diatom species are known to form a dense mat at the sediment/water interface and thus prevent any continued reworking of the sediment.

Planktonic species are considered as part of the oceanic or neritic plankton. Oceanic (holoplanktonic) forms spend their entire existence in the open ocean and pass through the various phases of their life cycle in that environment. Neritic forms are found in association with coastlines. Since such forms frequently pass through a benthic stage, these coastal regions must possess rather shallow shelves.

Hendey (1964) divides neritic plankton into three categories: holoplanktonic, meroplanktonic and tychopelagic. The **holoplanktonic** existence was considered previously and is mentioned here only because some neritic plankton appear to be holoplanktonic. **Meroplanktonic** species live close to the coastline in the plankton but spend part of their existence in bottom sediments, probably as resting spores. It is sometimes difficult to differentiate holoplanktonic and meroplanktonic species. **Tychopelagic** species probably spend most of their lives on the bottom (Fig. 5).

Frequently a diatom assemblage far from

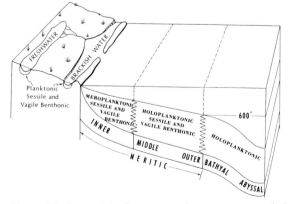

Fig. 5. Marine and fresh-water environments occupied by diatoms. (From Wornardt, 1969).

the present coastline may contain representatives of both the neritic and oceanic plankton as well as a number of sessile and vagile forms. This is understandable since bottom currents or storms may dislodge shallow-water forms and transport them many kilometers from their original habitat.

Factors affecting distribution

Factors which control diatom distribution vary with habitat. Temperature and salinity appear to be the principal factors controlling distribution of marine diatoms. Thus, diatoms are relatively easy to culture, although frequently it is difficult to reconstruct diatom paleoecology.

MAJOR MORPHOLOGICAL GROUPS

Diatoms take on a great many shapes and thus it is difficult to present a simple picture of the major morphological groups. Most diatomists follow the lead of their nineteenth century predecessors in recognizing two major divisions — the Centrales and the Pennales. The Centrales may be circular, oblong, hemicircular, triangular, or quadrangular, but the surface structures are arranged with reference to some central or near-central point. The Pennales, on the other hand, are elongate with major structures at approximately right angles from a median line which runs parallel to the long axis.

The Pennales can further be divided into those forms which possess a true raphe, or cleft, and those that do not. Any further morphological breakdown is largely based on such features as shell geometry and surface structures. Although the many different kinds of shapes, geometries and structures in the diatom present a somewhat confusing picture, N. Ingram Hendey (1964) has succeeded in assigning most diatoms to one of seven "shape groups". These seven minor groups can be lumped together into two major groups: (1) those whose valves usually have a raphe or pseudoraphe; and (2) those whose valves are without a raphe or pseudoraphe. In the first category, Hendey recognized four "shape groups": the linear diatoms, the cuneate diatoms, the cymbiform diatoms and the carinoid diatoms.

Linear diatoms have one long axis (the apical axis) and two shorter axes (the transapical and pervalvar axes). They are isopolar, symmetrical about the apical axis (with a few exceptions) and posses a clear silica area running down the center of the valve parallel to the apical axis. If this area is free of any structure it is termed the pseudoraphe. If, however, it possesses a V-shaped slit, usually seen as a narrow line under the light microscope, running parallel to the apical axis, it is termed the raphe. Linear diatoms which contain a raphe also have a central nodule and, at the apical ends, polar nodules. Surface markings, in the form of ridges, costae, or pores, may radiate away from the raphe or pseudoraphe (Fig. 6).

Cuneate diatoms differ from the linear forms in their symmetry. Although cuneate forms are symmetrical about the apical axis,

they are asymmetrical about the transapical axis. They may have a raphe or a pseudoraphe (Fig. 7A).

Symmetry is also a major distinguishing characteristic in the **cymbiform diatoms**. In this group, the asymmetry is about the apical axis, while it is symmetrical about the transapical axis. The valves are isopolar and may have a raphe or a pseudoraphe, depending upon the genus (Fig. 7B).

The last group, the **carinoid diatoms**, are based upon the placement and structure of the raphe. Symmetry and valve shape are relatively unimportant. The raphe is usually on a raised keel and the group may be subdivided into those genera that have one raphe and those genera having two raphes; in both subgroups the raphe is situated along the margin of the valve (Fig. 8).

In the second category, Hendey recognized three "shape groups". **Discoid diatoms** have a relatively simple structure, are circular in valve view and rectangular in girdle view. The valve surface may be flat, concave, convex or concavoconvex. Possible surface structures on the valve face include: areolae, pores, ribs, spines, etc. (Fig. 9A).

The **gonioid diatoms** are usually angular in outline but, otherwise, are difficult to characterize. The shape may be circular, elliptical or triangular. Since a definitive description

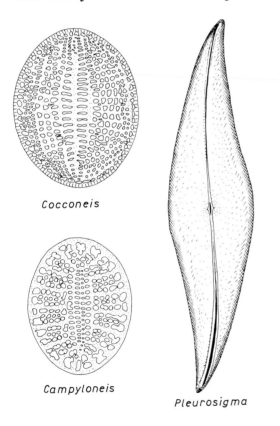

Cocconeis

Campyloneis

Pleurosigma

B

Fig. 6. Linear diatoms. Diagnosis: One long axis (apical axis) and two shorter axes (transapical and pervalvar axes). Isopolar, symmetrical about the apical axis with clear silica area running down center of valve parallel to apical axis. May contain raphe, pseudoraphe, central nodule, polar nodule, ridges, costae and pores.

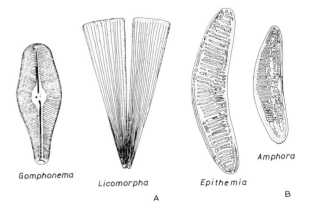

Gomphonema

Licomorpha

Epithemia

Amphora

A   B

Fig. 7. A. Cuneate diatoms. Diagnosis: Symmetrical about the apical axis and asymmetrical about the transapical axis. May have a raphe or pseudoraphe. B. Cymbiform diatoms. Diagnosis: Asymmetrical about the apical axis. Valves are isopolar and may have a raphe or pseudoraphe.

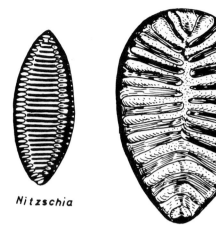

Nitzschia

Surirella

Fig. 8. Carinoid diatoms. Diagnosis: Based on placement and structure of raphe. Raphe placed on raised keel near valve margin. Symmetry and valve shape relatively unimportant.

of this group is difficult, the reader is referred to Fig. 9B for sample shapes.

Solenoid diatoms feature exceptional development of the girdle band. The girdle length may be many times the valve diameter, resulting in a frustule with a tubular shape. In valve view, the cell is usually circular or elliptical in outline. However, because of the expanded girdle, members of this group are also always cemented on a glass slide with the girdle view up (see Fig. 9C).

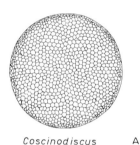

Coscinodiscus   A

B

Biddulphia

Isthmia

Chaetoceras

Triceratium

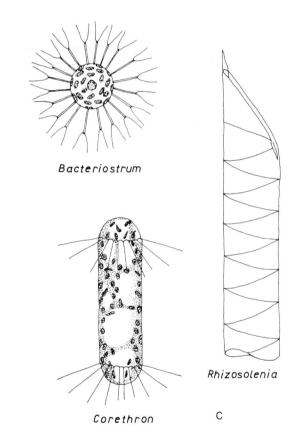

Bacteriostrum

Corethron   C

Rhizosolenia

Fig. 9. A. Discoid diatoms. Diagnosis: Circular in valve view and rectangular in girdle view. Valve surface may be flat, concave, convex or concavo-convex. Surface structures on valve face include: areolae, pores, ribs, spines, etc. B. Gonioid diatoms. Diagnosis: Usually angular in outline. Shape may be circular, elliptical or triangular. C. Solenoid diatoms. Diagnosis: Unusual development of girdle band. Girdle length many times valve diameter resulting in tubular-shaped frustule.

EVOLUTIONARY TRENDS

The earliest diatoms date from the middle Cretaceous, although there are unsubstantiated reports of their occurrence in the Jurassic and even in the late Paleozoic. Since these Cretaceous diatoms occur in assemblages of numerous species it has been argued that diatoms really developed in late Paleozoic or early Mesozoic time and either existed as naked protoplasm or were simply not fossilized. Isolated, heavily silicified diatom valves found in early to middle Cretaceous sediments suggest that selective solution is the final arbiter of diatom first appearance rather than any biochemical changes.

The Centrales are the only group of diatoms to occur in the Cretaceous and earliest Paleocene with pennate forms first appearing in the late Paleocene and increasing somewhat in abundance during the Eocene. Tappan and Loeblich (1971) pointed out that the expansion of the Centrales had stabilized by Eocene time and, except for a burst in the Miocene, their diversity was essentially unchanged throughout the Tertiary. The Pennales, on the other hand, have undergone an exponential increase since at least Miocene time and, at present, far surpass the Centrales in numbers of living species. The centric forms tend to dominate the plankton while the pennate forms are more abundant in the benthic realm.

Therefore, it is not surprising that the earliest fresh-water diatoms (Eocene) are pennate forms. Indeed, at present most fresh-water diatoms, including benthics and quasi-planktonics, are pennate forms. Diatoms were probably introduced into the fresh-water environment via brackish-water littoral forms.

Unfortunately, we do not see a progressive increase in complexity of the diatom valve with time. Cretaceous forms show as much variation in shell structure as do the Tertiary forms. Because of this, it is difficult to use "stage of evolutionary development" as a gross-stratigraphic tool.

Reasons for evolutionary change

In considering benthic and fresh-water diatoms it is relatively easy to account for evolutionary change. At the very least we can in-voke geographic isolation followed by progressive changes in the composition of the gene pool to explain any character shifts. With planktonic communities, many species appear to be restricted to specific water masses, thus achieving some degree of geographic isolation.

What we are at a loss, however, to explain is the evolution of one species into another in what was apparently the same water mass. We cannot invoke geographical isolation nor can we see any measurable ecological reason why there should be this time-sequential grading of one species into another. We are led to suspect that some property of the environment (vitamin composition, trace-element abundance, minor though long-term changes in temperature, salinity, or upwelling) has promoted a character shift.

Many species whose evolution can be documented, last a relatively short time in the geologic record. For example, from the late Miocene to Recent of the equatorial Pacific we can observe the evolution of six species. One of these survives to the present while the other five become extinct. Most of them become extinct within a half million years of their first appearance. This fact suggests that there was something inherently unstable in the new species or that they evolved in response to some transitory (in a geological sense) phenomenon in the environment. The fact that we know so little about the multitude of environmental influences operating on the diatoms leaves us relatively free to speculate on the cause of their evolution.

PALEOECOLOGY

Prior to discussing the paleoecological importance of diatoms, several words of caution are necessary. Because planktonic diatoms are built to remain in suspension, they can easily be transported laterally by both bottom and surface currents. This may result in an anomalously high concentration of planktonic forms where none existed in the local, living assemblage, the **biocoenose**. Burckle and Biscaye (1971) found that in the Antarctic high southern latitude neritic and oceanic diatoms are carried northward by Antarctic Bottom Waters as far as the equatorial regions (Fig. 10). The possibility of such displacements

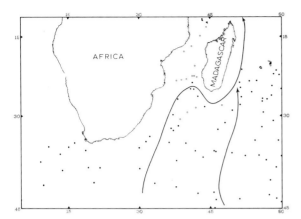

Fig. 10. Distribution of displaced Antarctic diatoms (solid circles) in surface sediments of the southwest Indian Ocean. The arrows indicate the probable avenues by which Antarctic Bottom Water moves northward.

must be seriously considered in any paleoecological reconstruction.

Secondly, we should keep in mind that selective solution both in the water column and in the sediments may significantly alter the diatom death assemblage, **the thanatocoenose**. In the equatorial Pacific the solution rate may remove more than 80% of the biocoenose and in those sections where monospecific oozes are recovered we suspect that solution has removed everything else.

A final point to keep in mind is that the word "paleoecology" when applied to diatoms should be read as quantitative paleoecology. In order to achieve maximum results the floras must be subjected to some sort of mathematical treatment which may range from extremely simple to algebraically complex.

Quantitative paleoecology

The initial step in quantitative paleoecology is counting and categorizing (into taxa) a minimum number of diatoms per sample. Statistics have shown us that no fewer than 300 individuals per sample ought to be counted. Where species diversity is high, the count should be increased to that level where no additional species are encountered. The resulting diatom spectrum showing the relative percent abundance for each species versus time is useful in demonstrating time-sequential changes in the diatom flora.

However, some diatomists prefer to integrate such a spectrum into a single curve and thereby determine the relative importance of various ecological parameters upon the changing diatom community. To achieve this end, several avenues are open. The simplest is the T_d value:

$$T_d = \frac{X_w}{X_c + X_w} \times 100$$

where X_w = total count of warm-water diatom species, and X_c = total count of cold-water diatom species. T_d stands for diatom temperature and is used in a relative sense in reference to cold or warm assemblages.

In some cases we already know the temperature preferences of the diatoms with which we are dealing. In others, it may be necessary to determine the constituency of a diatom assemblage and its relationship to latitude, water mass and temperature. To divide diatom assemblages along a north—south gradient into a number of species groups we can use a recurrent group analysis:

$$\alpha_{ab} = \frac{J_{AB}}{\sqrt{N_A N_B}} - \frac{1}{\sqrt{N_B}}$$

where N_A = total number of occurrences of species A, N_B = total number of occurrences of species B, and J_{AB} = joint occurrences of species A and B. When $N_B \geqslant N_A$, then pairs of species are considered to show affinity (i.e., they belong to the same group) when $\alpha_{ab} \geqslant 0.5$. This permits the eventual definition of a number of species groups all of whose members show affinity for one another.

In addition to these relatively simple statistical techniques a number of more sophisticated mathematical tools are available to the student. Cluster analysis may be used to define fossil associations and the reader is encouraged to become familiar with the treatment of this subject given in Michener and Sokal (1957). A final method, and one which will come into increasing use in the future is the Biological Response Model of Imbrie and Kipp (1971). This method uses Factor Analysis to group species relative to some ecologic parameters. With planktonic species the principal paleoecological parameter sought is temperature. Paleoecological equations developed by Imbrie also permit the reconstruction of other parameters such as salinity and nutrient supply, but these factors usually can be related to temperature.

Pleistocene temperatures

Since most paleoecological work deals with the Pleistocene, we are naturally most interested in temperature change during that period. Kanaya and Koizumi (1966), for example, used recurrent group analysis and T_d values in reconstructing late Pleistocene temperature

change in the North Pacific. Although there was little time control, they were able to document at least three warm intervals during the late Pleistocene. T_d values have been used in the equatorial Pacific (Fig. 11). You can readily see that a substitution is necessary in the original equation since cold-water forms are not present. However, if we substitute cosmopolitan species for cold-water ones we find that the equatorial assemblage can be broken down into two groups: a tropical/subtropical group and a cosmopolitan group. As shown in Fig. 11 there are significant changes in the assemblages, changes which essentially agree with climatic oscillations found by other methods.

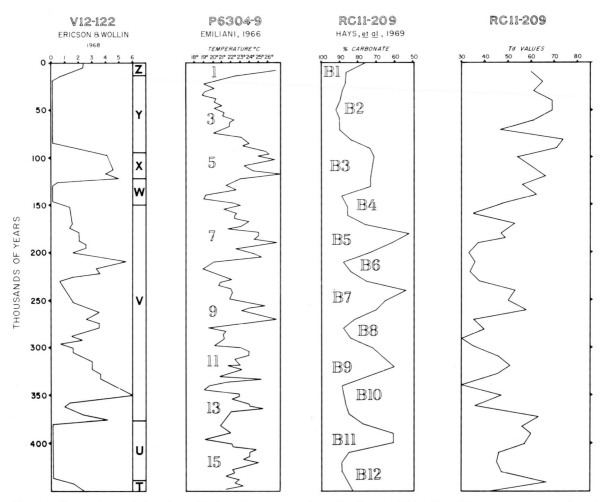

Fig. 11. Comparison of T_d values obtained from a core from the Equatorial Pacific with temperature data based upon Foraminifera (Ericson and Wollin, 1968), $^{18}O/^{16}O$ isotope ratios (Emiliani, 1966) and abundance of biogenic calcium carbonate (Hays et al., 1969).

Valve size and paleoecology

You will recall that Belayeva noted that certain species show a marked increase in size (in this case, valve diameter) over areas of strong upwelling. This size increase appears to be related to the nutrient supply which in turn is governed by the rate of upwelling.

In 1954, Kolbe performed a simple but time-consuming experiment. In a deep-sea core from the eastern equatorial Pacific (an area of upwelling), Kolbe measured the valve diameters of 300 specimens per sample of *Coscinodiscus nodulifer* A. Schmidt, a species which ranges from approximately 10 μm to over 100 μm in diameter. His work showed good correlation between maximum diameter values and glacial periods. If, as Arrhenius (1952) proposed, circulation increased during glacial periods and if vigorous circulation means more vigorous upwelling, the diatom valve size would be expected to increase during glacials.

Brackish, fresh-water and benthic diatoms in paleoecology

Marine benthic and brackish-water communities can also be utilized in paleoecological reconstruction. One must repeat the caution, however, that currents, particularly bottom currents, may displace these diatoms from their normal habitat. Sedimentary structures may provide some clue as to whether or not displacement has taken place. The condition of the diatom valve may also be a key, for if they are broken and partially dissolved they may very well have been displaced. On the other hand, we find in the South Atlantic, well-preserved benthic diatoms of Eocene age from depths of approximately 900 m on the Argentine continental slope — clear evidence, it would seem, of considerable subsidence along this slope since the Eocene.

We can use brackish-water diatoms to reconstruct sea-level changes in the recent past as well as any significant changes in temperature or water chemistry. Many of these species have very narrow temperature, salinity and chemical tolerances. Changes in any parameter of the environment, therefore, lead to a change in the diatom community. By careful analysis of the diatom biocoenosis we should be able to reconstruct shoreline (sea-level) changes and, thereby, reconstruct past climates.

Fresh-water diatoms can indicate the sources of continental detritus found in the marine environment. For instance, along the southwest edge of the Sahara a long wet season alternates with a much shorter dry season. Intermittent lakes, usually containing abundant diatoms, dry out during the dry season and the ever present Trade Winds carry lake sediment out to sea where it is incorporated into marine sediments. By studying paleomagnetically dated deep-sea cores from the equatorial Atlantic we can determine the time of first influx of wind-blown continental detritus into the Atlantic (about 1.2 m.y. B.P.) as well as changes in volumes of continental detritus. Of special interest is the observation that *Stephanodiscus astraea* (Ehrenberg) Grunow was a common constituent of equatorial Atlantic sediment from about 1.2 m.y. B.P. to approximately 400,000 years B.P. Such a convergence of evidence and circumstance suggests that this section of North Africa was continually humid prior to 1.2 m.y. B.P. and that, prior to that time, these lakes were perennial.

Diatom abundance and diversity

Two additional measurements which may have some paleoecological significance are total diatom abundance per gram of sediment and diatom diversity. Diatom numbers per gram of sediment have been much used by modern-day diatomists as an indication of past changes in surface productivity. The diatoms in an aliquot of the original sample are counted and then, by simple ratio, numbers per gram are determined. The problem arises in the method used — no two diatomists employ the exact same methods in sample preparation and sample counting. Therefore, the results usually differ by a great deal.

Diatom diversity has also been used as an indicator of surface productivity. The usual method of determining diversity is to determine the numbers of species in a random count of several hundred specimens. This method has been found unreliable because it does not account for the relative abundance of individual species. In our work we have found good results using the Shannon-Wiener Information Measure as a Diversity Index:

$$H_s = - \sum_{i=1}^{S} P_i \ \log_e P_i$$

where S = number of species in a random count of 300 individuals;

P_i = the relative abundance of the ith species measured from 0 to 1.0;

$\log_e P_i$ = the logarithm of P_i to the base e.

A high value for H indicates a very diverse assemblage while a 0 or low value indicates a monospecific or near monospecific assemblage. In the equatorial Pacific we find high diversity values during a glacial period which may result from increased circulation vigor during cold periods resulting in increased surface nutrients and productivity.

As mentioned earlier, diatoms are easily displaced by currents and, as an example, the action of Antarctic Bottom Waters in carrying high-latitude diatoms toward the equator is noted. It can be seen that the transportability of the diatom valve can be turned into a plus factor and we might use the distribution of Antarctic diatoms in surface sediments to trace the northward flow of Antarctic Bottom Water. Fig. 10 (p. 253) shows the distribution of these diatoms in surface sediments of the southwest Indian Ocean. One can clearly trace the influence of bottom water to the south and east of Malagasy. Similar distributions are observed in other parts of the Southern Hemisphere.

BIOSTRATIGRAPHY

The fundamental tenet of biostratigraphy is that species evolve and become extinct at specific times and can thus be used to subdivide strata.

Because the appearance of benthic diatoms, like any other benthic organism, can be time-transgressive, true time-stratigraphic correlations are best drawn using planktonic diatoms. There are a number of species which can be used to define and differentiate the Cretaceous and Paleogene. The Cretaceous may be defined by several species of *Triceratium*, *Coscinodiscus*, *Aulacodiscus* and *Bid-*

dulphia, which appear to be restricted to the Mesozoic. The "crisis of life" which occurred at the Mesozoic/Cenozoic boundary also affected the diatoms because there is a difference between the late Cretaceous and Paleogene forms. The true extent of this difference has not yet been determined because of the paucity of Paleogene diatomites. However, the Paleogene witnessed the first appearance of the Pennales.

The Eocene may be defined by several species of the genus *Pyxilla* as well as *Cyclotella hannae* Kanaya, *Coscinodiscus oblongus* Greville and several species of *Triceratium*. *Pyxilla* spp. had a rather broad distribution during the Eocene, occurring in the South Atlantic, the Southwest Pacific, the California sections and the Norwegian Sea. Some of these forms extend upward into the early Oligocene. However, little is known about the Oligocene floras, particularly the younger part. Burckle (1972) described a high-latitude Oligocene flora from the North Atlantic but reported low-diversity assemblage dominated by the genus *Stephanopyxis*.

In Miocene and younger sediments, we have more vertical continuity and thus a greater opportunity to define biostratigraphic zones (Fig. 12). Hans-Joachim Schrader (1973) has defined 25 diatom zones in the North Pacific for the interval from the middle Miocene to the Recent, while Japanese workers have named 7 zones for the same time interval on the Japanese mainland. Since both continental and deep-sea Neogene sections may be diatomaceous, we can correlate the two.

In general, we can characterize the diatom assemblages of marine Neogene sections as follows: The early Miocene in low latitudes can be defined by the occurrence of such forms as *Coscinodiscus lewisianus* Greville, *C. lewisianus* var. *similis* Rattray, *Rhaphidodiscus marylandicus* Christian and *Coscinodiscus praepaleaceus* Schrader. The last mentioned species is found as a constituent of high-latitude diatom assemblages as well. In intermediate latitudes (northern and southern Italy, Maryland and California), *C. lewisianus* var. *lewisianus* and *similis*, *C. praepaleaceus* and *R. marylandicus* are found. The late early Miocene and early middle Miocene may be characterized by the problematical diatom

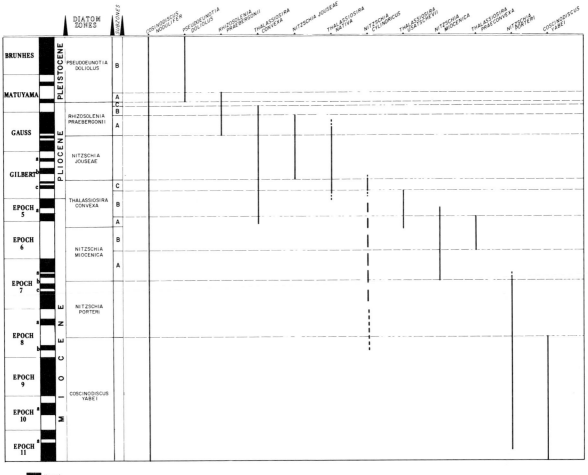

Fig. 12. Ranges of selected diatom species from the late Miocene to Recent of the Equatorial Pacific. (From Burckle, 1972).

Macrora stella (Azpeitia) Hanna. This form is found most abundantly in intermediate latitudes but has also been found in the equatorial regions as well as higher-latitude sections of Japan.

The base of the middle Miocene is defined by the first appearance of *Annelus californicus* Tempere and Peragallo. This is an extremely important species, particularly useful in circum-Pacific correlations. It occupies a very short time range during the early middle Miocene and is found in both continental marine and deep-sea sections.

In equatorial regions a number of species are found in association with one another and do not range above the middle Miocene. Such associations, therefore, may be used to define the middle Miocene and include *Coscinodis-*

cus lewisianus Greville, *Craspedodiscus coscinodiscus* Ehrenberg, *Actinocyclus ingens* Rattray, *Cestodiscus pulchellus* Greville, and *C. pulchellus* var. *maculatus* Kolbe. Besides these, there are a number of forms which provide specific information on stratigraphic levels within the middle Miocene.

An important first appearance of *Denticula nicobarica* Grunow occurs in the early middle Miocene, followed by the first occurrence of *Denticula hustedtii* Simonsen and Kanaya in the middle middle Miocene. The middle middle Miocene is also characterized by the first occurrence of plicated species of the genus *Coscinodiscus*. Members of this group characterize the late middle Miocene and early middle Miocene of many low- and intermediate-latitude sections.

Ranging concurrently with the early part of the range of the plicate *Coscinodiscus*, is the species *Mediaria splendida* Sheshukova-Poretzkaya. This form is found in intermediate and high latitudes and provides an important link in correlating equatorial and high-latitude sections.

The late Miocene is characterized in equatorial regions by such forms as *C. paleaceus* (Grunow) Rattray and *C. yabei* Kanaya which both disappear around the middle of the late Miocene. Two *Denticula* species (*D. nicobarica* Grunow and *D. hustedtii* Simonsen and Kanaya) disappear near the middle/late Miocene boundary but are found higher up in the section in the higher latitudes. *Actinocyclus ingens* Rattray is another form whose last occurrence is time-transgressive from low to high latitudes.

The latest Miocene is characterized by a number of evolutionary first appearances. *Nitzschia miocenica* Burckle, for example, evolves from its ancestor *N. porteri* Frenguelli, and ranges to just before the Miocene/Pliocene transition. *Thalassiosira convexa* Muchina evolves before the Miocene/Pliocene transition and ranges to the latest Pliocene (Fig. 12). Both these species are fairly cosmopolitan and were distributed from the equator to about 40° north and south of the equator. An equally important species with a much broader distribution is *Nitzschia reinholdii* Kanaya and Koizumi. This form evolved during the latest Miocene and ranged into the middle Pleistocene

High northern latitudes may be characterized during the late Miocene by the first appearance of *Rhizosolenia barboi* Brun, *Den-*

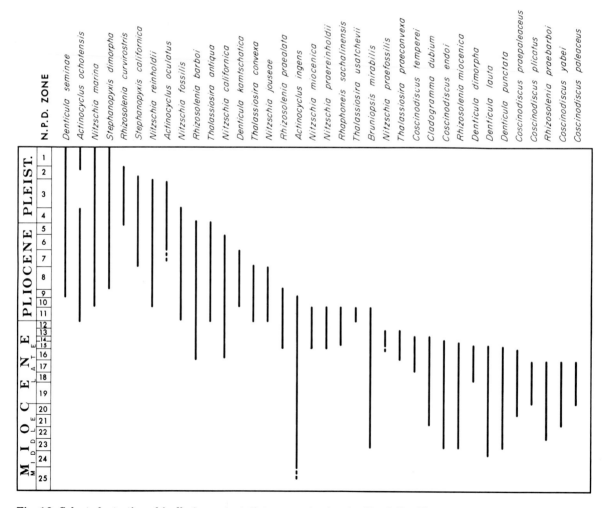

Fig. 13. Selected, stratigraphically important diatom species for the North Pacific.

ticula kamtschatica Sabelina and *Denticula seminae* (Semina) Simonsen and Kanaya (Fig. 13).

During the Pliocene a number of first and last appearances provide us with stratigraphic markers. *Nitzschia jouseae* Burckle makes an evolutionary first appearance during the early Pliocene and disappears during the middle late Pliocene. This form is abundant in equatorial sediments but has been found as far away as 40° of latitude from the equator. The early/ late Pliocene boundary is the approximate position of the first appearance of *Rhizosolenia praebergonii* Muchina. This species ranges to just after the Pliocene/Pleistocene transition and is generally restricted to low latitudes (Fig. 14).

In high northern latitudes of the Pacific region, a number of forms define the Plio-

cene: *Actinocyclus ochotensis* Jousé, *Thalassiosira punctata* Jousé, *Rhizosolenia barboi* Brun to mention a few.

The base of the Pleistocene is marked by the first evolutionary appearance of *Pseudoeunotia doliolus* (Wallich) Grunow. It is found mainly in equatorial regions but is also present in intermediate latitudes as far away as 40° from the equator. By early Pleistocene times an essentially modern diatom flora had developed in most marine environments of the world.

In the high northern latitudes, we observe a distinct provincialism. In the North Pacific, the flora is dominated by *Denticula*, several species of *Rhizosolenia*, several species of *Thalassiosira* and the genus *Coscinodiscus*. In the North Atlantic, *Denticula* is not present and the assemblages appear to be dominated

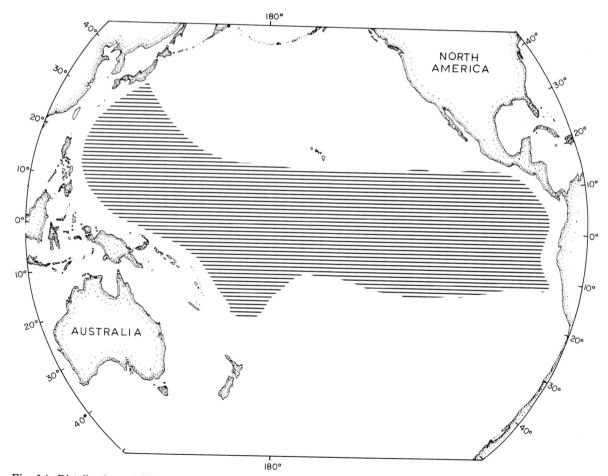

Fig. 14. Distribution of *Rhizosolenia praebergonii* Muchina late Pliocene/earliest Pleistocene sediments of the Pacific. (From Burckle, 1972.)

by *Thalassiosira gravida* Cleve on the one hand and *Thalassionema nitzschioides* Grunow on the other.

The Pleistocene diatoms of the Antarctic region have been studied by a number of workers (Jousé and others, 1963, Kozlova, 1964; Donahue, 1970 to mention a few) (Fig. 15). Jousé and others (1963) defined a number of abundance zones based upon the alternation in abundance of Antarctic and Subantarctic forms.

In addition to these, there are numerous diatoms which serve as indicators to the Pleistocene and are unique to the Antarctic region. These include at least six species of the genus *Nitzschia* (of which *N. kerguelensis* (O'Meara) Hasle is the most abundant), *Eucampia balaustium* Castracane, *Coscinodiscus lentiginosis* Janisch, *C. margaritae* Frenguelli and *Thalassiosira gracilis* (Karsten) Hustedt.

APPENDIX

Brief description of key Neogene diatoms (Multiple photographs show different views of same species, or at different focal levels)

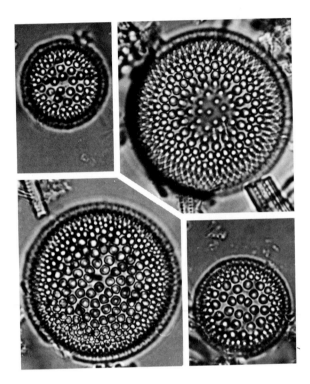

Actinocyclus ingens Rattray. Valve flat to concavo-convex. Dispersed rows of areolae radiate from central area. Pseudonodule small, indistinct, surrounded by small poroids. X 750.

DIATOM ZONES

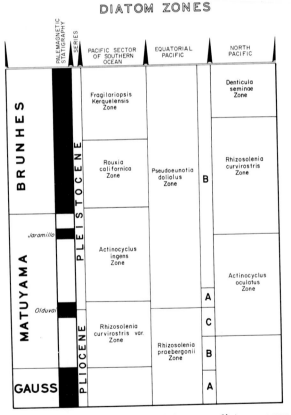

Fig. 15. Late Pliocene and Pleistocene diatom zones for the Pacific. (From Donahue, 1970, and Burckle, 1972.)

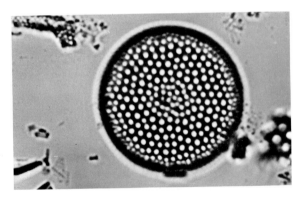

Actinocyclus ochotensis Jousé. Valve discoid, flat, usually with hyaline central area. Areolae usually fasciculate, interrupted by a narrow hyaline zone near the margin. X 1000.

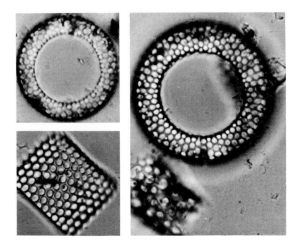

Annellus californicus Tempère and Peragallo. Tubular valve, usually with large central area absent. Areolae coarse. × 750.

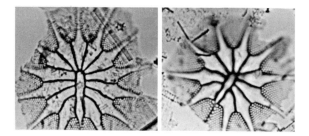

Asteromphalus hiltonianus (Greville) Ralfs. Valve discoid, flat. Central hyaline surface covers nearly half of valve surface. From 10 to 12 hyaline rays extend from central area to valve margin. Gently curved lines extend from the central area toward the valve margin. × 750.

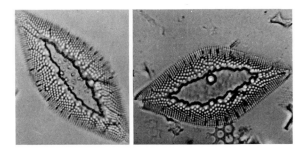

Cestodiscus peplum Brun. Valve ovoid, flat, usually with a large ovoid central area. Areolae cover about one-third of the valve around the periphery. × 750.

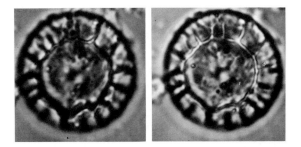

Cladogramma dubium Lohmann. Valve circular, convex. Central area clear. Margin characterized by ribs which extend about one-third distance toward central area. × 1000.

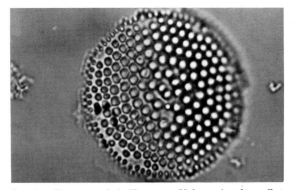

Coscinodiscus endoi Kanaya. Valve circular, flat. Areolae radiate from central area and are loosely fasciculate. A central nodule is present as is a broad hyaline ring around the margin. × 1000.

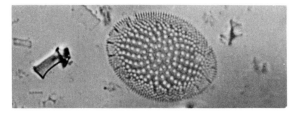

Coscinodiscus lewisianus Greville. Valve ovoid, flat to slightly convex. Large areolae spiral out from central area. Areolae smaller and more tightly spaced around margin. × 1000.

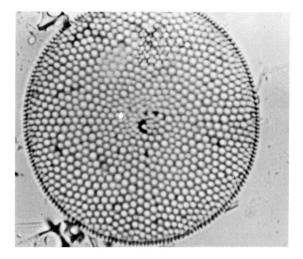

Coscinodiscus nodulifer A. Schmidt. Valve circular, flat. Areolae fasciculate, increasing slightly in diameter toward the margin. A small nodule is present near the center of the valve. × 1000.

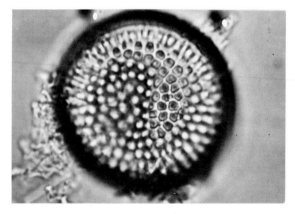

Coscinodiscus plicatus Grunow. Valve circular, broken by a tangential plication; one side convex, the other side concave. Areolae cover entire valve, increasing in size slightly toward margin and then decreasing at the margin. × 1000.

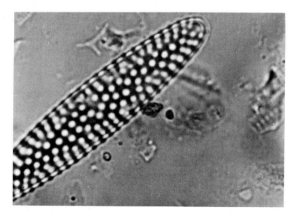

Coscinodiscus praepaleaceus Schrader. Valve ovoid, flat. Transapical ribs extend to middle of valve where they are joined by zig-zag rib arrangement. × 1000.

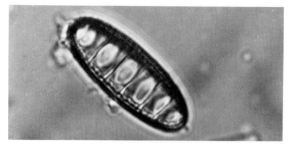

Denticula dimorpha Schrader. Valve linear-elliptical with bluntly rounded ends. No secondary pseudosepta. Almost hyaline valve, with strongly silicified thickenings which protrude into the interior of the cell. × 1000.

Denticula hustedtii Simonsen and Kanaya. Valve elliptical to elliptical-linear with bluntly rounded ends. Secondary pseudosepta present. Valve is finely punctate with punctae in decussate pattern. × 1000.

Denticula kamtschatica Sabelina. Valve linear-elliptical with broadly rounded ends. Secondary pseudosepta, when present, found only near valve ends. × 1000.

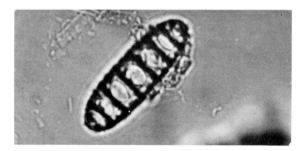

Denticula lauta Bailey. Valve linear-elliptical to elliptical with bluntly rounded ends. Secondary pseudosepta, when present, found only near valve ends. Punctae present in decussate pattern. × 1000.

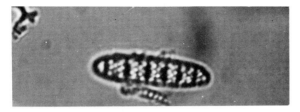

Denticula nicobarica Grunow. Valve narrow, linear to linear-elliptical. Secondary pseudosepta not present. Punctae large, in quincunx pattern. × 1000.

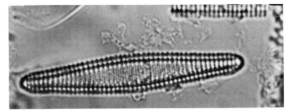

Nitzschia fossilis (Frenguelli) Kanaya. Valve elliptical with slightly rounded apices. Valve margins slightly convex. Intercostal membranes with two rows of decussate punctae. × 750.

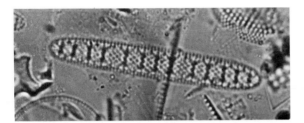

Denticula punctata Schrader. Valve linear to linear-elliptical with broadly rounded ends. Secondary pseudosepta sometimes present. Coarsely punctate. × 750.

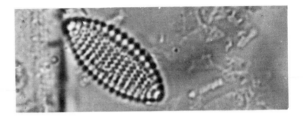

Nitzschia jouseae Burckle. Valve linear-lanceolate with subacute apices and convex margins. Intercostal membrane contains double row of decussate punctae. Valve margin well developed. × 1000.

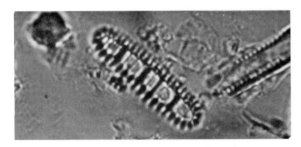

Denticula seminae (Semina) Simonsen and Kanaya. Valve linear to elliptical with broadly rounded ends. Secondary pseudosepta present. Finely punctate in decussate patterns. × 1000.

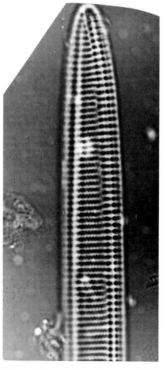

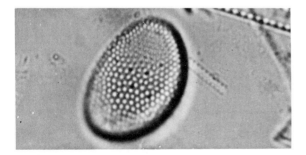

Hemidiscus cuneiformis Wallich. Valve flat with dorsal margin gently to strongly convex and ventral margin flat to gently concave. Areolae usually small, arranged in fascicules. × 1000.

Nitzschia marina Grunow. Valve linear to linear-elliptical with broadly rounded ends. Valve margins parallel to gently convex. Intercostal membrane with two rows of punctae. × 750.

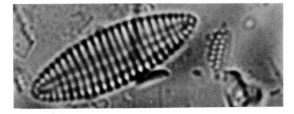

Nitzschia miocenica Burckle. Valve flat, linear-lanceolate with gently rounded apices. Two lines run down center of valve, some 2—3 μm apart. Intercostal membrane with two rows of small punctae. × 1000.

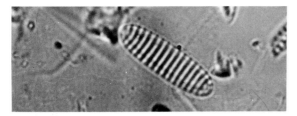

Nitzschia porteri Frenguelli. Valve small linear-elliptical with broadly rounded ends. Costae close together and intercostal membrane with rows of small punctae. × 1000.

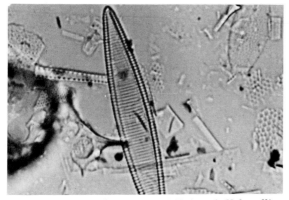

Nitzschia reinholdii Kanaya and Koizumi. Valve elliptical with slightly convex margins. Costae close together. Intercostal membrane with rows of small decussate punctae. × 1000.

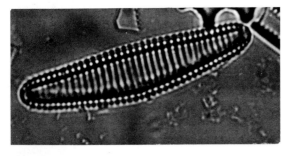

Pseudoeunotia doliolus (Wallich) Grunow. Valve heteropolar with dorsal margin gently convex and ventral margin gently concave. Costae close together. Intercostal membrane with rows of small decussate punctae. × 1000.

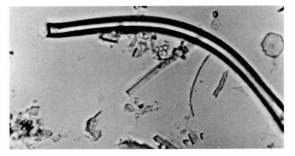

Rhizosolenia barboi Brun. Curving cylindrical tube terminated at the apical end by two small spines. Spine is present at point of maximum curvature. × 1000.

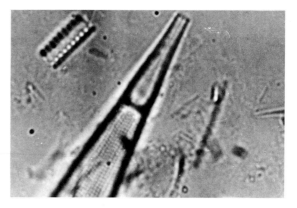

Rhizosolenia bergonii Peragallo. Gently tapering tube terminating at the apical process. Calyptra is finely punctate. × 1000.

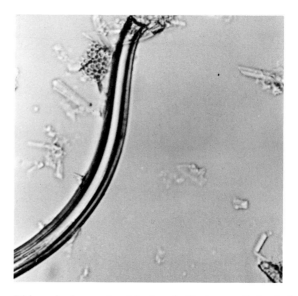

Rhizosolenia curvirostris Jousé. Similar to *R. barboi* except that it is larger and lacks the spine at the point of maximum curvature. × 1000.

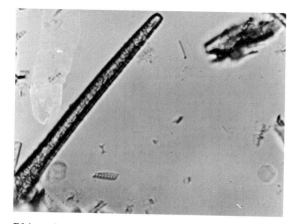

Rhizosolenia praealata Schrader. Valve cylindrical, with rows of radial punctae extending up to the middle of the apical process. Apical process is without spines. × 1000.

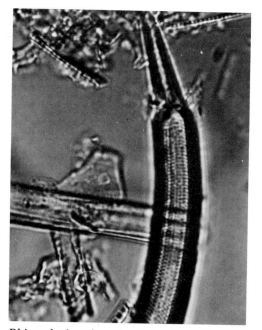

Rhizosolenia miocenica Schrader. Valve cylindrical, tapering toward the apex. Apical process with lines of punctae. Apical spine present. × 750.

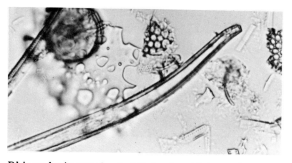

Rhizosolenia praebarboi Schrader. Valve cylindrical, straight or gently curved. Apical process hyaline. No apical spine. × 750.

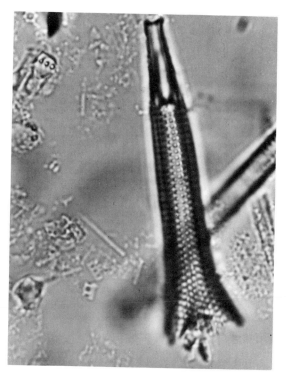

Rhizosolenia praebergonii Muchina. Valve cylindrical, robust, strongly areolate. Apical process thick, flaring out at the apex. × 1000.

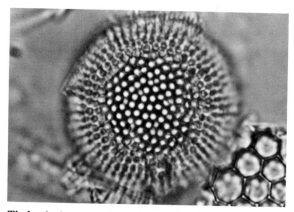

Thalassiosira convexa Muchina. Valve circular, strongly convex. Areolae coarse, frequently fasciculate. Well-developed margin. × 1000.

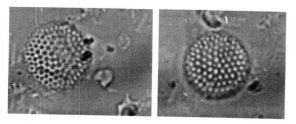

Thalassiosira praeconvexa Burckle. Valve small, convex. Fine areolae are frequently in fascicules. × 1000.

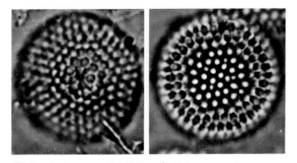

Thalassiosira usatchevii Jousé. Valve strongly convex. Coarse areolae are nonfasciculate. Well-defined margin. × 1000.

SUGGESTIONS FOR FURTHER READING

Hendey, N.I., 1964. An introductory account of the smaller algae of British coastal waters. *Fish. Invest. Ser. 4, Part 5, Bacillariophyceae (Diatoms)*, pp. 1—317 (London). [Although an account of diatoms in the northeast Atlantic, this book has an excellent introduction to diatom taxonomy, biology, ecology and methodology.]

Hustedt, F., 1927—1964. Die kieselalgen Deutschlands, Österreichs und der Schweiz mit Berücksichtigung der übrigen Länder Europas sowie der angrenzenden Meeres-Gebiete. In: *Dr. Rabenhorst's Kryptogamen-Flora von Deutschland, Oesterreichs und der Schweiz*. 7, I, Sect. 1—5, pp. 1—920. (1927—1930); II, Sect. 1—5, pp. 1—736 (1931—1937); III, Sect. 1—3, pp. 1—556 (1961—1964). Leipzig [A "source book" in living marine and fresh-water diatoms. Morphology and ecology are stressed.]

Koizumi, I., 1975. Neogene diatoms from the Western Margin of the Pacific Ocean, Leg 31, Deep Sea Drilling Project, *Initial Rep. Deep Sea Drilling Project*, 31: 779—819. [A key paper stressing biostratigraphic correlation of Equatorial and North Pacific diatom zonations.]

Patrick, R. and Reimer, C., 1966. The diatoms of the United States, 1. *Monogr. Acad. Nat. Sci. Phila.*, 13: 1—688. [A book on fresh-water diatoms with excellent introductory chapters on various aspects of diatom biology and distribution.]

Simonsen, R. (Editor), 1972. First symposium on Recent and fossil marine diatoms. *Nova Hedwigia*, 39: 1—294. [Some fifteen contributions on various aspects of fossil and living marine diatoms.]

CITED REFERENCES

Arrhenius, G., 1952. Sediment cores from the East Pacific. *Swed. Deep-Sea Expedition (1947—1948), Rep.*, 5: 1—89.

Blow, W.H., 1969. Late Middle Eocene to Recent planktonic biostratigraphy. In: P. Brönnimann and H.H. Renz (Editors), *Proc. First Int. Conf. Planktonic Microfossils, Geneva, 1967*. Brill, Leiden, 1: 199—421.

Burckle, L.H., 1971. Correlation of Late Cenozoic marine sections in Japan and the Equatorial Pacific. *Trans. Proc. Paleontol. Soc. Jap., N.S.*, 82: 117—128.

Burckle, L.H., 1972. Late Cenozoic planktonic diatom zones from the eastern Equatorial Pacific. In: R. Simonsen, (Editor), *Symposium on Recent and Fossil Marine Diatoms. Nova Hedwigia*, 39: 217—256.

Burckle, L.H., 1975. Diatom evidence bearing on the correlation of the Relizian stage, California. *Pac. Geol.*, 9: 33—34.

Burckle, L.H. and Biscaye, P., 1971. Sediment transport by Antarctic Bottom Water through the eastern Rio Grande

Rise. *Annu. Meet. Geol. Soc. Am.*, pp. 518—519 (abstract).

Donahue, J., 1970. *Diatoms as Quaternary Biostratigraphic and Paleoclimatic Indicators in High Latitudes of the Pacific Ocean*. Thesis, Columbia University, New York, pp. 1—230.

Emiliani, C., 1966. Paleotemperature analysis of the Caribbean cores P6304-8 and P6304-9 and a generalized temperature curve for the past 425,000 years. *J. Geol.*, 74: 109—126.

Ericson, D. and Wollin, G., 1968. Pleistocene climates and chronology in deep-sea sediments. *Science*, 162: 1227—1234.

Hays, J.D., Saito, T., Opdyke, N.D. and Burckle, L.H., 1969. Pliocene-Pleistocene sediments of the Equatorial Pacific: Their paleomagnetic, biostratigraphic and climatic record. *Geol. Soc. Am. Bull.*, 80: 1481—1514.

Hendey, N.I., 1964. Introductory account of the smaller algae of British coastal waters. *Fish. Invest., Ser. 4, Part 5, Bacillariophyceae (Diatoms)*, pp. 1—317 (London).

Imbrie, J. and Kipp, N., 1971. A new micropaleontological method for quantitative paleoclimatology: application to a late Pleistocene Caribbean core. In: K.K. Turekian (Editor), *Late Cenozoic Glacial Ages*. Yale University, New Haven, Conn., pp. 71—181.

Jousé, A.P., Koroleva, G.S. and Nagaeva, G.A., 1963. Stratigraphic and paleogeographic research in the Indian Ocean sector of the Antarctic. *Okeanol. Issled.*, 8: 137—160 (in Russian).

Kanaya, T. and Koizumi, I., 1966. Interpretation of diatom thanatocoenosis from the North Pacific applied to a study of core V20-130 (studies of a deep-sea core V20-130, Part IV). *Tohoku Univ. Sci. Rep., Ser. 2, Geol.*, 37: 89—130.

Kilham, P., 1971. A hypothesis concerning silica and the fresh-water planktonic diatoms. *Limnol. Oceanogr.*, 16: 10—18.

Koizumi, I., 1968. Tertiary diatom flora of Oga Peninsula, Akita Prefecture, northeast Japan. *Tohoku Univ. Sci. Rep., Ser. 2, Geol.*, 40: 170—240.

Koizumi, I., 1972. Marine diatom flora of the Pliocene Tatsunokuchi Formation in Fukushima Prefecture. *Trans. Proc. Paleontol. Soc. Jap., N.S.*, 86: 340—359.

Kolbe, R.W., 1954. Diatoms from equatorial Pacific cores. *Swed. Deep-Sea Expedition (1947—1948), Rep.*, 6: 183—196.

Kozlova, O.G., 1964. *Diatoms of the Indian and Pacific Sectors of the Antarctic*. Acad. Sci. U.S.S.R., Inst. Oceanol., Israel Program for Scientific Translations, Jerusalem, pp. 1—191.

McCollum, D.W., 1973. Neogene genus *Trinacria* as a stratigraphic marker in southern ocean sediments. *Antarct. J. U.S., p.* 198.

Mitchener, C.U. and Sokal, R.R., 1957. A quantitative approach to a problem in classification. *Evolution*, 11: 130—162.

Nakaseko, K., Koizumi, I., Sugano, K. and Maiya, S., 1972. Microbiostratigraphy of the Neogene formations in the Nadaura area, Toyama Prefecture, Japan. *Geol. Soc. Jap., Bull.*, 78: 253—264.

Schrader, H., 1973. Cenozoic diatoms from the Northeast Pacific, Leg 18. In: *Initial Report of the Deep See Drilling Project*, Leg 18: 673—797.

Simonsen, R., 1969. Diatoms as indicators in estuarine environments. *Veröff. Inst. Meeresforsch. Bremerhaven*, 11: 287—292.

Simonsen, R. and Kanaya, T., 1961. Notes on the marine species of the diatom genus *Denticula* Kutz. *Int. Rev. Gesamten Hydrobiol.*, 46: 498—513.

Tappan, H. and Loeblich Jr., A.R., 1971. Geologic implications of fossil phytoplankton evolution and time-space distribution. In: R. Kosanke and A.T. Cross (Editors), *Symposium on Palynology of the Late Cretaceous and Early Tertiary. Geol. Soc. Am., Spec. Pap.* 127: 247—340.

Wornardt, W.W., 1969. Diatoms, past, present and future. In: P. Brönnimann and H.H. Renz (Editors), *Proc. First Int. Conf. Planktonic Microfossils, Geneva, 1967*. Brill, Leiden, 2: 690—714.

11

SILICOFLAGELLATES AND EBRIDIANS

BILAL U. HAQ

INTRODUCTION

Silicoflagellates and ebridians are silica-secreting marine microplankton. Recent revival of interest in the history of ocean basins and paleoclimates has enhanced their importance in paleooceanographic interpretations, especially in those areas where calcareous microfossils are either relatively scarce or completely lacking (e.g., in higher latitudes and deeper waters).

Together with rare siliceous dinoflagellates, silicoflagellates and ebridians are the only silica-secreting flagellates. The first appeared in the early Cretaceous and have survived with little change in their basic morphology to the present day. In present-day seas silicoflagellates constitute a relatively minor part of the total phytoplankton populations.

Fossil silicoflagellates and ebridians are found together and have been traditionally studied together. Besides the differences in their skeletal morphology there are also some other basic differences between them. Whereas silicoflagellates are cosmopolitan and photosynthetic (autotrophic), ebridians are found more commonly in temperate to cold water and do not possess chromatophores and are thus heterotrophic.

HISTORY OF STUDY

Like most other groups of microfossils, silicoflagellates were first observed by the German biologist C.G. Ehrenberg, who described fossil and living silicoflagellates between 1837 and 1840. He also recorded the first fossil ebridian in 1844 and Schumann later reported living ebridians from modern seas in 1867. After being variously assigned to diatoms and radiolaria, silicoflagellates were separated into the Order Silicoflagellata by Bogart in 1890 on the basis of the presence of a single flagellum. The materials collected during various oceanographic expeditions in the late nineteenth and early twentieth centuries were a great impetus to the study of silicoflagellates and ebridians. Taxonomic and systematic studies continued through to the 1960's particularly by many workers in Western Europe and Russia (for a summary of previous works, see Loeblich and others, 1968). Recent studies of silicoflagellates have been mainly directed towards their utility in biostratigraphy and paleoclimatology. In spite of their slow evolutionary rates, silicoflagellates have been successfully applied in biostratigraphy in the late Cenozoic, especially in high latitudes where most other microfossil groups are of limited or no help.

BIOLOGY OF SILICOFLAGELLATES

The silicoflagellates are minute (20—50 μm, rarely up to 100 μm), unicellular, marine flagellates with a siliceous skeleton. The major part of the clear, hyaline, protoplasm is enclosed within the skeleton, with only an outer layer (including pseudopodia, some granules and chromatophores) and a single flagellum extending beyond the skeletal lattice. The cell contains an oval to subcircular nucleus and many yellowish or greenish chromatophores necessary for photosynthesis (Fig. 1).

The skeleton

The silicoflagellate skeleton is composed of opaline silica. In various species it may vary in shape from a simple ring-like structure

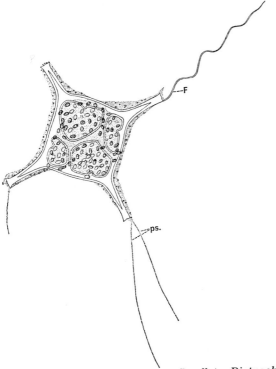

Fig. 1. Living cell of the silicoflagellate *Dictyocha fibula* Ehrenberg. Abapical view showing single flagellum, pseudopodia, protoplasm enclosing the skeleton, chromatophores and granules. (After Marshall, 1934, and Loeblich and others, 1968.) Length of the skeleton is about 40 μm.

with spines to elongated geometrical forms and relatively more complex dome-shaped structures. However, a **basal ring** is common to all species on which more complex structures may be built up on one side thus making

most species asymmetrical about their basal rings and giving the whole skeleton a somewhat hemispherical shape with a radial symmetry. The skeletal tubular rods are hollow (Fig. 2A) and their surfaces are generally smooth. However, under higher magnifications the surface of some species is seen to be rough, reticulate or with ridges parallel to the long axis of the rods (Fig. 2B).

In culture studies wide variations in the skeletal morphology have been observed within the same species (Van Valkenburg and Norris, 1970). Biologists stress the plasticity of the silicoflagellate skeleton and the use of caution in defining fossil taxa (see section on Major Morphologic Groups).

Nutrition

The presence of chromatophores suggests that silicoflagellates are autotrophic, manufacturing their own food through photosynthesis. The possibility of their capturing food particles with pseudopodia is also likely, but has not been observed in culture studies. Blue-green algal symbionts have been observed associated with some species in culture and these may also help in nutrition (Norris, 1967).

Reproduction

Silicoflagellates reproduce by vegetative division. In the rare observations made on

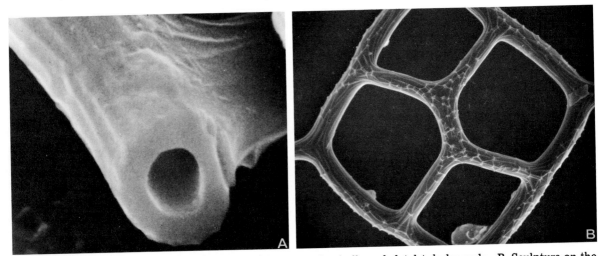

Fig. 2. A. Scanning electron micrograph at × 12,800 showing hollow skeletal tubular rods. B. Sculpture on the surface of *Dictyocha fibula* Ehrenberg. × 1,600. (After Wornardt, 1971.)

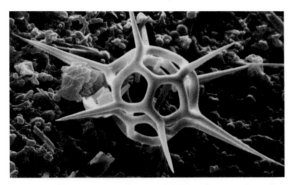

Fig. 3. A double skeleton of *Distephanus speculum* (Ehrenberg) Haeckel, first stage in the vegetative reproductive cycle. × 1000. (Photo: H. Okada.)

silicoflagellates, protoplasm is seen to extrude towards the basal ring, where a new skeleton forms as a mirror image of the parent (Fig. 3) (Gemeinhardt, 1930). Although other modes of reproduction have not been observed in silicoflagellates, it is possible that as in other phytoflagellates sexual reproduction (i.e., fusion of gametes and subsequent division) does take place.

ECOLOGY

Silicoflagellates are cosmopolitan. Like other marine microplankton the make-up of assemblages and abundance of silicoflagellates is closely related to water mass properties (mainly temperature and salinity), and the availability of nutrients.

Temperature seems to be a major controlling factor, imparting a latitudinal differentiation in the assemblages. For example, in the South Atlantic Gemeinhardt (1934) found the genus *Dictyocha* to be restricted mainly to the low and mid latitudes (10°C and above) and the genus *Distephanus* to high latitudes (20°C and below).

Availability and concentrations of nutrients are important for the numerical abundance of taxa. In the case of silicoflagellates concentration of silica for skeletal construction is also an important factor. Since silica levels are generally low in surface waters, where photosynthetic silicoflagellates must remain in the euphotic zone, areas of upwelling where both nutrient and silica-rich deeper waters are brought to the surface are ideally suited for silicoflagellate abundance (Lipps, 1970). The

fluctuations in the intensity of upwelling would thus affect the numerical abundance of silicoflagellate taxa. As changes in upwelling are intimately related to changes in the atmospheric and oceanic circulation, this makes silicoflagellates a potentially useful tool in the study of paleoecology (paleoproductivity and paleoclimatology) in the ancient sediments.

MAJOR MORPHOLOGICAL GROUPS

Silicoflagellates are protists and claimed both by zoologists (as protozoans) and botanists (as algae). However, due to their autotrophic phytoplanktonic nature it is more logical to regard them as planktonic algae. The algalogists include them in the Order Siphonotestales, Subclass Silicoflagellatopheidae, Class Chrysophyceae and Phylum Chrysophycophyta.

Various descriptive terms used for silicoflagellates are shown in Fig. 4 (also see Bukry, 1976, for terminology). Silicoflagellates are either studied in smear-slides made from raw sediment samples or they are extracted by dissolving and chemically disaggregating all of the sediment content except the siliceous fraction which can then be concentrated. For details of the chemical extraction method the reader is referred to Mandra and others (1973).

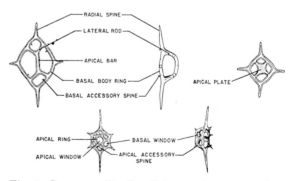

Fig. 4. Common silicoflagellate terminology used to describe the skeleton (after Mandra and Mandra, 1972).

The criterion of silicoflagellate basic morphology (see Skeleton, p. 267) helps in distinguishing silicoflagellates from other taxa erroneously described as such in the paleontological literature. Lipps (1970) has presented a critical review of silicoflagellate taxonomy based on an analysis of their functional morphology. He concludes that there are eight generic groups that can be accepted as genuine silicoflagellates. Other forms do not conform to the basic morphology of the silicoflagellates. These show more affinity to either diatoms or radiolaria. In this chapter several major silicoflagellate morphological groups are recognized.

Genuine silicoflagellates

(1) The **dictyochids**. Typified by the genus *Dictyocha* Ehrenberg (Fig. 5A, B). The basal ring in the species of this group is quadrate with spines or knobs at each corner. The central structure in various species is a variation of a diagonal apical bar connected to two short lateral bars on either side. In the genus *Hannaites* Mandra the corners of the quadrate basal ring are produced into rounded protuberances (Fig. 5C).

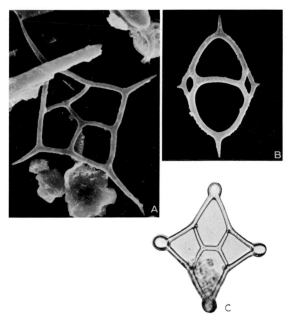

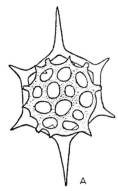

Fig. 5. Dictyochids. A. *Dictyocha fibula* Ehrenberg. × 1,140. B. *Dictyocha ausonia* Deflandre. × 570. (After Wornardt, 1971.) C. *Hannaites quadria* Mandra. × 340. (After Perch-Nielsen, 1975a.)

(2) The **distephanids**. Typified by the genus *Distephanus* Stohr. The basal ring is generally hexagonal with spines at each corner. A smaller apical hexagonal ring is connected to

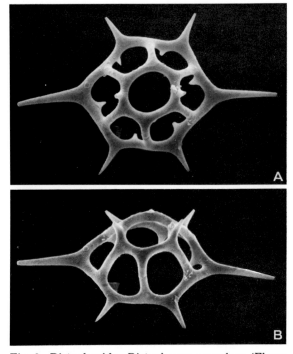

Fig. 6. Distephanids. *Distephanus speculum* (Ehrenberg) Haeckel. A. Abapical view. B. Side view of the same specimen. × 1,500. (After Wornardt, 1971.)

the basal ring between corners by lateral bars (Fig. 6).

(3) The **cannopilids**. Typified by the genus *Cannopilus* Haeckel. There is a great morphological variation within this group. The basal ring varies from a distephanid-type hexagonal ring with spines to a circular ring with numerous spines. The central structure is more complicated than distephanids and is usually dome-shaped, with variations from

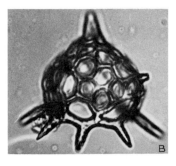

Fig. 7. Cannopilids. A. *Cannopilus hemisphaericus* (Ehrenberg) Haeckel. × 610. (After Schulz, 1928.) B. *Cannopilus depressus* (Ehrenberg) Perch-Nielsen. × 580. (After Perch-Nielsen, 1975a.)

regularly spaced lateral bars meeting a symmetrical, complex, apical ring to a more irregular asymmetrical central structure. (Fig. 7.)

(4) The **mesocenids.** Typified by the genus *Mesocena* Ehrenberg. This is the simplest of the silicoflagellate morphological groups, consisting of basal rings without any central structure. In some cases the rings of mesocenids are similar to basal rings of other genera in the above three morphologic groups, with the difference that apical central structures either failed to form or are missing. (Fig. 8.)

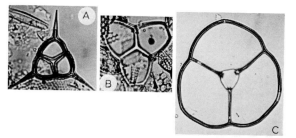

Fig. 9. Corbisemids. A. *Corbisema triacantha* (Ehrenberg) Hanna. × 480. B. *Corbisema bimucronata* Deflandre. × 480. C. *Corbisema geometrica* Hanna. × 300. (After Perch-Nielsen, 1975a.)

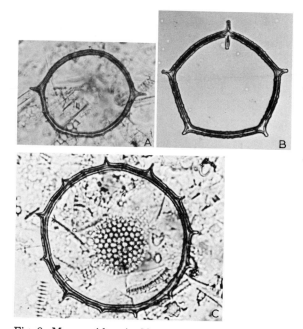

Fig. 8. Mesocenids. A. *Mesocena diodon* Ehrenberg. × 410. B. *Mesocena elliptica* (Ehrenberg) Ehrenberg. × 410. C. *Mesocena circulus* (Ehrenberg) Ehrenberg. × 650. (After Perch-Nielsen, 1975a.)

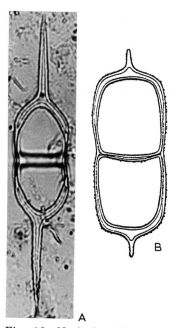

Fig. 10. Naviculopsids. A. *Naviculopsis biapiculata* (Lemmermann) Frenguelli. × 800. (After Perch-Nielsen, 1975a.) B. *Naviculopsis rectangularis* (Schulz) Frenguelli (after Frenguelli, 1940).

(5) The **corbisemids.** Typified by the genus *Corbisema* Hanna. The basal ring is trigonal, with or without spines. A simple central structure is connected to the basal ring by lateral bars. (Fig. 9.)

(6) The **naviculopsids.** Typified by the genus *Naviculopsis* Frenguelli. Two-sided basal rings range from long and narrow to short and broad in various species, pointed and spinose at both ends. An arched central bar spans the central area which may or may not bifurcate near the basal ring. (Fig. 10.)

Genera incertae sedis

Forms traditionally included among silicoflagellates but not showing the basic silicoflagellate morphology and with greater affinity to either diatoms or radiolaria are:

Genus *Vallacerta* Hanna. A pentagonal basal ring with spines and a convex, sculptured, disc-like central structure cover the central area. It is described only from Upper Cretaceous rocks. (Fig. 11A.)

Genus *Lyramula* Hanna. These Y- or U-shaped hollow rods may represent incomplete rings of silicoflagellates or they may be fuse·

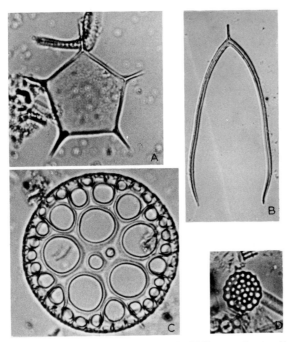

Fig. 11. Genera incertae sedis. A. *Vallacerta hortonii* Hanna. × 720. B. *Lyramula furcula* Hanna. × 430. C. *Rocella gemma* Hanna. × 660. D. *Pseudorocella corona* Deflandre. × 660. (A,B: after Perch-Nielsen, 1975a; C,D: after Perch-Nielsen, 1975b.)

setae as seen in some diatoms. (Fig. 11B.)

Genera *Rocella* Hanna and *Pseudorocella* Deflandre (Figs. 11C, D). These are circular discs with numerous circular holes and lacking spines. In *Rocella* the holes increase in size towards the center of the disc which is almost flat. In *Pseudorocella* the pores of the central disc are about the same size and the disc is slightly convex; the margin is flat and without pores. *Rocella* includes a single species (*R. gemma* Hanna); and *Pseudorocella*, two species (*P. barbadiensis* Deflandre and *P. corona* Deflandre). These forms most likely represent valves of diatoms (Lipps, 1970).

Other genera included with silicoflagellates have even less or no affinity to true silico-flagellates. These forms include: *Clathrium* Frenguelli (?fragments of diatom frustule); *Clathropyxidella* Deflandre (probably an ebridian); *Cornua* Schulz (at least some species are fragments of ebridians, others may be fragments of metazoan spicules); *Lutetianella* Filipescu and *Nothyocha* Deflandre (both radiolarian fragments); *Pseudomesocena* Hovasse (= *Planicircus* Frenguelli) (probably a fragment of radiolarian skeleton).

PALEOCLIMATOLOGY

Since Gemeinhardt's (1934) observation that *Dictyocha* and *Distephanus* replace each other in dominance in warm and cold waters, these silicoflagellate genera have often been used in paleoclimatic interpretations. This paleoclimatic method involves a comparison of the *Dictyocha/Distephanus* ratios from Recent (core-top) sediments with the mean annual surface temperatures of the water masses above. A temperature vs. generic-ratio curve can then be constructed and used in down-core paleoclimatic interpretations.

Mandra (1969) first applied this method to obtain a temperature vs. generic-ratio curve and inferred relatively warm temperate climatic conditions for the late Eocene off the Antarctic Peninsula. Jendrzejewski and Zarillo (1972) found a close correlation between the frequencies of *Dictyocha* abundance and warm-water foraminifera and between *Distephanus* and cold-water foraminifera in a subantarctic deep-sea core in the southeastern Indian Ocean. On the basis of dictyochid abundances they recognized five warm climatic peaks in the late Pleistocene.

Ciesielski and Weaver (1974) used the *Dictyocha/Distephanus* ratios from six deep-sea cores to infer several temperature oscillations and a dramatic change in the thermal structure of the Southern Ocean. They concluded that temperatures in the Antarctic in the early Pliocene (between 4.3 to about 3.95 m.y. B.P.) were much higher than today (as much as $10°C$) so as not to be able to support an extensive ice sheet. Such an ice sheet must have developed or re-established itself after 3.95 m.y. B.P., reaching its present dimensions between 3.85 and 3.80 m.y. B.P. These authors also attempted an independent test of Mandra's (1969) temperature vs. generic-ratio curve and concluded that although paleoclimatic inferences based on this curve would be accurate qualitatively, the temperature values inferred from this curve would be anomalously high for their data from the early Pliocene. They suggested an alternative curve, more compatible with their data.

Recently Poelchau (1974) has suggested that the relationship between surface temperatures and the *Dictyocha/Distephanus* ratios is not a simplistic one. Such varied factors as differential grazing by predators, differences and fluctuations in the productivity of the water masses and endemism amongst species play important roles in determining the composition of assemblages.

In spite of these complicating factors, the silicoflagellates still remain an important tool for paleoclimatic and paleoproductivity studies, admittedly, perhaps for more qualitative interpretations.

GEOLOGICAL DISTRIBUTION OF SILICOFLAGELLATES

The earliest known silicoflagellates are from the early Cretaceous. Rare fragments of a probable corbisemid have been reported from early Cretaceous strata (Deflandre, 1950; Tynan, 1957). In the late Cretaceous dictyochids, corbisemids and *Vallacerta* become fairly common. The genera *Vallacerta* and *Lyramula* are restricted to the late Cretaceous. All other silicoflagellate groups appeared in the Tertiary: the naviculopsids in the early Paleocene, the distephanids in the middle Eocene, and the cannopilids in the Oligocene. The genus *Hannaites* has so far been reported from the Eocene only. All major genera (except *Vallacerta*, *Hannaites*, and *Corbisema*) continue to the Recent.

Silicoflagellate zonation

Attempts at biostratigraphic zonation using silicoflagellate taxa have been frustrated to some extent by the slow evolutionary rate of this group. Although new taxonomic data are still forthcoming from less well-studied parts of the stratigraphic column (the Mio-Pliocene has been preferentially studied, thus explaining a high number of species recorded from this interval), a general assessment of evolutionary rates can be made from the degree of morphological variety in the already described taxa. Thus, evolutionary rates (origination vs. extinction of species) are relatively low in the late Cretaceous, rising sharply but not markedly in the Paleocene, offset by the appearance of new lineages such as the distephanids and the naviculopsids. The

rates rise relatively slightly once again in the Eocene and then drop again in the Oligocene. From the middle Miocene to Pliocene the evolutionary rates are high relative to the preceding early Tertiary and the Pleistocene.

It is obvious that biostratigraphic subdivisions (zones) are bound to be of long duration throughout the late Cretaceous and early Tertiary. However, even long zones, in areas where other microfossil groups are absent, can prove useful. We can, on the other hand, hope that relatively more refined (short duration) zonations for the Mio-Pliocene interval will eventually be available.

It is beyond the scope of an introductory text such as this to include all the existing biostratigraphic zonation schemes. The interested reader is referred to the following papers for further perusal of the subject: Martini (1971), Ling (1972), Bukry and Foster (1973), and Bukry (1976) for zones of tropical to subtropical areas, and to Martini (1972, 1974), Ling (1970), Bukry (1974), Bukry and Foster (1974), Perch-Nielsen (1975a, 1976) and Ciesielski (1975) for zones of non-tropical and high latitude areas.

EBRIDIANS

These siliceous, marine planktonic flagellates are within the same size range as silicoflagellates. In contrast to silicoflagellates the living ebridian cell possesses two unequal flagella, it lacks chromatophores and has an internal skeleton (Fig. 12). The skeleton is also of solid construction rather than tubular and its composition suggests affinity to the Radiolaria. According to Loeblich and others (1968) the possession of a dinoflagellate-like nucleus and two unequal flagella in ebridians suggests a closer affinity to siliceous dinoflagellates and should thus be considered a separate class within the Division Pyrrhophyta.

Biology and ecology of ebridians

An ebridian cell is illustrated in Fig. 12. The two unequal flagella are attached to the large (dinokaryon) nucleus. The skeleton is entirely within the protoplasm. Ebridians most probably reproduce both vegetatively and sexually. In vegetative reproduction, first

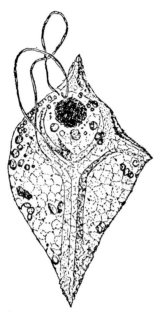

Fig. 12. Cell components of an ebridian, *Hermesinum adriaticum* Zacharias showing the two inequal flagella attached to the large dinokaryon nucleus. (After Hovasse, 1934.)

individuals. Sometimes the daughter skeleton may not separate and a double skeleton may result. Sexual reproduction is also suspected because occasionally in fossil forms a lorica surmounting two or three normal ebridian specimens have been observed (Loeblich and others, 1968).

Ebridians are heterotrophic and diatoms form their major food source. They are generally less common and less diverse than silicoflagellates. They are relatively more common only in cold to temperate waters. No substantial work has so far been done on the geographical distribution of ebridians.

Common ebridian genera and geological occurrence

The ebridian skeleton has triradial or tetraxial symmetry. In many cases the skeleton may not develop completely, or may occur in fragmentary forms as fossils. This adds considerably to the complexity of ebridian taxonomy.

a daughter skeleton is formed within the parent cell and then the nucleus and other cell contents divide to give rise to two separate

The most common ebridians are the thecate type, with a geological record going back to Paleocene. They are most numerous in Eocene and Miocene. The more commonly

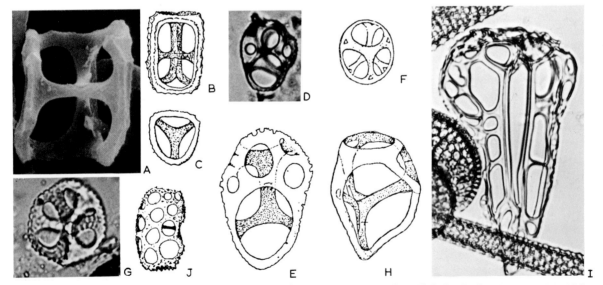

Fig. 13. Common ebridian genera. A—C. *Ammodochium*. A. *A. rectangulare* (Schulz) Deflandre. × 2,400. (After Perch-Nielsen, 1975b.) B,C. *A. prismaticum* Hovasse; B, lateral view; C, apical view. × 950. (After Hovasse, 1932a.) D,E. *Craniopsis*. D. *C. octo* Hovasse. × 800. (After Perch-Nielsen, 1975b.) E. *C. octo* Hovasse. × 1,400. (After Deflandre, 1934.) F. *Ebria antiqua* Schulz. × 850. (After Schulz, 1928.) G. *Ebriopsis crenulata* Hovasse. × 800. (After Perch-Nielsen, 1975b.) H. *Hermesinella transversa* Deflandre. × 1,400. (After Deflandre, 1934.) I. *Hermesinum geminum* Dumitrica and Perch-Nielsen. × 600. (After Perch-Nielsen, 1975b.) J. *Pseudoammodochium dictyoides* Hovasse, lateral view. × 950. (After Hovasse, 1932b.)

occurring genera are: *Ammodochium* (Paleocene—Miocene) (Fig. 13A—C); *Craniopsis* (Eocene) (Fig. 13D—E); *Ebria* (Miocene—Recent) (Fig. 13F); *Ebriopsis* (Paleocene—Miocene) (Fig. 13G); *Hermesinella* (Eocene—Recent) (Fig. 13H); *Hermesinum* (Paleocene—Recent) (Fig. 13I) and *Pseudoammodochium* (Paleocene—Miocene) (Fig. 13J).

Most non-thecate ebridians are incompletely formed or fragments of complete skeleton of thecate ebridians (see Loeblich and others, 1968, for illustrations). The commonly occurring non-thecate genus *Carduifolia* (Paleocene—Miocene) (Fig. 14) that is often included amongst ebridians is now considered to be a siliceous endoskeletal dinoflagellate.

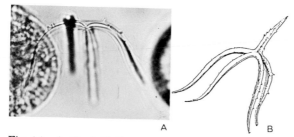

Fig. 14. A. *Carduifolia gracilis* Hovasse. × 540. (After Perch-Nielsen, 1975b.) B. *C. apiculata* Hovasse. × 1,350. (After Hovasse, 1932b.)

SUGGESTIONS FOR FURTHER READING

Lipps, J.H., 1970. Ecology and evolution of silicoflagellates. *Proc. N. Am. Paleontol. Conv.*, G: 965—993. [A good general introduction to the subject and a critical review of silicoflagellate functional morphology and validity of various genera.]

Loeblich III, A.R., Loeblich, L.A., Tappan, H. and Loeblich Jr., A.R., 1968. Annotated index of fossil and Recent silicoflagellates and ebridians with descriptions and illustrations of validly proposed taxa. *Geol. Soc. Am., Mem.* 106: 1—319. [Brief introductions to silicoflagellate and ebridian biology, history of study and geological distribution is followed by descriptions of taxa in the original language in which these taxa were first described. Original illustration of species are also included in plates. Includes comprehensive bibliography until 1965.]

CITED REFERENCES

Bukry, D., 1974. Stratigraphic value of silicoflagellates in non-tropical regions. *Geol. Soc. Am. Bull.*, 85: 1905—1906.

Bukry, D., 1976. Silicoflagellate and coccolith stratigraphy, southeastern Pacific Ocean, Deep Sea Drilling Project. *Init. Rep. D.S.D.P.*, 34: 715—735.

Bukry, D. and Foster, J.H., 1973. Silicoflagellates and diatom stratigraphy, Leg 16, Deep Sea Drilling Project. *Init. Rep. D.S.D.P.*, 16: 815—871.

Bukry, D. and Foster, J.H., 1974. Silicoflagellate zonation of Upper Cretaceous to Lower Miocene deep-sea sediments. *J. Res. U.S. Geol. Surv.*, 2: 303—310.

Ciesielski, P.F., 1975. Biostratigraphy and paleoecology of Neogene and Oligocene silicoflagellates from cores recovered during Antarctic Leg 28, Deep Sea Drilling Project. *Init. Rep. D.S.D.P.*, 28: 625—691.

Ciesielski, P.F. and Weaver, F.M., 1974. Early Pliocene temperature changes in the Antarctic Seas. *Geology*, 2: 511—515.

Deflandre, G., 1934. Nomenclature du squelette des Ebriacées et description de quelques formes nouvelles. *Ann. Protistkd.*, 4: 75—96.

Deflandre, G., 1950. Contribution à l'étude des silicoflagellidés actuels et fossiles. *Microscopie*, 2: 72—108; 191—210.

Frenguelli, J., 1940. Consideraciones sobre los silicoflagelados fósiles. *Rev. Mus. La Plata, Ser. 2*, 2: 37—112.

Gemeinhardt, K., 1930. Silicoflagellatae. In: L. Rabenhorst, *Kryptogamen-Flora von Deutschland, Österreich und der Schweiz*, 10. Akademische Verlagsgesellschaft, Leipzig, pp. 1—87.

Gemeinhardt, K., 1934. Die Silicoflagellaten des südatlantischen Ozeans. *Wiss. Ergebn. Dtsch. Atlantischen Exped. "Meteor", 1925—1927*, 12: 274—312.

Hovasse, R., 1932a. Note préliminaire sur les Ebriacées. *Soc. Zool. Fr. Bull.*, 57: 118—131.

Hovasse, R., 1932b. Troisième note sur les Ebriacées. *Soc. Zool. Fr. Bull.*, 57: 457—476.

Hovasse, R., 1934. Ebriacées, Dinoflagellés et Radiolaires. *C.R. Acad. Sci. Paris*, 198: 402—404.

Jendrzejewski, J.P. and Zarillo, G.A., 1972. Late Pleistocene paleotemperature oscillations defined by silicoflagellate changes in a subantarctic deep-sea core. *Deep-Sea Res.*, 19: 327—329.

Ling, H.Y., 1970. Silicoflagellates from central North Pacific core sediments. *Bull. Am. Paleontol.*, 58: 85—129.

Ling, H.Y., 1972. Upper Cretaceous and Cenozoic silicoflagellates and ebridians. *Bull. Am. Paleontol.*, 62: 135—229.

Mandra, Y.T., 1969. Silicoflagellates: A new tool for the study of Antarctic Tertiary climates. *J. Antarct. Res.*, 4: 172—174.

Mandra, Y.T. and Mandra, H., 1972. Paleoecology and taxonomy of silicoflagellates from an Upper Miocene diatomite near San Felipe, Baja California, Mexico. *Occ. Pap. Calif. Acad. Sci.*, 99: 1—35.

Mandra, Y.T., Brigger, A.L. and Mandra, H., 1973. Chemical extraction techniques to free fossil silicoflagellates from marine sedimentary rocks. *Proc. Calif. Acad. Sci.*, 39: 273—284.

Marshall, S.M., 1934. The Silicoflagellata and Tintinnoinea. *Br. Mus. (Nat. Hist.), Great Barrier Reef Exped. 1928—1929, Sci. Rep.*, 4: 623—664.

Martini, E., 1971. Neogene silicoflagellates from the equatorial Pacific. *Init. Rep. D.S.D.P.*, 7: 1696—1703.

Martini, E., 1972. Silicoflagellate zones in the late Oligocene and early Miocene of Europe. *Senckenberg. Lethaea*, 53: 119—122.

Martini, E., 1974. Silicoflagellate zones in the Eocene and early Oligocene. *Senckenberg. Lethaea*, 54: 527—532.

Norris, R.E., 1967. Algal consortiums in marine plankton. In: V. Krishnamurthy (Editor), *Seminar Sea, Salt and Plants*. Central Salt and Marine Chem. Res. Inst., Bhavnagar, India, pp. 178—189.

Perch-Nielsen, K., 1975a. Late Cretaceous to Pleistocene silicoflagellates from the southern SW Pacific, D.S.D.P. Leg 29. *Init. Rep. D.S.D.P.*, 29: 677—721.

Perch-Nielsen, K., 1975b. Late Cretaceous to Pleistocene archaeomonads, ebridians, endoskeletal dinoflagellates and other siliceous microfossils from the subantarctic S.W. Pacific, D.S.D.P. Leg 29. *Init. Rep. D.S.D.P.*, 29: 873—907.

Perch-Nielsen, K., 1976. New silicoflagellates and a silicoflagellate zonation in north European Paleocene and Eocene diatomites. *Bull. Geol. Soc. Den.*, 25: 27—40.

Poelchau, H.S., 1974. *Holocene Silicoflagellates of the North Pacific, Their Distribution and Use for Paleotemperature Determination*. Thesis, University of California at San Diego, 165 pp.

Schulz, P., 1928. Beiträge zur Kenntnis fossiler und rezenter Silicoflagellaten. *Bot. Arch.*, 21: 225—292.

Tynan, E.J., 1957. Silicoflagellates of the Calvert formation (Miocene) of Maryland. *Micropaleontology*, 3: 127—136.

Van Valkenburg, S.D. and Norris, R.E., 1970. The growth and morphology of the silicoflagellate *Dictyocha fibula* Ehrenberg in culture. *J. Phycol.*, 6: 48—54.

Wornardt, W.W., 1971. Eocene, Miocene and Pliocene marine diatoms and silicoflagellates studied with the scanning electron microscope. *Proc. II Planktonic Conf., Rome*, 2: 1277—1300.

PHOSPHATIC
MICROFOSSILS

Priniodid conodont element with a basal body from
the Ordovician of central Australia. × 140.

12

CONODONTS AND OTHER PHOSPHATIC MICROFOSSILS

KLAUS J. MÜLLER

INTRODUCTION

Throughout evolution both animals and plants have sometimes used phosphatic material for construction of their hard parts. In the early Cambrian phosphatic matter (apatite) seems to have been one of the main materials used for the construction of hard parts in many phyla of the animal kingdom. From late early Cambrian to late Ordovician it was replaced gradually by calcareous material, mainly calcite. Since that time onwards the number of taxa as well as individuals with phosphatic hard parts have decreased steadily up to the Recent. An exception to this general trend is the most advanced phylum, the Chordata.

Phosphatic microfossils are widely distributed in Paleozoic and early Mesozoic sediments and are valuable index fossils for parts of these eras.

Extraction methods

Phosphatic microfossils can be isolated by washing from shales as well as by etching from limestones with diluted acetic or formic acids. Hence indurated limestones which otherwise have been regarded as unfossiliferous have yielded identifiable microfossils. The effect of metamorphism on such fossils and their resistance to various extraction media are shown in Fig. 1. As they occur in smaller numbers than calcareous microfossils, a relatively large sample is required for their concentration. For details of preparation techniques the reader is referred to the *Treatise on Invertebrate Paleontology* (Hass, 1962, pp. 4—5).

COMMON PHOSPHATIC MICROFOSSILS

Conoidal shells (Hyolithelminthes)

These conical tubes 5—15 mm in length with circular to elliptical cross-sections have tips tapered and/or curved. These may be adorned externally by a sculpture of transverse ridges and/or longitudinal striae. In some cases a lid (operculum) closed the opening of the cavity in which the animal lived. Their systematic relationship is unknown. *Hyolithelminthes* range from early Cambrian to Ordovician and in early Cambrian they were locally rock-forming (Fisher, 1962). Two examples, *Hyolithellus* and *Lapworthella*, are shown in Fig. 2.

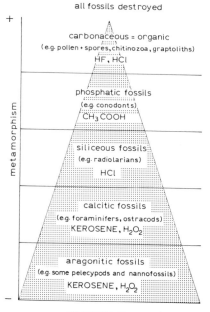

Fig. 1. Fossil resistance to metamorphism and separating agents used for each group. Phosphatic microfossils are among the most resistant to metamorphism.

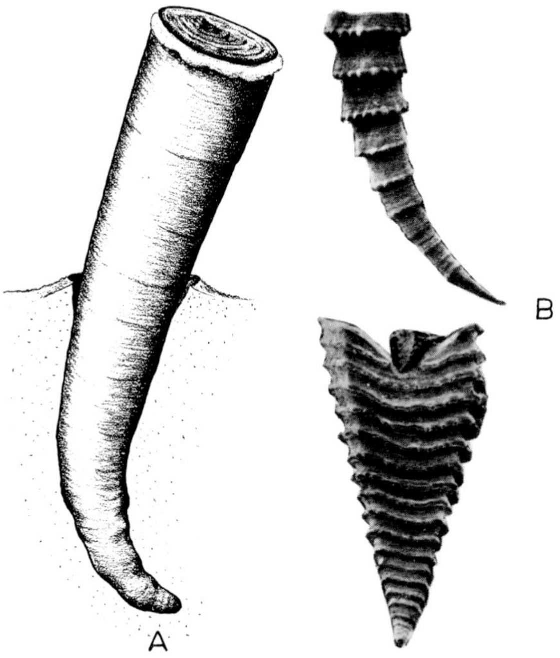

Fig. 2. Conoidal shells. Common in Cambrian. A. *Hyolithellus*. Lower Cambrian. × 5. B. *Lapworthella*. Lower Cambrian. × 20, × 25. (A after Fisher, 1962; B after Matthews, 1973.)

Horny brachiopods

Microscopic brachiopods with phosphatic skeleton were widespread in the Cambrian and early Ordovician. They are usually studied together with other brachiopods by macropaleontologists (Fig. 3A).

Horny ostracode-like arthropods (Phosphatocopina)

Phosphatic ostracode-like Arthropoda are restricted to the Cambrian (Müller, 1964). An example, *Vestrogothia*, is shown in Fig. 3B.

Remains of vertebrates

Bone fragments, spines, scales and teeth of vertebrates can accumulate locally in bone beds. In spite of their fragmentary nature they are sometimes useful index fossils, in particular around the Silurian—Devonian boundary. Several types are shown in Fig. 4.

CONODONTS

By far the most widespread and biostrati-

Fig. 3. A. Example of small horny brachiopods. Upper Cambrian, Utah. × 7. B. Horny ostracode, *Vestrogothia spinata* Müller. Restricted to Cambrian. × 100.

graphically important group of phosphatic microfossils are the conodonts.

These are minute (200 μm to 6 mm) isolated hard parts of unknown organisms. They have been described from rocks of Cambrian to Triassic age in which they are the most important single group of microfossils for biostratigraphic zonation. Their utility as index fossils is enhanced by their diverse morphologies and wide distributions which make them applicable for correlation on a worldwide basis.

History of conodont study

Conodonts were first observed by C.H. Pander in 1856. He described a number of species and coined the term "conodont" to denote the tooth-like morphology of these

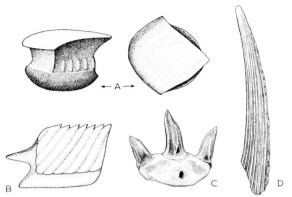

Fig. 4. Examples of vertebrate remains. A. Scale of *Thelodus* (Silurian). × 35. B. Scale of *Palaeoniscus* (Permian). × 3.5. C. Tooth of *Cladoselache* (Upper Carboniferous). × 40. D. Spine of *Climatius* (Silurian). × 1.3. (A and B after Gross, 1966.)

fossils. Conodont research continued at a slow pace during the remainder of the nineteenth and early twentieth century. However, a great impetus came in the nineteen-twenties and early thirties, when Ulrich and Bassler (1926), and Branson and Mehl (1933—34) described rich assemblages of these fossils from North America and established many new taxa.

The discovery in 1934 (Schmidt, Scott) of disjunct elements grouped in natural assemblages in still-water sediments required a fundamental revision of conodont classification; a problem which has been only partially solved. Further research in the nineteen-thirties to seventies has extended their stratigraphic range from Lower Ordovocian to Lower Cambrian (uppermost Precambrian ?) and from Upper Carboniferous to Upper Triassic. Conodonts recorded from the Upper Cretaceous and Jurassic are now considered reworked.

Composition

Conodonts consist of an organic matrix in which crystallites of apatite similar to the mineral francolite are imbedded. Well preserved material has a whitish-amber colour and is translucent. However, they are usually changed by diagenesis or natural staining to grey, black or brown and become opaque.

Morphology

Most conodonts are easily differentiated by gross outer features.

Morphologically, four main groups of conodonts can be distinguished (see Fig. 5).

(1) *Simple cones*: formed by a single tooth, or **denticle**. Various genera can be recognized from gross features in cross-section. These are important time markers for the Cambro-Ordovician and became extinct in the Devonian.

(2) *Bar-type conodonts*: thin bars, with or without a bent shaft which is commonly branched. Although conspicuous and common in many associations, these have proved to be the most conservative taxa from the evolutionary viewpoint. Thus, the group includes long-ranging 'form-species' with stable morphologies which are of little use as index fossils.

(3) *Blade-type conodonts*: elongate, laterally compressed units formed by a row of denticles which are fused except at their tips. They include index fossils from Silurian to Triassic.

(4) *Platform conodonts*: these have most probably evolved from the bar- and blade-type conodonts by the development of the broad flanges into plates. Platforms are the most highly differentiated elements, and many of them are excellent index fossils from Ordovician to Triassic.

Terminology

The important descriptive terms on conodonts are illustrated in Fig. 5. In the description of isolated conodont elements, an indication of orientation is useful; this however does not necessarily denote their former orientation inside the conodont animal. Denticles are considered to point towards the "posterior" end, and in most cases orientation can be ascertained using this character. "Upper side" is a designation for the side with the tips of denticles, its opposite being the "lower side". The terms oral and aboral are misleading and should be avoided since it has not been proved that conodonts functioned as teeth or had a position in a mouth.

The "inner side" is the concave side of the longitudinal axis along which in the platform-type conodonts the larger portion of the platform is developed; its opposite is termed the "outer side".

In single-cone types the symmetry, position

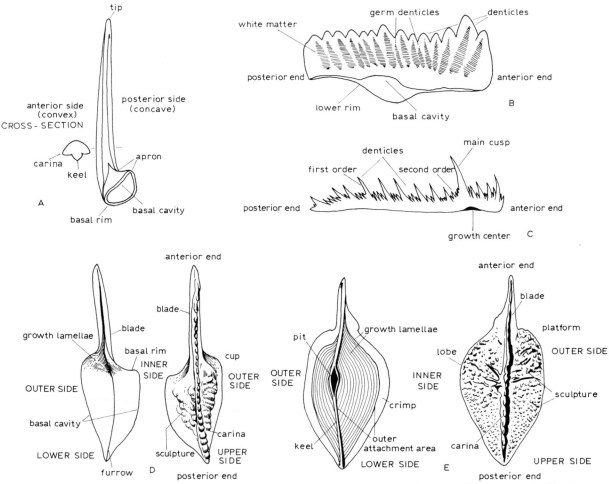

Fig. 5. Morphological terminology of conodonts. A. Single-cone type. B. Blade type. C. Bar type. D. Platform type (*Gnathodus*). E. Platform type (*Polygnathus*).

and development of **keels** and **carinae** on the sides are of systematic value. They are observed in cross-sections of denticles above the **apron**. The flaring portion above the **basal rim**, and its development and shape is also of taxonomic importance.

Germ denticles is a term for structures which have been suppressed during ontogeny due to overgrowth by adjacent structures. They can be observed in translucent fossils in transmitted light.

In platform-type conodonts a **blade** is the edge-like free portion at the anterior end. In many cases it integrates into a carina on the upper side of the platform. Protrusions on one or both sides of the platform are named **lobes**.

On the lower side of platforms, such as in *Polygnathus* (see Fig. 5) the main elements

are: (1) the **pit** around the growth center; (2) the attachment area where the edges of **growth lamellae** are visible and to which the basal organ is attached; (3) the **crimp**, or the portion outside the basal organ which does not expose the edges of growth lamellae; and (4) the keel in the center, which corresponds to the carina at the upper surface.

The lower side of some platforms, such as *Gnathodus* (see Fig. 5) is entirely excavated, and exposes the edges of growth lamellae on the entire surface. It is termed a **cup**. In the center a furrow, which is homologous to the keel, is developed.

Internal structures

The study of internal structures is important for suprageneric systematics as well as for investigations of the nature and function of conodonts. Initially, a nucleus was formed by a lamella. In subsequent stages this center is covered from outside by growth lamellae. In the earliest conodonts (suborder Paraconodontida), the growth lamellae are always open at the "upper side" (Fig. 6). In the suborder Conodontiformes, the upper lamellae are uninterrupted and at the lower side of the conodont a basal plate which is composed of much finer crystallites is developed (Fig. 7). It is not as sturdily built as the rest of the conodont and in many instances is not preserved.

The lower side of most conodonts is excavated. Two taxonomically significant structures can be distinguished:

(1) *The basal cavity*: an excavation which increases in size throughout growth of the specimen, its size being therefore dependent on the size of the specimen (Fig. 8).

(2) *The pit*: which formed when growth proceeded laterally from or above the initial point of secretion. The size does not increase throughout ontogenetic growth and ceases after a number of growth lamellae have been formed. Large specimens may thus have small pits (Fig. 9).

The ontogeny of individual conodonts can be studied in thin sections. Initial stages of all types correspond to the simple-cone type. In compound conodonts the various denticles present have developed from this simple structure through the thickening of the growth lamellae in well-defined directions of prevalent growth and by condensing them in intermediate directions. This simple manner of developing highly complicated shapes is called **anisometric growth**. Commonly, the direction of accelerated growth changes during ontogeny, and may lead to the development of peculiar shapes which are of taxonomic importance.

Structures within the denticles have been named "**white matter**", because in translucent fossils they appear white under reflected light (Fig. 10). In transmitted light, however, they appear as dark areas.

The upper surface of most conodonts is adorned with a sculpture of nodes or ridges. In addition, very small ridges may be present on the upper surface and are arranged in striae, polygons, etc. Because of their small size, the use of scanning electron microscope is necessary for study of these features which may prove useful in taxonomy (Fig. 11).

Conodont assemblages

Hinde (1879), while observing conodonts

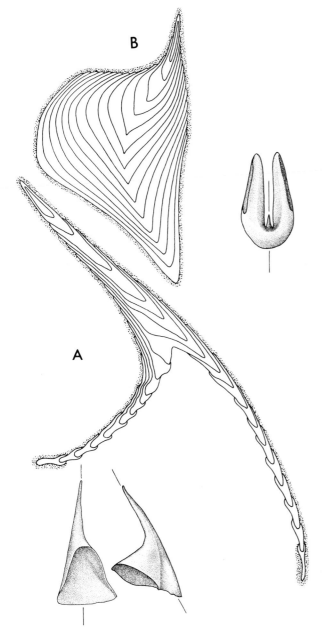

Fig. 6. Longitudinal sections through growth centers of earliest conodonts. A. *Furnishina furnishi* Müller, Upper Cambrian of Sweden. × 100. B. *Westergaardodina bicuspidata* Müller, Upper Cambrian of China. × 150. (After Müller and Nogami, 1971.)

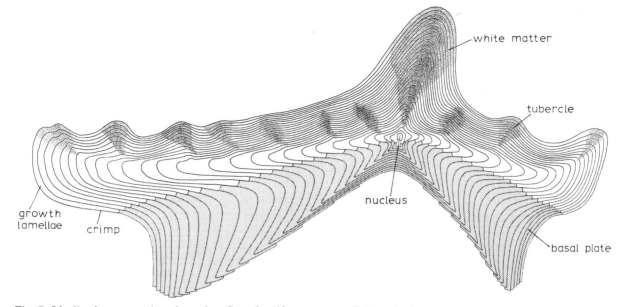

Fig. 7. Idealized cross-section through a Conodontiformes genus *Palmatolepis*.
Width of platform is developed by anisometric growth: lamellae are more widely spaced in this direction than in the other × 200. (After Müller and Nogami, 1971, modified from Gross, 1966.)

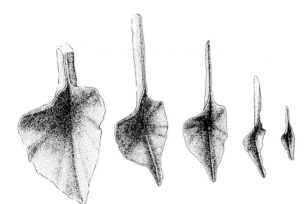

Fig. 8. Examples of conodont "basal-cavity" in various growth stages of *Gnathodus texanus* Roundy. × 25. (After Clark and Müller, 1968.)

on bedding planes of shales, realized that several conodonts with different shapes had been parts of the same animal. Such "natural" associations of conodonts are found in undisturbed sediments deposited in quiet waters, allowing the fossils to remain in their "natural" position (Fig. 12). These "associations" demonstrate that: (1) Many, if not all, of the elements of an assemblage occur in pairs, containing "right" and "left" specimens, indicating the conodont animal was bilaterally symmetrical. (2) Several pairs of one sort can be associated with one or more pairs of another sort. (3) The composition of the various

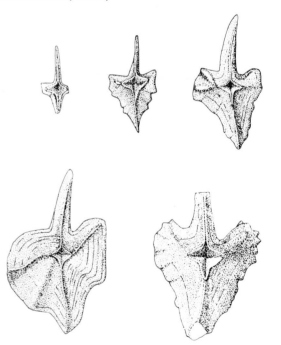

Fig. 9. Examples of conodont "pit" in specimens of *Ancyrodella rotundiloba* (Bryant). × 30. (After Müller and Clark, 1967.)

assemblages is diverse and the types of disjunct elements vary widely between different systematic units. Therefore, the composition of these assemblages is of significance for suprageneric taxonomy.

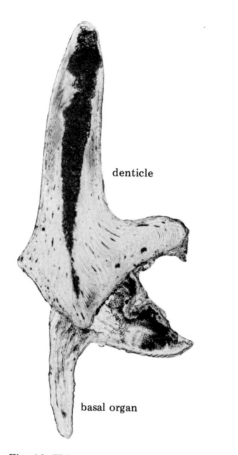

denticle

basal organ

Fig. 10. Thin section of a blade-conodont *Bryantodus inaequalis* Branson and Mehl, with basal organ, showing white matter, occurring as fine bubbles (seen as dark patches in transmitted light). × 145.

Such assemblages have been studied from very few localities. However, statistical methods using disjunct elements from etched residues allow theoretical constructions of a possible conodont apparatus (see Sweet and Bergström, 1972).

Function

Originally Pander (1856) regarded conodonts as teeth [conodont = cone-tooth (Greek)]. Although this name was based on an erroneous assumption, it remains valid. Our present state of knowledge shows that conodonts cannot have functioned as teeth because:

(1) They were built by outer apposition, and were therefore imbedded in soft tissue during their entire growth. Their internal position is also supported by their continuous growth pattern.

(2) They never show signs of functional wear that is observable on the chewing organs of vertebrates or annelids.

The shape and arrangement of conodont elements in the apparatuses suggest that they could have been supporting organs or possibly grasping, sieving or screening devices.

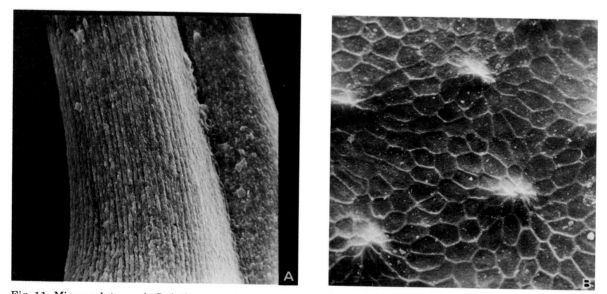

Fig. 11. Microsculpture. A. Striation on a single cone-type genus *Acontiodus*. × 560. B. Reticulation on platform-type genus *Idiognathodus*. × 580.

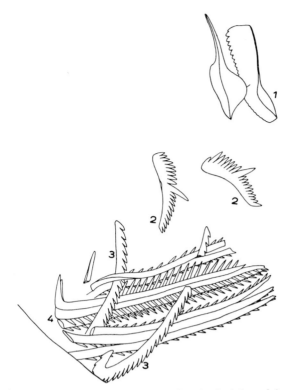

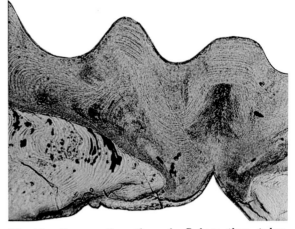

Fig. 13. Cross-section through *Polygnathus tuber-culatus* Hinde, showing resorption of white matter and growth lamellae which were later overgrown again. × 210. (From Müller and Nogami, 1971.)

Fig. 12. Conodont assemblage (*Gnathodus*) found *in situ* on a bedding plane. × 16. (After Schmidt and Müller, 1964.)

Histologic examination of conodonts has shown that resorption of a large portion of the conodont which was subsequently overgrown by regeneration is common and can be observed, sometimes repeatedly, on nearly every specimen (Fig. 13). From this evidence we believe that besides functioning as a tissue support, the conodont may have served as an organ for the temporary deposition of phosphatic matter, which might have been utilized later, either to form another conodont element in the same animal, or for other purposes.

Paleoecology

The conodont animal is believed to have been marine and free-swimming. Many species are distributed worldwide and have been found in areas as widely apart as North America, Europe, and Australia. For a long time it was believed that all representatives of the group were pelagic and drifted from the open seas towards the various continents. This is now considered true for many genera and species. Recent paleobiogeographical investigations have demonstrated the use of conodont assemblages for paleoecological reconstructions and faunal province delimitations.

The mode of life of chaetognath worms has been suggested as an ecologic model for the conodonts. Living chaetognaths are among the most common of planktonic animals, floating or swimming with tides and currents in the water without relation to the bottom. Chaetognaths show a distinct vertical stratification, many species being restricted to distinct depth zones (Hyman, 1959). The conodonts also seem to show in some cases such vertical stratification. Recognition of such depth zones is, however, difficult in fossil populations. Two theoretical models for such interpretations of conodonts are presented in Fig. 14.

Systematic position

Isolated conodonts are widespread and abundant. Nevertheless, their systematic position is still an open question. At various times different authors have classified conodonts under Cnidaria, Mollusca (Gastropoda and Cephalopoda), Annelida, Arthropoda, Chordata and even algae. However, none of these classifications seemed factually justified.

Although the main phylogenetic develop-

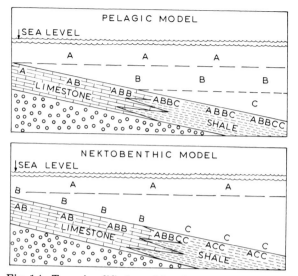

Fig. 14. Two simplified models of conodont habitat. A, B, and C represent different groups of genera with a simple two-component facies change and the resulting generic associations in the substrate. The pelagic model portrays a depth stratification. Here more diverse faunas are produced in deeper water. The nektobenthic model shows a widespread pelagic group (A) but stresses that most conodonts are laterally segregated (with some overlap) into offshore communities. (After Barnes and Nowland, 1975.)

ments occurred in the early Paleozoic, the conodont animal was highly differentiated and neither its histology, external morphology nor the range of occurrence coincides with any other living or fossil group.

Evolutionary trends

The evolutionary trends in conodonts are not well understood. All were originally formed from organic matter in which apatite crystallites have been imbedded. The earliest conodonts are largely composed of organic matter and sometimes retain their shape even when their small amount of phosphate is etched out with hydrochloric acid. In the most primitive groups (e.g., *Furnishina* and *Westergaardodina*, see Fig. 6), the separation between the conodont proper and the basal organ is not yet developed and the entire conodont has a uniform structure, which probably corresponds to the basal organ of later conodonts.

Various general trends can be observed in the development of conodont morphology. Single cones are the most primitive conodonts. In individual specimens there is a general

trend towards a restriction of the basal cavity.

In different lines of evolution, compound shapes developed from single-cone types. The earliest example is *Westergaardodina* on which the flanges on the side of the denticle have been bent upwards.

Conodonts are particularly well suited for the study of phylomorphogenesis. As do cephalopods and some foraminifera, they contain stages of their early ontogenetic development. These can be studied in thin sections.

Major conodont elements

In the following pages key conodont genera have been illustrated and described in stratigraphic order.

Cambrian

Furnishina Müller: Single-cone conodonts with extended basal cavity of triangular or quadrangular cross-section. (Fig. 15A.)
Westergaardodina Müller: Tri- or bicuspidate units with basal cavity either centrally located, or separated into two lateral cavities. (Fig. 15B.)

Ordovician

Scolopodus Pander: Single-cone conodont with costate and grooved sides. (Fig. 16A.)
Acontiodus Pander: Single-cone conodont with well-

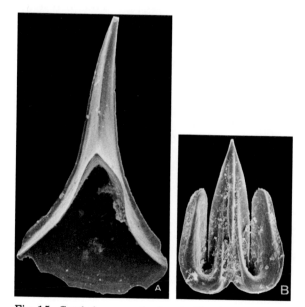

Fig. 15. Cambrian conodonts. A. *Furnishina furnishi* Müller. × 45. B. *Westergaardodina tricuspidata* Müller. × 75.

Fig. 16. Ordovician conodonts. A. *Scolopodus rex*, Lindström. × 45. B. *Acontiodus* sp. × 40. C. *Paroistodus parallelus* Lindström. × 40. D. *Panderodus simplex* (Branson and Mehl). × 90. E. *Cordylodus angulatus* Pander. × 60. F. *Pygodus anserinus* Lamont and Lindström. × 55. G. *Amorphognathus* sp. × 60. H. *Belodina compressa* (Branson and Mehl). × 60.

---

developed carina on one side. (Fig. 16B.)
Paroistodus Lindström: Single cone with sharp angle above apron on posterior side. (Fig. 16C.)
Panderodus Ethington: Single cone with conspicuously asymmetrical cross-section with deep basal cavity. (Fig. 16D.)
Cordylodus Pander: Unit with main cusp and posteriorily extended bar. Phylogenetically, it originated from a primitive single cone conodont. (Fig. 16E.)

Pygodus Lamont and Lindström: Modified single-cone type with reduced basal cavity and a well-developed plate-like side which is adorned with ribs and knots. (Fig. 16F.)
Amorphognathus Branson and Mehl: Platform type of irregular shape, with 4—6 lobes. (Fig. 16G.)
Belodina Ethington: Jaw-like, compound conodont with main denticle at tip and secondary denticles along one side. (Fig. 16H.)

Silurian

Icriodina Branson and Branson: Similar to the Devonian genus *Icriodus*. (Fig. 17A.)
Apsidognathus Walliser: Short, free blade; platform extended and sculptured. Large basal cup. (Fig. 17B.)
Pterospathodus Walliser: Similar to *Spathognathodus*,

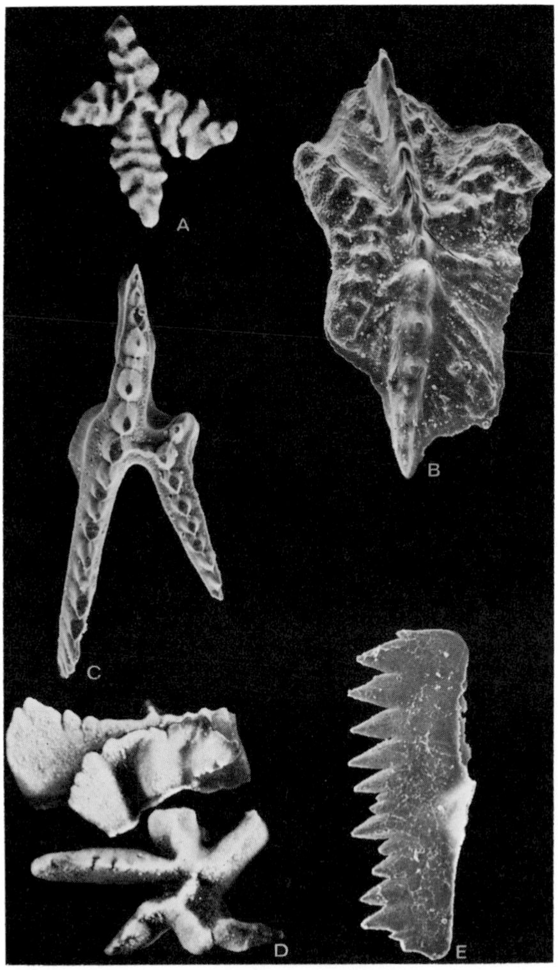

Fig. 17. Silurian conodonts. A. *Icriodina irregularis* Branson and Branson. × 30. (After Rexroad, 1967.) B. *Apsidognathus tuberculatus* Walliser. × 55. C. *Pterospathodus amorphognathoides* Walliser. × 60. D. *Kockelella variabilis* Walliser. × 23. (After Walliser, 1964.) E. *Spathognathodus steinhornensis* Ziegler. × 60.

but with a well-developed bifurcated lateral branch. (Fig. 17C.)

Kockelella Walliser: Similar to, and derived from, *Spathognathodus* by development of one or two branches. Large basal cavity. (Fig. 17D.)

Spathognathodus Branson and Mehl: Blade-type conodont with centrally located flanging extension above basal cavity. Long-ranging genus. (Fig. 17E.)

Devonian

Icriodus Branson and Mehl: Platform with three rows of nodes, in some cases fused to form transverse ridges; steep-sided, large basal cup beneath entire unit. (Fig. 18A.)

Polygnathus Hinde: Platform conodont with a rather well-developed free blade and a large, sculptured platform. Lower side with pit, keel and crimp. (Fig. 18B; see also Fig. 5.)

Ancyrognathus Branson and Mehl: Originated from *Polygnathus* by development of a lateral carina, pointing towards the posterior end. (Fig. 18C.)

Ancyrodella Ulrich and Bassler: Originated from *Polygnathus* by development of two lateral carinae pointing towards anterior end. (Fig. 18D.)

Palmatolepis Ulrich and Bassler: Platform, in most cases with one lateral lobe. Prominent node (azygous node) above growth center. Lower side with pit, outer attachment area and crimp. Index fossils in Upper Devonian. (Fig. 18E.)

Bispathodus Müller: Similar to *Spathognathodus*, but blade type with a double row of denticles. Index fossils in Upper Devonian—lowermost Carboniferous. (Fig. 18F.)

Lower Carboniferous

Siphonodella Branson and Mehl: Similar to *Polygnathus*, with conspicuous rostral ridges on platform behind the free blade. Lower side with pit. (Fig. 19A.)

Pseudopolygnathus Branson and Mehl: Similar to *Polygnathus*, with elongated, pit-like basal excavation. Most species with sturdy transverse ridges and/or nodes. (Fig. 19B.)

Scaliognathus Branson and Mehl: Anchor-shaped platform-type unit; lower side with pit. (Fig. 19C.)

Gnathodus Pander: Long free blade and well-developed platform with carina; basal cavity extends over entire lower side of the platform (cup). (Fig. 19D.)

Upper Carboniferous

Adetognathus Lane: Blade located at the side of the platform. Center of the platform trough-like. Lower side with basal cavity. (Fig. 20A.)

Idiognathodus Gunnel: Well-developed free blade; platform with conspicuous transverse ridges. Flaring basal cup extends over entire lower side of platform. (Fig. 20B.)

Fig. 18. Devonian conodonts. A. *Icriodus* sp. × 45. B. *Polygnathus asymmetricus* Bischoff and Ziegler. × 30. C. *Ancyrognathus iowaensis* Youngquist. × 45. D. *Ancyrodella rotundiloba* (Bryant). × 45. E. *Palmatolepis hassi* Müller and Müller. × 45. F. *Bispathodus costatus* (Branson). × 45.

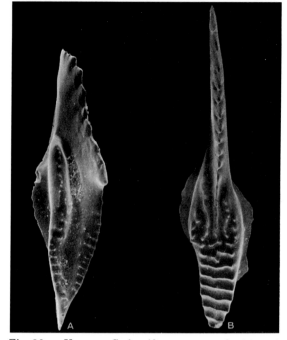

Fig. 20. Upper Carboniferous conodonts. A. *Adetognathus giganteus* (Gunnel). × 55. B. *Idiognathodus* sp. × 55.

Fig. 19. Lower Carboniferous conodonts. A. *Siphonodella duplicata* (Branson and Mehl). × 50. B. *Pseudopolygnathus* sp. × 65. C. *Scaliognathus anchoralis* Branson and Mehl. × 60. D. *Gnathodus semiglaber* Bischoff. × 60.

Fig. 21. A Permian conodont. *Gondolella rosenkrantzi*. × 70. (After Bender and Stoppel, 1965.)

Permian

Gondolella Stauffer and Plummer: Gondola- or boat-shaped platform conodonts; rounded termination on one side and sharp on the other. Upper surface with carina which is commonly flanked with furrows. (Fig. 21.)

Triassic

Epigondolella Mosher: Similar to *Gondolella*, but

with free anterior carina and node-like projections along the margin of the platform. (Fig. 22A.)
Neogondolella Bender and Stoppel: Similar to *Gondolella*, but probably from a different lineage. (Fig. 22B.)
Neospathodus Mosher: Similar to *Spathognathodus*, but with a basal cavity located at or near one end of the unit. (Fig. 22C.)

Conodont stratigraphy

The ranges of important conodont genera are given in Table I. As the conodont succes-

TABLE I

Stratigraphic range of important and index conodont genera

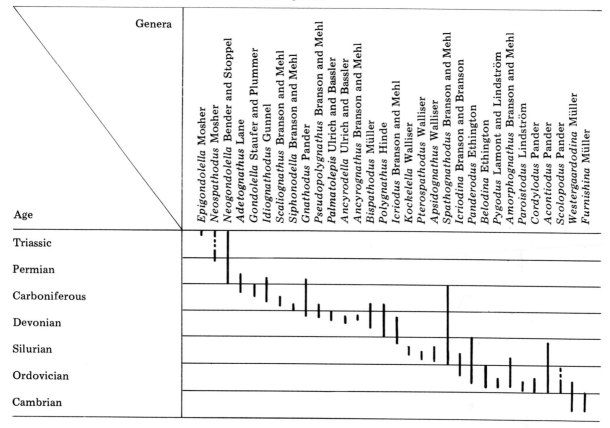

sion is fairly well known from many parts of the world, it is suitable for worldwide correlation. Cambrian conodont faunas, although widespread, are not as good zonal markers as most later ones; important examples are *Furnishina* and *Westergaardodina*. In the Tremadocian (early Ordovician) a diversification occurred which reached its peak shortly afterwards (around Arenigian). Single-cone types like *Scolopodus*, *Paroistodus*, *Acontiodus* and *Panderodus* were then prevalent. Further, *Cordylodus* as well as the first platform types, such as *Amorphognathus* and *Pygodus* have proved here to be good zonal fossils. Also various faunal province associations have been recognized in the Ordovician (Barnes and others, in Rhodes, 1973).

Although eleven zones have been established in Europe, North America, and Asia, using highly differentiated forms like *Hadrognathus* and *Pterospathodus*, the diversity of conodonts as well as their abundance

decreases again in the Silurian. The long-ranging genus *Spathognathodus* appears during the Silurian and later gives rise to many important descendents.

In the early and middle Devonian, species of *Icriodus*, *Spathognathodus*, and *Polygnathus* are prevalent. In the late Devonian the conodonts reach their second acme of development. New genera like *Palmatolepis*, *Ancyrodella*, and *Ancyrognathus* have developed from the *Spathognathodus—Polygnathus* stock, and are widely and abundantly distributed. A highly refined zonal scheme for the Devonian has been worked out particularly for Central Europe and North America (see Sweet and Bergström, 1971).

In the early Carboniferous, conodonts were still widespread and abundant, the most important genera being *Siphonodella* and *Gnathodus*. *Scaliognathus* which lived in an off-shore facies makes an excellent zonal fossil. In the late Carboniferous, *Idio-*

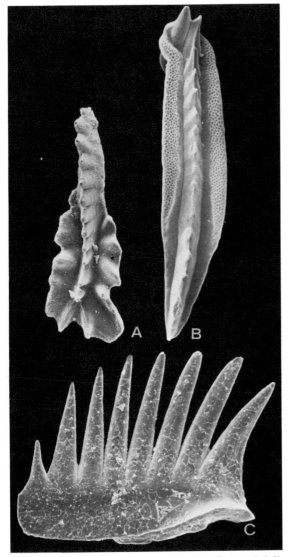

Fig. 22. Triassic conodonts. A. *Epigondolella abneptis* (Huckriede). × 90. B. *Neogondolella mombergensis prava* Kozur. × 90. C. *Neospathodus cristagalli* (Huckriede). × 150.

gnathodus, *Adetognathus* and the early gondolellids are index fossils. However due to reduced diversities, the zonal succession is no longer as refined as that for the previous periods.

In the Permian, *Idiognathodus*, *Streptognathodus*, *Gondolella* and *Neospathodus* yield a sequence of index species, but at this time conodonts underwent a crisis and almost became extinct.

In the early Triassic, the conodont stock regained their footing and became more widespread again (*Neogondolella*, *Epigondolella*), only to disappear in the late Triassic.

SUGGESTIONS FOR FURTHER READING

Hass, W.H., Rhodes, R.H.T., Müller, K.J. and Moore, R.C., 1962. Conodonts. In: R.C. Moore (Editor), *Treatise on Invertebrate Paleontology, part W*. Geol. Soc. Am., New York, N.Y., and Kansas Univ. Press, Lawrence, Kansas, pp. 3—69. [A revised edition of this treatise volume is in preparation.]

Lindström, M., 1964. *Conodonts*. Elsevier, Amsterdam, 196 pp. [General introduction.]

Lindström, M. and Ziegler, W. (Editors), 1972. *Symposium on Conodont Taxonomy*. Geologica et Palaeontologica, Sonderband 1, Marburg, 158 pp.

Müller, K.J. and Nogami, Y., 1972. Growth and function of conodonts. In: *Proc. 24th Int. Geol. Congr., Montreal, 1972, Sect. 7*, pp. 20—27. [Reviews histology.]

Pander, C.H., 1856. *Monographie der fossilen Fische des silurischen Systems der russisch-baltischen Gouvernements*. Königl. Akad. Wiss., St. Petersburg, 91 pp. [The earliest and classical work on conodonts.]

Rhodes, F.H.T. (Editor), 1973. Conodont Paleozoology. *Geol. Soc. Am., Spec. Pap.*, 141: 296 pp.

Seddon, G. and Sweet, W.C., 1971. An ecologic model for conodonts. *J. Paleontol.*, 45: 869—880. [Paleoecological model of conodont animal.]

Sweet, W.C. and Bergström, S.M. (Editors), 1971. Symposium on Conodont Biostratigraphy. *Geol. Soc. Am. Mem.*, 127: 499 pp. [Up-to-date information on biostratigraphy of conodonts by 30 contributors.]

Bibliographies

Ash, S.R., 1961. Bibliography and index of conodonts, 1949—1948. *Micropaleontology*, 7: 213—244.

Fay, R.O., 1952. Catalogue of conodonts. *Univ. Kansas Paleontol. Contrib., Vertebrata*, 3: 206 pp.

Ellison Jr., S.P., 1962. Annotated bibliography, and index, of conodonts. *Univ. Texas, Publ.*, Nr. 6210: 128 pp.

Ellison Jr., S.P., 1963. Supplement to annotated bibliography, and index, of conodonts. *Texas J. Sci.*, 15: 50—67.

Ellison Jr., S.P., 1964. Second supplement to annotated bibliography, and index, of conodonts. *Texas J. Sci.*, 16: 216—242.

Ellison Jr., S.P., 1967. Third supplement to annotated bibliography, and index, of conodonts. *Texas J. Sci.*, 19: 5—34.

Ziegler, W. (Editor), 1974, 1975. *Catalogue of Conodonts*. Vol. 1, 504 pp. Vol. 2, 404 pp. (further volumes forthcoming). [Well-illustrated compilation of index conodonts.]

CITED REFERENCES

Barnes, C.R. and Nowland, G.S., 1975. Conodonts: A thing of the past and of the future. *Geosci. Can.*, 2(2): 85—89.

Bender, H. and Stoppel, D., 1965. Perm-Conodonten. *Geol. Jahrb.*, 82: 331—364. [pl. 14—16.]

Branson, E.B. and Mehl, M.G., 1933—34. Conodont studies, 1—4. *Mo. Univ. Stud.*, 8: 349 pp. [pl. 1—28.]

Clark, D.L. and Müller, K.J., 1968. The basal opening of conodonts. *J. Paleontol.*, 42: 561—570.

Fisher, D.W., 1962. Small conoidal shells of uncertain affinities. In: R.C. Moore (Editor), *Treatise on Invertebrate Paleontology, Part W*. Geol. Soc. Am., New York, N.Y., and Kansas Univ. Press, Lawrence, Kansas, pp. 99—143.

Gross, W., 1966. Kleine Schuppenkunde. *Neues Jahrb. Geol. Paläontol. Abh.*, 125: 29—48.

Hinde, G.L., 1879. On conodonts from the Chazy and Cincinnati group of the Cambro-Silurian, and from the Hamilton and Genesee-shale divisions of the Devonian, in Canada and the United States. *Geol. Soc. Lond., Q.J.*, 35(art. 29): 351—369. [pl. 15—17.]

Hyman, L.H., 1959. *The Invertebrates*, 5. McGraw-Hill, New York, N.Y., 783 pp.

Matthews, S.C., 1973. Lapworthellids from the Lower Cambrian *Strenuella* Limestone at Comley, Shropshire. *Paleontology*, 16: 139—148. [pl. 8—9.]

Müller, K.J., 1964. Ostracoda (Bradorina) mit phosphatischen Gehäusen aus dem Oberkambrium von Schweden. *Neues Jahrb. Geol. Paläontol. Abh.*, 121: 1—46.

Müller, K.J. and Clark, D.L., 1967. Early late Devonian conodonts from the Squaw Bay Limestone in Michigan. *J. Paleontol.*, 41: 902—919. [pl. 115—118.]

Müller, K.J. and Nogami, Y., 1971. Über den Feinbau der Conodonten. *Mem. Fac. Sci., Kyoto Univ., Ser. Geol. Mineral.*, 38(1): 1—87. [pl. 1—22.]

Rexroad, C.B., 1967. Stratigraphy and conodont paleontology of the Brassfield (Silurian) in the Cincinnati Arch Area. *Geol. Surv. Ind., Bull.*, 36: 1—64. [pl. 1—4].

Schmidt, H., 1934. Conodonten-Funde in ursprünglichem Zusammenhang. *Paläontol. Z.*, 16: 76—85.

Schmidt, H. and Müller, K.J., 1964. Weitere Funde von Conodonten-Gruppen aus dem oberen Karbon des Sauerlandes. *Paläontol. Z.*, 38(3/4): 105—135.

Scott, H.W., 1934. The zoological relationships of the conodonts. *J. Paleontol.*, 8: 448—455.

Sweet, W.C. and Bergström, S.M., 1972. Multielement taxonomy and Ordovician conodonts. *Geologica et Palaeontologica, Sonderband* 1: 29—42.

Ulrich, E.O. and Bassler, R.S., 1926. A classification of the tooth-like fossils, conodonts, with descriptions of American Devonian and Mississippian species. *U.S. Natl. Mus. Proc.*, 68: 1—63. [pl. 1—11.]

Walliser, O.H., 1964. Conodonten des Silurs. *Abh. Hess. Landesamt. Bodenforsch., Wiesbaden*, 41: 106 pp. [pl. 1—32.]

ORGANIC-WALLED MICROFOSSILS

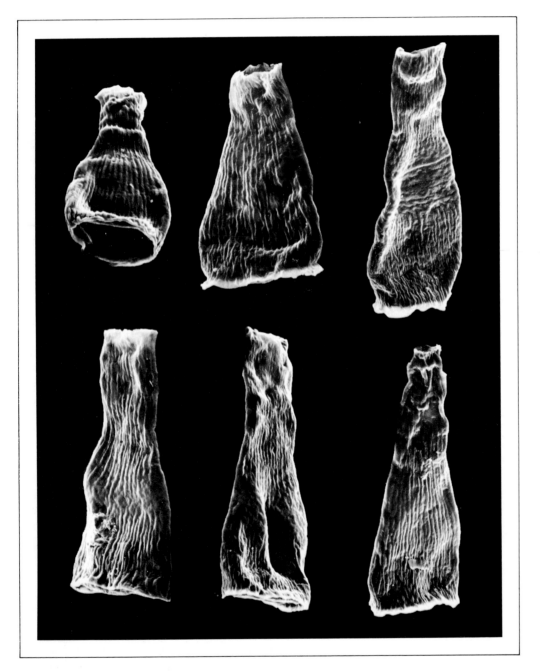

Scanning electron micrographs reveal different patterns or ornamental ribbing in closely related chitinozoans of the genus *Cyathochitina*. From type Caradoc (Upper Ordovician), England. × 180.

13

DINOFLAGELLATES, ACRITARCHS AND TASMANITIDS

GRAHAM L. WILLIAMS

INTRODUCTION

Dinoflagellates

In modern times man has become increasingly aware of a natural phenomenon called "red tide" and the disastrous effect which it can have on the ecology of marine coastal areas. Red tides are "blooms" or high concentrations (up to 6×10^6 organisms per liter of water) of small unicellular algae called dinoflagellates. In some areas, presumably where the water is richer in nutrients, the dinoflagellates multiply so rapidly and become so concentrated that their distinctive cell pigments impart a red hue to the water. Red tides can be a serious threat to life, because the contained dinoflagellates secrete a lethal toxin known as "paralytic shellfish poison". Other marine dinoflagellates are well known to man because they are responsible for most of the luminescence in the seas, such as that often seen in a ship's wake or in the breaking surf at night.

Living dinoflagellates are unicellular biflagellate algae ranging in size from 5 μm to 2 mm. They constitute the Division Pyrrhophyta (from the Greek *pyrrhos* = flame-colored, and *phyta* = plants) of the Algae. They have a varied habit; some are planktonic organisms in marine and fresh-water environments, where they are an important part of the food chain; but a few are marine sand dwellers, and some are symbionts or parasites. Most dinoflagellates are autotrophic: they contain chromatophores and carry out photosynthesis. Some are heterotrophic: they are devoid of chromatophores and feed like animals by ingesting their foods; either holozoic (ingesting whole food particles) or saprophytic (absorbing dissolved food material as parasites).

Dinoflagellates have a relatively simple life cycle, including a vegetative stage, and an encysted stage or resting cyst. The resting cyst is believed to be composed of a resistant organic substance similar to sporopollenin and is fossilizable. Sporopollenin derives its name from the fact that it was discovered as the outer coating in the walls of spores and pollen in terrestrial plants. It is now believed that all fossil dinoflagellates are sporopollenin cysts. Fossil dinoflagellates are common in marine sediments of Permian to Recent age.

Acritarchs

Acritarchs, like fossil dinoflagellates, are microscopic fossil cysts, but it cannot be determined from what type of ancient plankton they were derived. The majority lack diagnostic features which might identify them as dinoflagellate cysts, but otherwise are similar to them. Their wall is also believed to be composed of sporopollenin.

Tasmanitids

Tasmanitids are hollow spheres with a thick punctate sporopollenin wall. Living representatives are the cysts of pelagic chlorophyllous algae.

Hystrichospheres

Prior to 1963, all types of spiny organic-walled microplankton were called "hystrichospheres" or "hystrichosphaerids" and these terms were used freely in the older literature. However, these terms are no longer commonly

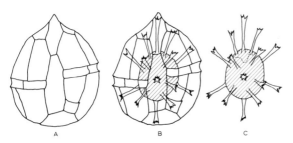

Fig. 1. Development of a "hystrichosphere" type of resting cyst within the thecate vegetative stage of a dinoflagellate. A. Thecate vegetative stage with *Gonyaulax* tabulation. B. Formation of the resting cyst within the theca. C. Resultant destruction of the theca leaving a typical hystrichosphere [*Oligosphaeridium complex* (White) Davey and Williams].

used because it was realized that many "hystrichospheres" are fossil dinoflagellate cysts (Fig. 1). All the "hystrichospheres" have now been reattributed either to the dinoflagellates, acritarchs or tasmanitids.

Fossil dinoflagellates, acritarchs, tasmanitids, spores and pollen are found in association in most marine sediments and are studied by palynologists. These microfossils must all be extracted from sediments by dissolving the mineral component away with hydrochloric and hydrofluoric acids. Collec-

tively, these microfossils are called "palynomorphs". The dinoflagellates, acritarchs and tasmanitids are also sometimes referred to as "acid-insoluble or organic-walled microplankton" for the same reason.

DINOFLAGELLATES

Within the algal Division Pyrrhophyta (the dinoflagellates) are four classes: the Ebriophyceae, the Ellobiophyceae, the Desmophyceae and the Dinophyceae (Table I). The Class Dinophyceae includes the Orders Dinophysiales (Fig. 2A), Gymnodiniales (Fig. 2B) and Peridiniales or thecate dinoflagellates (see Figs. 4, 6, 7). These three orders are the only dinoflagellates known from the fossil record.

Morphology

The dinoflagellates are unicellular organisms in which the **protoplast**, the cell as distinct from the cell wall, consists of a denser outer region harboring the chromatophores, when present, and an inner area containing the nucleus and vacuoles (Fig. 3). The chromatophores are commonly yellow or brown and often impart a distinctive hue to the individual

TABLE I

Classification of dinoflagellates

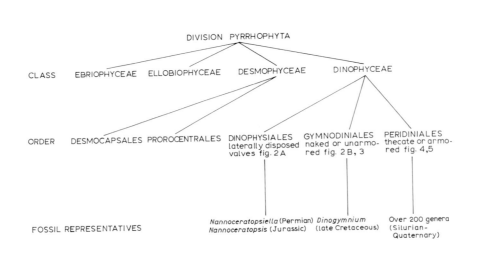

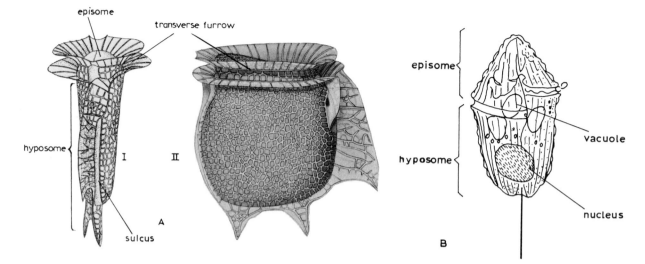

Fig. 2. Characteristics of the Dinophyceae. A. *Dinophysis collaris* Kofoid and Michener, of the Order Dinophysiales: I, ventral view; II, right lateral view. × 665. (From Kofoid and Skogsberg, 1928.) B. *Gymnodinium*, × 650. (From Lebour, 1925.)

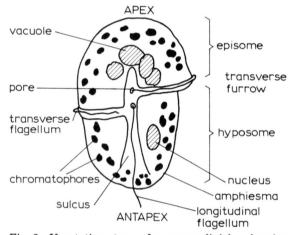

Fig. 3. Vegetative stage of a gymnodiniale, showing various descriptive features. This is colloquially known as a "naked" dinoflagellate or gymnodinioid stage.

cell. Food is stored as starch or fat, the latter sometimes being brightly colored. The large vacuoles, **pusules**, are connected to the exterior by a delicate canal and probably function in part as excretory organs. The protoplast possesses two flagella (longitudinal and transverse) arising from flagellar pores in the cell wall, and whose location is important in classification. In the vicinity of the flagellar pores the **cytoplasm**, all the protoplasm of the cell excluding the nucleus, is less viscous and often produces pseudopodia. **Ocelli**, light receptors, are found in some marine forms.

The Dinophyceae are characterized by their cell-wall morphology, the diagnostic motile or vegetative stage possessing an external, protective cellular covering, the **amphiesma**, and having two flagella of unequal length. In some species the amphiesma is composed of a series of plates and is termed the **theca**.

The whiplike **longitudinal flagellum** is directed backwards along a longitudinal furrow, the **sulcus**, which is a groove in the ventral surface of the organism. The **transverse flagellum** encircles the cell in a transverse or spiral groove, which is called the **cingulum** (see Fig. 3). The sulcus and cingulum intersect near the midventral point.

The cingulum divides the cell into an anterior area, **episome**, and a posterior area, the **hyposome**. The sulcus lies wholly or partially in the hyposome. The anterior portion of the episome is the **apex**, the posterior portion of the hyposome is the **antapex** (see Fig. 3).

The theca is composed of a latitudinal series of articulated plates, the number and position of which constitute the **tabulation** which is of major importance in classification. Kofoid (1907, 1909) proposed a system of nomenclature for the plates of the Peridiniales that has been widely adopted. The primary series of plates include the **apicals, precingulars, cingulars, postcingulars, antapicals** and **sulcals** (Table II). The shape of individual

TABLE II

Alphanumeric system of plate designation in dinoflagellates (after Kofoid, 1907, 1909)

Vegetative stage, plate series	Encysted stage, paraplate series	Location	Symbol	Individual plate or paraplate	
				position	symbol
Apical	apical	touching the apex but not the cingulum or paracingulum		second apical	2′
Anterior intercalary	anterior intercalary	between the apicals and pre-cingulars but not touching the apex or cingulum (paracingulum); can be ventrally or dorsally located	a	third anterior intercalary	3a
Precingular	precingular	immediately anterior to the cingulum or paracingulum and not touching the apex	″	fifth precingular	5″
Cingular	paracingular or girdle	cingulum or paracingulum	c or pc	fourth cingular or paracingular	4c 4pc
Postcingular	postcingular	immediately posterior to the cingulum or paracingulum and not touching the antapex	″ ′	sixth postcingular	6″ ′
Posterior, intercalary	posterior, intercalary	between the postcingulars and antapical(s) and not touching the cingulum (paracingulum) or antapex; usually adjacent to the sulcus or parasulcus	p	first posterior intercalary	1p
Antapical	antapical	touching the antapex but not the cingulum or paracingulum	″ ″	first antapical	1″ ″
Sulcal	parasulcal	occupying a position on the sulcus or parasulcus	s ps	anterior sulcal parasulcal	1s 1ps

plates can vary as in *Peridinium* (Fig. 4B). Secondary series of plates, termed **intercalaries**, are often present. These plates never touch the apex, antapex or cingulum. The series and number of plates can be expressed as the **plate formula**, using Kofoid's alphanumeric system (Table II).

Individual plates can be smooth or ornamented and are frequently **areolate, pitted,** and/or **porate**. Adjacent plates are firmly cemented together along their margins where they overlap in a tile-like fashion. Separation along the sutures, which sometimes remain visible, can be effected by sodium hypochlorite. Growth occurs in the sutural areas between plates, in the form of intervening bands, the **intercalary bands** (Fig. 4A).

The majority of fossil dinoflagellates have affinities with the modern Order Peridiniales (see Table I). Over 200 fossil genera are included in this order and most exhibit features that suggest they are related to one or other of the extant genera *Peridinium* (Fig. 4A,B), *Gonyaulax* (Fig. 6) or *Ceratium* (Fig. 7). In modern *Peridinium*, for example (Fig. 5), the shape of both the first apical (1′), and the second anterior intercalary plate (2a) may be 4-, 5- or 6-sided (Fig. 4B). This is also true of fossil genera such as *Wetzeliella* (see Fig. 11) and *Deflandrea*. Modern *Peridinium* cysts often have an intercalary archeopyle, excystment aperture, and this also is characteristic for peridinioid fossil forms.

All species of modern *Gonyaulax* (Fig. 6)

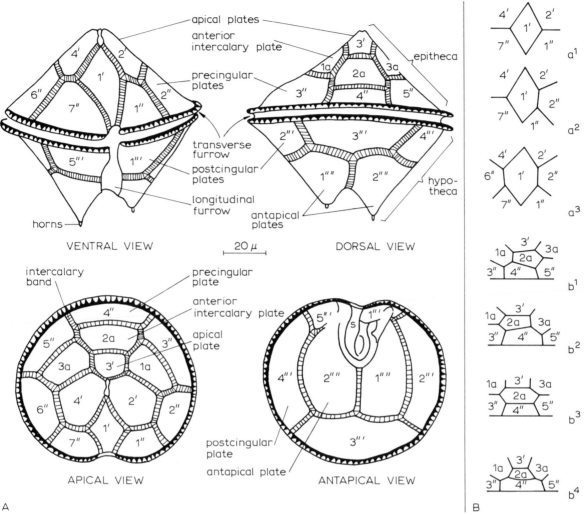

Fig. 4. A. Theca of *Peridinium leonis* Pavillard, showing tabulation and the alphanumeric symbols for denoting the individual plates. The tabulation or plate formula for this specimen may be expressed, thus, 4′, 3a, 7″, 5″′, 2″″. This means that there are four apical plates, three anterior intercalary plates, seven precingular plates, five postcingular plates, and two antapical plates. Note that the ventral, dorsal, apical and antapical views are always drawn as though they are the upper surface. S = sulcus. (After Lebour, 1925). B. Variation in shape of plates in *Peridinium*. a. Rhombic apical plate (1′) and the pattern of its relationship with bordering plates in *Peridinium*: a¹, Orthoperidinium (ortho); a² Metaperidinium (meta); a³, Paraperidinium (para). b. The second anterior intercalary plate (2a) in the various sections of Orthoperidinium and Metaperidinium. b¹ and b², penta; b³, hexa; b⁴, quadra. (From Graham, 1942.)

possess six precingular and one antapical plate(s). Fossils that are related to *Gonyaulax* form a major part of the fossil dinoflagellates and they include *Gonyaulacysta* (see Figs. 9 and 18), *Spiniferites* (see Fig. 23C), and *Oligosphaeridium* (see Fig. 15B, 19B). Archeopyle formation in *Gonyaulax*-related fossils often results from the loss of apical or precingular paraplates and never from the loss of intercalary paraplates as in peridinioid dinoflagellates.

The genus *Ceratium* (Fig. 7) is one of the most widely distributed of modern dinoflagellates, with species abundant in both fresh water and salt water. The presence of a postcingular horn is a very diagnostic feature of *Ceratium*. The cingulum is almost circular and the sulcus lies to the left of the midventral line. Fossil genera with similar features include *Odontochitina* (see Fig. 27C) and *Xenascus* (see Fig. 27B).

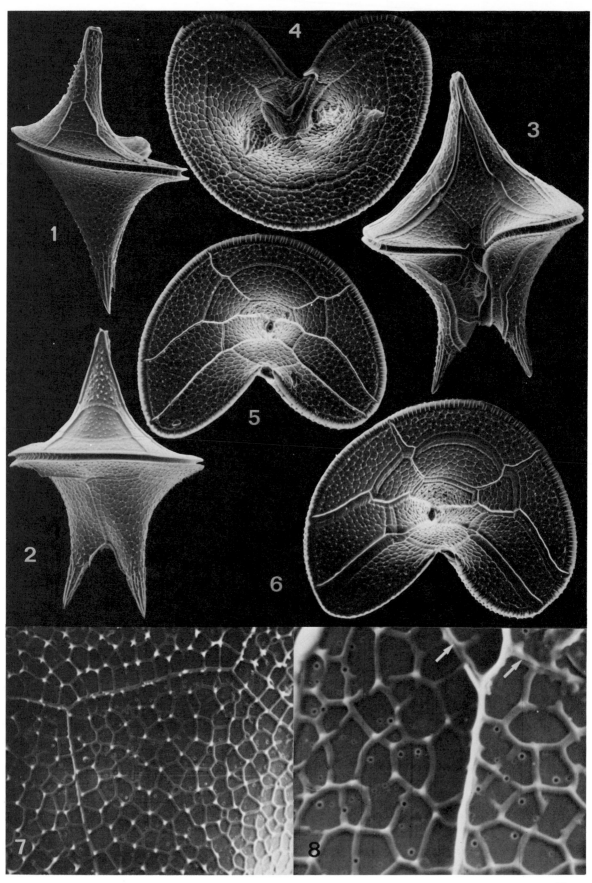

Fig. 5

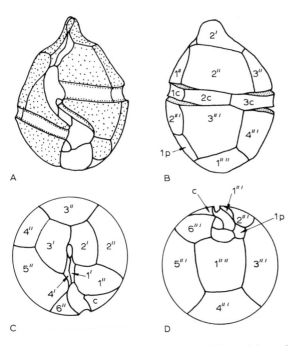

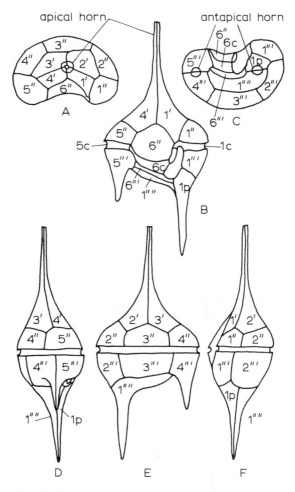

Fig. 6. Theca of *Gonyaulax spinifera* (Claparède and Lachmann) Diesing, showing tabulation. The tabulation may be expressed, thus, 4', 6", ?6c, 6" ', 1" ", 1p. c = cingular; p = posterior intercalary. A. Ventral view. B. Lateral view. C. Apical view. D. Antapical view. (From Wall and Dale, 1970.)

Fig. 7. *Ceratium* sp., showing tabulation of the theca or vegetative stage. The tabulation may be expressed, thus, 4', 6", 5—6c, 6" ', 1p, 1" ". A. Apical view. The circle denotes the apical horn. B. Ventral view. Note the offset sulcus. C. Antapical view. The circles denote the antapical horns. D. Right lateral view. E. Dorsal view. F. Left lateral view. (Figure courtesy W.R. Evitt.)

Ecology

The ecology of living dinoflagellates is very imperfectly known and we can only generalize about distribution patterns, with data based almost exclusively on the motile stage of oceanic species. Dinoflagellates are found in all environments from fresh water to open oceanic. The motile stages of autotrophic forms live in the photic zone. Species can be divided ecologically according to temperature, into **eurythermal** or temperature-tolerant species which are cosmopolitan, and **steno-thermal** or temperature-sensitive species which usually are restricted to warmer waters.

The distribution of the genus *Ceratium* is controlled in part by phosphate content of the sea water and ocean current systems.

The abundance of dinoflagellates fluctuates seasonally. Commonly, they multiply most

Fig. 5. SEM micrographs of the theca of the living species *Peridinium grande* (reproduced from Gocht and Netzel, 1974, by courtesy of the authors). 1. Young theca. × 290. 2. Theca with narrow dorsal intercalary bands. × 290. 3. Theca with broad ventral intercalary bands. × 300. 4. Theca with narrow antapical intercalary bands. × 380. 5. Theca with narrow apical intercalary bands. × 345. 6. Theca with broad apical intercalary bands. × 350. 7 and 8. Enlargement to show surface relief and pores. 7. Dorsal view, hypotheca of a young specimen, showing sutural ridges and reticulum. × 750. 8. Close-up of another specimen; sporadic separation of sutural ridges indicated by arrows. × 1900.

rapidly in late spring or summer, after the maximum diatom concentration and when nutrients are most abundant. During this period the concentrations of dinoflagellates, which can be as high as 100 million organisms per liter of water, often imparts a red or yellow coloration to the sea water.

Changes in salinity influence dinoflagellate concentrations. For instance, most species of *Ceratium* appear to be **stenohaline**, unable to tolerate variations in salinity. Although genera, such as *Gymnodinium* and *Peridinium* can range from fresh water to the open ocean, species or variants tend to be restricted to either oceanic, neritic, brackish, or fresh-water environments.

In modern seas the only phytoplankton more abundant than the dinoflagellates are the diatoms. Together with other phytoplankton, dinoflagellates are the vital first link in the food chain of the sea. They provide food for larger protozoans, such as Foraminifera and for metazoans, including the largest mammal ever to have lived, the blue whale.

Life cycle

There appear to be three distinct phases in the dinoflagellate life cycle. The most frequently encountered stage is the vegetative stage or motile unicell. This can reproduce asexually by fission. A second phase results from sexual reproduction, as in two species of *Ceratium*, when the fusion of two unequal gametes results in a **zygotic cyst** which is usually short-lived. The resting cyst, called a **dinocyst**, is the third distinct phase and is produced inside the cellulose wall of the vegetative stage. Resting cysts are now known in over forty species of dinoflagellates.

The two dominant cyst-forming groups are the *Gonyaulax* and the *Peridinium*, which also seem to dominate the fossil record. Members of both the Gymnodiniales and Peridiniales may form cysts at various times of the year, but **encystment** occurs primarily in association with dinoflagellate blooms (Fig. 8). A cyst is formed within the amphiesma, which sooner or later decays or sloughs off, leaving the nonmotile cyst to gradually sink to the bottom. Evitt and Davidson (1964) observed a "typical" hystrichosphere cyst within the theca of the species *Gonyaulax digitalis*, demonstrating the relationship of the cyst to the theca and proving that many of the spiny organisms collectively known as hystrichospheres are in reality dinoflagellate cysts.

The period of cyst dormancy is variable. Many cysts are protective and permit the protoplasm to survive adverse winter conditions.

Since encystment is partially controlled by temperature, cysts that go below 200 m, rarely germinate. Areas particularly favorable to germination are continental shelf environments where the water depth does not exceed 40 m. It is possible, however, that some heterotrophic dinoflagellates, which are found at depths of 400 m, have the ability to encyst. The majority of cysts, however, are produced by species concentrated in the inner neritic zone.

The release of the protoplasm from the cyst takes place through an excystment aperture, the **archeopyle** (see Fig. 17). In *Spiniferites bentori* (see Fig. 8), the initial free-swimming stage released from the cyst is surrounded by a very delicate membrane and possesses a longitudinal flagellum. This **uniflagellate stage** gives way to a **gymnodinioid stage** during which a transverse flagellum is developed and starts functioning immediately. In the ensuing stage the organism assumes a form resembling the vegetative stage. This is succeeded by the thecate stage in a few individuals. Almost identical thecae have been produced from the induced excystment of some morphologically dissimilar hystrichospheres. It is also possible that morphologically dissimilar vegetative stages produce identical cysts.

FOSSIL DINOFLAGELLATES

All fossil dinoflagellates are now believed to be cysts. They are present throughout the Mesozoic and Cenozoic and have been found in Silurian rocks, approximately 400 million years old. They are predominantly marine, although records of dinocysts in fresh-water deposits are known from the late Mesozoic and Cenozoic. They are most abundant in clays, shales, siltstones or mudstones.

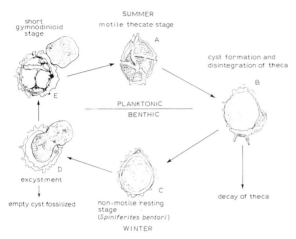

short
gymndinioid
stage

SUMMER
motile thecate stage

A

cyst formation and
disintegration of theca

B

E

PLANKTONIC
BENTHIC

D

excystment

empty cyst fossilized

non-motile resting
stage
(Spiniferites bentori)

WINTER

C

decay of theca

Fig. 8. Encystment—excystment cycle of *Gonyaulax digitalis* (Pouchet) Kofoid. The resting cyst, the dinocyst, is called *Spiniferites bentori* (Rossignol) Sarjeant. (From Wall and Dale, 1968.)

Most dinocysts are from 25 to 250 μm in overall diameter and may be spherical to ellipsoid to elongate. Many have the same diagnostic peridinioid outline as *Lejeunia* (see Fig. 12A). Others resemble *Gonyaulax*. Such species possess the characteristic dinoflagellate features such as tabulation, a cingulum or sulcus. Since these are cysts the prefix para is applied to several of these morphologic terms so that we speak of **paratabulation, paracingulum, parasulcus** and **paraplates**. In several dinocysts with **spines** or **processes**, such as *Hystrichosphaeridium* (see Fig. 10), the dinoflagellate affinities are indicated by the archeopyle, which permits orientation, and the position of the processes. Horns, which can be over 100 μm in length but are generally less than 50 μm, are invaluable in establishing orientation (see Fig. 14).

Fossil dinoflagellates were first discovered by the German biologist C.G. Ehrenberg in 1838. He observed two distinct types of organisms in thin sections of Upper Cretaceous flint. The first type was almost identical to the vegetative stage of Recent species of dinoflagellates that Ehrenberg had previously studied, hence he catalogued them as species of the genus *Peridinium*. The second type of fossil had a spherical body and numerous radiating spines. These spinose fossils later became known as "hystrichospheres", but Ehrenberg thought that they were siliceous zygospores (a body produced by the fusion of two similar gametes) of the fresh-water algal desmid genus *Xanthidium*. Mantell (1845) later proved conclusively that the "xanthidia" of Ehrenberg were of organic composition and in a later publication created the genus *Spiniferites* for them. His work was largely overlooked until the last decade, but was vindicated by recent recognition of the validity of his chosen generic name *Spiniferites*.

Studies of fossil dinocysts were very superficial until in 1933 O. Wetzel published a monograph of late Cretaceous microplankton. Later workers, notably Georges Deflandre and Lejeune Carpentier, expanded these studies.

The affinities of many of the hystrichospheres and the relationship of the dinocyst to the life cycle of the dinoflagellates remained enigmas until the studies of W.R. Evitt, D. Wall and their collaborators. Evitt (1961) was the first to call attention to the regular nature of openings in fossil dinoflagellate cyst walls.

Concentration techniques

Techniques for the extraction of dinocysts and acritarchs from consolidated and unconsolidated sediments are given by Gray (1965) and Barss and Williams (1973). Clays, silts, mudstones, sandstones, limestones and shales can be broken down by treating initially with hydrochloric acid. This removes the carbonates. After washing, hydrofluoric acid is added to remove the silicates. Further concentration can be achieved by sieving, chemical oxidation or heavy-liquid separations in which the organic fraction (including the dinocysts) floats when centrifuged in liquids of specific gravity of 1.6—2.5.

Morphology of the cyst

Cyst types

The dinocyst may closely resemble its corresponding vegetative stage or be totally dissimilar. The presence or absence of processes enables us to recognize two major types of cysts, the **proximate** and **chorate cysts**. Proximate cysts (Fig. 9) are devoid of processes, and the shape of the cyst closely approaches that of a theca. It presumably formed in close contact with the theca.

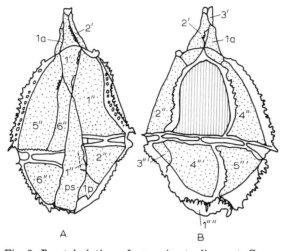

Fig. 9. Paratabulation of a proximate dinocyst, *Gonyaulacysta* cf. *jurassica* (Deflandre) Norris and Sarjeant, from the late Jurassic of England. A. Ventral view. B. Dorsal view, *ps* = parasulcus. The shaded area denotes the opening or archeopyle which has been formed from the loss of paraplate 3". Note that Kofoid's alphanumeric system has been adopted for denoting the cyst paratabulation.

Chorate cysts possess a main body bearing strongly developed ornamentation or processes, as in *Hystrichosphaeridium* (Fig. 10).

A third major cyst type, the **cavate cyst** (Fig. 11), has inner and outer bodies that are only in limited contact, with intervening spaces between them. This group includes

many cysts of the peridinioid type such as *Deflandrea* (see Figs. 25D and 26C). These cyst types are useful for morphologic distinction, but are only descriptive terms and should not be taken to indicate genetic affinities.

Shape and wall structure

Dinocysts are variable in shape. They may be spherical to ovoidal to ellipsoidal to elongate or peridinioid. Original flattening in a dorso-ventral plane is found in many genera, for example *Cyclonephelium*.

The wall of the dinocyst consists of one or more microscopic layers that form one or more bodies whose enclosure, one by another, isolates one or more cavities. By combining one of four prefixes designating spatial relationship with one of three suffixes, which refer either to the wall layer, the three-dimensional structure formed by that wall layer, or the enclosed cavity, twelve descriptive terms can be formed (Table III, Fig. 12). Thus, in a dinocyst with one recognizable wall layer, the **autophragm**, the body formed by this is the **autocyst** and the enclosed cavity is the **autocoel** (Fig. 12A). In bilayered forms there is one inner cavity, the **endocoel**, and one to several outer cavities or **pericoels** (Fig. 12B). The endocoel is completely closed off

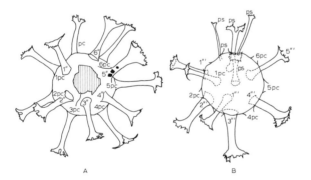

Fig. 10. Paratabulation of a chorate cyst, *Hystrichosphaeridium tubiferum* (Ehrenberg) Deflandre, as inferred from the number and position of the intratabular processes. A. Upper surface showing apical tetratabular archeopyle (shaded), precingular processes (1" etc.) and paracingular processes (*pc*). B. Lower surface seen through the upper, showing postcingular processes (1‴ etc.), parasulcal processes (*ps*), single antapical process (1‴‴), and posterior intercalary processes (*p*). This is a chorate cyst. Early Tertiary, England.

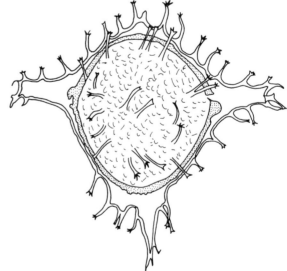

Fig. 11. *Wetzeliella lunaris* Gocht, a cavate cyst. Early Tertiary, England. × 480.

TABLE III

Wall terminology for dinocysts

	Meaning	Combination	Definition
Prefix			
auto-	single	autophragm autocyst autocoel	single wall single body single cavity
endo-	inner	endophragm endocyst endocoel	inner wall inner body inner cavity
meso-	middle	mesophragm mesocyst mesocoel	middle wall middle body middle cavity
peri-	outer	periphragm pericyst pericoel	outer wall outer body outer cavity
Suffix			
-phragm	wall	autophragm endophragm mesophragm periphragm	single wall inner wall middle wall outer wall
-cyst	body	autocyst endocyst mesocyst pericyst	single body inner body middle body outer body
-coel	cavity	autocoel endocoel mesocoel pericoel	single cavity inner cavity middle cavity outer cavity

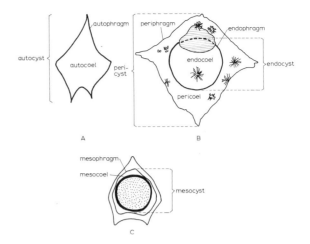

Fig. 12. Body, wall and cavity terminology in dinocysts: A. *Lejeunia* sp. B. *Rhombodinium glabrum* (Cookson) Vozzhennikova. C. *Deflandrea* sp.

divides the cyst into the **epicyst** and **hypocyst**. (These last two terms are the cyst equivalents of epitheca and hypotheca, respectively.)

Some cysts have a smooth surface; others possess ornamentation which can be minor elements such as granules or major elements like horns, septa or processes. A dinocyst never has more horns than its corresponding thecate stage and in both the maximum is five. Horns may be apical, antapical or lateral only (Fig. 14) and are useful criteria for the orientation of the dinocyst.

Processes are essentially columnar or spine-like (Fig. 15), whereas septa are membraneous, linear projections arising perpendicularly from the outer wall layer (see Figs. 9 and 18A). Processes may be located on or within paraplate boundaries. **Process complexes** are processes on individual paraplates that are united proximally, distally, or along their length (Fig. 16).

All the processes of a dinocyst may be similar (see Fig. 19B) or they may be differentiated. Process differentiation is not haphazard and a particular paraplate series often is characterized by a diagnostic type of process. In other taxa some of the paraplates, commonly the cingular, are devoid of processes (Fig. 15B). The preferred orientation of septa and processes often permits one to determine paratabulation in dinocysts. Process arrangements, shapes and terminations are important morphologic features in generic and specific classification.

from the outside if the archeopyle is not developed.

Paratabulation and ornamentation

In living species, a cyst often shows the same tabulation as the theca from which it was derived and if so, this is called reflected tabulation or paratabulation in the cyst to distinguish between cyst and theca. The alphanumeric system used to describe the thecal plates is, however, also used for the paratabulation in cysts. Similarly, the "plates" in cysts are called **paraplates** here and the principal paraplate series and intercalaries in cysts are shown in Fig. 13 and Table II. The cingular paraplates in cysts collectively form the **paracingulum** or **girdle**. The paracingulum

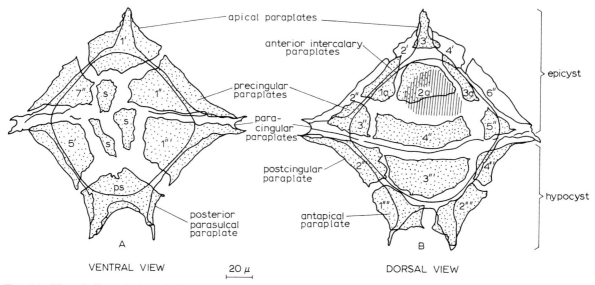

Fig. 13. *Wetzeliella reticulata* Williams and Downie, a dinocyst from the early Eocene of England, showing the "peridinioid" paratabulation characteristic of this genus. The archeopyle which is formed through the loss of the second anterior intercalary (2a) is shaded. The stippled intratabular areas represent the simulate process complexes. *s*, parasulcal paraplate. A. Ventral view. B. Dorsal view.

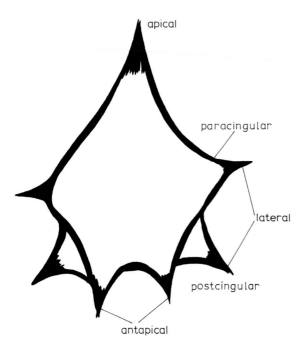

Fig. 14. Diagrammatic illustration of cardinal positions from which horns can originate in a dinocyst. The maximum number of horns on any one individual is five. Dorsal view.

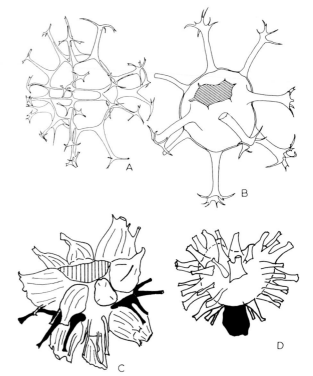

Fig. 15. Process types in chorate dinocysts. A. *Spiniferites ramosus* (Ehrenberg) Loeblich and Loeblich. Dorsal view. × 420. B. *Oligosphaeridium complex* (White) Davey and Williams, with one intratabular process per paraplate. × 330. C. *Hystrichokolpoma eisenacki* Williams and Downie, with process differentiation. × 330. D. *Diphyes colligerum* (Deflandre and Cookson) Cookson, with enlarged antapical process (black). × 540.

Archeopyle

Evitt (1961) was the first to realize the importance of the regular-shaped opening that is found in many dinocysts. He proposed the

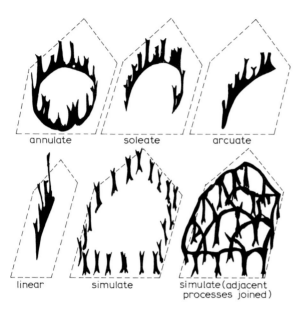

Fig. 16. Intratabular process complexes.

term archeopyle (from the Greek *arche* = old, and *pyle* = gate or orifice). It usually results from the complete or partial removal of one or more than one paraplate, the **operculum**. In some dinocysts the archeopyle is formed from rupture of the "suture" between adjacent paraplates and there is no operculum.

Archeopyles are of five major categories; apical, intercalary, precingular, combination and miscellaneous (Fig. 17). The archeopyle is a genetically determined feature and is a constant in all known species.

Major lineages and their stratigraphic distribution

All known fossil dinocysts are included in the Class Dinophyceae (see Table I). Two

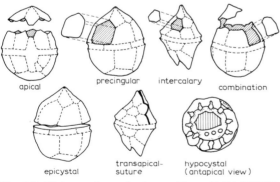

Fig. 17. Major archeopyle types. From Evitt (1967) and Wall and Dale (1970).

genera are assigned to the Order Dinophysiales and one to the Gymnodiniales. The majority of over two hundred genera belong to the Order Peridiniales. At the generic and specific level there are two distinct and independent classification schemes, the first based on fossil material and evolved by palynologists, the second based on the living vegetative stage and developed by biologists. A unified system based on all stages of the life cycle is still many years away.

In this chapter the fossil dinocysts have been placed into seven lineages (Table IV) based on paratabulation and archeopyle formation. The most important lineages are the *Gonyaulacysta* (see Figs. 18—24) which includes most of the proximate or chorate cysts, the peridinioid (see Figs. 25, 26) which is predominantly cavate cysts, the ceratioid (see Figs. 27, 28), and the *Cyclonephelium* (see Figs. 29, 30).

The lineage with the longest stratigraphic range is the *Gonyaulacysta* group which has a known range of Silurian to the present day. Dinoflagellates have not been described from Devonian or Carboniferous strata. A few specimens, including forms that appear to be related to *Nannoceratopsis* (see Fig. 31) are known from the Permian. In the Triassic, members of the *Gonyaulacysta* lineage with apical archeopyles appear.

Dinocysts with spines first appear in the Lias (early Jurassic). The spines may have been a flotation device, and/or a defensive mechanism. The middle Jurassic was a period of carbonate deposition in many regions. Recovery of dinocysts from these carbonate deposits has been poor, so knowledge is scanty. The *Gonyaulacysta* lineage predominates and dinocysts with intratabular processes first appear. *Deflandrea*-like species and *Spiniferites* (see Fig. 15A) occur first in the late Jurassic. During this Period the *Gonyaulacysta* lineage reaches its acme in number of species (Fig. 18A, B).

In the early Cretaceous, the *Gonyaulacysta* lineage is characterized by the appearance of thicker-walled species, which represent adaptation to a neritic environment. There is a marked influx of new species of chorate and cavate cysts. In the late Cretaceous the chorate and cavate cysts, most of

TABLE IV

Stratigraphic ranges of the seven dinocyst lineages

Age	Ceratioid	Cyclonephelium	Dinogymnium	Gonyaulacysta	Nannoceratopsis	Peridinioid	Tuberculodinium
Quaternary				│		│	│
Tertiary — Neogene				│		│	│
Tertiary — Paleogene		│		│		│	│
Cretaceous — late	│	│	│	│		│	
Cretaceous — early	│	│		│		│	
Jurassic — late	│	│		│	│	│	
Jurassic — middle		│		│	│	│	
Jurassic — early				│	│	│	
Triassic				│	┊		
Permian				│			
Carboniferous							
Devonian					│		
Silurian							

which belong to the *Gonyaulacysta* or peridinioid lineages respectively (Figs. 19,20, 25C, D), are abundant. This part of the geologic column is characterized by the presence of the *Dinogymnium* (see Fig. 32) lineage.

During the Paleogene evolution seems to accelerate, some lineages becoming extinct, others such as the *Tuberculodinium* lineage (Fig. 33) appearing for the first time. The *Gonyaulacysta* and peridinioid lineages are still dominant (see Figs. 21, 22, 26), although many species die out. The Neogene represents a decline phase (see Figs. 23, 24).

Williams (1977) has proposed thirty concurrent range zones and twelve subzones, named for index species, for the Mesozoic and Cenozoic. A distinctive suite of species characterizes each zone or subzone, which is usually cosmopolitan in the Mesozoic and early Tertiary and regional in the late Tertiary. These zones are being used to date the rocks in both surface sections and exploratory wells.

In the Quaternary there are only a few species of dinocysts, so relative abundances have been used for biostratigraphic zonation.

It is now known that over forty extant species possess an encysted stage. Wall (1971) defined the four major groups that have living species that produce resting cysts. These are the *Gonyaulax* group, the *Peridinium* group, the *Ceratium* group and the *Pyrophacus* group. The abundance in Recent seas of dinocysts with a gonyaulacacean or peridiniacean paratabulation is paralleled in the fossil

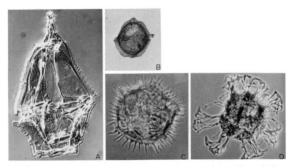

Fig. 18. Late Jurassic—early Cretaceous species of the *Gonyaulacysta* lineage. A. *Gonyaulacysta jurassica* (Deflandre) Norris and Sarjeant, late Jurassic. × 440. B. *Scriniodinium crystallinum* (Deflandre) Klement, late Jurassic. Archeopyle precingular. × 150. C. *Ctenidodinium elegantulum* Millioud, early Cretaceous. × 350. D. *Systematophora schindewolfi* (Alberti) Downie and Sarjeant, Early Cretaceous. × 725.

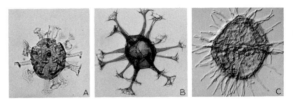

Fig. 19. Late Cretaceous species of the *Gonyaulacysta* lineage. A. *Hystrichosphaeridium salpingophorum* Deflandre emend. Davey and Williams. Operculum of the apical archeopyle is slightly displaced. × 212. B. *Oligosphaeridium pulcherrimum* (Deflandre and Cookson) Davey and Williams. Oblique apical view with apical archeopyle visible. × 275. C. *Hystrichodinium pulchrum* Deflandre. Dorso-lateral view. × 375.

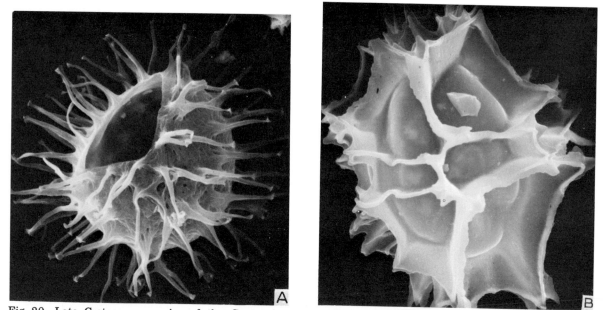

Fig. 20. Late Cretaceous species of the *Gonyaulacysta* lineage. A. *Exochosphaeridium bifidum* (Clarke and Verdier) Clarke et al. Dorso-lateral view, archeopyle is precingular. × 750. B. *Spiniferites cingulatus* (O. Wetzel) Sarjeant. Lateral view with paracingulum visible. × 800.

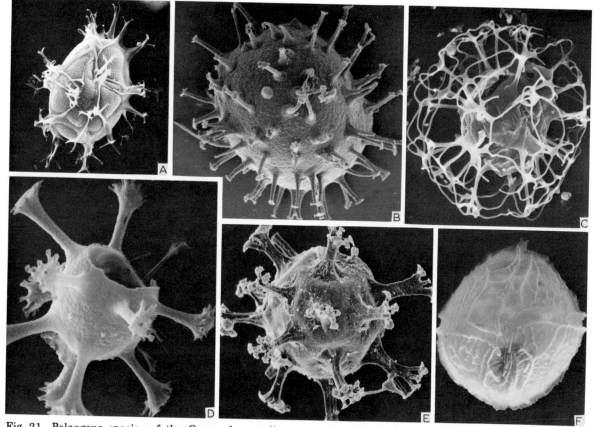

Fig. 21. Paleogene species of the *Gonyaulacysta* lineage. A. *Spiniferites* sp. Ventral view showing parasulcus. × 740. B. *Hemicystodinium zoharyi* (Rossignol) Wall. × 830. C. *Cannosphaeropsis* sp. Dorsal view showing processes united by trabeculae. × 830. D. *Perisseiasphaeridium* sp. Dorsal view showing apical archeopyle surrounded by large precingular processes. × 830. E. *Homotryblium plectilum* Drugg and Loeblich. × 540. F. *Gonyaulacysta giuseppei* (Morgenroth) Sarjeant. Ventral view showing parasulcus. × 830.

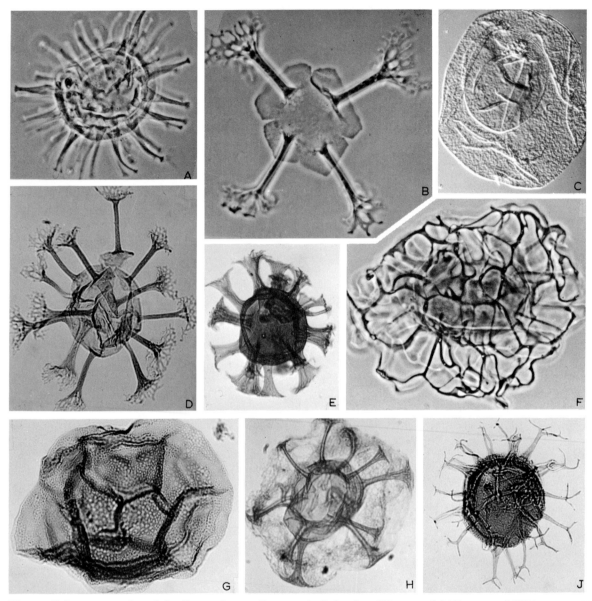

Fig. 22. Paleogene species of the *Gonyaulacysta* lineage. A. *Polysphaeridium pastielsi* Davey and Williams. × 680. B. *Areosphaeridium diktyoplokus* Eaton. Operculum of the apical archeopyle (cf. Fig. 22D). N 723. C. *Thalassiphora pelagica* (Eisenack) Eisenack and Gocht. × 255. D. *Areosphaeridium diktyoplokus* Eaton. Apical archeopyle partly visible at top. × 467. E. *Cordosphaeridium gracile* (Eisenack) Davey and Williams. Archeopyle precingular. × 420. F. *Adnatosphaeridium multispinosum* Williams and Downie. Processes united distally by traeculae. × 765. G. *Heteraulacacysta fehmarnensis* Lentin and Williams. Antapical view. × 680. H. *Membranilarnacia ursulae* Morgenroth. The processes are united distally by the perforated ectophragm. × 510. J. *Spiniferites ramosus* subsp. *granosus* (Davey and Williams) Lentin and Williams. × 510.

record. *Ceratium*, often dominant today, especially in oceanic environments, is relatively uncommon in the fossil state.

Resting cysts described for the modern Gymnodiniales resemble some of the acritarchs. It may be that many acritarchs are cysts of naked dinoflagellates, and are the an-

cestral stock of the thecate dinoflagellates. The only known fossil representative of the Gymnodiniales is the Cretaceous genus *Dinogymnium*. This possesses a phragma resistant to acetolysis and an archeopyle, which suggests it is a cyst. The similarity to the extant genus *Gymnodinium* may indicate

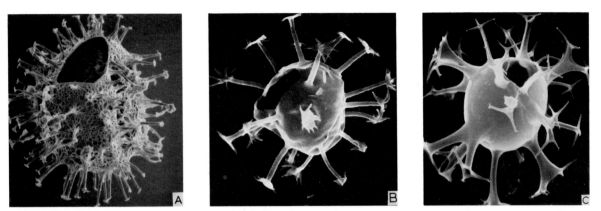

Fig. 23. Neogene examples of the *Gonyaulacysta* lineage. A. *Operculodinium israelianum* (Rossignol) Wall. Dorso-lateral view, the archeopyle is precingular. × 1000. B. *Hystrichosphaeridium pseudorecurvatum* Morgenroth. The archeopyle is precingular. × 1125. C. *Spiniferites pseudofurcatus* (Klumpp) Sarjeant. Dorsal view showing pre-cingular archeopyle. × 525.

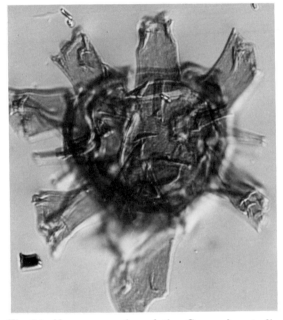

Fig. 24. Neogene species of the *Gonyaulacysta* lineage, *Hystrichokolpoma rigaudae* Deflandre and Cookson. The large processes are precingular, post-cingular and antapical. The slender processes are on the paracingulum. The very large antapical process is at the bottom. × 750.

a close relationship or it may be another example of homeomorphy.

Evolution

Population sizes of organisms do not fluctuate haphazardly. There is a recognizable population production cycle of species in algal cultures. An initial lag phase, in which the cells may undergo a size increase but do not reproduce, gives way to the exponential growth phase, a period of maximum cell division and hence increase of population. The third phase is the balanced growth phase in which the population remains stationary presumably when it has reached maximum concentration. The cycle terminates in the death phase, when there is a rapid population decline (Fig. 34). Zeuner (1946) believed that all organisms have a similar cycle in their evolution. A lag phase with relatively insignificant diversification is followed successively by an explosive phase (in which the influx of new species exceeds extinction), a climax or stationary phase, and a decline phase (in which the influx of new species is exceeded by extinction). This cycle appears to have been time controlled, lasting for approximately 50 m.y. regardless of the organisms involved.

Tappan and Loeblich (1970, 1972) have applied Zeuner's cycle to the phytoplankton evolution in the geologic column. They postulate that the cycle is particularly applicable to organisms that reproduce asexually. Also they believe that the explosive phase is characterized by small dinocysts, the decline phase by large dinocysts. The lag phase is recognized in the Silurian to early Jurassic. This is followed by an explosive phase in the early and middle Jurassic (24 m.y.) terminating in a climax and succeeded by a late Jurassic to early Cretaceous decline. A

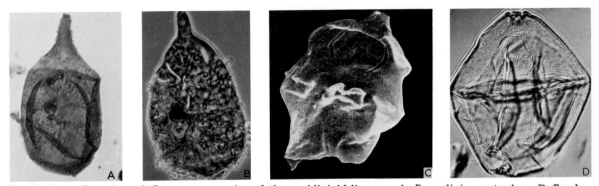

Fig. 25. Late Jurassic and Cretaceous species of the peridinioid lineage. A. *Pareodinia ceratophora* Deflandre, late Jurassic. × 650. B. *Imbatodinium* sp., early Cretaceous. This differs from *Pareodinia* only in possessing processes. × 640. C. *Australiella victoriensis* (Cookson and Manum) Lentin and Williams, late Cretaceous. Archeopyle intercalary with operculum remaining attached along posterior margin. × 640. D. *Deflandrea cretacea* Cookson, late Cretaceous. The operculum of the intercalary archeopyle remains attached along its posterior margin. × 805.

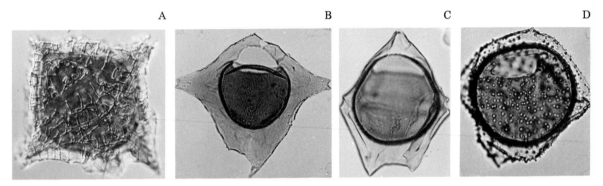

Fig. 26. Paleogene examples of the peridinioid lineage. A. *Wetzeliella tenuivirgula* subsp. *crassoramosa* (Williams and Downie) Lentin and Williams. The simulate process complexes are particularly noticeable on the epicyst. × 940. B. *Rhombodinium glabrum* (Cookson) Vozzhennikova. Dorsal view, archeopyle intercalary. × 300. C. *Deflandrea phosphoritica* Eisenack. Archeopyle intercalary. × 300. D. *Wetzeliella condylos* Williams and Downie. Dorsal view, archeopyle intercalary. × 350.

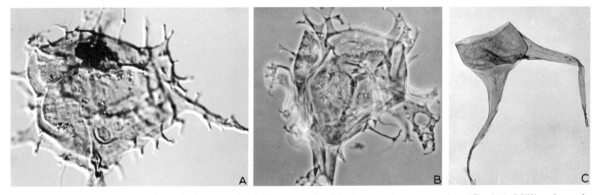

Fig. 27. Cretaceous examples of the ceratioid lineage. A. *Phoberocysta neocomica* (Gocht) Millioud, early Cretaceous. Dorsal view with apical archeopyle visible. × 670. B. *Xenascus ceratioides* (Deflandre) Lentin and Williams, late Cretaceous. Archeopyle apical. × 370. C. *Odontochitina operculata* (O. Wetzel) Deflandre and Cookson, late Cretaceous. Archeopyle apical. × 200.

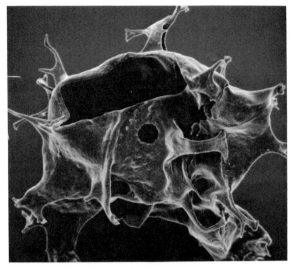

Fig. 28. Electron micrograph of the late Cretaceous species of the ceratioid lineage, *Xenascus ceratioides* (Deflandre) Lentin and Williams, showing the apical archeopyle towards the top. (cf. with Fig. 28B). × 1000.

second cycle, characterized by a separate evolutionary group commenced its explosive phase in the early Cretaceous and continued for 25 m.y. before it reached its climax. The late Cretaceous marked a decline phase. A third explosive phase commencing in the late Paleocene and extending into the early Eocene was followed by a decline phase, which extends to the Recent (Fig. 35). If we assume a cycle occupies 50 m.y. it is possible that Recent dinoflagellates are either experiencing or approaching the evolutionary phase of a fourth cycle.

The death phase of one cycle tends to be succeeded by the explosive phase of another cycle and these explosive cycles often seem to be related to maximum development of epicontinental seas. Other major environmental changes may also act as triggering mechanisms.

From the data generated by Sarjeant

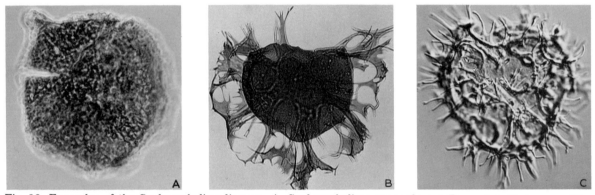

Fig. 29. Examples of the *Cyclonephelium* lineage. A. *Cyclonephelium vannophorum* Davey, late Cretaceous. The operculum of the apical archeopyle is slightly displaced. × 600. B. *Cyclonephelium ordinatum* Williams and Downie, Paleogene. The apical archeopyle permits orientation of the specimen. × 485. C. *Areoligera senonensis* Lejeune-Carpentier, Paleogene. The annulate process complexes delineate the paratabulation. × 485.

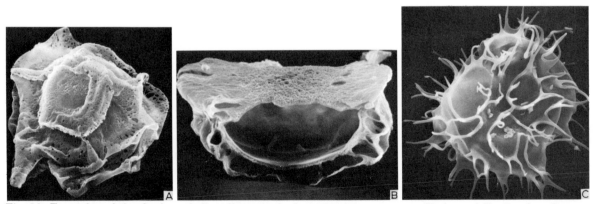

Fig. 30. Examples of the *Cyclonephelium* lineage. A. *Senoniasphaera protrusa* Clarke and Verdier, late Cretaceous. Dorsal view with apical archeopyle at top; the longer antapical horn is at the left. × 800. B. *Chiropteridium aspinatum* (Gerlach) Brosius, Paleogene. Apical view looking into the apical archeopyle. × 800. C. *Areoligera senonensis* Lejeune-Carpentier, Paleogene. Dorsal view showing annulate process complexes. × 800.

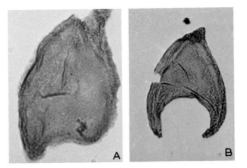

Fig. 31. Examples of *Nannoceratopsis* lineage. A. *Nannoceratopsis gracilis* Alberti, early Jurassic. The break in slope denotes the paracingulum. × 600. B. *Nannoceratopsis pellucida* Deflandre, late Jurassic. The two long curved horns are antapical. × 325.

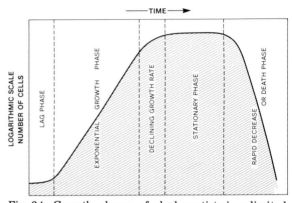

Fig. 34. Growth phases of algal protists in a limited culture, from inoculation to death. (From Tappan and Loeblich, 1970.)

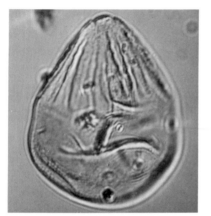

Fig. 32. Example of the *Dinogymnium* lineage, *Dinogymnium euclaensis* Cookson and Eisenack, late Cretaceous. × 1,350.

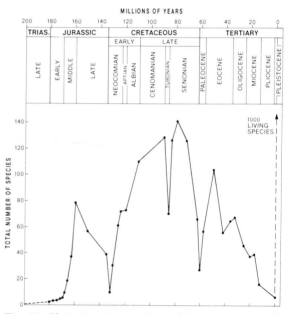

Fig. 35. Variations in number of dinocyst species in the Mesozoic—Cenozoic. Points denote end of period for which species were totaled. (From Tappan and Loeblich, 1970.)

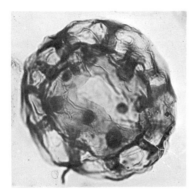

Fig. 33. Example of the *Tuberculodinium* lineage, *Tuberculodinium vancampoae* (Rossignol) Wall, Neogene. Archeopyle hypocystal. × 500.

(1967), Tappan and Loeblich estimate that a new fossil dinocyst genus appeared every 1.4 m.y. and that the average duration of a genus was 42.4 m.y. Including the number of modern genera lowers the post-Paleozoic evolution rate to one genus every 460,000 years. Such a rapid turnover suggests that they are ideally suited for biostratigraphic zonation.

Phyletic trends

Primitive features of paratabulation appear to be observable in some lineages. In the peridinioid lineage, genera with up to six anterior intercalaries, as *Pareodinia* (see Fig. 25A) are present in the early Jurassic and this apparently is a primitive feature. Other features of paratabulation are remarkably constant. The tendency for the first post-cingular to be reduced and incorporated into the sulcal region is found in the *Gonyaulacysta* lineage throughout its known stratigraphic range and is possibly also a feature of *Ceratium*. Eaton (1971) has recognized an evolutionary lineage in some dinocyst species from the Eocene of southeast England (Fig. 36). Such lineages enhance the value of dinocysts in biostratigraphic studies.

The recognition of evolutionary trends in archeopyle types also must be largely con-jectural, although Lister (1970b) has recognized a sequential development. In the most primitive stage, the excystment aperture is defined by a suture that has no surface manifestation before dehiscence. Thus, the trans-apical suture present in *Peridinium limbatum* (see Fig. 17) is a primitive feature and perhaps this species is a long-ranging stock. A more advanced stage is represented by species with an archeopyle margin that is not yet polygonal. Species with a polygonal archeopyle margin represent the most advanced stage in the development of the archeopyle. This last category would presumably include apical archeopyles although apical archeopyles are not known in present-day *Gonyaulax* species.

Biogeography and paleoecology

The distribution pattern of only one genus, *Ceratium*, is known in any detail. Unfortu-

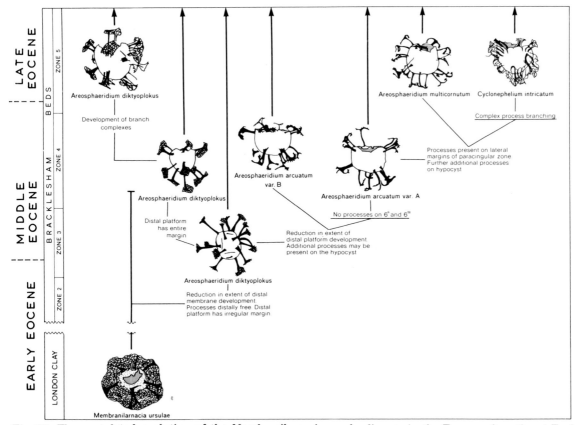

Fig. 36. The postulated evolution of the *Membranilarnacia ursulae* lineage in the Eocene of southeast England. (From Eaton, 1971.)

nately, marine species of this genus do not appear to produce cysts. Studies of the ecology of the approximately forty species which encyst and belong to either the *Gonyaulax*, *Peridinium* or *Pyrophacus* groups are very few. A pioneer study of the distribution of modern dinocysts was completed in 1971 by D.B. Williams, who plotted the distribution of cysts in Recent North Atlantic sediments (Fig. 37). He found that, although the distribution of most of the species was cosmopolitan, relative abundances varied and apparently were climatically controlled.

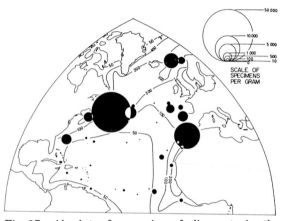

Fig. 37. Absolute frequencies of dinocysts in the North Atlantic. The size of the circle is proportional to the number of cysts per gram of sediment. Contours show estimated productivity of the surface waters. (From Williams, 1971.)

In studies of marine Quaternary sediments, data on dinocyst associations are facilitating surprisingly accurate conclusions on paleoclimatology and paleoenvironments. Wall and Dale (1968) recognized five species associations in the Pleistocene section penetrated by the Ludham borehole (Norfolk, England). These associations were interpreted as denoting alternating temperate (interstadial or interglacial), and subarctic or colder-water associations.

In pre-Quaternary sediments dinocyst associations have been studied in the early Eocene of southern England by Downie and others (1971). They recognized three associations, the gonyaulacacean, peridiniacean, and acritarch associations, which are often related to

facies (Fig. 38). These are interpreted as indicating open-marine (gonyaulacacean), littoral (acritarch), and lagoonal or brackish-water environments (peridiniacean). In the marine environments the gonyaulacacean forms become more abundant at the expense of the peridiniacean cysts; the reverse is true in the subsaline environments. This has made it possible to recognize the transgressive-regressive phases of cycles of sedimentation in the early Eocene.

In 1965, Vozzhennikova suggested that morphological differences in Mesozoic dinocysts could be used to predict depositional environments. Thick-walled dinocysts were concentrated in the littoral zone, while thin-walled forms with processes were largely confined to open-marine environments.

The relative abundances of palynomorphs, including dinocysts, can be used to predict the location of ancient shorelines. A simple increase in dinocysts and decrease in spore-pollen abundances ideally should indicate distance from shoreline, although too many deviations exist in nature for such an empirical situation to be common.

Reworked dinocysts can indicate proximity to shoreline. Derived dinocysts and acritarchs, primarily from the Middle and Upper Jurassic, are common in the Lower Eocene (London Clay) sediments of southeast England where their state of preservation is indistinguishable from that of the indigenous species. The one observable difference is that while the indigenous specimens contain pyrite inclusions, the reworked forms do not. The percentage of reworked dinocysts and acritarchs peaks in the vicinity of the presumed shoreline and declines markedly at the locality farthest from the shoreline. This method may have application in studies where derived material can be identified.

The belief has grown that fossil dinocysts are only found in marine sediments. Some modern cysts are produced, however, by fresh-water species while fresh-water sediments over 30 m.y. old contain dinocysts. Evidence is accumulating that fresh-water dinocysts existed in the late Cretaceous, so that the presence of dinocysts does not indicate an exclusively marine environment.

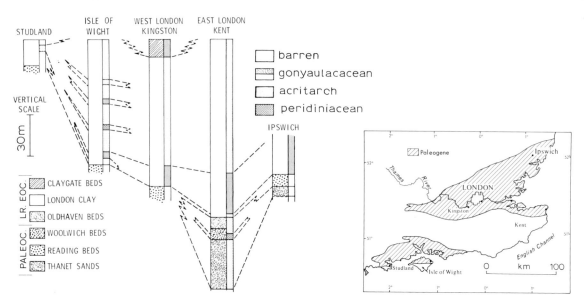

Fig. 38. Correlation of the gonyaulacacean, peridiniacean and acritarch associations in the Paleocene—early Eocene of southeast England. In each section lithology is to the left and the association to the right. Dashed lines indicate correlation between beds. Truncated lines depict lack of correlative beds. (After Downie and others, 1971).

ACRITARCHS

Introduction

The recognition of the relationships of many hystrichospheres to the dinoflagellates led Evitt in 1963 to erect the group Acritarcha (from the Greek *acritos* = uncertain, and *archae* = origin) to include those hystrichospheres whose affinities are unknown or uncertain. Both acritarchs and dinocysts have an organic wall and are concentrated by the same techniques in the laboratory.

There are various opinions concerning the nature and systematics of the Acritarchs. Downie and others (1963) proposed that the acritarchs are a heterogeneous mixture of fossil eggs, cysts, or tests of unicellular or multicellular animals or plants. They believed the majority are unicellular algae. Eisenack (1963) argued that the hystrichospheres (including the acritarchs) are a homogeneous group, in which shelled stages alternated with naked stages, and were the ancestral stock from which the dinoflagellates, and in particular the armored forms, developed.

Cramer and Diez de Cramer (1972) believed that acritarchs probably had algal affinities and were predominantly colonial. Lister (1970b) has concluded that the majority of Lower Paleozoic acritarchs are the cysts of unicellular phytoplankton. Those cysts possessing excystment apertures, but without tabulation, are probably the cysts of naked dinoflagellates (Gymnodiniales). In fact, the extant genus *Gymnodinium* produces spinose cysts which look very much like acritarchs. These cysts lack paratabulation and clearly defined archeopyles, but some species appear to release protoplasm from their apical or cingular areas. Cysts morphologically similar to the acritarchs also occur in various modern algal phyla apart from the Pyrrhophyta.

Acritarchs, first recorded in the Precambrian, became the dominant phytoplankton group in Precambrian and Paleozoic communities but decreased in importance in the Mesozoic and Cenozoic. It is probable that modern phytoplankton species which can be attributed to the Class Prasinophyceae, the green algae, would if found as fossils, be called acritarchs.

Morphology

The **test** (shell) of an acritarch usually consists of a spherical, ovoidal or triangular central body, the **vesicle** (Fig. 39), from which projections may arise in the form of spines, processes, or linear membranes, **septa**. Most genera are unicellular, but colonial multicellular forms are known. The wall of the vesicle may consist of one to three layers (Fig. 39). In two-layered walls the outer wall is not in contact with the vesicle cavity and may be in complete or partial contact or separated from the inner wall by a cavity analogous to a pericoel in dinocysts.

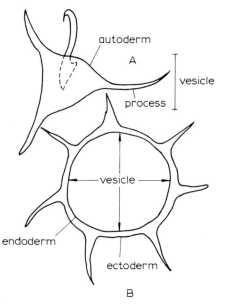

Fig. 39. Morphology of an acritarch. A. *Veryhachium europaeum* Stockmans and Willière, with a single wall layer. This species belongs to the subgroup Polygonomorphitae. B. Hypothetical acritarch of the subgroup Acanthomorphitae.

Sculptural elements and processes

There are two major types of outgrowths of the vesicle wall: sculptural elements (smaller than 5 μm) and processes. Sculptural elements do not considerably modify the vesicle shape, but are usually solid. Processes exceed 5 μm in length. They can vary in number from one to several hundred, be hollow or solid, and branched or unbranched (Fig. 40). From the orderly arrangement of the processes Lister (1970a) recognized paratabulation in acritarchs.

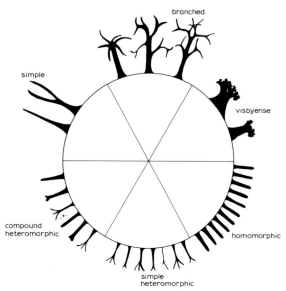

Fig. 40. Process types in acanthomorphitic acritarchs (from Cramer, 1970a).

Excystment apertures

The most frequently described excystment aperture in acritarchs is the **pylome**, which is a more or less circular opening (Fig. 41). Six types of pylomes have been recognized according to their shape and margin. The excystment aperture is always located on the **epicyst**.

Major morphologic groups

Downie and others (1963) proposed a suprageneric classification for the acritarchs, erecting "subgroups", whose names are based on morphologic features. Taxonomically useful morphologic features for classification of genera and species are: (1) the form of the vesicle; (2) the ornamentation and its areal distribution on the vesicle wall; and (3) the process type and formula, and the sculpture.

Most Precambrian acritarchs belong to the subgroup Sphaeromorphitae (Fig. 42F), which includes forms with a spherical or subspherical test with a single wall layer. The surface of the wall may be smooth or ornamented. Pylomes have been observed in some species. The Sphaeromorphitae are probably algae, though a few may be benthic bacteria.

The Disphaeromorphitae (Fig. 42P) possess a spherical to ellipsoidal central capsule enclosed within a spherical to ellipsoidal test.

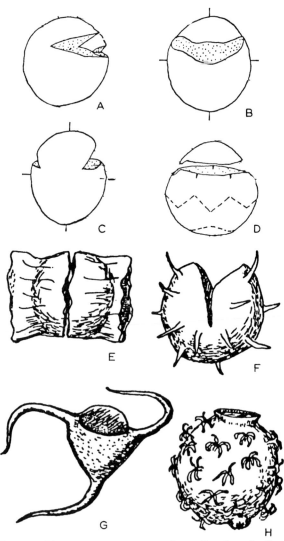

Fig. 41. Excystment apertures in acritarchs: A—C. Opening by cryptosuture. A. Lateral view. B. Dorsal view. C. Ventral view. D. Opening by obvious suture and with detached operculum. E. Median split of central body in *Riculasphaera*. F. Partial rupture of vesicle wall in *Micrhystridium*. G. Epityche or out-folded flap formed by an elongate slit in the vesicle wall in *Veryhachium*. H. Pylome with bordering rim and plug-like operculum in *Polyancistrodorus*. This genus also possesses an antapical pseudopylome. (A—D after Lister, 1970b; E—H after Loeblich and Tappan, 1969.)

The surface of the test wall may be smooth or granular. This subgroup had a period of exponential evolution in size in the late Pre-cambrian where they sometimes attained an overall diameter of 250 μm. The Precambrian genera are probably resting cysts of unicellular algae.

The most variable morphologic develop-ment is seen in the subgroup Acantho-morphitae (Figs. 42B, L, Q). These are cysts possessing a spherical, ellipsoidal or ovoidal vesicle which may be single or double-walled and bear processes (Fig. 43). The processes may show a regular arrangement in relation to the excystment aperture or test axis, or appear to lack an apparent distribution pattern. Included within this subgroup are all the Paleozoic genera previously regarded as hystrichospheres.

The Polygonomorphitae possess a polygonal vesicle of one wall layer with up to eight hollow processes (Fig. 42E). Only two genera are included in this subgroup whose retention is of doubtful value.

Acritarchs with an elongate to fusiform test and lacking an inner capsule are classified as the subgroup Netromorphitae (Figs. 42D, 42J and 44). The wall is single-layered. One or more processes may be developed as exten-sions of one or both poles of the vesicle, their number and disposition being of importance in generic classification. Excystment apertures have not been observed.

Acritarchs with a spherical to elongate vesicle and marked bilateral symmetry, in that the surface ornamentation is more pronounced at the "poles" than in the "equatorial" region, comprise the subgroup Diacromorphitae (Fig. 42K). An excystment aperture has been ob-served in some species. The affinities of this subgroup are unknown.

Members of the Herkomorphitae (Fig. 42H) superficially resemble a proximate dinocyst. They possess a spherical to ellipsoidal test whose surface is divided into polygonal fields of variable shape and size by septa, ridges or processes. A paracingulum has not been ob-served. Some forms referred to the Herko-morphitae are probably paratabulate dino-cysts seen in polar or oblique view and in which the paracingulum has been obscured.

Characteristic of the Pteromorphitae is the genus *Pterospermopsis* (Fig. 42I) which has a spherical to ellipsoidal vesicle surrounded "equatorially" by a flange.

The Prismatomorphitae (Fig. 42G) are acri-tarchs with a prismatic to polyhedral vesicle with more or less sharp edges, often produced into a distinct flange, entire or serrate, with or

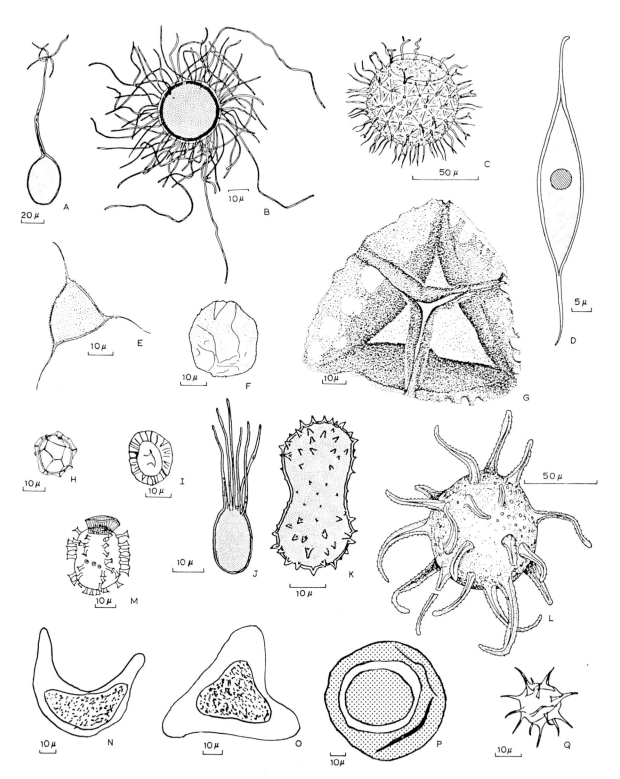

Fig. 42

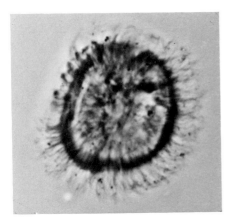

Fig. 43. *Comasphaeridium* cf. *cometes* Staplin, Jansonius and Pocock, an acritarch of the subgroup Acanthomorphitae. × 1,350.

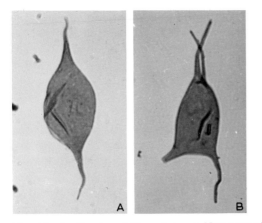

Fig. 44. Acritarchs of the subgroup Netromorphitae. A. *Leiofusa jurassica* Wall, Paleogene. × 1,100. B. *Domasia* sp., Paleogene. × 1,100.

without projections at the angles. They have a restricted stratigraphic range of Devonian to Triassic. Excystment apertures have not been described for this subgroup.

In the subgroups Stephanomorphitae (Fig. 42M), Dinetromorphitae (Fig. 42N) and Platymorphitae (Fig. 42O), dinoflagellate affinities have either been proven or are indicated.

Stratigraphic distribution

The earliest known organic microfossils are spheroidal or ellipsoidal granular unicells recorded by Schopf and Barghoorn (1967) from the Precambrian Fig Tree Group of South Africa; they are believed to be over 3.1×10^9 years old. The species *Archaeosphaeroides barbertonensis* is similar to modern coccoid blue-green algae. Barghoorn and Tyler (1965) erected the genus *Huroniospora* (Fig. 45A) for similar microfossils from the Gunflint Formation, United States (1900 m.y. old). Many of the Precambrian genera have a wide geographic distribution.

Palynologically, there is little difference between Precambrian and Lower Cambrian assemblages. Spiny acritarchs have been recorded from Lower Cambrian rocks in Russia and Scotland. Also appearing in the Lower Cambrian are genera that exhibit bilateral symmetry in the distribution of their ornamentation. In the Middle Cambrian species diversity increases markedly, but the Ordovician appears to represent the acme of development of the acritarchs in numbers and species diversity.

Fig. 42. Acritarch species characteristic of the various subgroups: A. *Deunffia ramusculosa* Downie, subgroup Netromorphitae. Silurian. B. *Comasphaeridium williereae* (Deflandre and Deflandre) Cramer, subgroup Acanthomorphitae. Silurian. C. *Cymatiogalea stelligera* Gorka, subgroup Herkomorphitae. Ordovician. D. *Leiofusa fusiformis* (Eisenack) Eisenack, subgroup Netromorphitae. Silurian. The shaded area is the pylome. E. *Veryhachium calandrae* Cramer, subgroup Polygonomorphitae. Silurian. F. *Leiosphaeridia faveolata* (Timofeev) Downie and Sarjeant, subgroup Sphaeromorphitae. Cambrian. G. *Polyedryxium trifissilis* Deunff, subgroup Prismatomorphitae. Devonian. H. *Cymatiosphaera eupeplos* (Valensi) Deflandre, subgroup Herkomorphitae. Early Jurassic. I. *Pterospermopsis* cf. *helios* Sarjeant, subgroup Pteromorphitae. Paleogene. J. *Polydeunffia eisenacki* Cramer, subgroup Netromorphitae. Silurian. K. *Lophodiacrodium pepino* Cramer, subgroup Diacromorphitae. Silurian. L. *Baltisphaeridium spinigerum* Gorka, subgroup Acanthomorphitae. Ordovician. M. *Stephanelytron redcliffense* Sarjeant, formerly of the subgroup Stephanomorphitae, now known to be a dinocyst. Late Jurassic. N. *Wallodinium lunum* (Cookson and Eisenack) Lentin and Williams, subgroup Dinetromorphitae, probably a dinocyst. Cretaceous. O. *Trigonopyxidia ginella* (Cookson and Eisenack) Lentin and Williams, subgroup Platymorphitae, probably a dinocyst. Cretaceous. P. *Disphaeria macropyla* Cookson and Eisenack, subgroup Disphaeromorphitae. Late Cretaceous. Q. *Micrhystridium stellatum* Deflandre, subgroup Acanthomorphitae. Paleogene. (A, B, D, E, J and K from Cramer, 1970a; C and L from Gorka, 1969; F from Martin, 1969; G from Deunff, 1971; H from Wall, 1965; M from Sarjeant, 1961; and N, O and P from Cookson and Eisenack, 1960).

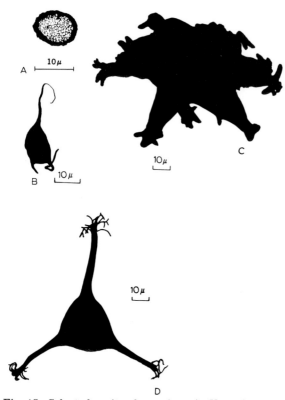

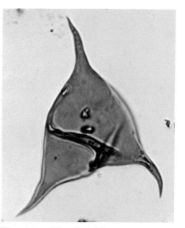

Fig. 46. *Veryhachium europaeum* Stockmans and Willière. A Paleogene acritarch of the subgroup Polygonomorphitae. × 1000.

Fig. 45. Selected acritarch species. A. *Huroniospora microreticulata* Barghoorn, an organic-walled microfossil over 1900 m.y. old. Precambrian. B. *Domasia liassica* Wall, subgroup Netromorphitae. Paleogene. C. *Multiplicisphaeridium radicosum* Loeblich, subgroup Acanthomorphitae. Ordovician. D. *Dateriocradus polydactylus* Tappan and Loeblich, subgroup Polygonomorphitae. Silurian. (A from Barghoorn and Tyler, 1965; C from Loeblich, 1970; and D from Tappan and Loeblich, 1971).

There is a major influx of new species in the early Silurian. The distinctive genus *Domasia* (Figs. 44B and 45B) is common in the Silurian. Acanthomorphitic acritarchs are also abundant. The acritarchs of the late Silurian of England show a more or less progressive vertical change in the assemblage. Whether this is stratigraphically or ecologically controlled is not known. There is a major decline towards the end of the Silurian which continues throughout the Devonian and Carboniferous. In the late Paleozoic the acritarch assemblages are dominated by small forms of the genera *Veryhachium* (Fig. 46) and *Micrhystridium*. Throughout the Mesozoic—Cenozoic, acritarchs play an insignificant part in phytoplankton assemblages. Genera present include *Baltisphaeridium*, *Coma-*

sphaeridium (Fig. 43), *Domasia*, *Leiofusa* (Fig. 44A), *Micrhystridium* and *Veryhachium* (Fig. 46).

Evolutionary trends

In the acritarchs size seems to have followed a definite evolutionary trend that parallels that of other organisms. The increase in overall size in the acanthomorphitic acritarchs continues into the Ordovician after which, in the Silurian and Devonian periods, there is a gradual, continuing reduction.

Processes develop from few to many, many to few, simple to branched, homomorphic to heteromorphic. The oldest recorded acanthomorphitic acritarch has simple processes, which are expanded proximally and are conical or subconical. This primitive stage gives way to forms with longer more slender processes in the Cambrian. In the Ordovician, complexly branched processes are developed in forms with conical or subconical processes as in *Multiplicisphaeridium radicosum* (see Fig. 45C). During the Silurian, forms with complexly branched, slender processes become more common in the acanthomorphitic acritarchs. The *Veryhachium* complex also shows a tendency to develop branched processes as in *Dateriocradus* (Fig. 45D). Evolutionary trends in vesicle outline, wall structure, or surface ornamentation have not yet been recognized.

Paleoecology

That most acritarchs are probably the tests of planktonic marine organisms is suggested by the widespread distribution of some species and their occurrence in sediments containing other marine fossils. A few Holocene species have been recorded from fresh-water deposits, so it is feasible that fresh-water Paleozoic, Mesozoic and Cenozoic forms existed. However, if the majority of acritarchs are the cysts of unicellular phytoplankton, their abundance in sediments can generally be taken to indicate a marine environment.

Use of acritarchs to predict specific marine environments was largely initiated by Staplin (1961) in a study of Upper Devonian rocks of western Canada. He recognized three distinct associations of acritarchs whose distribution patterns were related to proximity to reefs (Fig. 47). Simple spherical forms (Association 1) had a lateral distribution ranging from beds interfingering with reef carbonates to off-reef areas. Forms with thin spines (Association 2) also were widespread, but were infrequent in sediments within 1.6 km of the reefs. Thick-spined and polyhedral forms (Association 3) were only found, with two exceptions, at distances greater than 6.5 km from the reefs. The absolute abundance of all species increased with increasing distance from the reefs. Staplin reasoned that the optimum en-

vironment for the acritarchs, that is the motile phase, was in the quiet, deeper water of the off-reef areas. He also believed that concentrated acritarch assemblages denoted major marine currents.

The stratigraphic distribution of acritarchs in the Lias (early Jurassic) of England and Wales is environmentally controlled, with many common long-ranging species showing significant fluctuations in abundance. Wherever paleogeographic data indicate an inshore, basinal environment, the assemblages are dominated by members of the Acanthomorphitae, while a richer assemblage including members of the Polygonomorphitae and the Netromorphitae is present in the open-sea environment. Forms with long processes are concentrated in regions of quiet deposition, whereas those with reduced processes tend to be dominant in turbulent environments associated with sandstone deposition. This accords with the previous findings of Staplin.

From such observations, Wall (1965) could distinguish the transgressive—regressive phases of the Liassic cycles of sedimentation using acritarch assemblages. During the early and late phase of the cycle, species diversity is low with a single species being dominant in some samples; with the return of more open marine conditions, there is an influx of acritarchs and an increase in species diversity.

Cramer and Diez de Cramer (1972) in a study of a tidal flat area in northwest Spain, noted the sensitivity of certain acritarch taxa to turbidity as expressed in quantitative fluctuations in their abundances in sandstone and shale. These authors have also delineated five acritarch associations, whose distribution patterns appear to be climatically controlled and hence can be correlated with paleolatitudes in the Silurian. Many Silurian acritarchs are worldwide in their distribution. The species characterizing the associations, however, have a restricted lateral occurrence. All the Silurian samples analyzed were from epicontinental sea or shelf deposits, and this led Cramer (1970a) to believe that the associations were climatically controlled. Assuming continental drift occurred, the palynofacies can be related to paleolatitudinal zones with warmer to cooler zones being represented by transition from one facies group to another.

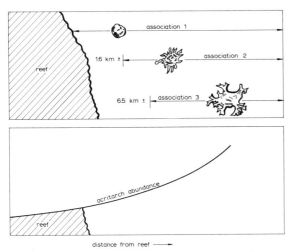

Fig. 47. Diagrammatic representation of distribution of acritarchs in the vicinity of a reef in the late Devonian of central Alberta, Canada (from Staplin, 1961).

From his studies, Cramer (1970b) has calculated the direction of drift of the reassembled Atlantic continents during the Wenlockian to be N45°W plus or minus 10° to 20° and the rate of continental drift as 2 cm per year.

TASMANITIDS

Introduction

The tasmanitids are unicellular, relatively thick-walled organic fossils characterized by the presence of numerous, fine, radial canals, which may pass partially or completely through the wall. They are predominantly spheroidal, more rarely ovoidal, reniform, biconvex or disc-shaped; the last three shapes probably resulting from compression. They vary in size from less than 100 to more than 600 μm. Their known stratigraphic range is Cambrian to Miocene. The family Tasmanaceae of 29 genera and over 100 species includes the genus *Tasmanites* (see Fig. 51B) after which it is named.

Historical background

The Tasmanaceae derive their name from a deposit of "tasmanite" or "white coal", an impure coal of Permian age found in Tasmania. Tasmanite is largely composed of *Tasmanites*, thick-walled spherical bodies to which Newton assigned the generic name in 1875. He observed numerous pits in the surface wall, did not see a triradiate mark as found on many spores, and so he correctly interpreted them as algal. Prior workers had postulated that they were seed spore cases of plants.

The affinities of *Tasmanites* lie apparently with the Prasinophyceae (green algae). Wall in 1962 noted the morphological similarity between *Tasmanites* (Fig. 48A) and the living species *Pachysphaera pelagica* of the Prasinophyceae (Fig. 48B). *Pachysphaera* is a pelagic organism with the characteristics of a vegetative and encysted stage. Mary Parke (1966) has successfully cultured *Pachysphaera marshalliae* and determined its life cycle (Fig. 49). The *Pachysphaera* cyst is developed from a motile swarmer. When the cyst is fully grown (100 to 175 μm in diameter), the nu-

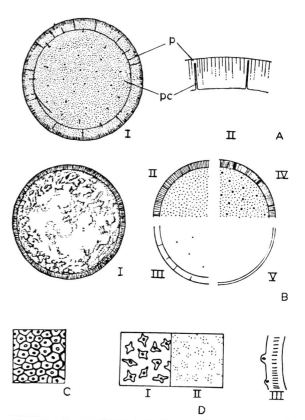

Fig. 48. Comparison of the wall structure of *Tasmanites* with that of Recent plankton. A. *Tasmanites* cf. *tardus* Eisenack, early Jurassic. I. Surface view. × 500. II. Optical section of wall. × 1,300. *p* = punctae; *pc* = pore canals. B. Recent plankton of the class Prasinophyceae. I—III. *Pachysphaera pelagica* Ostenfeld. I. Optical section. II. Wall structure showing numerous regularly arranged pores. III. Wall structure of specimen with few pores. IV. Wall structure of *Pachysphaera* sp. with fine punctae and coarser pore canals. V. Wall structure of *Halosphaera minor* Ostenfeld. All × 330. (From Wall, 1962.) C. *Tytthodiscus* sp., a tasmanitid; outer surface of wall. Paleogene. D. *Crassosphaera stellulata* Cookson and Manum, a tasmanitid. I. Outer surface of wall. II. Inner surface of wall. III. Optical section of wall. Paleogene.

Coscinodiscus paleaceus (Grunow) Rattray. Valve ovoid, flat with large, well-developed areolae. × 1000.

cleus then undergoes division. The resulting swarmers develop flagella and are released through a linear fissure in the cyst wall. The motile swarmers then repeat the cycle. If *Tasmanites* has a comparable life cycle to *Pachysphaera* the variation in cyst diameter could be explained thus. The tasmanitids are believed to have affinities with the Prasinophyceae.

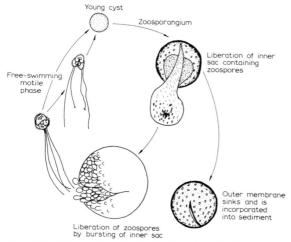

Fig. 49. Life cycle of the living alga *Pachysphaera marshalliae* Parke, a member of the subgroup Prasinophyceae. (From Sarjeant, 1970; redrawn after Jux, 1969, incorporating extra detail from Parke, 1966.)

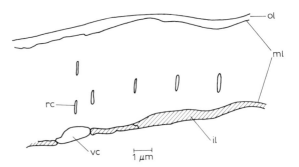

Fig. 50. *Tasmanites huronensis* (Dawson) Schopf, Wilson and Bentall. Schematic optical section of wall. *il* = inner wall layer; *ml* = middle wall layer; *ol* = outer wall layer; *rc* = radial canal; *vc* = vesicular cavity. (After Jux, 1968.)

Morphology

The tasmanitid wall is yellow to reddish brown to dark brown. It is usually 5 to 20 µm thick and is composed of two or three layers (Fig. 50), although the outer layer is rarely preserved. The middle wall layer forms the bulk of the wall, and there may also be a spongy or almost fibrous inner layer. Radial canals characterize the middle wall and it is possible to divide many of the tasmanitids into species on the basis of the arrangement and development of these radial canals.

Pylome

Some species of *Tasmanites* have a circular pylome complete with operculum. Others possess sutures or splits (Fig. 51B) which presumably delineate a line of weakness in the cyst. The closely related genus, *Pachysphaera*, develops a linear suture in the cyst wall prior to the escape of the motile swarmers. Comparison with *Pachysphaera* suggests that the "split" in *Tasmanites* is the excystment aperture, although splits can often result from compression.

Major morphologic groups

The tasmanitids are characterized by the possession of radial and pore canals. The

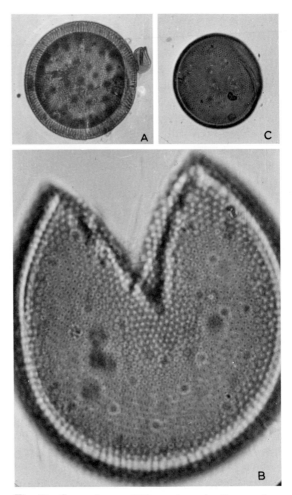

Fig. 51. Some tasmanitid species. A. *Crassosphaera hexagonalis* Wall, Paleogene. The wall is 5—10 µm thick. × 300. B. *Tasmanites suevicus* Eisenack, Paleogene. × 300. C. *Crassosphaera hexagonalis* Wall, Paleogene. The hexagonal wall structure is visible on the surface. × 1,250.

structure, number, and arrangement of these is important in classification. *Tasmanites* (see Fig. 48A) possesses radial and pore canals, which may form distinctive patterns on the outer surface of the wall. It has a known range of Ordovician to Miocene. *Tytthodiscus* (see Fig. 48C) from the Tertiary of California has a wall composed of radially oriented hexagonal elements each of which is penetrated by a tubule. The radial canals are arranged in a diagnostic pattern. *Crassosphaera* (Figs. 48D and 51A, C) is considered to differ from *Tytthodiscus* by not possessing a segmented wall. However, the radial canals do show an orderly arrangement. Other genera are separated primarily on the basis of wall thickness and surface ornamentation.

Stratigraphic distribution

The Tasmanaceae have a known stratigraphic range of Cambrian to Recent if we include the living genus *Pachysphaera* in this family. They become increasingly common in the Ordovician and Silurian when species with and without pylomes were extant. Most of these early Paleozoic forms are assigned to the genus *Tasmanites*. In some Devonian black shales they are the only microfossils present and form a large percentage of the rock. There are few records of tasmanitids in the Carboniferous. In the Permian they are often found in very high concentrations as in the "white coal" or "tasmanite" of Tasmania.

The Tasmanaceae range through the Mesozoic—Cenozoic, exhibiting increasing species diversity, but decreasing in abundance especially in the late Tertiary. Since the tasmanitids are very resistant to weathering and oxidation, they are frequently present as reworked material in younger sediments. One of the earliest records of their occurrence was as Devonian contaminants in the water supply of Chicago. Their presence in Mesozoic and Cenozoic sediments must be analyzed carefully to determine whether they are indigenous or reworked.

Evolutionary trends

The tasmanitids are a remarkably stable group and exhibit only one observable evolu-

tionary feature, the tendency for the radial canals and pore canals to show a more orderly arrangement with time, which attains a peak in *Crassosphaera stellulata* (see Fig. 48D). The radial canals may have functioned as a hydrostatic mechanism for the daily vertical movement of the tasmanitids, although if, as seems probable, they were cysts, this would not be a necessary function on a daily basis.

Paleoecology

The tasmanitids appear to have been cysts of planktonic organisms living in a restricted marine environment, sometimes forming marine peats. The enormous proliferation necessary to produce these marine peats must represent "algal blooms". It is believed that such deposits were formed in a littoral environment where fresh-water run-off had lowered the salinity. In the Paleozoic of the Sahara the subsaline environment favorable to tasmanitids resulted from the inflowing of meltwaters from surrounding glaciers. The common occurrence of *Tasmanites* in euxinic (restricted circulation) black shales suggests that the environment of deposition was a death trap for them. Alternatively, it may be that *Tasmanites* cysts can survive in anaerobic toxic environments which restrict or inhibit the existence of other organisms.

SUGGESTIONS FOR FURTHER READING

Downie, C. and Sarjeant, W.A.S., 1966. The morphology, terminology and classification of fossil dinoflagellate cysts. In: R.J. Davey, C. Downie, W.A.S. Sarjeant and G.L. Williams, *Studies on Mesozoic and Cainozoic dinoflagellate cysts. Bull. Br. Mus. Nat. Hist. (Geol.)*, Suppl. 3: 10—17. [Modern concepts of dinocyst morphology with particular regard to wall layers and the informal groups including chorate, proximate and cavate cysts.]

Evitt, W.R., 1967. Dinoflagellate studies II. The archeopyle. *Stanford Univ. Publ., Geol. Sci.*, 10(3): 1—83. [The standard reference on archeopyles in dinocysts.]

Evitt, W.R., 1969. Dinoflagellates and other organisms in palynological preparations. In: R.H. Tschudy and R.A. Scott (Editors), *Aspects of Palynology*. Wiley-Interscience, New York, N.Y., pp. 439—479. [A brief discussion of the morphology of modern dinoflagellates and the vegetative stage is followed by a relatively detailed section on the morphology of fossil dinocysts.]

Fritsch, F.E., 1965. Class VI. Dinophyceae (Peridinieae). In: F.E. Fritsch, *Structure and Reproduction of the Algae*. Cambridge University Press, London, pp. 664—720. [Discusses the classification, morphology and reproduction of modern dinophyceae. Presents a detailed, but slightly outdated, account, from the biological viewpoint.]

Lister, T.R., 1970. The acritarchs and chitinozoa from the Wenlock and Ludlow Series of the Ludlow and Millichope areas, Shropshire. *Palaeontogr. Soc. (Monogr.)*, 124: 1—100. [The terminology, including a glossary, classification and biological affinities of the acritarchs is outlined. A detailed section deals with "excystment openings". The theory that acritarchs are the resting cysts of "naked" dinoflagellates is forcefully presented. This publication is recommended for an insight into recent theories concerning these organisms.]

Loeblich Jr., A.R. and Loeblich III, A.R., 1966. Index to the genera, subgenera, and sections of the Pyrrhophyta. *Stud. Trop. Oceanogr. Miami*, No. 3: 94 pp. [The introductory chapter outlines the suprageneric classification, habitat, morphology and life cycle of the dinoflagellates. It is a concise account which includes details garnered from many sources.]

Loeblich III, A.R., 1970. The amphiesma or dinoflagellate cell covering. *N. Am. Paleontol. Conv., Chicago, 1969*, Proc. G, pp. 869—929. [A detailed section on the composition and structure of the cell wall of the vegetative stage and resting cyst, and cell division and a comprehensive suprageneric classification.]

Muir, M.D. and Sarjeant, W.A.S., 1971. An annotated bibliography of the Tasmanaceae and of related living forms (Algae: Prasinophyceae). In: S. Jardiné (Editor), *Les acritarches*. Editions C.N.R.S., Paris, pp. 56—117. [Rather heavy for the student but a useful and unique work on the tasmanitids.]

Sarjeant, W.A.S., 1969. Microfossils other than pollen and spores in palynological preparations. In: G. Erdtman (Editor), *Handbook of Palynology: Morphology, Taxonomy, Ecology*. Munksgaard, Copenhagen, pp. 165—208. [The morphology, habitat and life cycle of modern dinoflagellates is reviewed. The morphology of the dinocysts is categorized according to proximate, chorate and cavate cysts, and the various subdivisions. The section on the acritarchs and tasmanitids is presented in a very readable style.]

Sarjeant, W.A.S., 1970. Xanthidia, Palinospheres and 'Hystrix': A review of the study of fossil unicellular microplankton with organic cell walls. *Microscopy, London*, 31: 221—256. [An interesting account of the history of the study of fossil dinocysts with brief biographical sketches of many of the earlier scientists who studied these organisms. Also discusses vegetative stage, cyst relationship, acritarchs and tasmanitids.]

Sarjeant, W.A.S., 1974. *Fossil and Living Dinoflagellates*. Academic Press, London, 182 pp. [Morphology, life cycle and ecology of living dinoflagellates is discussed in considerable detail followed by the history of study and morphology of fossil dinoflagellates. The philosophy and problems of classification of living and fossil dinoflagellates accompany an Appendix, given over solely to one of the existing classifications for dinocysts. Preparation techniques are also discussed in an Appendix.]

Wall, D. and Dale, B., 1968. Modern dinoflagellate cysts and evolution of the Peridiniales. *Micropaleontology*, 14: 265—304. [The vegetative stage, cyst relationship is now known for over 30 species of modern dinoflagellates. The concept of lineages, as based primarily on archeopyle type, is a new field of study.]

CITED REFERENCES

Barghoorn, E.S. and Tyler, S.A., 1965. Microorganisms from the Gunflint chert. *Science*, 147: 563—577.

Barss, M.S. and Williams, G.L., 1973. Palynology and nannofossil processing techniques. *Geol. Surv. Can., Pap.*, 73-26: 1—25 (pls. 1—2).

Cookson, I.C. and Eisenack, A., 1960. Microplankton from Australian Cretaceous sediments. *Micropaleontology*, 6: 1—18.

Cramer, F.H., 1970a. Distribution of selected Silurian acritarchs; *Rev. Esp. Micropaleontol.*, Spec. No. 1: 1—202 (pls. 1—23).

Cramer, F.H., 1970b. Middle Silurian continental movement estimated from phytoplankton-facies transgression. *Earth Planet. Sci. Lett.*, 10: 87—93.

Cramer, F.H., 1971. A palynostratigraphic model for Atlantic Pangea during Silurian time. *Bur. Rech. Géol. Min., Mém.* 73: 229—235.

Cramer, F.H. and Diez de Cramer, M.d.C.R., 1972. North American Silurian palynofacies and their spatial arrangement: Acritarchs. *Palaeontographica, Abt. B*, 138: 107—180 (pls. 31—36).

Deunff, J., 1971. Le genre *Polyedryxium* Deunff. In: S. Jardiné (Editor), *Les Acritarches*. Editions C.N.R.S., Paris, pp. 17—49.

Downie, C., Evitt, W.R. and Sarjeant, W.A.S., 1963. Dinoflagellates, hystrichospheres, and the classification of the acritarchs. *Stanford Univ. Publ., Geol. Sci.*, 7(3): 1—16.

Downie, C., Hussain, M.A. and Williams, G.L., 1971. Dinoflagellate cyst and acritarch associations in the Paleogene of southeast England. *Geosci. Man*, 3: 29—35 (pls. 1—2).

Eaton, G.L., 1971. A morphogenetic series of dinoflagellate cysts from the Bracklesham Beds of the Isle of Wight, Hampshire, England. In: A. Farinacci (Editor), *Proc. Second Planktonic Conf., Roma, 1970*. Edizioni Tecnoscienza, Rome, 1: 355—379 (pls. 1—4).

Ehrenberg, C.G., 1838. Über das Massenverhältnis der jetzt lebenden Kiesel-Infusorien und über ein neues Infusorien-Conglomerat als Polirschiefer von Jastraba in Ungarn. *Abh. Preuss. Akad. Wiss.*, 1836: 109—135 (pls. 1—2).

Eisenack, A., 1963. Hystrichosphären. *Biol. Rev.*, 38: 107—139 (pls. 2—4).

Evitt, W.R., 1961. Observations on the morphology of fossil dinoflagellates. *Micropaleontology*, 7: 385—420 (pls. 1—9).

Evitt, W.R., 1963. A discussion and proposals concerning fossil dinoflagellates, hystrichospheres and acritarchs, I and II. *Proc. Natl. Acad. Sci., Wash.*, 49: 158—164; 298—302.

Evitt, W.R., 1967. Dinoflagellate studies. II. The archeopyle. *Stanford Univ. Publ., Geol. Sci.*, 10(3): 1—83 (pls. 1—9).

Evitt, W.R., 1975. *Manual for a Teaching Conference on Fossil Dinoflagellates*. Private publication, Stanford University, Calif.

Evitt, W.R. and Davidson, S.E., 1964. Dinoflagellate studies. I. Dinoflagellate cysts and thecae. *Stanford Univ. Publ., Geol. Sci.*, 10(1): 1—12 (pl. 1).

Fritsch, F.E., 1965. *The structure and Reproduction of the Algae*. Cambridge University Press, Cambridge, 791 pp.

Gocht, H. and Netzel, H., 1974. Rasterelektronenmikroskopische Untersuchungen am Panzer von *Peridinium* (Dinoflagellata). *Arch. Protistenkd.*, 116: 381—410.

Górka, H., 1969. Microorganismes de l'Ordovicien de Pologne. *Palaeontol. Pol.*, No. 22: 102 pp.

Graham, H.W., 1942. Studies in the morphology, taxonomy and ecology of the Peridiniales. *Carnegie Inst. Wash. Publ.*, No. 542: 129 pp.

Gray, J. (coord.), 1965. Techniques in palynology; Part III. In: B. Kummel and D.M. Raup (Editors), *Handbook of Paleontological Techniques*. Freeman and Co., San Francisco, Calif., pp. 469—706.

Jux, U., 1968. Über den Feinbau der Wandung bei *Tasmanites* Newton. *Palaeontographica, Abt. B.*, 124: 112—124.

Jux, U., 1969. Über den Feinbau der Zystenwandung von *Pachysphaera marshalliae* Parke, 1966. *Palaeontographica, Abt. B*, 125: 104—111.

Kofoid, C.A., 1907. The plates of *Ceratium* with a note on the unity of the genus. *Zool. Anzeiger*, 32: 177—183.

Kofoid, C.A., 1909. On *Peridinium stecni* Jorgensen, with a note on the nomenclature of the skeleton of the Peridinidae. *Arch. Protistenkd.*, 16(1): 25—47.

Kofoid, C.A. and Skogsberg, T., 1928. The Dinoflagellata: the Dinophysoidae. *Harvard Univ., Mus Comp. Zool., Mem.*, 51: 766 pp.

Lebour, M.V., 1925. *The Dinoflagellates of Northern Seas*, Mar. Biol. Assoc., U.K., London, 250 pp.

Lister, T.R., 1970a. The method of opening, orientation and morphology of the Tremadocian acritarch, *Acanthodiacrodium ubui* Martin. *Yorks. Geol. Soc., Proc.*, 38(1): 47—55 (pl. 5).

Lister, T.R., 1970b. A monograph of the acritarchs and chitinozoa from the Wenlock and Ludlow Series of the Ludlow and Millichope areas, Shropshire. *Palaeontogr. Soc. (Monogr.)*, part 1: 1—100 (pls. 1—13).

Loeblich III, A.R., 1970. The amphiesma or dinoflagellate cell covering. *N. Am. Paleontol. Conv., Chicago, 1969, Proc. G*, pp. 869—929.

Loeblich Jr., A.R. and Tappan, H., 1969. Acritarch excystment and surface ultrastructure with descriptions of some Ordovician taxa. *Rev. Esp. Micropaleontol.*, 1: 45—57.

Mantell, G.A., 1845. Notes of a microscopical examination of the Chalk and Flint of southeast England with remarks on the Animalculites of certain Tertiary and modern deposits. *Ann. Mag. Nat. Hist.*, 16: 73—88.

Martin, F., 1969. Les Acritarches de l'Ordovicien et du Silurien Belges. *Inst. R. Sci. Nat. Belg., Mem.*, 160: 175 pp.

Newton, E.T., 1875. On "Tasmanite" and Australian "White Coal". *Geol. Mag., Ser. 2*, 2(8): 337—342 (pl. 10).

Parke, M., 1966. The genus *Pachysphaera* (Prasinophyceae). In: H. Barnes (Editor), *Some Contemporary Studies in Marine Science*. Allen and Unwin, London, pp. 555—563.

Sarjeant, W.A.S., 1961. Microplankton from the Kellaways Rock and Oxford Clay of Yorkshire. *Palaeontology*, 4: 80—118.

Sarjeant, W.A.S., 1967. The stratigraphical distribution of fossil dinoflagellates. *Rev. Palaeobot. Palynol.*, 1: 323—343.

Schopf, J.W. and Barghoorn, E.S., 1967. Alga-like fossils from the Early Precambrian of South Africa. *Science*, 156: 508—512.

Staplin, F.L., 1961. Reef controlled distribution of Devonian microplankton in Alberta. *Palaeontology*, 4: 392—424 (pls. 48—51).

Tappan, H. and Loeblich Jr., A.R., 1970. Geobiologic implications of fossil phytoplankton evolution and time-space distribution. In: R.M. Kosanke and A.T. Cross (Editors), *Symposium on Palynology of the Late Cretaceous and Early Tertiary. Geol. Soc. Am., Spec. Pap.*, 127: 247—340.

Tappan, H. and Loeblich Jr., A.R., 1971. Surface sculpture of the wall in Lower Paleozoic acritarchs. *Micropaleontology*, 17: 385—410.

Tappan, H. and Loeblich Jr., A.R., 1972. Fluctuating rates of protistan evolution, diversification and extinction. In: B.L. Mamet and G.E.C. Westermann (Editors), *Sect. 7, Paleontology, Int. Geol. Congr., 24th, Montreal, 1972*, pp. 205—213.

Vozzhennikova, T.F., 1965. Vvedenie v izuchenie iskopaemykh peridineevykh vodorosley. (Introduction to the study of fossil peridinian algae.) *Akad. Nauk S.S.S.R., Sib. Otd., Inst. Geol. Geofiz., Tr.*, 156 pp.

Wall, D., 1962. Evidence from Recent plankton regarding the biological affinities of *Tasmanites* Newton 1875 and *Leiosphaeridia* Eisenack 1958. *Geol. Mag.*, 99: 353—362 (pl. 17).

Wall, D., 1965. Microplankton, pollen, and spores from the Lower Jurassic in Britain. *Micropaleontology*, 11: 151—190 (pls. 1—9).

Wall, D., 1971. Biological problems concerning fossilizable dinoflagellates. *Geosci. Man*, 3: 1—15 (pls. 1—2).

Wall, D. and Dale, B., 1968. Early Pleistocene dinoflagellates from the Royal Society Borehole at Ludham, Norfolk. *New Phytol.*, 67: 315—326 (pl. 1).

Wall, D. and Dale, B., 1970. Living hystrichosphaerid dinoflagellate spores from Bermuda and Puerto Rico. *Micropaleontology*, 16: 47—58 (pl. 1).

Wetzel, O., 1933. Die in organischer Substanz erhaltenen Mikrofossilien des baltischen Kreidefeuersteins, mit einem sedimentpetrographischen und stratigraphischen Anhang. *Palaeontographica*, 77: 147—186; *Palaeontographica*, 78: 1—110 (pls. 1—7).

Williams, D.B., 1971. The occurrence of dinoflagellates in marine sediments. In: B.M. Funnell and W.R. Riedel (Editors), *Micropalaeontology of the Oceans*. Cambridge University Press, London, pp. 231—243.

Williams, G.L., 1977. Biostratigraphy and palaeoecology of Mesozoic—Cenozoic dinocysts. In: A.T.S. Ramsay (Editor), *Oceanic Micropalaeontology*. Academic Press, London.

Zeuner, F.E., 1946. *Dating the Past, An Introduction to Geochronology*. Methuen and Co., London, 444 pp. (24 pls.).

14

SPORES AND POLLEN IN THE MARINE REALM

LINDA HEUSSER

INTRODUCTION

In contrast to the other microfossils discussed in this book, pollen and spores of plants are an allochthonous marine micropaleontologic group. That is, unlike foraminiferans or radiolarians, pollen and spores are not marine organisms, but are the products of continental vegetation, and might best be considered as a biogenous component of fine-grained (5—150 μm) detrital or terrigenous marine sediment.

Interest in marine palynology, the study of pollen and spores in present-day oceans and ocean sediments, has grown concomitantly with the mid-twentieth century development of oceanography. Although the stratigraphic and ecologic potential of terrestrial microfossils in marine sediments was recognized when pollen and spores were first described from marine cores, the full potential has yet to be exploited due to lack of meaningful data. Muller's (1959) classic work on the Orinoco Delta, South America, was the first comprehensive study of the distribution of marine palynomorphs in space and time.

Subsequent studies of a similar nature have been conducted on the continental shelf and in relatively restricted areas such as the Gulf of California and the Sea of Okhotsk (Koreneva, 1957; Cross and others, 1966). When Stanley reviewed the field of marine palynology in 1969, less than fifty papers had been published by about half as many authors, most of which dealt with pollen in surface sediment or in cores with sediment of Quaternary age.

Pollen

Pollen grains are the male reproductive bodies (microgametophytes) in seed plants (Fig. 1). They originate on the anthers of flowering plants, or angiosperms (like oaks, grass, roses) and in the microsporangia of gymnosperms (pines, firs, spruces). After pollination (transfer of the pollen grains to the female reproductive part) and fertilization (fusion of the sperm or male gamete with the egg or female gamete), the embryo develops within the ovule which becomes a seed. Coniferous pollen types were first described from the Pennsylvanian and angiosperm pollen from the early Cretaceous.

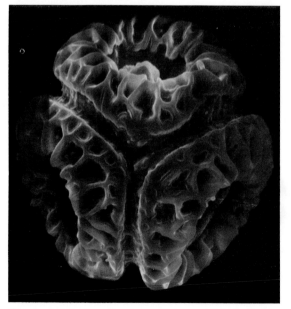

Fig. 1. A tetrad pollen type, *Drimys winteri*, showing three of the four components. × 1800.

Spores

Spores produced by "lower plants", such groups as algae, fungi, mosses, and ferns, are some of the earliest preserved remains of plants (Fig. 2). The Silurian marks the appearance of the first accepted spores with triradiate sutures. Spore production in primitive vascular plants, such as the club moss, may be homosporous, that is, a single type of spore is formed; or heterosporous. Heterosporous plants, those that produce male microspores and female megaspores, are known since the late Devonian. (In a strict sense pollen grains are microspores and are analogous to the microspores of ferns.) In homosporous plants the spore germinates and develops into either bisexual gametophytes or into two plants one of which produces male gametophytes and one female gametophytes.

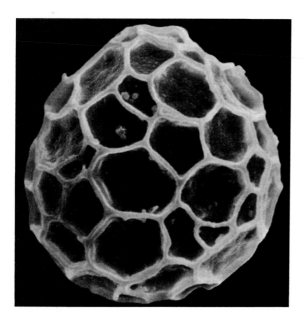

Fig. 2. A monad spore, *Lycopodium gayanum.* × 1500.

CHARACTERISTICS

Distinguishing criteria for pollen

The reader should consult one or more of the texts listed in "Suggestions for further reading" for a more extensive account of pollen and spore morphology and taxonomy. We will touch upon these aspects only briefly.

Pollen grains (and spores) are distinguished by the aspects of morphology — size, shape, apertures, surface sculpture — and wall structure.

Size

The size of the majority of pollen grains is between 20 and 80 μm, with rare forms less than 10 μm to more than 200 μm. Approximate size is relatively constant and useful in identification; however, it should be noted that size variations are caused by a number of factors, including preparatory technique. For example, mounting in glycerine gelatin tends to cause swelling of the pollen grains; the same grains embedded in silicone oil are about 30% smaller.

Shape

A wide variety of radially symmetrical shapes exist, of which variations of the rotational ellipsoid are more common. The relation between the polar and equatorial dimensions forms the basis for definition of commonly used shape classes, some of which are illustrated in Fig. 3. Variations of the ellipsoid form are frequently encountered.

Bilaterally symmetrical convex pollen grains are produced principally by two groups, the monocotyledons (lilies) and the gymnosperms. The latter produce vesiculate grains, those with sacci (bladders or wings) (Fig. 3F).

Apertures

The number and type of apertures (openings or thinning of part of the exine) form the primary basis for pollen differentiation. Generally, isodiametric apertures are called **pores**, and elongate or furrow-like apertures are called **colpi** (Fig. 4). Pollen grains may have no apertures (**inaperturate**), single apertures (**monoporate** or **monocolpate**), multiple pores (**diporate, triporate, stephanoporate,** or **periporate**), multiple colpi (**dicolpate, tricolpate, stephanocolpate, heterocolpate,** or **syncolpate**), or multiple pores and colpi (**dicolporate, tricolporate, stephanocolporate, pericolporate**).

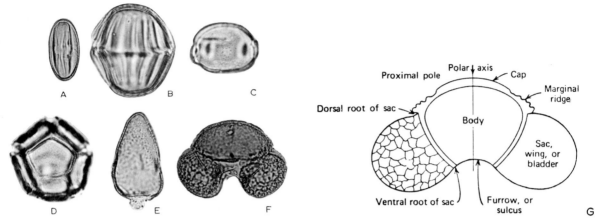

Fig. 3. Some common shapes of pollen grains. A. Prolate. B. Subspheroidal. C. Oblate. D. Pentagonal. E. Irregular. F. Vesiculate. G. Descriptive terms. (All figures approximately × 330, except D: × 890.)

Sculpture

Perhaps of greatest diagnostic value in the identification of pollen grains are surface sculpture and wall structures. Some twelve kinds of sculpture are recognized: these include types that are smooth, pitted, grooved, or ones that exhibit more or less isodiametric or elongate elements (see Fig. 4).

Wall structure

Sculptural elements are developed in the outer structural layer of the pollen grain, the **exine**, which is frequently divided into an inner homogeneous layer, the **endexine**, and an outer layer, the **ektexine** (Fig. 5). The exine is composed of a complex of minerals and organic compounds which include **sporopollenin**, a "natural plastic" which is highly resistant to degradation. The **intine** or interior wall of the pollen grain which encloses the protoplasm or living nucleus is largely composed of cellulose and other substances which are readily destroyed and thus are rarely fossilized.

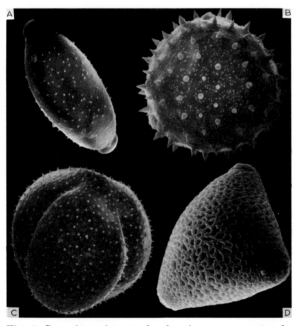

Fig. 4. Scanning micrographs showing some apertural types and sculpture on pollen grains. A. Diporate, microechinate grain of *Embothrium coccinineum*. × 1400. B. Tricolporate, echinate grain of *Corynabutilon vitifolium*. × 750. C. Triporate, foveolate—echinate grain of *Cereus chiloensis*, × 750. D. Triporate, echinate heterobrochate grain of *Lomatia ferruginea*. × 1500.

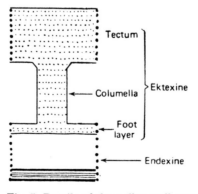

Fig. 5. Details of the pollen wall structure.

Spore characteristics

Spores are spheroidal, tetrahedral, or elongate (generally biconvex or planoconvex) and when mature occur both singly and in tetrads. The size of late Cenozoic spores is comparable to that of pollen; older spores may be as large as 2 mm. Surface sutures, **scars** or **laesurae**, are usually single (**monolete**), or triradiate (**trilete**), although they may be absent (**alete**) (Fig. 6). The spore wall, like that of the pollen grain, is composed of intine and exine. An outer layer or envelope called the **perine** appears in certain spores (Fig. 7). Many spores are decorated with sculptural elements in the same manner as pollen grains. Variations in sculpture, as well as structure, are more prevalent in trilete spore groups than in monolete groups. Envelopes enclosing spores to varying degrees are characteristic of many Paleozoic groups. **Crassitudes** (equatorial envelopes), **sacci** (wings), extensions of the tectum (fused and unfused) are commonly developed in Paleozoic and Mesozoic spores.

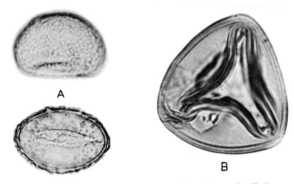

Fig. 6. Major spore types. A. Monolete. B. Trilete. × 500.

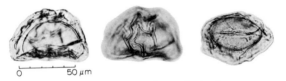

Fig. 7. Loosely folded outer layer (prine) on a monolete spore, *Athyrium alpestre*.

Methods

Identification of pollen and spores requires a high quality compound binocular microscope equipped with an oil immersion objective (× 950—1500 magnification) and an authenticated reference collection of pollen and spores from the area being studied with which comparisons can be made.

Preparation techniques designed to extract and concentrate the pollen by chemical or physical methods vary according to the matrix and personal preference of the palynologist. Many extraction techniques are covered by Gray (1965) in the *Handbook of Paleontological Techniques*.

Obviously sampling size and intervals will depend on the purpose of the study, the type of sediment, and the rate of sedimentation. Generally significant numbers of palynomorphs can be extracted from 3 to 5 cm³ of organic terrigenous lutites. Up to 10—20 cm³ of less organic terrigenous sediment may have to be processed to obtain counts of more than 100 pollen grains. Traverse and Ginsburg (1966), for example, processed 25 g of calcareous sands from the Great Bahama Bank. As plants are sensitive indicators of terrestrial ecologic change, it is desirable to sample at the closest intervals possible. In some areas, such as Santa Barbara Basin and Saanich Inlet, Vancouver, British Columbia, our samples cover about 10-year intervals and the presence of varved sediments may permit the sampling of yearly intervals.

In sampling marine sediments for pollen analysis, it is extremely important to obtain uncontaminated material. Many marine samples and cores are stored at temperatures which permit the growth of fungi on the outer surface of the sediment. External surfaces of cores may also be contaminated by pollen from the air. By avoiding material from the external surfaces, the possibility of contamination can be reduced.

FACTORS AFFECTING THE NATURE OF POLLEN IN THE MARINE ENVIRONMENT

The nature of the relationship between pollen in marine sediment and the source of the pollen, the species of the plant communities, is complex and can be envisioned as a function of a number of interrelated processes (ecologic, penecontemporaneous, and diagenetic).

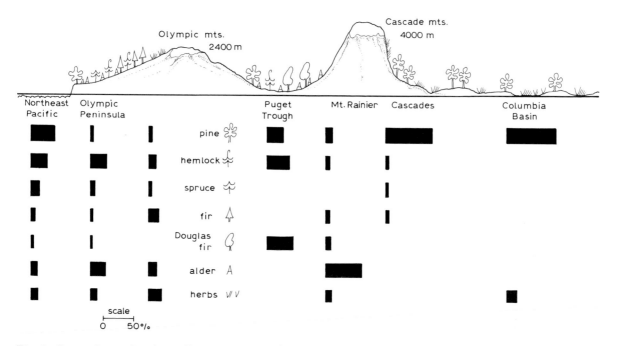

Fig. 8. Comparison of major pollen groups in surface sediments from the northeast Pacific and from a transect across Washington. Data from the core tops are the author's. Data from western Washington were derived from Florer (1972), Hansen (1947, 1949), and Heusser (1969, 1973). The cartoon illustrates the dominant vegetation of different elevations of the western Olympic Peninsula, Puget Trough, Mt. Rainier and the Cascades, and the Columbia Basin (Franklin and Dyrness, 1969).

Ecologic processes

Among the ecologic processes which delimit the plant communities from which pollen and spores are derived are climate, topography, edaphic or local factors such as soil and ground water, fire, and man. Plant communities are characterized by distinctive pollen assemblages. Although the relationship between the abundance of plants in the ecotone and the abundance of pollen is not always a simple one-to-one ratio*, in many geographic regions the pollen assemblages can be used to distinguish the broad features of the vegetation. For example, in eastern North America, Davis and Webb (1975) show close geographical correlation between contemporaneous pollen rain and the distribution of modern vegetation. The correlation between modern pollen rain and plant assemblages of western Washington is illustrated in Fig. 8.

---

*Pollen produced by individual species in a plant community ranges from one million grains/flower in pine to ten grains/flower in beech (Erdtman, 1969). Generally less pollen appears to be produced in restrictive environments, such as in tundra biomes, than in more temperate environments.

Penecontemporaneous processes

Diffusion, deposition, and destruction, the major penecontemporaneous processes act selectively and in essence screen the pollen and spore assemblage. Initial pollen dispersal from the plant to the surrounding environment is primarily **anemophilous** (by wind), **zoogamous** (animal) and **hydrophilous** (water); initial diffusion is comparatively unimpotant and dispersal distance short. Estimates vary from 10 to 150 km as a "natural" limit of primary diffusion, although long distance wind transport of up to 1500 km of small amounts of pollen has been reported. Some pollen may be wind-transported from coastal regions to the marine borderland. Estimates of coastal pollen rain range from 10^2 to 10^3 grains/cm^2 per year (Dyakowska, 1948). The nature and concentration of aeolian pollen would obviously depend on atmospheric circulation and on the aerodynamic properties of the pollen grains. Pollen and spore content of the atmosphere beyond the continental margins is low, on the order of $10-10^2$, and it would appear that aeolian transport would be

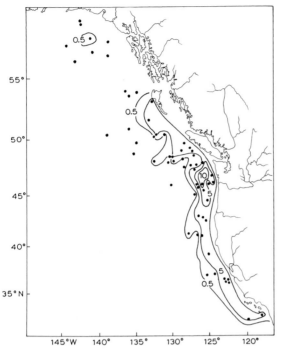

Fig. 9. Map showing distribution of pollen concentration (grains/cm³ in marine sediments) off NE Pacific coast of the United States. Contours are in grains × 10^3/cm³.

relatively unimportant in the dispersal of pollen to the deep sea. In some areas, however, as off western Equatorial Africa and in the Mediterranean, windblown pollen may form a significant proportion of marine palynomorphs.

Fluvial transport is probably the major means of pollen dispersal to the marine environment. Although long-term data are lacking, isolated observations on the amount of pollen in rivers and estuaries range up to 8000 grains/liter. In an extended study of a small catchment basin in England (Peck, 1973), comparison of the aerial and fluvial components of the pollen supply indicated that greater than 90% of the new pollen was derived from stream transport. Concentrations of pollen on the continental margins are higher adjacent to stream influx and lower off areas lacking permanent drainage. Pollen concentration in the northeast Pacific Ocean (Fig. 9) clearly reflects the influence of fluvial input of pollen. Contours of maximum amounts of pollen coincide with the distribution of the Columbia River plume, the major source of suspended sediment in the northeast Pacific. Secondary maxima occur off the outlet of the Sacramento and San Joaquin Rivers in California.

When pollen grains enter the water and are wetted, it appears that they are transported and deposited in the same manner as suspended particles of the size of fine silt and clay. Little data are available concerning the hydrodynamics of pollen grains, however Hopkins' (1950) observations suggest that hydrodynamic properties of pollen and spores may be species specific.

Analysis of suspended and surface sediments of rivers and estuaries indicate that fluvial pollen assemblages are similar to the regional vegetation of the entire drainage basin rather than to local environments. According to McAndrews and Power (1973) pollen in the surface sediments of Lake Ontario is generally uniform and reflects the deciduous and coniferous vegetation of the various streams which drain into the lake. Similar results were found by Groot (1966) in the Delaware River and by Thompson (1972) in the Raritan River in New Jersey, where the pollen content of the river, the estuary (Raritan Bay) and the atmosphere show a 67% similarity.

In the ocean, as in streams, pollen is believed to function as suspended, terrigenous particles. Overall, the amount of pollen in modern marine sediments is decreased seaward due to progressive lateral and vertical mixing with marine water of lower sedimentary particle concentration; however, the concentration of pollen grains in marine sediments is not necessarily in a simple linear relation to distance from land or to increased depth of water. In the northeast Pacific Ocean, pollen concentration is polymodal, being lower on the shelf, higher on the slope and rise, and decreases to minimal values in the basins. The quantity of pollen seems to be positively related to the distribution of lutites which bypass the outer shelf (Harlett and Kulm, 1973) and are transported to the slope by subsurface currents and particle settling from overlying waters (Baker, 1973). The general parallelism of the pollen concentration contours with marine isoclines in the northeast Pacific suggests that the pattern of pollen concentration is partly a function of distribution by surface currents. The complex distribution of pollen in shelf sediments of the Orinoco Delta was related by Muller (1959) to the influence of surface currents.

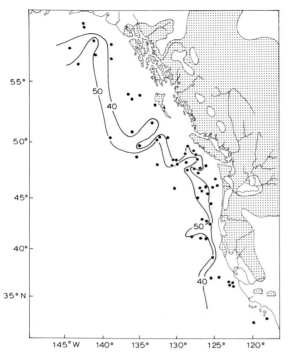

Fig. 10. Map showing distribution of relative abundance (percent of the pollen sum) of *Pinus*. Geographic extent of *Pinus* species from which pollen in marine sediment may be derived is indicated by shading.

The selective nature of marine transport has frequently been inferred from the decreased diversity of marine pollen assemblages compared to continental pollen assemblages and from the prominence of pine pollen in marine sediments. In core tops from the continental margin of the northwest Atlantic, about 90% of the pollen is pine, whereas on the adjacent coast the relative frequency of pine pollen ranges from 30 to 75% (Davis and Webb, 1975). At least 30% of all pollen in over 100 core tops from the northeast Pacific Ocean is pine (Fig. 10), both in samples adjacent to areas where pine is prolific and where pine trees and pollen are absent, such as the northern and southern extremities of the coast. This suggests transport by surface currents, the northwest flowing Alaska gyre, and the south flowing California current. Similar observations on the importance of surface currents on the distribution of pine pollen were made by Traverse and Ginsburg (1966) on the Great Bahama Bank. Even more striking is the uniform increase of pine pollen seaward, which apparently reflects the relative hydrodynamic efficiency of pine, although some differential resistance to destruction may also be a factor. The selective effects of marine transport on pinaceous pollen have been described by Cross and others (1966); Koreneva (1964) in a reconaissance study of the western Pacific Ocean, and by Lubliner-Mianowska in the Bay of Gdansk, Poland (1962).

Differential destruction of pollen and spores is not confined to the marine environment but is undoubtedly a selective factor from the time pollen grains are produced.

TABLE I

Comparative ranking of some major pollen types in relation to corrosion and oxidation susceptibility, sporopollenin content, and preservation ability (based on Havinga, 1964; and Sangster and Dale, 1964)

Pollen type	Corrosion susceptibility	Oxidation susceptibility	Sporopollenin content	Preservation ability
Conifers (pines, spruce, fir, etc.)	low	low	high	high
Spores (lycopods, polypods)				
Alder, birch, linden, hazel				
Oak, maple willow, poplar	high	high	low	low

Observations on chemical destruction of pollen and spores in terrestrial environments, such as lakes and bogs, and in the laboratory are summarized in Table I. Bacterial and fungal activity also causes differential destruction of palynomorphs. Different layers of the spore are selectively attacked by microbial activity. The susceptibility of pollen to chemical and microbial degradation may be related to differences in the chemical and mineral composition of the exine, to differences in the chemical structure of the sporopollenin, and/or to difference in the relationship between the sporopollenin and the mass of pollen.

Diagenesis

Oxidation and bioturbation, diagenetic processes which affect the distribution of

organic carbon in marine sediments, undoubtedly also affect the abundance of pollen and spores which are susceptible to degradation by oxidizing agents. Relatively high pollen concentrations and maximum organic concentrations appear related to minimal dissolved oxygen concentration in surficial marine sediments on the continental margin off Oregon and Washington. The unusually high number of pollen grains per gram of sediment found in the Black Sea by Traverse (1974) may be due, at least in part, to the euxinic character of the sediment. Pollen grains are apparently absent in ancient carbonate environments. Although this may be due to the adverse effect of alkalinity on pollen, as has been suggested, it may also be a function of oxidation and/or sedimentation. In the present carbonate sediments of the Bahamas, pollen is fairly abundant (greater than 100 grains/g) and maximum abundance is associ-

TABLE II

Stratigraphic distribution of some major spore and pollen groups (Figs. 1, 4: × 400; 2,3,8: × 325; 5, 6, 7, 9 : × 450).

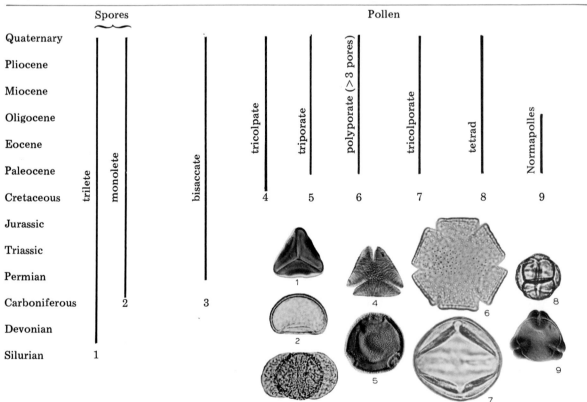

ated with fine-grained sediments (Traverse and Ginsburg, 1966).

During diagenesis pollen grains may be filled with opaque minerals, such as pyrite or manganese, or with oil. Over time the sporopollenin undergoes colorimetric evolution which can be detected through fluorescence or by reflective techniques. Pollen and spores stained with Safranin-0 show a gradation in stain acceptance which has been correlated with time by Stanley (1966b). Older grains, Paleozoic or Mesozoic in age, do not readily accept the stain and appear yellow or brown; Pleistocene grains accept stain and appear much like Recent pollen and spores. This coloric differentiation has been used by some palynologists to separate contemporaneous and reworked palynomorphs.

DISTRIBUTION OF POLLEN AND SPORES IN THE MARINE ENVIRONMENT

Horizontal distribution

Despite the complex nature of the processes affecting pollen and spores from the time they leave the plant until they are collected from the sediment, marine pollen assemblages do appear to reflect the vegetation from which they are derived. Comparison of the relative frequency of marine pollen in the northeast Pacific Ocean with the distribution of vegetation on the adjacent Pacific U.S. coast reveals a fairly good correlation of the dominant taxa. For example, western hemlock (*Tsuga heterophylla*), a prominent member of the temperate conifer forest, reaches optimal development in western Washington and coastal British Columbia, off which maximum percentages of western hemlock are found (Fig. 11A). Spruce (*Picea sitchensis*) becomes more important in southeastern Alaska, where high relative frequencies of marine spruce pollen occur (Fig. 11B). In California, *Sequoia* becomes a significant member of the temperate conifer forest and distribution of *Sequoia* on the continental margin is restricted to coastal California, as is oak (*Quercus* species) which although more widespread is relatively insignificant in the vegetation and pollen rain north of California (Fig. 11C and D).

Generally marine pollen assemblages appear to be an amalgam of coastal and regional vegetation, the extent of the region depending on the area of permanent drainage. Muller (1959) found a few grains of alder pollen from the Andes in sediments of the Orinoco Delta, and pollen in the eastern Mediterranean is apparently derived from coastal vegetation and from the Nile River, according to Rossignol-Strick (1973). The relative absence of representatives of the Compositae in marine sediments off the mouth of the Columbia River suggests that the pollen of these sagebrush herbs which cover vast areas of interior Oregon and Washington does not reach tributaries of the Columbia River. The lack of permanent drainage in arid northwest Africa has also been cited as the reason for the depauperate pollen content offshore (Koreneva, 1971).

Vertical distribution

The stratigraphic ranges of some major pollen groups are indicated in Table II.

Published studies of pre-Quaternary palynomorphs in cores from the present oceans are rare. Organic lutites from several cores in the Bahama Abyssal Plain which contained a rich Cretaceous pollen assemblage were described by Habib (1968), who also identified Cretaceous spores in cores from the North Atlantic off West Africa and north of Surinam, South America. Williams and Brideaux (1972) used distinctive Tertiary pollen assemblages to zone and correlate cores from the Grand Banks, Newfoundland. Pollen and spores of Tertiary age have also been recorded from the shelf off New Zealand, the Aleutian Abyssal Plain, the Black Sea, and the Gulf of Mexico.

Spores and pollen in Mesozoic and Tertiary marine sediments have been used primarily to determine the age of the polliniferous intervals. In the Gippsland Basin of southeastern Australia, late Cretaceous, Paleocene, and Eocene biostratigraphic zones based on spores and pollen were used by Stover and Evans (1973) and Stover and Partridge (1973) to correlate more than 40 offshore wells on the continental shelf. The zonation, an essentially

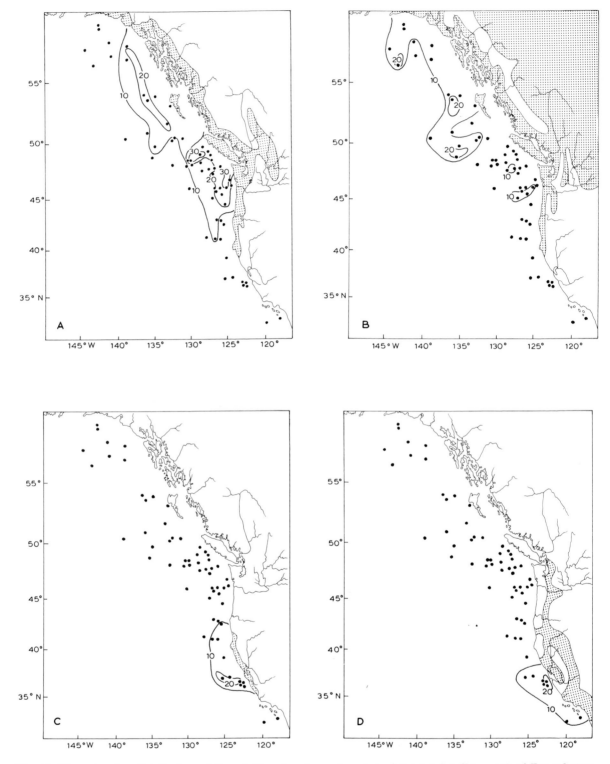

Fig. 11. Maps showing distribution of the relative abundance (percent of the total pollen sum) of *Tsuga hetero-phylla* (A), *Picea* (B), *Sequoia* (C) and *Quercus* (D). Geographic extent of *T. heterophylla, Picea, Sequoia* and *Quercus* species from which pollen in marine sediments may be derived is indicated by shading.

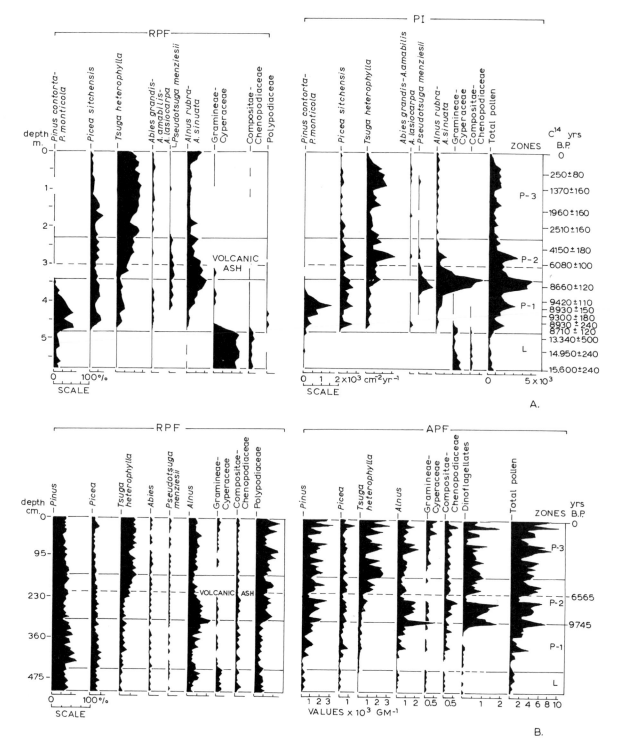

Fig. 12. A. Relative pollen frequency (RPF) and pollen influx (PI) diagrams of principal taxa in core HV-67 from the Hoh River Valley, western Olympic Peninsula. B. Relative pollen frequency (RPF) and absolute frequency (APF) diagrams of principal taxa in core 63-013 from the continental slope, northeast Pacific Ocean. (From Heuser and Florer, 1973.)

continuous, gradually changing sequence of pollen assemblages, was based on the changes of about 150 species and on the internal consistency of the palynomorph assemblages in each zone. Correlation of offshore wells with wells drilled in coals, interbedded sandstones, clays, and volcanics onshore was based on palynomorph stratigraphy.

Pollen analyses from a site drilled in the Ross Sea, Antarctica, suggest that southern beech forests similar to that presently growing in cool temperate South America and New Zealand persisted in the area through the early phases of glaciation in the late Oligocene (Kemp and Barrett, 1975). The Antarctic pollen assemblage is not unlike that from the Ninety-East Ridge in the Indian Ocean which is interpreted as reflecting flourishing island floras which were colonized by long distance dispersal mechanisms (Kemp and Harris, 1975).

Pre-Quaternary and Quaternary spores and pollen have been used as biostratigraphic tools. Quaternary palynomorphs can also be used as paleoclimatologic tools. Our initial studies of chronologically controlled cores from the northwest Atlantic and northeast Pacific suggest that marine pollen assemblages, at least for the past 70,000 years, are correlative with terrestrial pollen assemblages in the region from which the pollen is derived. Comparison of the pollen assemblages in a bog from western Washington (Fig. 12A) with those from a core on the continental slope offshore (Fig. 12B) shows the same sequence of pollen assemblage zones.

The late glacial vegetation of the western Olympic Peninsula (Fig. 12A) is composed primarily of grass (Gramineae) and sedge (Cyperaceae), zone L. Minor amounts of pine and herbs are present, and pollen influx (pollen grains/cm^2 per year) is low (less than 1000 grains/cm^2 per year). Arboreal succession during climatic amelioration is reflected in the pollen diagrams by the successive peaks in the profiles of pine, spruce, alder, and hemlock. The postglacial rise in pollen influx (to 5000 grains/cm^2 per year) is interpreted as reflecting increased pollen production during mild interglacial climate.

The concentration (pollen grains/g) of pollen in late glacial marine sediments is also low, less than 1000 grains/g. Grass and sedge are relatively more abundant in the lower part of the core (zone L) and successive peaks occur in the profiles of pine, spruce, alder, and hemlock, as in the core from continental Washington. An abrupt increase in the density of pollen ca. 10,000 years B.P. reflects the rise in continental pollen influx.

The profile of pine in zones P-2 and P-3 of the marine core differs from that of the same zones in the terrestrial core. In the Hoh River Valley pine decreases as the spruce—hemlock forest develops, whereas pine pollen remains relatively constant (approximately 30%) in the marine sediments. The higher amount of pine in zone P-2 and P-3 of core 63-013 may be due to the influx of pollen derived from the postglacial pine forest steppe vegetation of the Columbia River Basin, the principal source of sediments in the core, as well as to the selective effects of marine dispersion and destruction.

As pollen in a given level of a marine core is probably contemporaneous with other microfossils in that level, marine pollen provides a direct means of correlating marine and continental cores. In cores from the continental slope off western Washington state, the glacial—interglacial boundary interpreted from the change in pollen composition and from the abrupt increase in the frequency of pollen is the same as that determined by the sharp increase in dinoflagellates and from the change in the planktonic foraminiferan/ radiolarian ratio (Barnard, personal communication). Analysis of pollen, radiolaria, and oxygen isotopes in a core from the Pacific Ocean off southern Oregon shows that the response of these three variables to climatic change over the past 80,000 years is generally synchronous (Heusser and others, 1975). The continuous marine pollen record is correlative with pollen records of western Washington covering the same time interval.

Marine pollen, used with discretion, can provide information regarding the vegetation of the terrestrial source. The relative continuity of the marine pollen record is particularly important in areas where terrestrial records are interrupted, as in glaciated regions, or where the vegetation record is absent. In the Pacific Northwest, as in many parts of the world, relatively little information concerning early Pleistocene vegetation is available. First order analyses (40 cm sampling intervals) of a core from the Oregon slope (DSDP Site 175) suggest that pollen assemblages throughout the last 920,000 years are not unlike those of the last 30,000 years.

SUGGESTIONS FOR FURTHER READING

Faegri, K. and Iversen, J., 1975. *Textbook of Pollen Analysis.* Hafner, New York, N.Y., 295 pp. [New edition of the compact, information-filled basic palynology text. Includes keys to pollen identification.]

Kapp, R.O., 1969. *How to know Pollen and Spores.* Brown and Co., Dubuque, Iowa, 249 pp. [Inexpensive, illustrated key to major pollen and spore groups.]

Manton, A.A. (Editor), 1966. Marine palynology. *Mar. Geol.,* 4(6): 395—582. [Contains general articles as well as specific studies from various parts of the world.]

Tschudy, R.H. and Scott, R.A., 1969. *Aspects of Palynology.* Wiley, New York, N.Y., 510 pp. [An excellent compilation by fourteen authorities on various aspects of palynology ranging from the role of pollen and spores in the plant kingdom to the geologic record of palynomorphs.]

REFERENCES CITED

Baker, F.T., 1973. Distribution and composition of suspended sediment in the bottom waters of the Washington continental shelf and slope. *J. Sediment. Petrol.,* 43: 812—821.

Cross, A.T., Thompson, G.C. and Zaitzeff, J.B., 1966. Source and distribution of palynomorphs in bottom sediments, southern part of Gulf of California. *Mar. Geol.,* 4: 467—524.

Davis, R.B. and Webb III, T., 1975. The contemporary distribution of pollen in eastern North America: a comparison with vegetation. *Quaternary Res.,* 5: 395—434.

Dyakowska, J., 1948. The pollen rain on the sea and on the coast of Greenland. *Bull. Acad. Polon. Sci. Lettr.,* Sér. B, 1: 25—33.

Erdtman, G., 1969. *Handbook of Palynology.* Hafner, New York, N.Y., 486 pp.

Florer, L.E., 1972. Quaternary paleoecology and stratigraphy of the sea cliffs, western Olympic Peninsula, Washington. *Quaternary Res.,* 2: 202—216.

Franklin, J.F. and Dyrness, C.T., 1969. Vegetation of Oregon and Washington. *U.S. Dep. Agric. Forest Serv. Res. Pap.* PNW-80, 216 pp. (Portland, Ore.).

Gray, J., 1965. Techniques in palynology. In: B. Kummel and D. Raup (Editors), *Handbook of Paleontological Techniques.* Freeman and Co., San Francisco, Calif., pp. 471—708.

Groot, J.J., 1966. Some observations of pollen grains in suspension in the estuary of the Delaware River. *Mar. Geol.,* 4: 409—416.

Habib, D., 1968. Spores, pollen and microplankton from the Horizon Beta Outcrop. *Science,* 162: 1480—1481.

Hansen, H.P., 1947. Postglacial forest succession, climate, and chronology in the Pacific Northwest. *Trans. Am. Philos. Soc.,* 37: 130 pp.

Hansen, H.P., 1949. Pollen content of moss polsters in relation to forest composition. *Am. Midl. Nat.,* 42(2): 473—479.

Harlett, J.C. and Kulm, L.D., 1973. Suspended sediment transport on the northern Oregon continental shelf. *Bull. Geol. Soc. Am.,* 84: 3815—3826.

Havinga, A.J., 1964. Investigation into the differential corrosion susceptibility of pollen and spores. *Pollen Spores,* 6: 621—635.

Heusser, C.J., 1969. Modern pollen spectra from the Olympic Peninsula, Washington. *Bull. Torrey Bot. Club,* 96: 407—417.

Heusser, C.J., 1973. Modern pollen spectra from Mt. Ranier, Washington. *Northwest Sci.,* 47: 1—8.

Heusser, C.J. and Florer, L.E., 1973. Correlation of marine and continental Quaternary pollen records from the northeast Pacific and western Washington. *Quaternary Res.,* 3: 661—670.

Heusser, L.E., Shackleton, N.J., Moore, T.C. and Balsam, W.L., 1975. Land and marine records in the Pacific Northeast during the last glacial interval. *Geol. Soc. Am., Abstr. Progr.,* 7: 1113—1114.

Hopkins, J.S., 1950. Differential floatation and deposition of coniferous and deciduous tree pollen. *Ecology,* 31: 633—641.

Kemp, E.M. and Barrett, P.J., 1975. Antarctic glaciation and early Tertiary vegetation. *Nature,* 258: 507—508.

Kemp, E.M. and Harris, W.K., 1975. The vegetation of Tertiary islands on the Ninety-East Ridge. *Nature,* 258: 303—307.

Koreneva, E.V., 1957. Spore-pollen analysis of bottom sediments of the Sea of Okhotsk. *Tr. Inst. Okeanol. Akad. Nauk S.S.S.R.,* 22: 221—251.

Koreneva, E.V., 1964. Distribution and preservation of pollen in sediments in the western part of the Pacific Ocean. *Tr. Geol. Inst. Akad. Nauk S.S.S.R.,* 109: 1—88.

Koreneva, E.V., 1971. Spores and pollen in Mediterranean bottom sediments. In: B.M. Funnell and W.R. Riedel (Editors), *The Micropaleontology of Oceans.* Cambridge University Press, New York, N.Y., pp. 361—371.

Lubliner-Mianowska, K., 1962. Pollen analysis of the surface samples of bottom sediments in the Bay of Gdansk. *Acta Soc. Bot. Polon.,* 31: 305—312.

McAndrews, J.H. and Power, D.M., 1973. Palynology of the Great Lakes, the surface sediments of Lake Ontario. *Can. J. Earth Sci.,* 10: 777—792.

Muller, J., 1959. Palynology of recent Orinoco delta and shelf sediments. *Micropaleontology,* 5: 1—32.

Peck, R., 1973. Pollen budget studies in a small Yorkshire catchment. In: H.J.B. Birks and P.G. West (Editors), *Quaternary Plant Ecology.* Wiley, New York, N.Y., pp. 43—60.

Rossignol-Strick, M., 1973. Pollen analysis of some sapropel layers from the deep-sea floor of the eastern Mediterranean. *Init. Rep. Deep Sea Drill. Proj.,* XIII: 971—991.

Sangster, A.G. and Dale, H.M., 1964. Pollen grain preservation of under-represented species in fossil spectra. *Can. J. Bot.,* 42: 437—449.

Stanley, E.A., 1966a. The application of palynology to oceanography with reference to the northwestern Atlantic. *Deep-Sea Res.,* 13: 921—939.

Stanley, E.A., 1966b. The problem of reworked pollen and spores in marine sediments. *Mar. Geol.,* 4: 397—408.

Stanley, E.A., 1969. Marine Palynology. *Oceanogr. Mar. Biol. Annu. Rev.,* 7: 277—292.

Stover, L.E. and Evans, P.R., 1973. Upper Cretaceous—Eocene spore-pollen zonation, offshore Gippsland Basin, Australia. *Geol. Soc. Aust. Spec. Publ.,* 4: 55—72.

Stover, L.E. and Partridge, A.D., 1973. Tertiary and late Cretaceous spores and pollen from the Gippsland Basin, southeastern Australia. *R. Soc. Vic. Proc.,* 85(2): 237—286.

Thompson, D., 1972. *Paleoecology of the Pamlico Formation, St. Mary's County, Maryland.* Thesis, Rutgers University, New Brunswick, N.J.

Traverse, A., 1974. Palynologic investigation of two Black Sea cores. *AAPG, Mem.,* 20: 381—388.

Traverse, A. and Ginsburg, R.N., 1966. Palynology of the surface sediments of the Great Bahama Bank, as related to water movement and sedimentation. *Mar. Geol.,* 4: 417—459.

Williams, G.L. and Brideaux, W., 1972. Palynologic analyses of cored sediments from the Grand Banks, Newfoundland. *Am. Assoc. Stratigr. Palynol., Abstr. 3rd Meet.,* 136.

15

CHITINOZOA

J. JANSONIUS and W.A.M. JENKINS

INTRODUCTION

The Chitinozoa are a group of extinct microscopic animals whose systematic position is not known. They have hollow, organic-walled tests that are radially symmetrical about a central longitudinal axis and closed at one end; in short they look like minute bottles. Individual tests measure 50–2000 μm in length. They were named after the chitinoid appearance of their tests by their discoverer Alfred Eisenack.

The Chitinozoa evolved rapidly during the Ordovician, Silurian, and Devonian periods. They were exclusively marine and formed an important part of the small animal life in the oceans of the time. They are found today in organic residues from virtually all types of marine sedimentary rock and consequently are useful in stratigraphy for dating and correlating rocks of Ordovician, Silurian and Devonian age.

HISTORY OF STUDY

The Chitinozoa were named and first formally described by Eisenack in 1931, who established a system for classifying chitinozoans that is based upon morphology and does not necessarily reflect natural or evolutionary relationships within the group. This system of classification is widely used today.

For over forty years Eisenack continued to publish studies of Ordovician and Silurian chitinozoans from the Baltic region. In the 1950's, a series of papers described Devonian chitinozoans from Brazil and the United States.

In the early 1960's, the chitinozoan literature began to grow rapidly, and French paleontologists became the dominant contributors.

Much of their work was on subsurface material obtained in the course of exploring for oil and gas in North Africa. Research and publication continued to increase through the 1960's as a consequence of the petroleum industry's growing interest in chitinozoans as a useful means of interpreting subsurface geology particularly in the search for oil and gas, and their value in Lower Paleozoic stratigraphy became widely recognized.

In the late 1960's and 1970's studies were published detailing the stratigraphical ranges of chitinozoan species in Ordovician, Silurian and Devonian type sections, and other reliably dated outcrop sections in Europe and the Americas.

MORPHOLOGY OF THE CHITINOZOAN TEST

The chitinozoan test consists of two main parts, the **chamber** and the **oral tube**, which generally can be distinguished easily (Fig. 1).

Chamber

For descriptive purposes four parts of the chamber have been distinguished; these are the **shoulder, flanks, basal margin**, and **base**, the terms being more or less self-explanatory. The basal margin assumes a greater diversity of form than other parts of the chamber and is important in classification at both generic and specific levels. Three types of structure on the basal margin are:

(1) The **carina** (Fig. 2): a sharp outward extension from the chamber wall resembling a skirt or the brim of a hat.

(2) The **siphon** (Fig. 3): a hollow, open-ended tube extending from the aboral end. It is developed in some chitinozoans with two-layered test walls by the separation of the

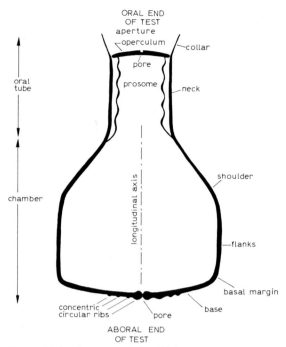

Fig. 1. Major features of the chitinozoan test.

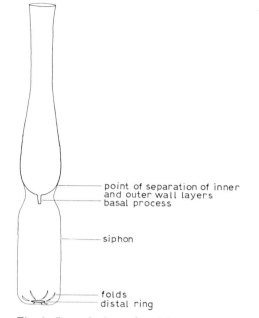

Fig. 3. Lateral view of a siphonate chitinozoan (diagrammatic). × 200.

Fig. 2. Types of carinae (stippled) in lateral view. × 125.

Fig. 4. Types of appendices (stippled) in lateral view. × 125.

inner and outer layers at the basal margin, to form, respectively, the base and the siphon.

(3) **Appendices** [sing. **appendix**] (Fig. 4): generally discrete processes suspended from the basal margin. They rise from the same part of the chamber as the carinae and siphons of other tests and encircle the base. Occasionally they coalesce distally. Appendices are hollow but their interiors do not open into the body cavity.

The base is the aboral surface of the chamber. The term "base" does not refer to the original life position and direction of growth of the living organism, which are not known. Orientation of the test is a matter of choice, although most illustrations show the wider part directed downward and the large opening uppermost. The base may be convex,

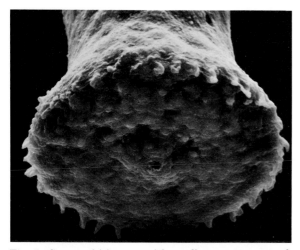

Fig. 5. *Coronochitina* sp., with small pore at center of base and many small appendices, in oblique aboral view. × 900. Ordovician, Britain.

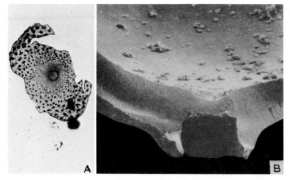

Fig. 6. A. Fragment of *Conochitina micracantha* Eisenack, in aboral view showing sealed pore at center of base. × 200. B. Fragment of a chitinozoan in lateral view showing, in section, the pore sealed off from the interior of the test by an inner layer of the test wall; a depression on the inner surface of the test marks the position of the pore. × 1000. Specimens from the subsurface Upper Ordovician of Mississippi and the Lower Devonian of New York, respectively.

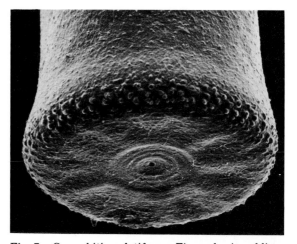

Fig. 7. *Conochitina latifrons* Eisenack, in oblique aboral view showing concentric ribbing around a central pore. × 1000. Upper Silurian, Gotland.

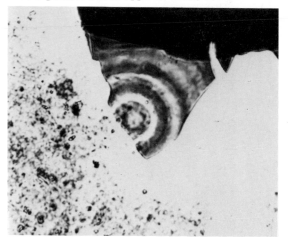

Fig. 8. Fragment of base in polar view showing concentric ribbing around a central pore. × 500. Ordovician, Arabia.

flat, concave or hemispherical, and may bear a stout process (see Fig. 33). In mature specimens of some species, the center of the base is perforated by a fine pore (Figs. 5—8), which may be closed on the inside by an inner layer of the test wall (Fig. 6). Concentric circular ribs often surround the pore (Figs. 7—8) and are thought to result from contact with the aperture of another test when, at some stage in the animal's life cycle, the test formed part of a chain of tests joined aperture-to-base. The pores may at some stage have provided for communication between the interiors of adjacent tests in chains.

Oral tube

The oral tube has two parts, the **collar** and the **neck** (Fig. 1). They may be distinguished from each other by a change in thickness of the test wall or by a change in the wall's profile. The collar encircles the **aperture** and sometimes terminates in a fringe of small spines. The neck is a hollow, more or less cylindrical tube, however, few chitinozoans lack necks and collars altogether.

An internal structure, the **prosome**, lies within the neck (Fig. 1). In most chitinozoans it is made up of a series of rings or discs and terminates orally in a disc-like **operculum** which may have sealed the contents of the test from the outside (Fig. 9). In some genera, however, particularly those without necks,

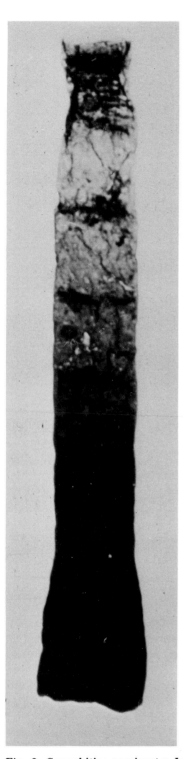

Fig. 9. *Conochitina parviventer* Jenkins, showing the rings or discs that make up the prosome, and the dark operculum sealing the aperture, in lateral view. × 450. Ordovician, Britain.

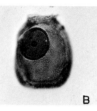
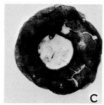
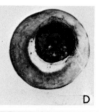

Fig. 10. Chitinozoan operculum. A, B. Two examples of *Desmochitina minor* Eisenack, in lateral view showing the operculum fallen into the chamber, × 250. C, D. Lenticular chitinozoans in oral view, showing (C) open aperture after the operculum has been lost, and (D) operculum lying partly across the aperture, × 200. E. Lenticular chitinozoan in oblique oral view showing operculum firmly in place. × 600. A, B and D from the Upper Ordovician of Oklahoma; C from the Ordovician of Britain; and E from the Lower Devonian of Pennsylvania.

the prosome lacks the rings or discs and consists solely of an operculum (Fig. 10). Sometimes an operculum shows a central pore and concentric ribbing that reflect the structure of the base to which it was once attached.

Test wall

In well-preserved material the wall of the chitinozoan test is translucent and amber-colored. Carbonized tests recovered from rocks altered by metamorphism are opaque, black and brittle.

Chitinozoan species with one, two and three wall layers are known (Laufeld, 1974) although in most species the wall is made up of two layers. The outer surface of the wall may be smooth or ornamented (Fig. 11).

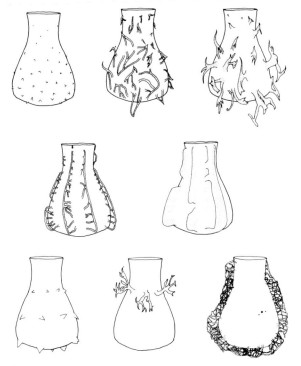

Fig. 11. Types of ornament (stippled) in lateral view. × 125.

Ornament generally is less strongly developed on the neck than on the chamber but there are important exceptions. Ornamental processes often stand in rows parallel with the length of the test (see Figs. 22, 28). Small ornamental processes are solid or have "spongy" interiors, whereas large processes

are largely hollow. Like appendices, large ornamental processes do not open into the body cavity, and the surface of the test lying within the base of a large process does not differ from the adjacent surface outside the base of the process. These observations have led Laufeld (1974) to conclude that the test wall was formed before the appendices or spines were secreted on it. Processes may have served to control buoyancy or as a means of attachment.

OCCURRENCE OF CHITINOZOAN TESTS

Most chitinozoans occur in the rock as single tests (Fig. 12A). Frequently, however, two or more tests are found joined together in linear chains (Fig. 12B). Such chains are known in all the common genera, though in some they may be rare. It is believed that the capacity to form chains is a fundamental chitinozoan characteristic, although most species have not yet been found preserved in this condition. In some species chains are very long and form spiral coils (Fig. 12C). Occasionally, the whorls of a spiral chain comprising several hundred tests have been found packed tightly together and enclosed in an organic pellicle called a **cocoon** (Fig. 12D).

Collection of samples

The concentration of chitinozoans in sediments generally is less than 20 tests per gram of rock and may be less than one test per gram. The greatest recorded concentration is 984 tests per gram in Silurian rocks on Gotland. Comparable figures for other organic-walled microfossils, such as acritarchs and spores, could be in the thousands per gram. Consequently, rock samples collected specifically for chitinozoan studies should be about 500 g.

Preparation of samples

Chitinozoans are prepared for microscopy in much the same way as other organic-walled microfossils. Care must be taken not to damage the relatively large and brittle tests which often bear delicate ornamentation. The

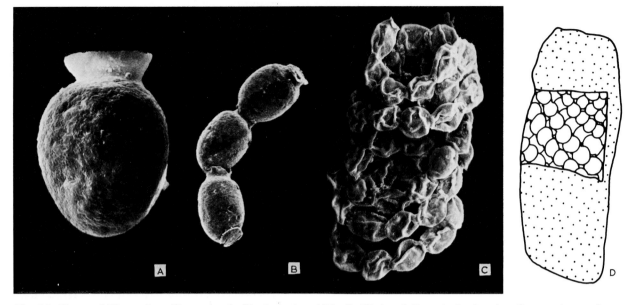

Fig. 12. *Desmochitina minor* Eisenack. A. Single test. × 400. B. Chain of three tests showing the apertures of two tests applied to the bases of their orally adjacent neighbours; the operculum of a fourth test remains firmly attached to the base of the test at lower left. × 160. C. Long spiral chain consisting of numerous tests. × 100. Ordovician, Britain. D. Cocoon (stippled).

rock sample should be broken into fragments roughly the size of centimeter cubes. The fine powder that results from fragmenting the sample usually contains a high proportion of broken, unidentifiable chitinozoans and is better discarded if quantitative studies are contemplated. After treatment with hydrochloric and hydrofluoric acids (respectively, to dissolve the carbonate minerals and to disaggregate and remove the clay minerals and quartz), chitinozoans can be further concentrated by sieving.

Chitinozoans should be mounted as nearly as possible in a single focal plane. For transmitted light microscopy a two-layer permanent mounting technique is preferable, by which the tests are closely applied to the cover slip and firmly held in the same plane.

For details of preparation techniques for Chitinozoa, see Jansonius (1970).

CLASSIFYING CHITINOZOANS

Classification within the Chitinozoa is based largely upon the gross morphology of the test. The genera so defined are arbitrary, artificial groupings, not necessarily reflecting the natural or evolutionary relationships between different kinds of chitinozoans. The original shape of a chitinozoan test requires careful interpretation because most tests have, to some extent, been distorted by compression during compaction of the sediment. Other diagenetic changes, particularly mineral recrystallization, also may alter the shapes of chitinozoan tests; and the various chemical treatments used to free chitinozoans from rock fragments may be responsible for slight changes in both shape and size.

Several characteristics used in defining genera are:

(1) The tendency for tests to occur in chains.
(2) The overall shape of the test.
(3) The nature of the basal margin. About one-third of the chitinozoan genera are based on this character.
(4) Internal structure. These may be more significant phylogenetically than is the overall shape of the test, but frequently they are not visible through the opaque test wall.
(5) Size.

Species are classified according to minor changes in test shape; according to the style, development and distribution of ornamental processes; according to minor differences in the character of the basal margin and its elaborations (carina, appendices); according to the character of the aperture and collar (for example, whether it flares or ends in a fringe of short spines); and according to the size and shape of the basal process. Size is also used in the diagnosis of species. Infra-red photography can also reveal features useful for classification.

Identification of chitinozoan species or the diagnosis of new species is, in general, relatively straightforward because of the general absence of "immature" forms dispersed in sediments and the fact that chitinozoan assemblages universally consist of relatively few species: about six to eight on average and rarely as many as twelve.

PALEOECOLOGY

Chitinozoans were marine. We know little about the factors that determined the composition of chitinozoan assemblages in different types of sedimentary rock. There is no general agreement as to whether chitinozoans were benthic or planktonic. On the one hand, their wide geographical distribution, rapid dispersal across northern Europe and North America, and occurrence in most types of sedimentary rock suggest that they were planktonic. On the other hand, their often very thick test wall and evidence that some forms were attached to the substrate suggest they were benthic. It is possible, however, that some chitinozoans were planktonic while others were benthic, or that the chitinozoan life cycle involved planktonic and benthic stages.

Recently, Laufeld (1974) has shown that the abundance of chitinozoans in the Silurian rocks of Gotland is inversely proportional to the quantity of calcium carbonate in the rock; and that chitinozoans were not adapted to life on reefs or in the turbulent waters associated with them. Many environmental factors, such as water temperature and salinity, undoubtedly affected the numbers of chitinozoans in Paleozoic seas. Sedimentary processes such as sorting also would have affected the numbers and the size (and therefore the number of taxa) of chitinozoans accumulating in sediments.

THE SYSTEMATIC POSITION OF THE CHITINOZOA

What is the systematic position of the Chitinozoa? For nearly half a century little progress has been made towards answering this question. Eisenack (1931) suggested that chitinozoans might be related to protozoans of the rhizopod order Testacea, which contains species with comparable chitinoid tests. Eisenack acknowledged, however, that members of the Testacea live today mainly in fresh water and that their tests are soluble in potassium hydroxide whereas those of chitinozoans are not. A year later (1932) he noted that the cellulose test and exclusively fresh-water habitat of the testacean genus *Trachelomonas* opposed any relationship between those flagellate protozoans and chitinozoans.

In 1955 Collinson and Schwalb considered the Chitinozoa to be an extinct order of marine protozoans, which they referred to the class Rhizopoda on account of the thick pseudochitinous tests and marine habitat.

Kozlowski (1963), however, was more hopeful and thought it unlikely that there could have existed during a large part of the Paleozoic Era (Ordovician to Devonian) organisms representing a type of animal totally unknown today. He therefore attempted to interpret the biological affinities of the Chitinozoa on the premise that clusters of tests in cocoons were hermetically sealed and that tests, therefore, corresponded to cysts or eggs. He noted that nothing like clusters of chitinozoans in a common envelope was known in the Protozoa and therefore rejected a relationship with this phylum.

Considering next the possibility that chitinozoans were metazoans, Kozlowski concluded that chitinozoan tests were morphologically and structurally more complicated than eggs of recent marine invertebrates and only remotely analogous to the eggs and egg capsules of existing animals.

In 1970, Jenkins suggested that the problem of chitinozoan affinities could be approached from two directions. Firstly, by a direct zoological approach one could attempt to relate the Chitinozoa to another group of living or fossil animals on the basis of anatomy, morphology and chemical composition. Secondly, one could study the fossil associations of chitinozoans, noting with what other groups of fossils chitinozoans were consistently found.

Following the second approach, Jenkins was struck by the association between chitinozoans and graptolites which suggested the existence of a genetic or ecological relationship between these two groups. He noted that an abundance of one was associated with an abundance of the other and, conversely, that strata containing no chitinozoans contained no graptolites. He admitted, however, that the techniques involving hydrochloric and hydrofluoric acids used to recover chitinozoans from sedimentary rocks could exaggerate the frequency of this association, particularly at the expense of associations between organic-walled fossils and those composed of acid-soluble mineral material.

The stratigraphic ranges of chitinozoans and graptolites correspond to a remarkable degree. In addition, the chitinozoan species at a given horizon almost invariably number from three to twelve and average about seven or eight. These figures correspond approximately to those for graptolites at productive horizons where a thorough search had been made for macrofossils.

The hypothesis that graptolites and chitinozoans were related genetically does not conflict with the limited available biochemical data. Analyses of graptolites (Foucart, in Kozlowski, 1966) and chitinozoans (Voss-Foucart and Jeuniaux, 1972) have shown that neither group of fossils contains chitin.

Graptolite reproduction is incompletely understood and an early part of the life cycle, following sexual reproduction, has not been recorded in the fossil record. The "missing" stage presumably precedes the prosicula which usually is less than 500 μm in length. If chitinozoans were indeed the pre-prosicular stages of graptolites it remains for future research to show how graptolites produced chitinozoans and how the latter gave rise to prosiculae.

EVOLUTION AND BIOSTRATIGRAPHY

Because of their abundance, rapid evolu-

tion and wide-spread occurrence in most types of sedimentary rocks, chitinozoans afford a means of correlation that is more frequently available, particularly in subsurface work, than do the relatively scarce and more facies-dependent macrofossils. Their scope in stratigraphy is even greater because they sometimes remain identifiable in strongly metamorphosed rocks.

Currently available information about the vertical distribution of chitinozoans permits their use in regional and intercontinental correlation at about the series level. In areas where the chitinozoan succession has been studied in detail, a much finer stratigraphic subdivision is generally possible.

The major changes in the Ordovician to Devonian chitinozoan succession are summarized in Fig. 35. They were not merely local events, but took place in the same order and at approximately the same time across an area which now extends from the southwestern United States to the eastern Baltic and southward into South America and North Africa. The Ordovician succession is described in greater detail than the Silurian and Devonian successions because it is more fully documented in terms of precisely dated sedimentary sequences.

In the Lower Ordovician chitinozoans are smooth-walled and often very large. A trend toward smaller size may be traced from the base of the system to the top.

Very early in the group's evolutionary history the siphon appeared. Early Ordovician faunas of Arenig to early Llanvirn age are often distinguished by chitinozoans with large siphons (Fig. 13). Thereafter, the structure decreased in size and became relatively rare. It persisted into early Caradoc time.

The carina also first appeared in the early Ordovician and carinate chitinozoans (Fig. 14) rapidly became numerically important elements in the fauna. Whereas siphonate chitinozoans became extinct in the middle Ordovician, carinate forms thrived and diversified moderately throughout the Ordovician Period. Chitinozoans with elaborate networks suspended from the basal margin were common in the Arenig but died out during the early Ordovician. Large, smooth-walled, cylindrical and elongate conical chitinozoans up to

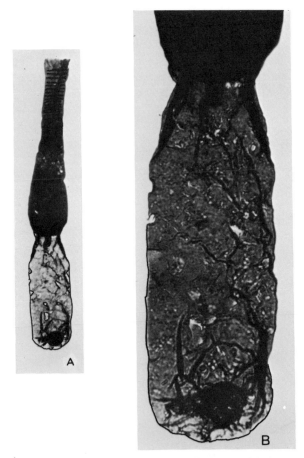

Fig. 13. *Siphonochitina formosa* Jenkins, with large siphon. A. Lateral view. × 200. B. Detail of siphon. × 650. Ordovician, Britain.

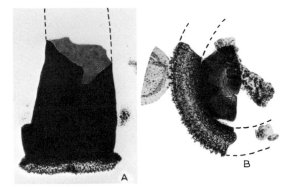

Fig. 14. *Anthochitina* sp., with wide spiny carina. × 150. A. Lateral view of incomplete test. B. Aboral view of incomplete test. Ordovician, Arabia.

1000 µm in length were common in the early and middle Ordovician becoming rare in the late Ordovician and early Silurian.

In the middle Ordovician, perhaps in Llandeilo time, a distinctive group of

lenticular chitinozoans appeared which, because they are wider than long, generally are seen in polar view (see Figs. 10C, D). The aperture is sealed by a disc-like operculum. Lenticular chitinozoans enclosed in translucent outer membranes also first occur at this time. Chitinozoans of this general type continued through the Ordovician, Silurian and Devonian and are found in the youngest known (Famennian—Strunian) chitinozoan assemblages. Also in the middle Ordovician, and possibly earlier, there appeared small, roughly spherical chitinozoans commonly joined together in chains. By Caradoc time, these had universally become important elements in chitinozoan assemblages.

During the latter part of the Ordovician Period chitinozoans developed some very elaborate features. Early in Caradoc time, the carina developed into a wide, flaring, skirt-like, translucent membrane, which is found more or less consistently throughout the Caradoc—Ashgill succession (Fig. 15). Shortly after this, during the middle Caradoc, species bearing appendices appeared at approximately the same time in North America, Britain and the Baltic region. At first the appendices were tiny processes (see Fig. 5) but they quickly became larger and more elaborate (Figs. 16, 17). Appendix-bearing chitinozoans strikingly distinguish most late Ordovician faunas from faunas occurring

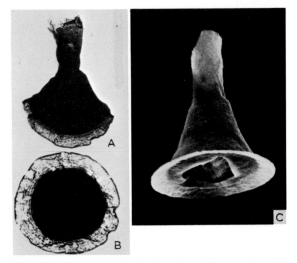

Fig. 15. *Cyathochitina kuckersiana* (Eisenack), with wide skirt-like carina. × 160. A. Lateral view. B. Aboral view showing concentric striations on carina. C. Oblique aboral view. Ordovician, Britain.

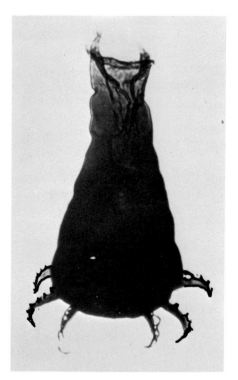

Fig. 16. *Ancyrochitina alaticornis* Jenkins, with perforate appendices, in lateral view. ×400. Ordovician, Britain.

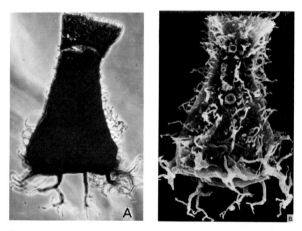

Fig. 17. *Ancyrochitina onniensis* Jenkins, with branching appendices and ornament of λ-spines, in lateral view. ×700. Ordovician, Britain. (A, phase-contrast illumination; B, scanning electron micrograph.)

Fig. 20. *Angochitina dicranum* Jenkins, with many long distally branching spines, in lateral view. ×750. Ordovician, Britain. ➝

18   19

Fig. 18. ?*Conochitina* sp., with minute, simple spines (>2 μm in length), in lateral view. ×200. Ordovician, Arabia.

Fig. 19. *Conochitina* sp., with many short simple spines, in lateral view. ×250. Ordovician, Britain.

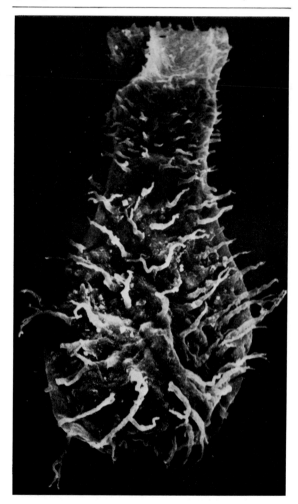

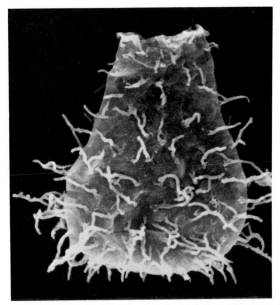

Fig. 21. *Kalochitina hirsuta* (Laufeld), with many λ-spines, in lateral view. × 600. Ordovician, Britain.

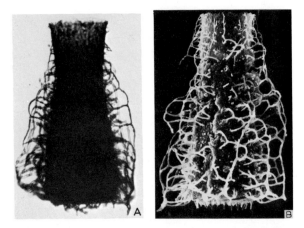

Fig. 22. *Hercochitina downiei* Jenkins, in lateral view showing longitudinal rows of spines connected at their tips by longitudinal bars, in lateral view. A. × 200. B. × 270. Ordovician, Britain.

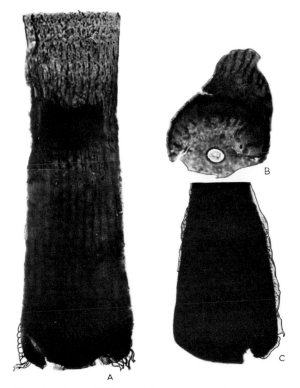

Fig. 23. *Hercochitina crickmayi* Jansonius, showing ornament arranged in longitudinal rows. A. Lateral view showing ornament visible through translucent test wall and in profile. × 325. B. Oblique aboral view showing ornament and pore at center of base. × 325. C. Lateral view showing fins in profile. × 250. A and B from the Upper Ordovician of Anticosti Island, Quebec; C from the Upper Ordovician of Oklahoma.

earlier in the period, but not until the Silurian did they reach their greatest diversity and abundance.

Up to late Caradoc time chitinozoan faunas had consisted very largely of smooth-walled forms. These now gave way to assemblages dominated by strongly and elaborately ornamented species. Small simple spines and small λ-spines are known as far back as the early Caradoc, but only in the late Caradoc did these become larger and more common (Figs. 18, 19). Shortly after distally branching

processes appeared in the late Caradoc (Fig. 20), λ-spines appeared simultaneously in several otherwise very different genera (Figs. 17, 21). There then appeared, also in the late Caradoc, chitinozoans furnished with longitudinal rows of spines connected at their tips by longitudinal bars (Fig. 22). In North America, chitinozoans with longitudinal ribs or fins (Fig. 23) are first known in Edenian (mid-Caradoc) rocks but have been recorded in somewhat older beds (basal Caradoc) in Sweden and Estonia. They are most common in Ashgill.

A striking style of ornament occurs in the latest Ordovician in eastern North America and Europe. It consists of closely spaced upright processes each of which divides into a number of arms. The arms of adjacent processes unite to form a reticulum which stands above the surface of the test wall (Fig. 24).

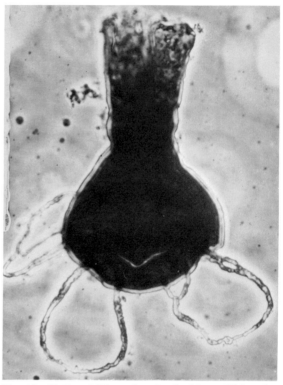

Fig. 25. *Clathrochitina sylvanica* Jenkins, with looping and anastomosing appendices, in lateral view. × 550. Uppermost Ordovician, Oklahoma.

Fig. 24. *Acanthochitina barbata* Eisenack, illustrating ornament (diagrammatic). A. Lateral view of test. × 200. B. Reconstruction of ornament, × 1000.

Acanthochitina barbata Eisenack bears this type of ornament and has a very short stratigraphical range across the Caradoc—Ashgill boundary.

At the very top of the Ordovician in North America, there occur cylindroconical chitinozoans with strongly developed looping and anastomosing appendices suspended from the basal margin (Fig. 25).

Typical Ordovician chitinozoans such as *Siphonochitina* (see Fig. 13), *Sagenachitina*, *Hercochitina* (Figs. 22, 23), *Acanthochitina* (Fig. 24), *Desmochitina minor* Eisenack (see Figs. 10, 12) and most carinate species (see Figs. 14, 15) died out at or just before the

Fig. 26. Two Silurian chitinozoans, *Sphaerochitina sphaerocephala* (Eisenack) (A), and *Angochitina echinata* Eisenack (B), in lateral view. × 425. The shape of the tests is characteristically Silurian. Upper Silurian, Gotland.

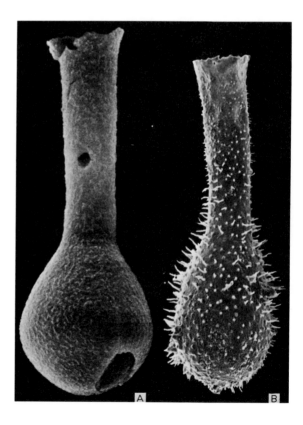

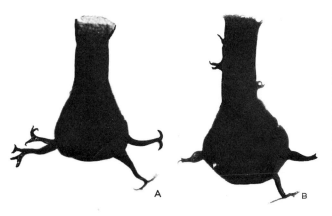

Fig. 27. *Ancyrochitina ancyrea* (Eisenack), in lateral view. × 250. A. Typical example with long, elegant appendices. B. Example with thick processes on neck. Silurian, Mississippi.

end of the period. They were replaced in the Silurian by a variety of species with spheroidal chambers and long necks (Fig. 26). Some of these species bear stout processes on the neck (Figs. 27, 28A), while others have large processes on the chambers arranged in longitudinal rows (Fig. 28C). Also in the Silurian are small cylindroconical species with dense ornaments of short processes (Fig. 29); lenticular chitinozoans with various types of surface texture (Fig. 30); long, thin, cylindrical or elongate conical forms with short, sturdy basal processes; and goblet-shaped species (Fig. 31) that frequently form long chains

Fig. 29. *Eisenackitina philipi* Laufeld, with dense ornament of short simple processes, in lateral view. × 600. Upper Silurian, Gotland.

(Middle Silurian—Lower Devonian). The large, cylindrical and elongate conical chitinozoans, which (see Fig. 35) characterize most Ordovician assemblages, become rarer during the Silurian and do not occur later.

In the Devonian, the numbers and variety of chitinozoans decreased. Typical were species with spheroidal chambers and distinct

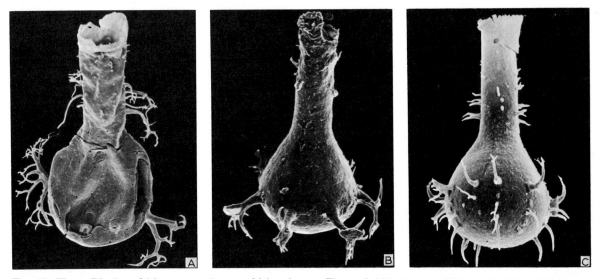

Fig. 28. Three Silurian chitinozoans, *Ancyrochitina desmea* Eisenack (A), *Ancyrochitina pedavis* Laufeld (B) and *Gotlandochitina villosa* Laufeld (C), in lateral view. × 400. The shape of these tests is characteristically Silurian. Upper Silurian, Gotland.

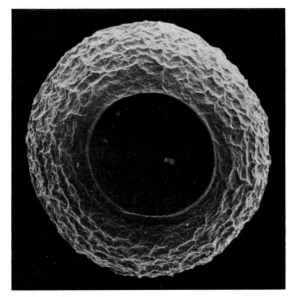

Fig. 30. *Desmochitina acollaris* Eisenack, with a distinctive surface texture and no collar, in oral view. × 1000. Middle Silurian, Gotland.

Fig. 31. *Margachitina margaritana* (Eisenack), in lateral view. × 600. Silurian, Mississippi.

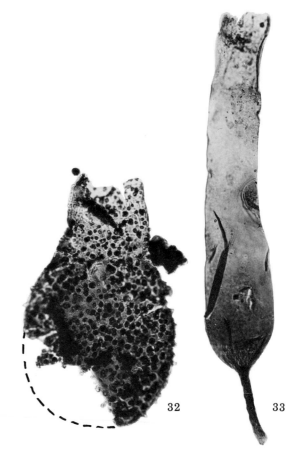

Fig. 32. *Angochitina* sp., covered with small nodes, in lateral view. × 250. Devonian, Mississippi.

Fig. 33. *Urochitina* sp., with large basal process, in lateral view. × 250.

necks, some covered with small nodes (Fig. 32); others with large coarsely branching spines; cylindroconical forms with well-developed processes confined to the basal margin, the collar, and sometimes the shoulders; cylindroconical species with strongly developed anastomosing appendices; species with relatively tall, slender chambers often forming long chains; lenticular chitinozoans (often relatively abundant); and elongate forms with long basal processes (Fig. 33).

Over short vertical intervals such as stages, chitinozoan faunas change more subtly than they do over the much greater intervals of the Ordovician, Silurian and Devonian systems. Populations of chitinozoan species from continuous sedimentary sequences sometimes reveal evolutionary lineages not unlike those in the Foraminifera. One such lineage is summarized in Fig. 34. It is from the lower Viola Limestone of southern Oklahoma and in chronostratigraphic terms it runs for about half a stage.

Evolutionary changes in the chitinozoan test from the Ordovician through the Devonian are summarized schematically in Fig. 35 (see pp. 356—357).

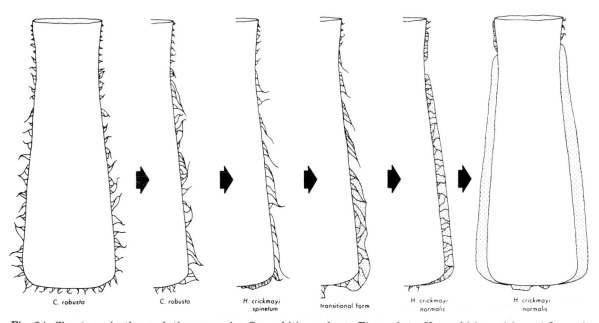

C. robusta   C. robusta   H. crickmayi   transitional form   H. crickmayi   H. crickmayi
                                        spinetum   normalis   normalis

Fig. 34. Six stages in the evolutionary series *Conochitina robusta* Eisenack to *Hercochitina crickmayi* Jansonius forma *spinetum* and *H. crickmayi* forma *normalis* present in the Ordovician, Oklahoma (diagrammatic). × 240.

SUGGESTIONS FOR FURTHER READING

Cramer, F.H., 1964. Microplankton from three Palaeozoic formations in the province of León, NW Spain. *Leidse Geol. Meded.*, 30: 253—261 (pls. 1—24). [Describes Devonian chitinozoans from Europe.]

Eisenack, A., 1968. Über Chitinozoen des baltischen Gebietes. *Palaeontographica*, 131A: 137—198 (pls. 24—32). [A comprehensive review of chitinozoans with particular reference to those of the Baltic.]

Jansonius, J., 1970. Classification and stratigraphic application of Chitinozoa. *Proc. N. Am. Paleontol. Conv., 1969, G.*, pp. 789—808. [Outlines methods of preparation, and system of classification based on internal and aboral structures. Complete bibliography through 1969.]

Jenkins, W.A.M., 1970. Chitinozoa. *Geosci. Man*, 1: 1—21. [A general introduction to the study of chitinozoans.]

Lange, F.W., 1967. Biostratigraphic subdivision and correlation of the Devonian in the Paraná Basin. In: J.J. Bigarella (Editor), *Problems in Brazilian Devonian Geology. Bol. Parana. Geocienc.*, 21/22: 63—98 (pls. 1—5). [Describes species with short vertical ranges and wide geographical distributions in the Devonian (Emsian—Frasnian) rocks of Brazil.]

Taugourdeau, P. and De Jekhowsky, B., 1960. Répartition et description des Chitinozoaires siluro-dévoniens de quelques sondages de la C.R.E.P.S., de la C.F.P.A. et de la S.N. REPAL au Sahara. *Rev. Inst. Fr. Pét.*, 15: 1199—1260 (pls. 1—13).

Taugourdeau, P., and others, 1967. *Microfossiles organiques du Paléozoique. Les Chitinozoaires (1). Analyse bibliographique illustrée.* Published for the Commission internationale de Microflore du Paléozoique by the Centre National de la Recherche Scientifique, Paris. [Gives worldwide stratigraphic ranges and illustrates holotypes of all known species.]

CITED REFERENCES

Collinson, C. and Schwalb, H., 1955. North American Paleozoic Chitinozoa. *Rep. Invest. Ill. State Geol. Surv.*, 186: 33 pp. (2 pl.).

Combaz, A., Calandra, F., Jansonius, J., Millepied, P., Poumot, D. and Van Oyen, F.H., 1967. *Microfossiles organiques du Paléozoique. Les Chitinozoaires (2). Morphographie.* Published for the Commission Internationale de Microflore du Paléozoique by the Centre National de la Recherche Scientifique, Paris, 43 pp.

Deflandre, G., 1945. Microfossiles des calcaires siluriens de la Montagne Noire. *Ann. Paléontol.*, 31 (1944—45): 39—75 (pls. 1—3).

Eisenack, A., 1931. Neue Mikrofossilien des baltischen Silurs, I. *Paläontol. Z.*, 13: 74—118 (pls. 1—5).

Eisenack, A., 1932. Neue Mikrofossilien des baltischen Silurs, II. *Paläontol. Z.*, 14: 257—277 (pls. 11—12).

Kozlowski, R., 1963. Sur la nature des Chitinozoaires. *Acta. Palaeontol. Pol.*, 8: 425—49.

Kozlowski, R., 1966. On the structure and relationships of graptolites. *J. Paleontol.*, 40: 489—501.

Laufeld, S., 1974. Silurian Chitinozoa from Gotland. *Fossils and Strata, 5.* Universitetsforlaget, Oslo, 130 pp.

Voss-Foucart, M.F. and Jeuniaux, C., 1972. Lack of chitin in a sample of Ordovician Chitinozoa. *J. Paleontol.*, 46: 769—70.

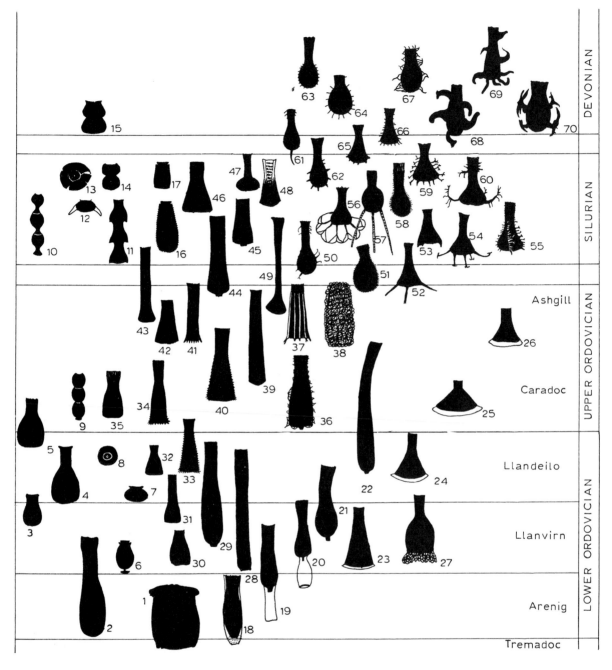

Fig. 35. Illustrating some of the major changes in the Ordovician—Devonian chitinozoan succession. Numbers on figures correspond to the following species:

1. *Ollachitina ingens* Poumot.
2. *Lagenochitina maxima* Taugourdeau and De Jekhowsky
3. *Lagenochitina brevicollis* Taugourdeau and De Jekhowsky
4. *Lagenochitina esthonica* Eisenack
5. *Lagenochitina baltica* Eisenack
6. *Desmochitina minor* Eisenack
7. *Hoegisphaera complanata* (Eisenack)
8. *Hoegisphaera bransoni* Wilson and Dolly
9. *Desmochitina nodosa* Eisenack
10. *Margachitina margaritana* (Eisenack)
11. *Linochitina cingulata serrata* Taugourdeau and De Jekhowsky

12. *Pterochitina perivelata* (Eisenack)
13. *Hoegisphaera glabra* Staplin
14. *Eisenackitina sphaerica* (Taugourdeau and De Jekhowsky)
15. *Eisenackitina bursa* (Taugourdeau and De Jekhowsky)
16. *Eisenackitina oblonga* (Taugourdeau and De Jekhowsky)
17. *Eisenackitina cylindrica* (Taugourdeau and De Jekhowsky)
18. *Siphonochitina veligera* (Poumot)
19. *Siphonochitina copulata* (Poumot)
20. *Siphonochitina formosa* Jenkins
21. *Eremochitina mucronata* Taugourdeau and De Jekhowsky
22. *Conochitina minnesotensis* (Stauffer)
23. *Cyathochitina calix* (Eisenack)
24. *Cyathochitina campanulaeformis* (Eisenack)
25. *Cyathochitina kuckersiana* forma *brevis* Eisenack
26. *Cyathochitina kuckersiana* (Eisenack)
27. *Sagenachitina striata* (Benoit and Taugourdeau)
28. *Rhabdochitina magna* Eisenack
29. *Eremochitina baculata* Taugourdeau and De Jekhowsky
30. *Conochitina simplex* Eisenack
31. *Conochitina primitiva* Eisenack
32. *Conochitina conulus* Eisenack
33. *Belonechitina wesenbergensis* (Eisenack)
34. *Belonechitina micracantha* (Eisenack)
35. *Conochitina aculeata* Taugourdeau
36. *Hercochitina downiei* Jenkins
37. *Hercochitina crickmayi* Jansonius
38. *Acanthochitina barbata* Eisenack
39. *Rhabdochitina hedlundi* Taugourdeau
40. *Belonechitina robusta* (Eisenack)
41. *Coronochitina coronata* (Eisenack)
42. *Conochitina turris* Taugourdeau
43. *Conochitina elegans* Eisenack
44. *Conochitina proboscifera* Eisenack
45. *Conochitina tuba* Eisenack
46. *Conochitina communis* Taugourdeau
47. *Sphaerochitina pistilliformis* (Eisenack)
48. *Sphaerochitina vitrea* Taugourdeau
49. *Sphaerochitina longicollis* Taugourdeau and De Jekhowsky
50. *Angochitina filosa* Eisenack
51. *Angochitina eisenacki* Bachmann and Schmid
52. *Ancyrochitina longicornis* Taugourdeau and De Jekhowsky
53. *Ancyrochitina diabolo* (Eisenack)
54. *Ancyrochitina ancyrea* (Eisenack)
55. *Gotlandochitina martinssoni* Laufeld
56. *Plectochitina carminae* Cramer
57. *Ancyrochitina nodosa* Taugourdeau and De Jekhowsky
58. *Angochitina capillata* Eisenack
59. *Gotlandochitina spinosa* (Eisenack)
60. *Ancyrochitina desmea* Eisenack
61. *Urochitina simplex* Taugourdeau and De Jekhowsky
62. *Angochitina crassispina* Eisenack
63. *Angochitina comosa* Taugourdeau and De Jekhowsky
64. *Angochitina mourai* Lange
65. *Ancyrochitina primitiva* Eisenack
66. *Ancyrochitina* sp.
67. *Angochitina* sp.
68. *Ramochitina magnifica* Lange
69. *Cladochitina biconstricta* (Lange)
70. *Angochitina devonica* Eisenack

GLOSSARY OF TERMS

A₁ GENERATION: multinucleate megalospheric foraminifera.

A₂ GENERATION: uninucleate megalospheric foraminifera.

ABDOMEN: in nassellarian radiolarians, the third (second post-cephalic) chamber.

ACANTHOPORE: spine-like structure in the skeleton of Paleozoic Bryozoa, with a clear axis of calcite and steeply inclined laminae surrounding it.

ACCESSORY APERTURES: test openings that do not lead directly into primary chambers but extend beneath or through accessory structures on planktonic foraminifera.

ACCESSORY ARCHEOPYLE SUTURE: a suture developed between adjacent paraplates which form the archeopyle margin, or within the operculum. Such sutures do not result in complete separation of the paraplates involved.

ADDUCTOR MUSCLE SCARS: imprints on the inner side of the outer lamella made by attachment of closing (adductor) muscle scars.

AGGLUTINATED: foreign particles bound together by cement into a test.

ALVEOLI: bubble-like liquid inclusions constituting a frothy mass in the ectoplasm of radiolarians.

AMPHIDONT: hinge with teeth in both valves.

AMPHIESMA: the external protective cellular covering surrounding the protoplast.

ANCESTRULA: fist-shaped individuals in bryozoan colony.

ANNULATE PROCESS COMPLEX: a series of intratabular processes arranged in a circle on an individual paraplate. The processes of an individual complex may be united proximally, along their length, or distally.

ANTAPEX: in a motile dinoflagellate, the area of the amphiesma at the posterior end of the hyposome or hypotheca, respectively. The corresponding area in the dinocyst is similarly named.

ANTAPICAL: adjectival form of antapex.

ANTAPICAL PARAPLATE: the paraplate, or one of the group of paraplates, occupying the antapex of a paratabulate dinocyst.

ANTAPICAL PLATE: the plate, or one of the group of plates, occupying the antapex of a tabulate motile dinoflagellate.

ANTERIOR INTERCALARY PARAPLATE: the paraplate, or one of the group of paraplates, in a paratabulate dinocyst, which lies between the apicals and precingulars, without touching the apex or paracingulum.

ANTERIOR INTERCALARY PLATE: the plate, or one of the group of plates, of a tabulate motile dinoflagellate, which lies between the apicals and precingulars, without touching the apex or paracingulum.

ANTERIOR SULCAL PORE: the pore from which the transverse flagellum arises.

ANTRUM: concave area in hollinacean ostracodes bordered by dolonal parts of adventral structures.

APERTURE (foraminifera): opening or openings from chamber of test to exterior.

APERTURE (pteropods): opening at last-formed margin of shell, providing outlet for the head-foot mass.

APERTURE (spore or pollen): an opening or modification in the exine where the pollen tube or developing prothallus develops.

APEX: in a motile dinoflagellate, the area of the amphiesma at the anterior end of the hyposome or hypotheca. The corresponding area in the dinocyst is similarly named.

APICAL ARCHEOPYLE: an archeopyle formed by the partial or complete loss of the single apical, or all the apical paraplates.

APICAL AXIS: the longitudinal axis of the ostracode valve.

APICAL PARAPLATE: the paraplate, or one of the group of paraplates, occupying the apex of a paratabulate dinocyst.

APICAL PLATE: the plate, or one of the group of plates, occupying the apex of a tabulate motile dinoflagellate.

APRON: flaring base of cone conodont.

ARCHEOPYLE: the excystment aperture in a dinocyst through which the dinoflagellate escapes during excystment.

ARCHEOPYLE SUTURE: the suture(s) developed in formation of the archeopyle.

ARCUATE: with a complete or partial arc.

ARCUATE PROCESS COMPLEX: a series of intratabular processes arranged in an arc on an individual paraplate. The processes of an individual complex may be united proximally, along their length, or distally.

AREAL APERTURE: aperture on face of final chamber of test.

AREAL BULLA: blister-like structure covering multiple areal apertures (e.g., *Globigerinatella*).

ARENACEOUS: composed of sand or other foreign particles.

AREOLAE: polygonal shaped or rounded cavities separated by siliceous partitions.

ARMORED: possessing a theca. Synonym of thecate.

ARTICULATED CORALLINES: Those coralline algae having segmented thalli.

ASCUS: hydrostatic organ in certain cheilostome bryozoa.

AUTECOLOGY: study of relationship of elements in a community ecology.

AUTOCOEL: the cavity enclosed by the autophragm in a dinocyst.

AUTOCYST: the body formed by the autophragm in a dinocyst.

AUTODERM: the one layer present in acritarchs with a single-layered wall.

AUTOPHRAGM: the one layer present in dinocysts with a single-layered wall.

AUTOTROPHIC: pertaining to a plant that is able to manufacture its own food by photosynthesis.

AUTOZOOID: feeding individual in a bryozoan colony.

AUXOSPORE: a special cell formed by the fusion of two protoplasts.

AVICULARIUM: specialized zooid in Cheilostomata with a mandible-like operculum.

AXIS: imaginary line around which spiral or cyclical shell is coiled, transverse to plane of coiling.

AXOFLAGELLUM: specialized pseudopod in some radiolarians resembling a bundle of axopods.

AXOPLAST: a special knot-like structure on the nuclear membrane of some radiolarians, the origin of axopodia.

AXOPODIA: long, slender, straight pseudopodia with a characteristic stiff axial filament.

BARS: straight, elongate skeletal elements in radiolarians connected at both ends.

BASAL CAVITY: a structure below the growth center of a conodont, it develops continuously with the conodont structure, forming e.g. a cup.

BASAL RIM: rim on the lower side of the conodont.

BATHYPELAGIC: open-sea organisms found between 1000 and 4000 m depth.

B-FORM: microspheric form.

BILAMELLAR: walls of each chamber consisting of two primarily formed layers.

BINARY NOMENCLATURE = BINOMIAL: two-named; in biology referring to the genus to which an organism belongs and its own species name.

BIOFACIES: a subdivision of a sedimentary unit based on its biologic characters (fossil, fauna, or flora).

BISACCATE: possessing two sacci.

BISERIAL: having chambers arranged in rows, two per whorl.

BIUMBILICATE: having central depression (umbilicus) on each side of test.

BLADE: laterally compressed structure usually bearing denticles: (a) in bar and blade conodonts divisible into posterior and anterior portions; (b) in platform conodonts it is the edge-like free portion of carina, not surrounded by platform.

BODY CAVITY: space between the two inner lamellae of the ostracode carapace, occupied by the ostracode body.

BROWN BODY: organic tissue in bryozoan zooid resulting from disintegration from polypide.

BUCCAL CAVITY: pharynx in pteropods.

BULLA (pl. bullae): structure that partially or completely covers primary or secondary apertures. May be umbilical, sutural, or areal in position and may have one or more accessory marginal apertures.

BY-SPINES: short, tiny spines on radiolarian skeletons; usually numerous and ordinarily one at each junction of bars.

CALYMMA: a relatively thick outer zone of the radiolarian ectoplasm, contains the alveoli.

CANCELLATE: having honeycomb-like surface.

CANALICULATE: possessing series of fine tubular cavities.

CAPPA (pl. cappae): the proximal portion of the corpus or body of pollen grains in the Pinaceae and Podocarpaceae.

CARAPACE: protective covering of the ostracode soft body, formed by two valves joined together along the dorsal margin. Mostly calcified, soft in some myodocopids and archeocopids; in some members of the last-named group probably slightly phosphatized in living animal.

CARDINAL MARGIN: part of the carapace along the dorsal margin occupied by hinge structures. Syn.: hinge margin.

CARINA (pl. carinae). Ridge or crest present on the side of single-cone element, also row of nodes or low denticles on the upper side of platform conodonts.

CAVATE: a dinocyst in which the endophragm and periphragm are generally not in contact and in which the endocyst and pericyst are therefore separated.

CCD: level below which all carbonate should go into solution due to partial pressure of CO_2 or the decreased alkalinity of the environment.

CENTRAL CAPSULE: the contents of a central part of a radiolarian cell (endoplasm) enclosed in a distinctive membrane; often roughly spherical, but may be variously shaped.

CENTRAL SECTION: slice bisecting central chambers of test.

CENTRIC: term used in an informal classification of diatoms. Centric diatoms are those forms which are disc-shaped or have morphological features radiating from a central area.

CEPHALIS: in nassellarian radiolarians, the tiny apical chamber; usually spherical to subspherical, but often elongate, and sometimes hidden within the thorax.

CHAMBER: test cavity and its surrounding wall, formed at single growth stage.

CHAMBERLET: subdivision of chamber produced by axial or transverse septula.

CHAROPHYTES: referring to algae or algal organs belonging to the phylum Charophyta.

CHLOROPHYLL: the green pigment found in the chloroplast, important for the absorption of light energy in photosynthesis.

CHLOROPLAST: specialized cytoplasmic body, containing chlorophyll, where reaction of starch or sugar synthesis occur.

CHORATE: a dinocyst which possesses processes and does not resemble the corresponding motile stage, which develops from it.

CHROMATOPHORE: (see chloroplast).

CINGULAR PARAPLATE: one of the paraplates forming the paracingulum in a paratabulate dinocyst.

CINGULAR PLATE: one of the plates forming the cingulum in a tabulate motile dinoflagellate.

CINGULUM: synonym of transverse furrow.

CLAVATE: club-shaped.

COCCOID: resembling a spherical body.

COCCOLITH: elliptical to circular calcite plates covering the cell of coccolithophores.

COCCOSPHERE: complete envelope of coccoliths around the cell of coccolithophores.

COLPUS (pl. colpi): a furrow or kind of elongate aperture with a ratio of length to width > 2; a sulcus.

COLUMELLA (pl. columellae): a column-like or rod-like structure that interconnects the inner part of the ektexine (foot layer) with the tectum or outer part; columellae occur in both tectate and semitectate structure. In pteropods solid or hollow pillar-surrounding axis of a coiled shell, formed by adaxial walls of whorls.

COMBINATION ARCHEOPYLE: an archeopyle which results from the loss of paraplates belonging to more than one series; e.g., in some dinocysts the archeopyle is formed by the loss of apical and anterior intercalary paraplates.

COMPOUND HETEROMORPHIC: when there are branched and unbranched processes in a single cyst.

CONCEPTACLE: a cavity in which the reproductive organs, such as sporangia, are developed.

CONISPIRAL: with spire projecting as cone or conoid.

CONTACT GROOVE: a groove along the extracardinal margin, into which the free edge of the opposite valve fits.

CONVERGENT EVOLUTION: the evolution of analogous morphologic features in unrelated taxa.

CORALLINE: referring to algae belonging to the family Corallinaceae (Rhodophyta).

CORPUS: the body or central part of vesiculate or saccate pollen.

CORTICAL SHELL: the main external shell of multi-shelled radiolarians.

COSTATE: having raised ridges or costae.

CRIBATE: perforated with round holes, sieve-like.

CRIMP: marginal band on the lower side of platform representing area covered by the last lamella accreted to conodont structure.

CRUMINA: brood pouch in the beyrichiomorphs, situated anteroventrally to centroventrally.

CRUSTOSE CORALLINE: referring to coralline algae having, mostly attached, crust-like thalli.

CRYPTOSUTURE: a suture in an acritarch which only becomes visible when dehiscence has commenced. Partly analogous to archeopyle suture in dinocysts.

CUP: greatly expanded cavity beneath the platform of certain conodonts (e.g. Gnathodus).

CYPHONAUTES: pelagic larva in some cheilostomatous bryozoa.

CYST: the spore or resting stage of unicellular algae.

CYSTOPORE: zooid in cystoporate bryozoans modified for supporting functions.

CYTOPLASM: all the protoplasm of a cell apart from the nucleus.

DENDRITIC: branched, tree-like.

DENTICLE: spine-like, needle-like, or sawtooth-like structure, similar to cusp but commonly smaller.

DEXTRAL: right-handed; term originally applied to any shell with aperture on observer's right when shell apex is directed upward, or with apparent clockwise coiling when viewed from above apex.

DIAGENESIS: all chemical, physical, and biological modifications undergone by a sediment after its initial deposition.

DIAGONAL SECTION: slice cutting axis of coiling obliquely.

DIATOMITE: a rock with diatoms as the most conspicuous component, generally thought of as a soft, fine-grained, porous, low-density material which is white or very light in color.

DICTYOSOME: a unit of Golgi apparatus several of which occur as discrete bodies in cells of invertebrates and plants.

DIMORPHISM: refers to two morphologically distinct phases or forms of a species.

DINOCYST: the resting cyst of a dinoflagellate. All fossil dinoflagellates appear to be dinocysts.

DIPLOID STAGE: having the chromosomes in pairs, the members of each pair being homologous, so that twice the *haploid* number is present.

DIPORATE: possessing two pores and no colpi.

DISTAL: that part of a spore or pollen grain that faces to the outside when a member of a tetrad.

DOLON: part of velar or histial structure which is modified in heteromorphs in comparison with that of adult tecnomorphs.

DOMICILIUM: part of the carapace exclusive of space enclosed by velar structures. In wider sense the part of the carapace enclosing the body of the animal, but excluding any kind of sculptural extensions of the valve.

DORSAL: the surface directly opposite to that containing the sulcus in a motile dinoflagellate, and the parasulcus in a dinocyst.

DORSAL: opposite to ventral side; spiral side of trochoid test.

DORSAL MUSCLE SCAR FIELD: imprints of muscles below the dorsum, including scars of different origin.

DUPLICATURE: in zoological works, synonymous with inner lamella. In paleontological literature, terms of recalcified part of inner lamella.

ECDYSIS: molting of body cover in arthropods.

ECHINATE: sculpture type of pointed elements at least 1 μm in height.

ECHINUS (pl. echini): component element of echinate sculpture.

ECTODERM: the outer layer present in acritarchs with bi-layered walls.

ECTOPHRAGM: the outermost layer present in dinocysts with three walls in which the other two walls are the endophragm and periphragm, respectively.

EKTEXINE: usually the outermost layer of the exine when both the ektexine and endexine, or innermost layer, are represented; see perine.

ELLIPSOIDAL: shape like an ellipsoid; a solid the plane sections of which are largely ellipses but also circles.

ENCYSTMENT: the enclosure of nucleus and cytoplasm within a resistant wall commonly as a response to changing environmental conditions or as a preliminary to cell division. The cyst is formed within the amphiesma, but may grow to greater dimensions after the latter has decayed.

ENDEXINE: innermost layer of the exine.

ENDOCOEL: the cavity enclosed by the endophragm in a dinocyst.

ENDOCYST: the body formed by the endophragm in a dinocyst.

ENDODERM: the innermost wall in acritarchs with two or more walls.

ENDOPHRAGM: the innermost wall in dinocysts with two or more walls.

ENDOPLASM: central part of cytoplasmic mass, commonly granulated.

ENTOSOLENIAN: having internal tube-like apertural extension.

EPICYST: the portion of a dinocyst corresponding to the epitheca of a motile dinoflagellate. In paratabulate dinocysts this is the area anterior to the paracingulum and consisting of the apical, anterior intercalary (if present) and precingular paraplates.

EPICYSTAL ARCHEOPYLE: an archeopyle involving the loss of the epicyst and in which the archeopyle suture runs immediately anterior to the paracingulum.

EPISOME: in naked or unarmored motile dinoflagellates, the portion of the amphiesma which is anterior to the cingulum.

EPITHECA: also called the epivalve. The larger of the two valves is the frustule, it fits over the smaller valve. In a thecate motile dinoflagellate, the portion of the theca which is anterior to the cingulum. It includes the apical, anterior intercalary (if present) and precingular plates.

EPITYCHE: an outfolded flap of the vesicle wall in acritarchs.

EQUATOR: imaginary line, situated approximately midway between the polar extremities, that encircles symmetrical spores and pollen.

EQUATORIAL AXIS: imaginary line, perpendicular to the polar axis, that extends to the equator; in radiosymmetric pollen and spores the axis is more or less equidistant, whereas in bilateral types a short axis is at right angles to a long axis.

EUCARYOTES: morphologically advanced organisms characterized by having the cellular organelles, including the nucleus, bounded by membranes.

EUPHOTIC ZONE: zone near surface of sea (roughly upper 100 m) into which sufficient light penetrates for photosynthesis.

EUTROPHIC: literally "well nourished". A condition where there are enough nutrients in the aquatic environment to support the population.

EVOLUTE: tending to uncoil; chambers non-embracing.

EXCYSTMENT: in dinoflagellates, the process of cyst abandonment, which may be as a response to improving environmental conditions or after completion of cell division. Excystment occurs through an archeopyle or other preformed opening; the cell initially has a soft flexible wall lacking any differentiation of plates, etc., but quickly develops a more rigid amphiesma.

EXCYSTMENT APERTURE: the aperture through which the dinoflagellate escapes on excystment. Partly synonymous with archeopyle.

EXINE: outermost resistant layer of the pollen or spore wall.

EXOPLASM: the outer zone of radiolarian cytoplasm, outside the central capsule.

EXTRACARDINAL MARGIN: part of the valve margin outside the cardinal (hinge) margin. Syn.: free margin.

EXTRAUMBILICAL APERTURE: opening in final chamber of test not connecting with umbilicus, commonly sutural midway between umbilicus and periphery.

FACIES: the general aspect, nature, or appearance of a sedimentary rock.

FEET: in nassellarian radiolarians, terminal spines, few in number and usually relatively robust, directed away from the apex, often homologs of dorsal and primary lateral spines.

FILAMENT: a slender row or line of cells.

FILOPODIA: thin, delicate, needle-shaped pseudopodia.

FISTULOSE: having tubular irregular growth.

FLAGELLATED: bearing a flagellum.

FLAGELLUM: a thin whip-like structure capable of undulatory movement.

FLANGE: ridge along the valve margin of some podocopids made by projection of outer lamella, on the outer side of the selvage.

FORAMEN (pl. foramina): opening between chambers located at base of septa or areal in position.

FREE MARGIN: see extracardinal margin.

FREE OPERCULUM: an operculum which is completely detached from the rest of the dinocyst.

FRONTAL MUSCLE SCARS: scars situated in front of the adductor muscle scars and above mandibular scars. In podocopids hitherto studied, they correspond to imprints of a mandibular muscle and muscle attaching itself to chitinous endoskeleton. In earlier literature, they were erroneously interpreted as antennal muscle scars.

FRUSTULE: the individual diatom cell consisting of the epivalve, the hypovalve and a connecting band.

GAMETANGIUM: an organ producing gametes (which form the zygote in the process of sexual reproduction).

GAMETE: reproductive cell capable of fusing in pairs to form new individual (zygote).

GAMONT: generation which forms gametes in sexual reproduction, commonly with megalospheric test (A-forms).

GERM DENTICLE: these are suppressed or aborted denticles, which could not develop into mature structures owing to crowded condition along growing edge of conodont.

GIRDLE (Diatoms): also called the connecting band or the girdle band. It holds the epivalve and hypovalve together. (Dinoflagellates: a term which can be synonymous with the transverse furrow or cingulum of a motile dinoflagellate or the paracingulum of a dinocyst. Its use is not recommended.)

GOLGI APPARATUS: local structure in cytoplasm of cells. Appears like a group of membranes constituting a number of flattened sacs aligned roughly parallel. Secretory function of wall material.

GONAL: the point of intersection of three or more paraplates, or of paraplates and paracingulum or parasulcus.

GONYAULACACEAN: dinoflagellates with the *Gonyaulax* tabulation, or dinocysts with the *Gonyaulacysta* paratabulation.

GRANULAR HYALINE WALL: perforate, lamellar test composed of granular calcite, seen between crossed nicols as multitude of tiny flecks of color.

GROWTH LAMELLAE: numerous thin layers or sheaths forming a conodont, and its basal body. They are accreted around a growth center.

GYMNODINIOID STAGE: the thin-walled uniflagellate stage which is formed soon after release of the protoplast from the dinocyst on excystment, and which rapidly gives away to a biflagellate stage.

GYROGONITES: the spiral-shaped calcified portion of the oogonium (charophyta).

HAPLOID STAGE: having a single set of unpaired chromosomes in each nucleus (e.g. in gametes).

HAPTONEMA: flagellar apparatus in coccolithophores.

HEIGHT: distance between two planes perpendicular to shell axis.

HEMISOLENIC: type of carapace closure having at each valve both salient (positive) and depressed (negative) elements.

HETEROCOCCOLITHS: coccoliths composed of crystallites of different shapes and sizes.

HETEROMORPH: term used in dimorphic species of extinct ostracodes for adult carapaces of that sex which differ more from the larval carapaces than those of the other sex. Considered to represent females.

HETEROMORPHIC: processes of more than one type in a single dinocyst or acritarch.

HETEROPHASIC: containing two or more distinct phases, e.g., alternation of generations produces heterophasic life cycle.

HETEROTROPHIC: organisms requiring organic food supply form its environment.

HINGE: dorsal part of the valves serving for their articulation.

HINGE LINE: line along which the valves articulate, seen in closed carapace.

HISTIUM: a frill-like adventral structure occurring dorsally from the velum.

HOLOCOCCOLITHS: coccoliths composed of crystals of same shape and size.

HOLOPLANKTON: organisms living their complete life cycle in the floating state.

HOLOSOLENIC: type of carapace closure having all the elements of both cardinal and extracardinal margins in the same valve either depressed or raised.

HOMOMORPHIC: when the processes on a dinocyst or acritarch are all of the one type, i.e. are all the same.

HORN: an outbulge or extension of the wall, or of its outer layer, in motile dinoflagellates or dinocysts. Horns may be apical, lateral or antapical in position. (In nassellarian radiolarians, a term sometimes applied to a stout apical spine.)

HYALINE AREA: usually refers to an area on the frustule which is composed of clear silica with no structures.

HYPOCYST: the portion of a dinocyst corresponding to the hypotheca of a motile dinoflagellate. In paratabulate dinocysts this is the area posterior to the paracingulum and consisting of the postcingular, posterior intercalary (if present) and antapical paraplates. Most or all of the parasulcus is located on the hypocyst.

HYPOCYSTAL ARCHEOPYLE: an archeopyle formed from the loss of some of the paraplates of the hypocyst.

HYPOSOME: in naked or unarmored motile dinoflagellates, the portion of the amphiesma which is posterior to the cingulum.

HYPOTHALLIUM: the basal or inner tissue in a coralline alga.

HYPOTHECA: also called the hypovalve. The smaller of the two valves in the frustule of a diatom. In a thecate motile dinoflagellate, the portion of the theca which is posterior to the cingulum. It includes the postcingular, posterior intercalary (if present) and antapical plates. Most or all of the sulcus is located on the hypotheca.

IMPERFORATE: without pores, used for porcelaneous tests and in describing ornamentation.

INAPERTURATE: without clearly defined apertures.

INFRALAMINAL ACCESSORY APERTURE: opening in planktonic foraminiferal test leading to cavity beneath accessory structures (bullae) at the margin of these structures.

INNER LAMELLA: thin layer enveloping the ostracode body in anterior, ventral and posterior parts of the carapace; chitinous, with marginal parts often calcified.

INTECTATE: without a tectum, that is, no ektexinous sculpture elements fused to form a membrane, or tectum, outside the endexine.

INTERCALARY: located either between the apical and the precingular or between the postcingular and antapical series of plates in the theca, or between the corresponding series of paraplates in the dinocyst.

INTERCALARY ARCHEOPYLE: an archeopyle formed from the loss of one, or more, of the anterior intercalary paraplates.

INTERCALARY BAND: the area between adjacent plates in peridinioid dinoflagellate thecae. Also found in cysts. It is commonly striate.

INTERGONAL: synonymous with parasutural.

INTRALAMINAL ACCESSORY APERTURE: opening in planktonic foraminiferal test leading through accessory structures (bullae) into cavity beneath but not directly into chamber cavity.

INTRAUMBILICAL APERTURE: opening of test located in umbilicus but not extending outside of it.

INTRATABULAR: within the parasutures of individual paraplates in dinocysts.

INVOLUTE: strongly overlapping; in enrolled forms, later whorls completely enclosing earlier ones.

ISOTONIC: solutions having the same osmotic pressure.

ITERATIVE EVOLUTION: the repeated evolution of a morphologic feature in related taxa.

KALYPTRA: the outer flocculant, granular, organic mantle found in some dinocysts such as the genus *Kalyptea*.

KENOZOOID: zooid in bryozoan colony modified for supporting functions.

LABYRINTHIC: having complex spongy wall with interlaced dendritic channels perpendicular to surface, characteristic of some agglutinated foraminifers.

LAMELLAR: composed of thin layers of aragonite or calcite, one layer being formed with addition of each new chamber and covering whole previously formed test.

LANCEOLATE: flat, narrow, and tapering to point.

LAST WHORL: in coiled shells, last-formed complete volution.

LATERAL PORE CANAL: tubular passage traversing more or less perpendicularly the outer lamella. Syn.: normal pore canal.

LATTICE, LATTICED, LATTICEWORK: referring to a type of wall structure in radiolarians consisting of a regular open two-dimensional network of connecting bars.

LENTICULINE: lens-shaped.

LIGAMENT: narrow stripe of cuticle uniting the two valves along the dorsal margin.

LIMBATE: referring to thickened border of a suture, may also be elevated.

LINEAR PROCESS COMPLEX: a series of processes arranged in a straight line, linked or unlinked proximally, along their length, or distally.

LINE OF CONCRESCENCE: proximal line of junction of the outer and inner lamellae.

LIP: elevated border of aperture, may be at one side of aperture or completely surround it.

LIST: low ridge along the valve margin of some podocopids, situated proximally to the selvage.

LOBE: major elevation usually best developed in the dorsal part of the ostracode valve and reflected on its internal surface. (In conodonts: lateral extension of platform or blade of a conodont.)

LOCULAR WALL: cell wall composed of two layers of silica separated by vertical sections, also made of silica.

LONGITUDINAL FLAGELLUM: the flagellum which arises from the posterior sulcal pore and is directed backwards along the longitudinal furrow or sulcus in a motile dinoflagellate.

LONGITUDINAL FURROW: the furrow on the ventral surface of the motile dinoflagellate which houses the longitudinal flagellum. It lies wholly or partially on the hyposome or hypotheca. Synonymous with sulcus.

LOPHOPHORE: circular or horshoe-shaped ring around mouth of bryozoans bearing ciliated tentacles.

LUNARIUM: horse-shoe shaped projection on proximal side of aperture in cystoporate bryozoa.

LYSOCLINE: level in the water column where significant amount of carbonate begins to be dissolved.

Ma: mega annum = million years B.P.

MAIN BODY: in chlorate dinocysts the central portion of the cyst from which the processes arise.

MAIN SPINES: conspicuous large, long spines on radiolarian skeletons; usually relatively few in number.

MANDIBULAR SCARS: imprints of chitinous support rods of the mandible, situated in front of the adductor muscle scars and below the frontal muscle scars.

MANTLE: the edge of the valve where it bends to join the connecting girdle.

MARGINAL PORE CANALS: tubules passing through the zone of concrescence. Synonymous with radial pore canals.

MATRIX: a thin inner zone of the radiolarian ectoplasm separating the calymma from the central capsule.

MEDIAN SECTION: slice in central sagittal position, perpendicular to axis of coilings.

MEDIAN SULCUS: sulcus corresponding to the attachment of adductorial muscles, designated as S_2.

MEDULLARY SHELL(S): the inner shell(s) of multi-shelled radiolarians.

MEGALOSPHERIC: having large proloculus, commonly representing gamont generation.

MERODONT: hinge with teeth in one valve only.

MEROPLANKTONIC: near-shore species which spend part of their existence in the pelagic realm and the remainder on the sea floor, probably as a resting spore.

MESOCOEL: the cavity, or cavities, between the endophragm and the mesophragm in a dinocyst.

MESOCYST: the body formed by the mesophragm in a dinocyst.

MESODERM: the middle wall in acritarchs with three walls.

MESOPHRAGM: the middle wall in dinocysts with three walls, in which the outer is the periphragm.

MESOPORE: modified zooid in trepostomatous Bryozoa.

METABOLISM; the process by which nutritive matter is built into living material.

MICROFLAGELLATES: tiny members of the group of flagellate protists.

MICROGRANULAR: microscopically granulose, referring to wall composed of minute calcite crystals, probably originally granular but possibly recrystallized; granules may be aligned in rows perpendicular to outer wall, resulting in fibrous structure.

MICROSPHERIC: having small proloculus, commonly agamont (schizont) generation, adult test large.

MICROSPORE: the small male spore in heterosporous ferns and fern allies, as in *Isoetes, Salvinia, Azolla*, and *Marsilea*; also, in microsporogenesis, the monoculeate forerunner of the pollen grain.

MILIOLINE: formed as in Miliolacea, commonly with elongate chambers, added in differing planes of coiling.

MISCELLANEOUS ARCHEOPYLES: archeopyles that cannot be placed into one of the four major categories which are the apical, intercalary, precingular and combination groups

MITOCHONDRIA: rod or thread-like microscopic bodies occurring in cytoplasm of every cell (except bacteria and blue-green algae).

MONAD: a pollen grain occurring singly, as opposed to pollen in tetrads or polyads.

MONOCOLPATE: with a solitary colpus.

MONOCOLPORATE: with one colpus interrupted by a pore.

MONOLETE: with a linear tetrad scar.

MONOPORATE: with one pore.

MOTILE DINOFLAGELLATE: a dinoflagellate that has not encysted, but is a functional component of the plankton, swimming, feeding and reproducing.

MULTILOCULAR: many-chambered test.

MUSCLE SCAR: marking on internal surface of the valve which indicates place of muscle attachment; it is depressed or raised and shows a different shell structure than the rest of the valve.

NONVASCULAR: referring to plants without conducting tissues or well-differentiated roots, stems, and leaves.

NORMAL PORE CANAL: see lateral pore canal.

NUCLEOLI: small dense bodies which contain ribose nucleoprotein.

NUCLEUS: spherical, compact mass of chromatin surrounded by membrane, lying within cytoplasmic body.

OBLATE: shape when the P:E ratio is 0.50:0.75.

OBLATE SPHEROIDAL: shape with a P:E ratio of 0.88:1.00.

OBLIQUE SECTION: slice through test cut in direction neither parallel to axis of coiling nor normal to it.

OLIGOTYPIC: containing one type only — assemblage of only one species.

OOGONIUM: the female reproductive organ of the Charophyta.

OOZES: ocean bottom sediments (usually deep) consisting mainly of skeletal material, e.g. diatom ooze, ratiolarian ooze, foram ooze, nannoplankton ooze.

OPERCULUM: the paraplate or group of paraplates, which are lost or partially detached in archeopyle formation and which are bounded by the principal archeopyle suture. Chitinous lid covering the aperture in cheilostomatous Bryozoa. Corneous or calcareous structure borne by foot of pteropods and serving for closure of aperture, wholly or partly.

ORIFICE: aperture or other opening in test.

OUTER LAMELLA: outer, usually well calcified layer, constituting the whole outer part of the ostracode valve. Along the extracardinal margin of the valve, it passes by a simple bend or by fusion into the inner lamella.

PALMATE: flat, resembling hand with outspread fingers.

PARACINGULAR: adjectival form of paracingulum.

PARACINGULUM: the area on the dinocyst analogous to the cingulum of the motile dinoflagellate. Paraplates may or may not be visible.

PARALLEL SECTION: slice through test in plane normal to axis of coiling but not through proloculus.

PARAPLATE: the cyst equivalent to a plate in the theca.

PARASULCAL: adjectival form of parasulcus.

PARASULCAL NOTCH: the re-entrant angle in the apical archeopyle margin which marks the posterior extension of the first apical paraplate. It is immediately anterior to the parasulcus.

PARASULCAL TONGUE: the posterior extension of the first apical paraplate in the operculum of an apical archeopyle.

PARASULCUS: the area on the dinocyst analogous to the sulcus or longitudinal furrow of the motile dinoflagellate. Paraplates may or may not be visible.

PARATABULATION: the pattern or arrangement of the constituent paraplates in a dinocyst. It is usually expressed as an alpha-numeric formula which gives the series of paraplates present, and the number of paraplates within each series.

P:E RATIO: the relationship between the lengths of the polar and equatorial axes in radiosymmetric grains.

PENNATE: term used in an informal classification of diatoms. Pennate diatoms are those forms which have features or structures running parallel to the transapical axis.

PENETABULAR: linear features which lie immediately interior to the margin of a paraplate.

PERFORATE PLATE: a type of wall structure in radiolarians in which a thin plate-like wall is perforated by pores, as opposed to a latticed wall which is constructed of discrete elements (bars).

PERICOEL: the cavity, or cavities, lying between the endophragm and periphragm, or mesophragm and periphragm, in a dinocyst.

PERICYST: the body formed by the periphragm in a dinocyst.

PERIDINIACEAN: dinoflagellates with the *Peridinium* tabulation, or dinocysts with the *Wetzeliella* paratabulation.

PERIDINIOID: the characteristic outline of a dinoflagellate of the genus *Peridinium*, with a pointed apex or apical horn and with two antapical horns. Also applied to motile dinoflagellates and dinocysts with a pentagonal outline, which may or may not be prolonged into one apical, two lateral (paracingular) and one or two antapical horns.

PERINE: an outer layer in the wall of certain spores, considered by some to be part of the ektexine and by others to be extra-exinous.

PERIPHRAGM: the outer wall in dinocysts with two walls. The outermost wall in dinocysts with three walls, of which the middle is the mesophragm. Next to the outer wall in dinocysts with three walls of which the outermost is the ectophragm.

PERIPORATE: with pores, exceeding three in number, arranged more or less equidistantly on the grain surface.

PERIPROLATE: shape when the P:E ratio is >2.

PERITHALLIUM: the upper or outer layer of tissue in a coralline alga.

PERVALVAL AXIS: also called the cell axis. The axis through the center point of the two valves.

PHAEODIUM: a mass of pigmented spaerules in the ectoplasm of one major group of radiolarians (*Tripylea* or *Phaeodaria*).

PHAGOTROPHY: ingestion of foreign food particles.

PHIALINE: having everted rim on apertural neck, as on neck of a bottle.

PHOTOSYNTHESIS: a process in which carbon dioxide and water are chemically combined to form carbohydrates, the energy for the process being sunlight.

PHRAGMA: the wall of a dinocyst. It can be composed of one, two, or more than two layers.

PIT: a structure around the growth center of a conodont which ceases to develop after it has attained a certain size.

PLANISPIRAL: coiled in single plane.

PLATE: one of the constituent and separable units of the theca.

PLATE FORMULA: the alpha-numeric representation of the plates in a thecate dinoflagellate. It shows the series of plates present, and the number of plates within each series, in an abbreviated formula. Thus $4'$, 1a, $6''$, 6c, $6'''$, $1''''$ means there are four apical, one anterior intercalary, six precingular, six cingular, six postcingular and one antapical plates in the taxon whose tabulation is given in the plate formula.

POLAR AXIS: imaginary line passing from pole to pole and constituting a major reference of symmetry.

POLLEN: the binucleate or trinucleate male reproductive body in seed plants.

POLYAD: grains united in multiples of more than four.

POLYHEDRON: a solid resulting from the intersection of many plane faces.

POLYMORPHISM: morphologically different forms of same species which may be result of different generations.

POLYPIDE: soft parts of individual in a bryozoan colony.

PORCELANEOUS: having calcareous, white, shiny, and commonly imperforate wall resembling porcelain in surface appearance; shows low-polarization tints between crossed nicols and has majority of crystals with *c*-axes tangential, or more rarely arranged radially; commonly brown in transmitted light.

PORE: a more or less isodiametric germinal aperture in spore or pollen with a length to width ratio of <2. Opening of pore canals in ostracodes.

PORE CANAL: tubular passage traversing the valve wall of ostracodes. The larger of the two types of pores which partially or completely penetrate the wall of a tasmanitid.

PORE (Foraminifera): found, slit-like or irregular openings ≃ 5—6 µm in size partially perforating the test wall of foraminifera.

PORTICUS (pl. portici): asymmetrical apertural flaps.

POSTCINGULAR: located between the cingulum and antapex in a motile dinoflagellate, and between the paracingulum and antapex in a dinocyst.

POSTCINGULAR HORN: a horn arising from a position on the hypotheca of a motile dinoflagellate immediately posterior to the cingulum, i.e. from a postcingular plate. Alternatively a horn arising from a position on the hypocyst of a dinocyst immediately posterior to the paracingulum, i.e. from a postcingular paraplate.

POSTCINGULAR PARAPLATE: one of the latitudinal series of paraplates lying immediately posterior to the paracingulum in a paratabulate dinocyst.

POSTCINGULAR PLATE: one of the latitudinal series of plates lying immediately posterior to the cingulum in a tabulate motile dinoflagellate.

POSTEQUATORIAL: in acritarchs one of the series of paraplates lying immediately posterior to the equator or "paracingulum". In dinocysts synonymous with postcingular.

POSTERIOR INTERCALARY PARAPLATE: the paraplate, or one of the group of paraplates, of a paratabulate dinocyst, which lies between the postcingulars and antapical(s), without touching the cingulum or antapex.

POSTERIOR SULCAL PORE: the pore from which the longitudinal flagellum arises.

PRECINGULAR: located between the apex and cingulum in a motile dinoflagellate, and between the apex and paracingulum in a dinocyst.

PRECINGULAR ARCHEOPYLE: an archeopyle formed from the loss of one, or more, of the precingular paraplates.

PRECINGULAR PARAPLATE: one of the latitudinal series of paraplates lying immediately anterior to the paracingulum in a paratabulate dinocyst.

PRECINGULAR PLATE: one of the latitudinal series of plates lying immediately anterior to the cingulum in a tabulate motile dinoflagellate.

PRE-EQUATORIAL: in acritarchs one of the series of paraplates lying immediately anterior to the equator or "paracingulum". In dinocysts synonymous with precingular.

PRIMARY APERTURE: main opening of test.

PRINCIPAL ARCHEOPYLE SUTURE: the suture developed between the operculum and archeopyle margin, or within the operculum when it results in complete separation of the portions.

PROCARYOTES: morphologically primitive organisms (bacteria and blue-green algae; mostly single cells or simple filaments) which do not have DNA separated from the cytoplasm by an envelope.

PROCESS: an essentially columnar or spine-like projection arising from the surface of a dinocyst or acritarch. Processes may be simple or intricately branched and interconnected. Processes are rarely, if ever, found in motile dinoflagellates.

PROCESS COMPLEX: the association of three or more adjacent intratabular processes to form a distinctly arranged and aligned group, often united, proximally, along their length and/or distally.

PROCESS FORMULA: the alpha-numeric representation of the processes in a dinocyst or acritarch. It shows the series of paraplates present, and the number of paraplates within each series, but only if such paraplates bear processes. Thus $4'$, $6''$, $6'''$, 1p, $1''''$ means there are four apical, six precingular, six postcingular, one posterior intercalary and one antapical processes in the taxon represented by this process formula. Since the paracingulars are devoid of processes they are not included.

PROLATE: shape with a P:E ratio of 1.32:2.

PROLOCUS (pl. proloculi). Initial chamber of foraminiferal test.

PROTOPLAST: the protoplasm of the cell excluding the cell wall.

PROXIMAL: side facing inward.

PROXIMATE: a dinocyst which in form and size closely resembles the corresponding motile stage, which develops from it upon excystment.

PSEUDONODULE: also called a pseudo-ocellus. A small clear area, usually on the periphery of the valve, slightly raised and frequently surrounded by small pores.

PSEUDOPODIA: cytoplasmic projections serving for locomotion, attachment, and capture of food in foraminifers.

PSEUDOPYLOME: a false pylome found in the antapical region of some acritarchs.

PSEUDORAPHE: a narrow, hyaline area running parallel to the apical axis. It is not true raphe because there is no cleft or notch on the valve.

PUNCTA (pl. punctae): holes or perforations.

PYLOME: the circular excystment opening found in many acritarchs.

RADIAL BEAM: a radial connecting bar in radiolarian skeletons.

RADIAL CANAL: the smaller of the two types of pores which partially or completely penetrate the wall of a tasmanitid.

RADIAL MICROSTRUCTURE: calcareous tests consisting of calcite or aragonite crystals with c-axes perpendicular to surface; under crossed nicols shows black cross with concentric rings of color mimicking negative uniaxial interference figure.

RADIAL PORE CANAL: see marginal pore canal.

RADIATE APERTURE: opening associated with numerous diverging slits.

RADIOLARIAN EARTH: a soft, fine-grained, porous rock similar to diatomite but with radiolarians dominant.

RADIOLARITE: a hard siliceous rock rich in radiolarian remains, term generally restricted to cherts or similar dense lithologies.

RADULA: ribbon-like band containing teeth in the pharynx of pteropods.

RAMUSCULOSUM TYPE: synonym of branched process.

RAPHE: a line or slit found in some pennate diatoms which runs parallel to the apical axis. In side view the raphe may be V-shaped or notched.

RECTILINEAR: growing in a straight line.

REFLECTED TABULATION: synonym of paratabulation.

RELICT APERTURES: short radial slits around umbilicus of test which remains open when umbilical portions of equatorial aperture are not covered by succeeding chambers.

RESTING CYST: synonym of dinocyst in dinoflagellate usage.

RETICULATE: ornamental ridges at surface of test or inner meshwork.

RHIZOPODIA: bifurcating and anastomosing pseudopodia.

ROSTRUM: a beak-like projection at the anterior carapace end of many myodocopids, overhanging a gap between the valves, called rostral incisure.

RUGOSE SURFACE: rough irregular ornamentation, may form ridges.

SACCATE: possessing sacci.

SACCUS (pl. sacci): air sac, bladder, or wing characteristic of vesiculate pollen types in which the ektexine and endexine have become separated; also hollow projections of the walls of some fern spores.

SAGITTAL RING: in nassellarians, a skeletal structure consisting of median bar and apical and vertical spines connected by an arched bar.

SAGITTAL SECTION: slice through test perpendicular to axis of coiling and passing through proloculus.

SCULPTURAL ELEMENT: a projection of less than 5 μm height on the outer surface of an acritarch.

SCULPTURE: ornamentation, or the external appearance of the ektexine without reference to its makeup or construction.

SECONDARY APERTURES: additional or supplementary openings into main chamber cavity; areal, sutural, or peripheral in position.

SECONDARY PIT CONNECTION: narrow passage connecting adjacent cells not belonging to the same filament in tissue of coralline red algae.

SELVAGE: chitinous fringe with calcified base, developed along the extracardinal margin and serving to seal the closed valves.

SEPTULUM (pl. septula): ridge extending downward from lower surface of spirotheca so as to divide chambers partially.

SEPTUM (pl. septa): partition between chambers, commonly consisting of previous outer wall or apertural face, may have single layer, be secondarily doubled enclosing canal systems or be primarily double. A membraneous, linear projection on the wall of a dinocyst or acritarch. Commonly parasutural in position.

SESSILE: attached, sedentary.

SEXUAL DIMORPHISM: difference in morphology of sexes of the same species.

SIEVE PLATE: a thin siliceous plate covering the areolae and perforated by fine pores.

SIEVE PLATE: minute discoidal plate with numerous circular, triangular, and polygonal micropores arranged in concentric rows, contained in pore canal of certain foraminifers.

SIMPLE HETEROMORPHIC: when there are branched processes only in a single cyst.

SIMULATE PROCESS COMPLEX: an arrangement of intratabular processes in the form of a closed polygon, linked or unlinked proximally, along their length, or distally. The complex is developed within, but parallel to the boundaries or parasutures of a paraplate. Partly synonymous with penetabular.

SINISTRAL: shell arranged as in mirror image of dextral (see dextral).

SIPHON: internal tube extending inward from aperture.

SOLEATE PROCESS COMPLEX: a series of intratabular processes arranged in a horse-shoe or crescent on an individual paraplate. The processes of an individual complex may be united proximally, along their length, or distally.

SPHEROIDAL: shape like a sphere with a P:E ratio of 0.88:1.14.

SPICULE: referring to a radiolarian skeleton or skeletal segment consisting of a simple association of radiating spines.

SPINOSE: having fine elongate solid spines on surface of test.

SPIRAL SIDE: part of test where all whorls are visible, also called dorsal side.

SPIRILLINE: planispiral nonseptate tube enrolled about globular proloculus.

SPIROTHECA: outer or upper wall of test in fusulinaceans.

SPIROUMBILICAL: interiomarginal aperture extending from umbilicus to periphery and onto spiral side.

SPONGY: referring to a type of wall structure in radiolarians consisting of an irregular three-dimensional system of interconnecting bars.

SPORANGIUM: a cell which eventually produces one or more spores.

SPORE: as used here, the reproductive body in the Pteridophyta that is asexual or sexual, isosporous plants producing asexual isospores and heterosporous plants forming male microspores and female megaspores.

STOLON: tube-like projections connecting chambers in orbitoids.

STREAMING: continuous linear motion in gel-like protoplasm, particularly pseudopodia.

STREPTOSPIRAL: coiled in several planes.

STRIATE: marked by parallel grooves or lines.

SUBOBLATE: shape with a P:E ratio of 0.75:0.88.

SUBPROLATE: shape when the P:E ratio is 1.14:1.33.

SULCAL PARAPLATE: a paraplate located within the parasulcus of a dinocyst.

SULCAL PLATE: a plate located within the sulcus of a motile dinoflagellate.

SULCAL PORE: one of the two pores located in the sulcus and from which arise either the transverse flagellum or the longitudinal flagellum.

SULCUS: synonym of longitudinal furrow in dinoflagellates. A trench or depression on the valve surface, roughly dorsoventrally oriented in ostracodes.

SUPPLEMENTARY APERTURES: secondary openings in test additional to primary aperture or completely replacing primary aperture.

SUPPLEMENTARY MULTIPLE AREAL APERTURES: subordinate openings in tests.

SUTURAL SUPPLEMENTARY APERTURES: relatively small sutural openings which may be single or multiple, with many openings along the sutures.

SUTURE: line of union between two chambers or between two whorls.

SWARMERS: flagellated gamete cells produced by protozoans.

SYMBIOSIS: life association mutually beneficial to both organisms; commonly refers to green or blue-green algae or yellow cryptomonads.

TABULATE: composed of plates.

TABULATION: the pattern or arrangement of the constituent plates in a tabulate motile dinoflagellate. It is usually expressed as an alpha-numeric formula which gives the series of plates present, and the number of plates within each series.

TANGENTIAL SECTION: slice through test parallel to axis of coiling or growth but not through proloculus.

TECNOMORPH: term used in dimorphic species of extinct ostracodes for larval carapaces and those adult carapaces which are essentially similar to the larval ones, generally interpreted as belonging to adult males.

TECTIN: organic substance having appearance of chitin but distinct chemically.

TECTUM: an outer layer of the wall formed by the distal fusion of ektexinous sculpture elements; grains with a more or less complete tectum are tectate, those with a partial covering of tectum are classified as semitectate.

TEST: shell or skeletal covering, may be secreted, gelatinous, chitinous, calcareous or siliceous; or composed of agglutinated foreign particles, or combination of these. In acritarchs all of the cyst, i.e. vesicle plus processes.

TETHYS: ancient circum-global equatorial seaway.

TETRAD: pollen or spore type in which the components are four in number; tetrads occur in the form of a tetrahedron (tetrahedral), rhomboid (rhomboidal).

TETRATABULAR: when there are four adjoining paraplates in an operculum, which is usually from an apical archeopyle.

THALLUS: plant body without true roots, stems, and leaves.

THANATOCOENOSES: death-assemblage accumulation of dead organisms which did not necessarily live together.

THECA: the body formed by the plates of a thecate or tabulate motile dinoflagellate.

THECATE: composed of plates which together form a theca; or possessing a theca.

THORAX: in nassellarian radiolarians, the second (first postcephalic) chamber or segment.

TISSUE: a group of cells of similar structure which performs a specialized function.

TOOTH: projection in aperture of test, may be simple or complex, single or multiple.

TOOTH PLATE: internal, apertural modification consisting of plate that extends from aperture through chamber to previous septal foramen. One side may be attached to chamber wall or base attached to proximal border of foramen.

TRANSAPICAL AXIS: the transverse axis of the valve.

TRANSAPICAL SUTURE: the excystment suture developed in peridinioid dinocysts with a transapical excystment aperture. It follows the ventral margins of paraplates $3''$, 1a, $3'$, 3a and $5''$.

TRANSVERSE FLAGELLUM: the flagellum which arises from the anterior sulcal pore and is directed laterally in an equatorial position along the transverse furrow or cingulum, in a motile dinoflagellate.

TRANSVERSE FURROW: the equatorially aligned furrow which almost completely encircles the amphiesma and which houses the transverse flagellum. It separates the episome or epitheca from the hyposome or hypotheca. Synonymous with cingulum.

TRICOLPATE: possessing three colpi.

TRICOLPORATE: with three colpi, each containing a pore; when the pores are rather indistinct, the type is referred to as tricolporoidate.

TRILETE: with a triradiate tetrad scar.

TRIMORPHISM: some megalospheric forms were plurinucleate and reproduced, producing a third generation.

TRIPOLI: a commercial term for soft, fine-grained porous material similar to or identical with diatomite or radiolarian earth.

TRIPORATE: with three pores.

TRISERIAL: chambers arranged in three columns, three chambers in each whorl.

TRISPINOSUM TYPE: a process that is simple, or bifurcate.

TROCHOID: Trochospiral, rotaloid, rotaliform; chambers coiled spirally, evolute on one side, involute on other.

TROCHOSPIRAL: trochoid, rotaliform; spirally coiled chambers, evolute on one side of test, involute on opposite side.

TUNNEL: resorbed area at base of septa in central part of test in many fusulinids, facilitating communication between adjacent chambers.

TYCHOPELAGIC: species which spend most or all of their lives on near-shore sea bottoms.

UMBILICAL SIDE: involute side in trochospiral forms, with only chambers of final whorl visible around umbilicus.

UMBILICAL TEETH: triangular modification of apertural lip, of successive chambers in forms with umbilical aperture giving characteristic serrate border to umbilicus.

UMBILICATE: having one or more umbilici.

UMBILICUS (pl. umbilici): space formed between inner margins of umbilical walls of chambers belonging to same whorl. Pteropods: cavity or depression formed around shell axis between faces of adaxial walls of whorls where these do not coalesce to form a solid columella; in conipiral shells its opening is at the base of shell.

UNARMORED: naked.

UNICELL: an organism consisting of a single cell.

UNIFLAGELLAGE STAGE: the protoplast which is surrounded by a delicate membrane and is released from the dinocyst on excystment. It is uniflagellate and lacks both a transverse furrow and a longitudinal furrow. It is the initial free-swimming stage which lasts approximately 15 min before giving way to the gymnodinioid stage.

UNILOCULAR: single-chambered.

UNIPARTITE HINGE: hinge not subdivided into median and terminal elements.

UNISERIAL: having chambers arranged in a single row.

VACUOLE: globular inclusion in cytoplasm; includes contractile vacuoles, food vacuoles.

VALVE: the largest component of a frustule. The two valves of a frustule fit over one another much like a pill box.

VALVE CAVITY: space between the outer and the inner lamellae of the valve.

VEGETATIVE DIVISION: asexual reproduction.

VEGETATIVE STAGE: all stages in the life cycle except the encysted stage or dinocyst. Partly synonymous with motile dinoflagellate.

VELUM: adventral elongate frill-like structure which parallels the extracardinal margins in some beyrichiomorphs.

VENTRAL: lower side of test, commonly used for umbilical side; opposite to dorsal; commonly apertural side. In dinoflagellates: the surface containing, the sulcus in the motile dinoflagellate, and the parasulcus in the dinocyst.

VENTRAL NOTCH: synonymous with parasulcal notch.

VENTRAL PARAPLATE: synonymous with sulcal paraplate.

VENTRAL TONGUE: synonymous with parasulcal tongue.

VESICLE: the test of an acritarch excluding the processes. Frequently called central body.

VESICULATE: possessing sacci.

VESTIBULE: space between the outer lamella and the calcified part.

VESTIBULUM (pl. vestibula): compartment situated between the exopore and endopore and resulting from the differentiation of the ektexine and endexine in the pore area.

VIBRACULARIA: zooid with the operculum modified into a long seta in cheilostomatous Bryozoa.

VIRGATODONT: terms used for hinges of Paleozoic ostracodes in which longitudinal grooves and bars prevail.

VISBYENSE TYPE: in acritarchs broad conical to subconical processes which distally are minutely branched.

VISCERAL MASS: in pteropods, mass in which internal organs are concentrated.

WATER MASS: a parcel of water of suboceanic scale with a distinctive set of physical—chemical characteristics usually defined on the basis of temperature and salinity.

WHORL: single turn or volution of coiled test (through 360°).

WINGS: in nassellarian radiolarians a term sometimes applied to conspicuous pre-terminal spines directed away from the apex, usually homologs of dorsal and primary lateral spines.

XANTHOSOME: small brown or yellowish, globular inclusions in cytoplasm.

ZOARIUM: term for the bryozoan colony.

ZONE OF CONCRESCENCE: zone in which the outer and inner lamellae fuse by their lateral surfaces. Syn.: zone of fusion.

ZOOECIUM: skeleton of individual zooid in bryozoan colony.

ZOOID: the single individual in a bryozoan colony.

ZOOSPORANGIUM: a body within which are produced asexual spores termed zoospores.

ZOOSPORE: a naked asexual spore possessing one or more flagella.

ZOOXANTHELLAE: symbiotic cells (usually yellow pigmented) of algal origin.

ZYGOTE: result of fusion of two gametes in process of sexual reproduction.

ZYGOTIC CYST: a cyst which contains a fertilized ovum.

INDEX

Abathomphalus mayaroensis, 67, 70
Abditoloculina, 145
A. pulchra, 133
Abies, 337
A. amabilis, 337
A. grandis, 337
A. lasiocarpa, 337
Absolon, A., 138
Abushik, A.F., 123, 247
Abyssocythere, 142, 145, 147
A. casca, 126
Acanthochitina, 352
A. barbata, 352, 357
Acanthoscarpha, 145
A. volki, 137
Acarinina, 68, 72
A. densa, 72
A. soldadoensis, 71
Acervulina, 40
Acetabularia, 174
Acicularia, 181, 185
Acontiodus, 283, 285, 289
Acontiodus spp., 286
Acritarchs, 293
— evolutionary trends, 320
— major morphologic groups, 316—319
— morphology, 316
— paleoecology, 321, 322
— stratigraphic distribution, 319, 320
Actinocyclus, 284
A. ingens, 257, 258, 260
A. ochotensis, 258, 259
A. oculatus, 258, 260
Actinocythereis, 147
Adams, C.G., 37, 76, 77
Adams, I., 83, 106
Adetognathus, 287
A. giganteus, 288
Adey, W.H., 176, 183, 186
Adnatosphaeridium multispinosum, 308
Adshead, P.C., 207, 209, 242
Aechmina, 145
A. bovina, 134
Agrenocythere, 142, 145, 147
A. spinosa, 126
Alabamina, 66
Albaillella, 233
Alexander, C.I., 110, 167
Alnus, 337
A. rubra, 337
A. sinuata, 337
Allemann, F., 167, 170
Alveolinella, 40
Amaurolithus, 94
A. delicatus, 94, 105
Ambocythere, 147
Ammobaculites, 27, 42, 67
Ammodiscus, 27, 57, 67
Ammodichium, 274, 275

A. rectangulare, 274
A. prismaticum, 274
Ammonia, 36, 37, 42
A. beccarii, 48
Amorphognathus, 286, 289
Amorphognathus spp., 286
Amphiroa, 185
Amphissites, 145
A. remesi, 139
Amphistegina, 28, 40
Amphizona, 145
A. asceta, 139
Amphora, 250
Amphorella calida, 162
Amplocypris recta, 115
Ancyrochitina alaticoris, 350
A. ancyrea, 353, 357
A. desmea, 353, 357
A. diabolo, 357
A. longicornis, 357
A. nodosa, 357
A. onniensis, 350
A. pedavis, 353
A. primativa, 357
Ancyrochitina sp., 357
Ancyrodella, 287, 289
A. rotundiloba, 282, 287
Ancyrognathus, 287, 289
A. iowaensis, 287
Andres, D., 129, 147
Angochitina capillata, 357
A. comosa, 357
A. crassispina, 357
A. devonica, 357
A. dicranum, 350
A. echinata, 352
A. eisenacki, 357
A. filosa, 357
A. mourai, 357
Angochitina sp., 354, 357
Anomalina, 45, 46, 47
Annelus californicus, 257, 261
Anthochitina sp., 349
Applin, E., 4
Apsidognathus, 286, 289
A. tuberculatus, 286
Aragonia, 61, 64
Archaeolithophyllum, 174, 178, 185
Archaeolithothamnium, 174, 176
Archaeosphaeroides barbertonensis, 319
Archaeotrypa, 196
Archaias, 30
Archimedes, 194, 197, 198
Areoligera senonensis, 311
Areosphaeridium arcuatum, 313
A. diktyoplokus, 308, 313
A. multicornatum, 313
Arkhangelskiella, 90
A. cymbiformis, 90
A. specillata, 90

Arnott, H.J., 173, 186
Arnold, 21
Arrhenius, G., 255, 266
Articulina, 29, 40
Artophormis barbadiensis, 238
A. gracilis, 238
Ash, S.R., 290
Astacolus, 60
Asterolampra, 247
Asteromphalus, 247
A. hiltonianus, 247, 261
Astrorhiza, 57
Athyrium alpestre, 330
Aubert, J., 45, 46, 64
Aubry, M.-P., 163, 170
Aulacodiscus, 256
Aurila, 147
Australiella victoriensis, 310
Avnimelech, M., 159

Bacteriostrum, 251
Bairdia, 124
Bairdiocypris prantli, 134
Baker, F.T., 332, 339
Balsam, W.L., 339
Bandy, O., 20
Barghoorn, E.S., 319, 320, 325, 326
Barker, R.W., 20, 77
Barnard, 338
Barnes, C.R., 285, 289, 290
Barrett, P.J., 338, 339
Barss, M.S., 301, 325
Barthel, K.W., 229, 242
Bassler, R.S., 134, 147, 149, 201, 279, 291
Bathysiphon, 24
Batostomella, 194, 196, 197
Bé, A.W.H., 22, 39, 76, 87, 106
Belayeva, T.V., 248, 255
Belodina, 286, 289
B. compressa, 286
Belonechitina micracantha, 357
B. robusta, 357
B. wesenbergensis, 357
Bender, H., 288, 290
Benson, R.H., 125, 126, 137, 140, 142, 147
Berger, W.H., 14, 15, 16, 51, 76, 77, 106, 107, 225, 230, 242
Berggren, W.A., 7, 12, 14, 16, 20, 45, 46, 48, 56, 76, 77
Bergström, S.M., 283, 289, 290, 291
Bernard, F., 80
Bernier, P., 165, 170
Berounella, 145
B. tricerata, 135
Beyrichia, 145
B. dactyloscopica, 115

Biddulphia, 251, 256
Bigenerina, 64
Bignot, G., 163, 170
Biology and evolution, 9—12
Biostratigraphy and biochronology, 5
Biscaye, P., 252, 266
Biscutum, 90
B. constans, 90
Bismuth, H., 170
Bispathodus, 287, 289
B. costatus, 287
Bjørklund, K.R., 214, 243
Blackites, 92
B. spinosus, 93
Black, M., 85, 106
Blow, W.H., 20
Blumenstengel, H., 134, 135, 137, 147
Bogart, 267
Boheminia, 145
B. extrema, 135
Bolivina, 38, 45, 47, 65
Bollia, 145
B. biocollina, 134
Bolli, H., 20, 77
Bonaduce, G., 149
Bonet, 163
Bonnevie, K., 158
Borza, K., 170
Bossellini, A., 228, 243
Braarudosphaera, 94
B. bigelowi, 94
B. discula, 94
Braarud, T., 80, 84
Brachycythere, 147
Bradleya, 145, 147
Br. dictyon, 126
Bradshaw, J.S., 35, 36, 37, 52, 77
Brady, G.S., 109, 147
Brady, H.B., 3, 20, 77
Bramlette, M.N., 80, 102, 106, 229, 243
Brandt, K., 213, 243
Branson, E.B., 279, 290
Brideaux, W., 335, 339
Bridge, J., 4
Brien, P., 201
Brigger, A.L., 275
Broinsonia, 90, 100
B. parca, 90, 103
Bronstein, 2, 5, 111, 147
Brood, K., 195, 201
Bruniopsis mirabilis, 258
Bryantodus inaequalis, 283
Bryozoa,
— biology, 190, 191, 192, 193
— ecology and paleoecology, 194—196
— geological occurrence and evolution, 196—201
— major divisions, 194
— skeletal structure, 193
Buge, E., 195, 201
Bukryaster, 95
B. hayi, 95
Bukry, D., 85, 86, 90, 92, 93, 95, 102, 105, 106, 269, 273, 275
Bulimina, 38
Buning, W.L., 76
Burckle, L.H., 252, 256, 257, 259, 260, 266
Bursa, A.S., 86, 106
Bythoceratina umbonata, 135, 145

Cachon, J., 206, 208, 243
Cachon, M., 206, 208, 243
Calcarina, 40
Calcareous algae,
— biogeography, 183, 184
— biology, 172—174

— ecology and paleoecology, 182, 183
— geologic distribution, 184, 185, 186
— history of study, 171, 172
— skeletal calcareous algae, 175—181
Calcareous nannoplankton,
— biogeography, 86, 87, 88
— biology, 80—86
— biostratigraphy, 102—105
— ecology, 86
— evolutionary trends, 97—99
— history of research, 79—80
— major morphological groups, 89—97
— paleobiogeography, 99—101
Calcicalathina oblongata, 103, 168
Calcispheres, 184
Callimitra agnesae, 211
Calpionellids,
— biology and paleoecology, 163, 164
— biostratigraphy, 167—170
— history of study, 162, 163
— major morphological groups, 164, 167
Calpionellids vs. tintinnids, 161, 162
Calocyclas hispida, 236
C. turris, 236
Calocycletta, 239
C. costata, 239
C. robusta, 239
C. virginis, 239
Calpionella, 162, 166, 167, 168
C. alpina, 161, 163, 164, 166, 167
C. elliptica, 161, 166, 167, 168
Calpionellites, 167, 168, 170
C. caravacaensis, 167
C. coronata, 166, 167
C. darderi, 161, 166, 167, 168, 170
Calpionellopsis, 167, 168, 170
C. oblonga, 161, 166, 167, 168, 170
C. simplex, 167, 168, 170
Calvert, S.E., 225, 243
Calyptrosphaera catillifera, 85
Campbell, A.S., 162, 242
Campbelliella striata, 165
Campyloneis, 250
C. laticonus, 237
Cannartus, 224
C. nanniferus, 237
C. prismaticus, 237
C. petterssoni, 237
C. tubarius, 237
C. violina, 237
Cannon, H.G., 110, 147, 148
Cannopilus, 270
C. depressus, 270
C. hemisphaericus, 270
Cannosphaeropsis sp., 307
Cardiniferella bowsheri, 139
Cardobairdia, 165
C. balconbensis, 134
Carduifolia, 275
C. apiculata, 275
C. gracilis, 275
Carpenteria, 40
Carpenter, W.B., 3
Casey, R.E., 209, 214, 243
Catalano, R., 170
Catapsydrax, 50, 68
C. dissimilis, 74
Catinaster, 96
Cavellina, sp., 144
C. missouriensis, 144
Cavolinia, 155, 157
C. gibbosa, 153, 156
C. globulosa, 153, 156
C. inflexa, 153, 156
C. longirostris, 153, 156
C. tridentata, 153, 156

C. uncinata, 153, 156
Cayeuxia, 185
Cellaria, 194, 197, 201
Centrocythere, 145, 147
C. denticulata, 136
Čepek, P., 105, 106
Ceramopora, 194, 197
Ceramoporella, 194
C. linstroemi, 198
Ceratium, 296, 297, 300, 306, 308, 313
Ceratium sp., 299
Ceratoikiscum, 223
Ceratolithus, 94
C. aff. acutus, 94
Cereus chiloensis, 329
Ceriopora, 194, 197, 198, 199
Cestodiscus peplum, 261
Chaetoceras, 251
Challenger, H.M.S., 3, 20, 80, 109
Cheetham, A., 195, 201
Chennaux, G., 162, 170
Chiasmolithus, 90, 97, 98
C. danicus, 97, 104
Chiastozygus, 92, 93
C. inturratus, 93
Chiloguembilina, 40, 68, 71
C. cubensis, 73
Chipping, D.H., 229, 243
Chiropteridium aspinatum, 311
Chitinodella, 164, 165, 166, 167, 168
C. boneti, 161, 166, 168
C. bermudezi, 161
C. cristobalensis, 161
Chitinozoa,
— classification, 346, 347
— evolution and biostratigraphy, 348—355
— history of study, 341
— morphology, 341—345
— occurrence, 345
— paleoecology, 347
— systematic position, 347, 348
Chora, 174
Chu Hsi, 2
Cibicidella, 24, 40
Cibicides, 24, 25, 42, 45, 47
Ciesielski, P.F., 272, 273, 275
Cifelli, R., 77
Cita, M.B., 51, 77
Citharina, 64
Cladochitina biconstricta, 357
Cladogramma dubium, 258, 261
Clark, D.L., 282, 290
Clathrium, 272
Clathrochitina sylvanica, 352
Clathropyxidella, 272
Clavofabella multidentata, 115
CLIMAP, 50
Climacocyclis elongata, 162
Climatius, 279
Clio, 155, 157
C. chaptali, 156
C. cuspidata, 153, 156
C. polita, 153, 156
C. pyramidata, 153, 156
Clodoselache, 279
Cloud, P.E., 184, 186
Clypeina, 185
Cnestocythere, 147
Coccolithus, 90
C. pelagicus, 82, 83, 84, 85
Cocconeis, 250
Collinson, C., 347, 355
Collins, R.L., 156, 159
Colom, H., 163, 164, 167, 170
Colomiella, 165

C. mexicana, 161, 165
C. recta, 161
Comasphaeridium, 320
C. cf. cometes, 319
C. williereae, 319
Conarachnium nigriniae, 261
Condonella bojiga, 162
Condonellopsis pacifica, 162
C. aculeata, 357
C. communis, 357
C. conulus, 357
C. elegans, 357
C. latifrons, 343
C. micracantha, 343
C. minnesotensis, 357
C. parviventer, 344
C. primativa, 357
C. proboscifera, 357
C. robusta, 355
C. simplex, 357
Conochitina sp., 350
C. tuba, 357
C. turris, 357
Conodonts,
— assemblages, 281, 282, 283
— composition, 279
— evolutionary trends, 285
— function, 283, 284
— history of study, 278, 279
— internal structures, 280, 281
— major conodont elements, 285—288
— morphology, 279
— paleoecology, 284
— systematic position, 284, 285
— stratigraphy, 288, 289, 290
— terminology, 279, 280
Conoidal shells, 277
Conusphaera mexicana, 103, 168
Cookson, I.C., 319, 325
Cooper, C.L., 139, 148
Corallina, 174, 178, 185
Corbisema, 271, 273
C. bimucronata, 271
C. geometrica, 271
C. triacantha, 271
Cordosphaeridium gracile, 308
Cordylodus, 286, 289
C. angulatus, 286
Cornua, 272
Corolla, 157
Corollithion, 92
C. exiguum, 93
Coronochitina coronata, 357
Coronochitina sp., 343
Coryell, H.N., 4, 5, 147
Corynabutilon vitifolium, 329
Coscinodiscus, 247, 251, 256, 257, 258, 259
C. endoi, 258, 261
C. lengtiginosis, 260
C. lewisianus, 256, 257, 261
C. margaritae, 260
C. nodulifer, 247, 255, 257
C. oblongus, 256
C. paleaceus, 258, 261
C. plicatus, 258, 262
C. praepaleaceus, 256, 258, 262
C. pulchellus, 257
C. temperei, 258
C. yabei, 257, 258
Coscinopleura, 194, 197, 200, 201
C. angusta, 200
Cox, A., 7
Coxliella coloradoensis, 162
C. helix, 162
Cramer, F.H., 315, 316, 319, 321, 325, 355

Craniopsis, 274, 275
C. octo, 274
Crassicollaria, 166, 167, 168
C. brevis, 166, 168
C. elliptica, 166
C. intermedia, 161, 166, 168
C. massutiniana, 166, 168
C. parvula, 161, 166, 168
Crassosphaera, 324
C. hexagonalis, 323
C. stellulata, 322, 324
Crepidolithus crassus, 103
Creseis, 152, 155, 157, 158
C. acicula, 153, 155
C. clava, 155
C. conica, 153
C. virgula, 153
Cretarhabdus, 91, 92
C. angustiforatus, 103
C. crenulatus, 92, 168
Cribrosphaera, 91, 92
Cribrosphaera ehrenbergi, 92
Cribrostomoides, 65
Cricosphaera, 86
C. carterae, 82
C. elongata, 84
Crisia, 194, 196, 197, 201
Croneis, C.G., 5
Cross, A.T., 333, 339
Cruciplacolithus, 90
C. tenuis, 91, 97, 104
Crucirhabdus primulus, 103
Cryptozoon proliferum, 175
Crystallolithus hyalinus, 83
Ctenidodinium elegantulum, 306
Cuneiphycus, 174, 185
Curfsina, 147
Curry, D., 159
Cushman Laboratory for Foraminiferal Research, 4
Cushman, J.A., 4, 20, 77
Cuvirina, 155, 156, 157
C. columella, 153, 155
Cyathochitina calix, 357
C. campanulaeformis, 357
C. kuckersiana, 349, 357
Cyclammina, 27
Cyclasterope fascigera, 129
Cyclococcolithus, 90
C. leptoporus, 79, 81, 91
Cycloclypeus, 41
Cyclocypris ovum, 138
Cyclonephelium, 302, 305, 306, 311
C. intricatum, 313
C. ordinatum, 311
C. vannophorum, 311
C. coronatus, 93
Cyclotella hannae, 256
Cymatiogalea stelligera, 319
Cymatiosphaera eupeplos, 319
Cymatocyclis situla, 162
Cymbalopora, 61
Cymbulia, 157
Cymopolia, 174, 180, 185
Cypridea granulosa, 138
Cyprideis, 147
C. torosa, 122
Cypridella sp., 129, 145
Cypridina mediterranea, 129
Cypridopsis vidua, 115, 116
Cypris pubera, 113
Cythere, 145
C. lutea, 136
Cythereis, 147
Cytherella, 145
C. posterospinosa, 118, 119

Cytherella sp., 114, 144
Cytheretta, 145, 147
C. gracillicosta, 137
Cytherelloidea, 124
C. chapmani, 139
Cytheridea, 145, 147
C. acuminata, 136
Cyttarocyclis magna, 162

Daday, 113
Daktylethra punctata, 85
Dale, B., 299, 305, 314, 325, 326
Dale, H.M., 333, 339
Danielopol, D.L., 121, 148
Danielopolina orghidani, 126
Darby, 132
D'Arcy Thompson, W., 211, 266
Darwinula stevensoni, 138
Dasycladus, 174
Dateriocradus, 320
Davidson, S.E., 300, 325
Davies, G.R., 182, 186
Da Vinci, Leonardo, 2
Davis, R.B., 331, 333, 339
Dawson, E.Y., 173, 186
Deep Sea Drilling Project, 4, 7, 20
Deflandrea, 296, 302, 305
D. cretacea, 310
D. phosphoritica, 310
Deflandrea sp., 303
Deflandre, G., 80, 163, 230, 243, 274, 275, 355
De Jekhowsky, B., 355
Dentalina, 66
Denticula, 258, 259
D. dimorpha, 258, 262
D. hustedtii, 257, 258, 262
D. kamtschatica, 258, 259, 262
D. lanta, 258, 262
D. nicobarica, 257, 258, 263
D. punctata, 258, 263
D. seminae, 258, 259, 260, 263
Dentostomina, 29, 30, 40
Desmochitina acollaris, 354
D. minor, 344, 346, 352, 356
D. nodosa, 356
Desmopterus, 157
Deunffia ramusculosa, 319
Deunff, J., 319, 325
Diacria, 155, 157
D. trispinosa, 153, 156
D. quadridentata, 153, 156
Diastopora, 190, 194, 197, 198
Dicloeopella borealis, 162
Dictyocha, 269, 270, 272, 273
D. ausonia, 270
D. fibula, 268, 270
Dictyococcites, 91
Dictyococcites sp., 91
Dictyocysta magna, 162
Dictyomitra multicostata, 262
Diez de Cramer, M.d.C.R., 315, 321, 322, 325
Dilley, F.C., 54, 76, 77
Dinoflagellates, 293
— biogeography and paleoecology, 313, 314
— cysts, 301, 302
— cyst terminology, 303, 304, 305
— ecology, 299
— evolution, 309, 311, 312
— fossil forms, 300, 301
— life cycle, 300
— major lineages and geologic distribution, 305—309
— morphology, 294—299
— phyletic trends, 313

Dinogymnium, 294, 306, 308, 312
D. eulaensis, *312*
Dinophysis collaris, 295
Diphes colligerum, 304
Diplopora, 185
Discoaster, 86, 94
D. barbadiensis, 97,
D. brouweri, 94, 105,
D. hamatus, 105,
D. lodoensis, 94, 104,
D. mirus, 94
D. mohleri, 97
D. multiradiatus, 94, 104
D. quinqueramus, 105
D. saipanensis, 94, 104
Discoasteroides, 94
D. kuepperi, 94, 97
Discorbinopsis, 37
Discorbis, 42
Discorhabdus, 90
D. rotatorius, 90
Disphaeria macropyla, 319
Distephanus, 269, 272, 273
D. speculum, 269, 270
Dollfus, G., 159
Domasia, 320
Donahue, J., 260, 266
Donze, P., 127, 148
D'Orbigny, A., 2, 19, 77
Dorcadospyris, 224, 239
D. alata, 239
D. ateuchus, 239
D. dentata, 239
D. forcipata, 239
D. papilio, 239
D. praeforcipata, 239
D. simplex, 239
Dorothia, 64
Douglas, R.G., 15, 17, 76, 77
Downie, C., 314, 315, 316, 324, 325
Drepanella, 145
D. crassinoda, 134
Drimys winteri, 327
Drooger, C.W., 66, 77
Dujardin, F., 2, 20
Dumitrică, P., 231, 243
Dunhum, J.B., 250, 263
Duturella ora, 162
Dyakowska, J., 331, 339
Dyrness, C.T., 331, 339

Eaton, G.L., 313, 325
Ebria, 275
E. antiqua, 274
Ebridians, 273
— biology and ecology, 273, 274
— common genera and geologic distribution, 274, 275
Ebriopsis, 275
E. crenulata, 274
Echinocythereis echinata, 126
Ecology: Paleoecology, 12—15
Edwards, A.R., 105, 106
Edwards, G., 20, 50, 77
Ehrenberg, C.G., 2, 79, 203, 205, 267, 301, 325
Eicher, D.L., 16, 162
Eiffellithus, 92, 93
E. turriseiffeli, 93, 103
Eisenack, A., 315, 319, 325, 341, 347, 355
Eisenackitina bursa, *357*
E. cylindrica, 357
E. oblonga, 357
E. philipi, 353
E. sphaerica, 357
Ellison Jr., S.P., 290

Ellisor, A., 4
Elofson, O., 110, 126, 148
Elphidium, 42
Embothrium coccinineum, 329
Emeis, J.D., 76
Emiliania, 91
E. huxleyi, 81, 82, 83, 84, 86, 91, 105
Emiliani, C., 20, 50, 77, 254, 266
Enay, R., 168, 170
Endothyra, 53
Enjumet, M., 206, 208, 211, 216, 242
Entomoconchus, 145
E. scouleri, 132
Entomoconchus sp., 132
Entosolenia, 24
Epigondollela, 288, 289, 290
E. abneptis, 290
Epimastopora, 185
Epistomina, 67
Epithemia, 250
Erdtman, G., 331, 339
Eremochitina baculata, 357
E. mucronata, 357
Ericson, D.W., 20, 77, 254, 266
Ethmodiscus rex, 248
Eucampia balaustium, 260
Euconchoecia chierchiae, 132
Eucypris clavata, 138
E. virens, 111
Eugonophyllum, 185
Euprimites, 145
E. effusus, 133
Eusyringium fistuligerum, 235
E. lagena, 235
Evans, P.R., 335, 339
Evitt, W.R., 299, 300, 301, 304, 305, 315, 324, 325
Exochosphaeridium bifidum, 307

Faegri, K., 339
Fares, F., 170
Farinacci, A., 88, 89, 106
Fasciculithus, 95, 97
F. schaubii, 95
F. tympaniformis, 95, 104
Favella helgolandica, 162
Fay, R.O., 290
Fenestella, 194, 197
F. pentagonalis, 196
Fillon, R., 27
Fischer, A.G., 243
Fisher, D.W., 277, 278, 290

Fistulipora, 194, 197, 198
Florer, L., 331, 337, 339
Foraminifera,
— apertures and openings, 32, 33
— biology, 21—26
— biostratigraphy and geochronology, 55—57
— cell contents, 21
— chamber shape and arrangement, 30—32
— ecology, 33, 35—44
— evolution, 57—66
— growth, 24, 25, 26
— history of study, 19—21
— iterative and convergent evolution, 65
— major morphological groups, 33, 34, 35
— movement, 24
— nutrition, 22, 24
— ornamentation, 33
— paleoecology, 44, 48—55
— phyletic trends, 65, 66
— pores, 32, 33

— range-charts, 69—75
— reproduction, 24
— systematic position, 21
— test mineralization, 26
— wall structure, 26—30
Forbes, 110
Forchheimer, S., 95, 106
Foreman, H.P., 216, 222, 223, 224, 230, 231, 262, 263
Forschia, 59
Fortey, R.A., 230, 263
Foster, J.H., 273, 275
Foucart, M.F., 348
Fragilariopsis kerguelensis, 260
Frenguelli, J., 271, 275
Franklin, J.F., 331, 339
Franklinella, 145
F. multicostata, 132
Frerichs, W.E., 65, 77
Friend, J.K., 216, 243
Fritch, F.E., 186, 324, 325
Frondicularia, 60, 67
Funnell, B.M., 213, 226, 230, 244
Furnishina, 285, 289
F. furnishi, 281, 285

Gaarder, K., 85, 106
Galloway, J.J., 4
Gardner, J., 76
Garrett, P., 184, 186
Garrison, R.E., 228, 243
Gartnerago, 90, 100
G. costatum, 90
Gartner, S., 85, 86, 97, 98, 102, 106
Gavellinella, 45, 66, 61, 66
Gebelein, C.D., 182, 186, 187
Gemeinhardt, K., 269, 272, 275
George, R.Y., 149
Gephyrocapsa, 91
G. oceanica, 82, 91, 105
Gervasio, A.M., 149
Geyssant, J., 168, 170
Gigantocypris, 124
G. muelleri, 129
Ginsburg, R.W., 172, 187, 330, 333, 335, 339
Givanella, 174, 175, 182, 184, 185
Gleba, 157
"*Globigerina*", 66, 68
Globigerina, 28, 39, 42, 65, 68
G. ampliapertura, 68, 73
G. angulisuturalis, 73
G. ciperoensis, 68
G. helvetojurassica, 64
G. nepenthes, 74
G. pachyderma, 40, 68, 75
G. quinqueloba, 158
G. sellii, 73
Globigerinatella, 68
Globigerinatheca, 32, 51, 68
G. index, 72
Globigerinelloides, 67
Globigerinita, 51
Globigerinoides, 22, 28, 68
G. fistulosus, 68, 75
G. primordius, 73, 74
G. ruber, 32
G. trilobus/sacculifer, 74
Globoquadrina, 57, 68
G. dehiscens, 68, 74
G. dutertrei, 158
Globorotalia, 28, 51
G. conomiozea, 68, 74
G. fimbriata, 68, 75
G. fohsi, 68, 74
G. inflata, 75
G. kugleri, 68, 73

G. margaritae, 75
G. menardii, 53, 68, 74, 75
G. miocenica, 53, 54, 75
G. multicamerata, 53, 54, 68
G. opima, 68, 73
G. tosaensis, 68, 75
G. truncatulinoides, 75
G. tumida, 75
Globotruncana, 40, 56, 64, 67, 48
G. angusticarenata, 67, 69
G. calcarata, 70
G. concavata, 70
G. contusa, 67, 68, 70
G. elevata, 70
G. fornicata, 70
G. gansseri, 67, 68
G. pseudolinneana, 67, 69
G. renzi, 67, 69
Glomar Challenger, D/V, 4
Glomospira, 27, 67
Glorianella, 145
G. vassoevichi, 135
Glyptopleura reniformis, 139
Gnathodus, 280, 287
G. semiglaber, 288
G. texanus, 282
Gocht, H., 299, 325
Goll, R.M., 214, 221, 243
Gomphonema, 250
Gondolella, 288, 289, 290
G. rosenkrantzi, 288
Goniolithon, 185
Gonyaulacysta,   297, 305, 306, 307,
    308, 309, 313
G. giuseppei, 307
G. cf. jurassica, 302
G. jurasica, 306
Gonyaulax, 294, 296, 297, 300, 301,
    306, 313, 314
G. digitalis, 300
G. spinifera, 299
Gordon, W.A., 77
Górka, H., 319, 325
Gotlandochitina martinssoni, 357
G. spinosa, 357
G. villosa, 353
Graham, H.W., 297, 325
Gramm, M.N., 119, 143, 144, 148
Gray, J., 301, 325, 330, 339
Greco, A., 136, 148
Grekoff, N., 147
Groot, J.J., 332, 339
Gross, W., 279, 290
Grunau, H.R., 242
Gründel, J., 146, 148
Gubkinella, 60, 64
Gymnodinium, 295, 308, 315
Gypsina, 177
Gyroidina, 66, 67

Habib, D., 335, 339
Hadrognathodus, 289
Haeckel, E., 3, 205, 206, 242
Haeldia, 145
Halimeda, 171, 174, 179, 180, 185
Hallam, A., 76
Hall, J., 175
Hallopora, 194, 197
Halosphaera minor, 322
Hammond, S.R., 231, 244
Hannaites, 270, 273
H. quadria, 270
Hansen, H.P., 331, 339
Hantkenina, 51, 68, 72
Haplophragmoides, 59
Haq, B.U., 12, 15, 16, 48, 77, 97, 99,
    100, 101, 102, 106

Harding, J.P., 112, 121, 148
Hare, P.E., 12, 17, 33, 36, 57, 77
Harlett, J.C., 332, 339
Harmelin, J., 195, 201
Harper, H.E., 224, 243
Harris, W.K., 338, 339
Hartmann, G., 125, 143, 147, 148
Hass, W.H., 277, 290
Hastigerina, 68
Havanardia, 145
H. havanensis, 137
Havinga, A.J., 333, 339
Hays, J.D., 231, 243, 254, 266
Hay, W.W., 80, 89, 102, 105, 106
Hazel, J.E., 141, 148
Healdia anterodepressa, 134
Heath, G.R., 225, 243
Hedbergella, 67, 69
H. trochoidea, 69
Hedley, R.H., 21, 76
Hedstroemia, 185
Helicosphaera, 91, 92, 97, 99
H. dinesenii, 92
H. recta, 92, 104
H. wilcoxonii, 92
Heliolithus, 95
H. kleinpelli, 95, 104
Helopora, 194, 197
H. lindstroemi, 197
Hemicystodinium zoharyi, 307
Hemicythere, 145
H. villosa, 137
H. folliculosa, 117
Hemidiscus cuneiformis, 247, 263
Henbest, L., 20, 77
Hendey, N.I., 246, 249, 266
Henryhowella, 145, 147
H. asperrima, 137
Hercochitina crickmayi, 351, 355, 357
H. downiei, 351, 357
Herman-Rosenberg, Y., 153, 159
Herman, Y., 158, 159
Hermesinella, 275
H. transversa, 274
Hermesinum, 275
H. adriaticum, 274
H. geminum, 274
Herodotus, 2
Herrig, E., 118, 119, 148
Hertwig, R., 203, 205
Heteraulacacysta fehmarensis, 308
Heterohelix, 40, 64, 66, 67
H. striata, 70
Heusser, C.J., 331, 337, 339
Heusser, L., 339
Hida, T.S., 158
Hinde, G.L., 281, 290
Hipponicharion, 145
H. loculàtum, 129
Hoegisphaera bransoni, 356
H. complanata, 356
H. glabra, 357
Hoeglundina, 42, 64
Hofmann, H.J., 175, 182, 186, 187
Holdsworth, B.K., 223, 230, 243
Holinella (Hollinella), 145
Hollande, A., 206, 208, 211, 216, 242
Hollinella bassleri, 133
Hollister, C.D., 14, 16
Holm, E.A., 242
Holodiscolithus macroporus, 85
Holoekiscum, 223
Homotrema, 40
Homotryblium plectilum, 307
Honjo, S., 14, 16, 87, 88, 100, 106
Hopkins, J.S., 332, 339
Hornera, 194, 197, 200, 201

H. striata, 200
Hornibrook, N. de B., 139, 148
Horny arthropods, 278
Horny brachiopods, 278
Hovasse, R., 274, 275
Howe, H.V., 126, 147, 148
Hulings, N.C., 123, 148
Hungarocypris madaraszi, 111, 113
Huroniospora, 319
Hussain, M.A., 325
Hussey, R.C., 133, 148
Hustedt, F., 245, 266
Huxley, T.H., 79
Hyalocylix, 155, 157
H. striata, 153, 155
Hyman, L.H., 201, 284, 290
Hyolites, 157
Hyolithellus, 277, 278
Hyolithelminthes, 277
Hyperammina, 27
Hystrichokolpoma rigandae, 309
Hystrichosphaeridium, 301, 302
H. eisenacki, 304
H. pseudorecurvatum, 309
H. pulchrum, 306
H. salpingophorum, 306
H. tubiferum, 302
Hystrichospheres, 293, 294

Icriodina, 286, 289
I. irregularis, 286
Icriodus, 286, 287, 289
Icriodus sp., 287
Idiognathodus, 283, 287, 289, 290
Idiognathodus sp., 288
Idmidronea, 194, 197, 198, 199
Ilyocypris, 117
I. bradyi, 117
Imbatodinium sp., 310
Imbrie, J., 50, 51, 52, 77, 253, 266
Isenberg, H.D., 86, 106
Issacharella zharnikovae, 144
Isthmia, 251
Isthmolithus, 89, 96
I. recurvus, 96, 104
Iversen, J., 339

Jaanusson, V., 121, 133, 148
Jahn, T.L., 22, 77
Jania, 174, 185
Jansonius, J., 346, 355
Jantzen, R., 106
Jenkins, D.G., 76
Jenkins, W.A.M., 347, 355
Jenrzejewski, J.P., 272, 275
Jepps, 21
Jeuniaux, C., 348, 355
Johansen, H.W., 176, 186
Johnson, D.A., 229, 231, 243
Johnson, J.H., 172, 186, 187
Johnson, T.C., 225, 243
JOIDES, 4
Jones, T.R., 3, 110
Jousé, A.P., 260, 266
Jung, P., 159
Jux, U., 323, 325

Kalochitina hirsuta, 351
Kamptner, E., 80
Kamptnerius, 90, 100
K. percivalli, 90
Kanaya, T., 254
Kapp, R.O., 339
Keega, 184
Keij, A.J., 148
Kellet, B., 134, 147
Kelts, K.R., 100, 107

Kemp, E.M., 338, 339
Kennett, J.P., 14, 17
Kesling, R.V., 132, 133, 148
Khabakov, A.P., 206, 242
Kilenyi, T.I., 127, 148
King, Jr., K., 12, 17, 33, 36, 57, 77
Kipp, N.G., 50, 52, 77, 253, 266
Kittl, E., 159
Kling, S.A., 203, 209, 214, 243
Knightina, 145
K. allerismoides, 139
Kniker, H., 4
Knoll, A.H., 224, 229, 231, 243
Kockelella, 287, 289
K. variabilis, 286
Kofoid, C.A., 162, 295, 296, 325
Koizumi, I., 254, 266
Kolbe, R.W., 255, 266
Koreneva, E.V., 333, 335, 339
Koroleva, G.S., 266
Kornicker, L.S., 129, 132, 148
Kozlova, O.G., 214, 260, 266
Kozlowski, R., 347, 348, 355
Kozur, H., 127, 133, 135, 142, 146, 148
Kristan-Tollmann, E., 139, 148
Krithe, 145, 147
K. undecimradiata, 136
Krömmelbein, K., 134, 148
Kruglikova, S.B., 214, 243
Kulm, L.D., 332, 339
Kummel, B., 16, 76, 172, 187

Lagena, 60
Lagenammina, 27
Lagenochitina baltica, 356
L. brevicollis, 356
L. esthonica, 356
L. maxima, 356
Lamarkina, 42
Lamprocyrtis haysi, 241
L. heteroporos, 241
L. neoheteroporos, 241
L. nigriniae, 241
Lange, F.W., 355
Langer, W., 129, 148
Laporte, L.F., 76
Laufeld, S., 345, 347, 355
Lapworthella, 277
Lebour, M.V., 295, 297, 325
LeCalvez, 20
Leedale, G.F., 82, 106
Le Hégarat, G., 168, 170
Leiofusa, 320
L. fusiformis, 319
L. jurassica, 319
Leiosphaeridia faveolata, 319
Lejeunia, 301
Lemoine, M., 172, 187
Lenticulina, 60, 64, 67
Leperditia, 129, 145
Leprotintinnus pellucidus, 162
LeRoy, D., 27
Leptocythere moravica, 116
L. pullucida, 136
Leviella dichotoma, 139
L. egorovi, 144
L. rudis, 139
Levinson, S.A., 129, 148
Lewin, J.C., 173, 187
Licomorpha, 250
Limacina, 157
L. bulimoides, 153, 154
L. helicina, 153
L. inflata, 153, 155
L. lesuerii, 153
L. retroversa, 153, 155
L. trochiformis, 153, 155, 158

Lindström, M., 290
Ling, H.Y., 244, 273, 275
Lingulina, 67
Linné, 2
Lintochitina cingulata, 356
Lipman, R.K.H., 206, 242
Lipps, J.H., 38, 39, 77, 269, 272, 275
Lisitzin, A.P., 226, 243
Lister, 21
Lister, T.R., 313, 315, 316, 317, 325, 326
Lithastrinus, 95
L. grilli, 95
Lithochytris archaea, 234
L. vespertilio, 234
Lithocyclia angusta, 236
L. aristolelis, 236
L. ocellus, 236
Lithophyllum, 172, 174, 185
Lithoporella, 177, 185
Lithostromation, 96
L. perdurum, 95
Lithothamnium, 174, 177, 184
Lituola, 27
Lituotuba, 57
Loculicytheretta, 147
Loeblich, Jr., A.R., 33, 77, 89, 106, 163, 164, 170, 266, 275, 309, 311, 317, 320, 325, 326
Loeblich, III, A.R., 267, 268, 273, 274, 275
Loeblich, L.A., 275
Loftusia, 61
Logan, B.W., 182, 187
Lohmann, G.P., 16, 101, 102, 106
Lohmann, H., 80
Lomatia ferruginea, 329
Londinia, 145
L. reticulifera, 120
Longispina, 145
L. oelandica, 129
Lophocythere, 145, 146
L. propinqua, 136
Lophodiacrodium pepino, 319
Lophodolithus, 91, 92
L. nascens, 92
Lorenziella hungarica, 161, 166
L. plicata, 166
Loxostomoides, 45
Loxoconcha, 145
L. bairdii, 136
Lubliner-Mianowska, K., 333, 339
Lunulites, 194, 197, 200, 201
L. saltholmiensis, 200
Luterbacher, H., 27, 64, 76
Lychocanoma bellum, 232
L. bipes, 238
L. elongata, 238
Lycopodium gayanum, 328
Lyramula, 271, 273
L. furcula, 272

MacIntyre, I.G., 176, 186
Macrocypris siliqua, 138
Macrodentina, 146
Macrora stella, 257
Mamet, B., 55, 77
Manawa, 145
M. tryphena, 139
Mandra, H., 269, 275
Mandra, Y.T., 269, 272
Manivit, H., 96, 105, 106
Mantell, G.A., 326
Manton, A.A., 339
Manton, I, 82, 106
Margachitina margaritana, 354, 356
Marginopora, 40

Marginulina, 60, 67
Marginulinopsis, 46
Marine diatoms,
— biology, 246—248
— biostratigraphy, 256—260
— ecology, 248—249
— evolutionary trends, 252
— history of study, 245
— major morphological groups, 249—251
— nutrition, 248
— paleoecology, 252—256
— reproduction, 248
Marinescu, F., 122, 149
Marshall, S.M., 268, 275
Marthasterites furcatus, 103
Martin, F., 319, 326
Martini, E., 102, 106, 273, 275
Martinsson, A., 133, 144, 148
Maslov, V.P., 172, 187
Mathews, S.C., 278, 290
Mauritsina, 147
Mayr, E., 10, 17
McAndrews, J.H. 332, 339
McIntyre, A., 87, 88, 100, 101, 106
McKenzie, D., 48, 77
McKenzie, K.G., 134, 140, 141, 148, 149
Mediaria splendida, 258
Mehl, M.G., 279, 290
Meisenheimer, J., 152, 159
Melobesia, 174, 185
Membranilarnacia ursulae, 308, 313
Membranipora, 194, 197, 201
Menzies, R.J., 142, 149
Merinfeld, E.G., 205
Mesocena, 271
M. circulus, 271
M. diodon, 271
M. elliptica, 271
Mesophyllum, 185
Mesotrypa, 194, 196, 197
Metacyclis annaluta, 162
Metrarabdotus, 194, 197, 201
M. helveticum, 201
Meyen, F.V.F., 203
Michener, C.U., 253, 266
Micrantholithus, 94
M. obtusus, 94
Micrhystridium, 317, 320
M. stellatum, 319
Micropaleontology,
— commercial, 4, 5
— future trends, 15, 16
— historical review, 2—4
Microrhabdulus, 89, 97
M. belgicus, 96
Micula, 89, 97
M. mura, 103
M. staurophora, 96, 103
Miniacina, 40
Miogypsina, 40, 66
Mizzia, 174, 185
Moelleritia moelleri, 129
Mohler, H.P., 102, 106
Monterey Formation, 229
Monticulipora, 194, 196, 197
Monty, C.L.V., 182, 187
Moore, Jr., T.C., 229, 239, 243, 339
Moore, R.C., 16, 147, 201, 290
Morozovella, 51, 68
M. angulata, 71
M. aragonensis, 72
M. lehneri, 72
M. subbotinae, 68, 72
M. uncinata, 71
M. velascoensis, 68, 71

Mosaeleberis, 145, 147
M. interruptoidea, 137
Mostler, H., 142, 148
Mucronella, 194, 197, 200, 201
M. hians, 200
Muir, M.D., 325
Muller, G.W., 109, 116, 129, 135, 149
Müller, J., 203, 205, 332, 335, 339
Müller, K.J., 278, 281, 282, 284, 290, 291
Müller, O.F., 109
Multiplicisphaeridium radicosum, 320
Murphy, M.A., 230, 243
Murray, J.W., 42, 76, 80
Mutilus, sp., 116
Myers, 2

Nagaeva, G.A., 266
Nagyella, 145
N. longispina, 135
Nannoceratopsiella, 294
Nannoceratopsis, 294, 305, 306, 312
N. gracilis, 312
N. pellucida, 32
Nannoconus, 89, 97
N. bucheri, 103
N. colomi, 168
N. elongatus, 96
N. globulus, 96
N. minutus, 96
N. trutii, 96
N. wassali, 96
Nannoconus sp., 96
Naviculopsis, 271
N. biapiculata, 271
N. rectangularis, 271
Neale, J.W., 110, 149
Nematopora, 194, 197
N. visbyensis, 198
Němejc, F., 172, 187
Neococcolithes, 92, 93
N. dubius, 93
Neocorbina, 67
Neocythere, 147
Neogloboquadrina acostaensis, 74
N. dutertrei, 12
Neogondolella, 288, 289, 290
N. mombergensis, 290
Neomeris, 174, 185
Neospathodus, 288, 289, 290
N. cristagalli, 290
Neothlipsura furca, 135
Netzel, H., 299, 325
Newton, E.T., 322, 326
Nigrini, C., 214, 229, 231, 241, 243, 244
Nisbet, E.G., 228, 244
Nitzschia, 251, 260
N. bicapitata, 247
N. californica, 258
N. cylindricus, 257
N. fossilis, 258, 263
N. jouseae, 257, 258, 259, 263
N. kerguelensis, 260
N. marina, 247, 258, 263
N. miocenica, 257, 258, 264
N. porteri, 257, 258, 264
N. praefossilis, 258
N. praereinholdii, 258
N. reinholdii, 258, 264
Nitzschia, sp., 247
Nodella, 145
N. svinordensis, 134
Nodibeyrichia tuberculata, 133
Nodosaria, 60
Noël, D., 106

Norris, R.E., 268, 275
Nothyocha, 272
Nowland, G.S., 285, 290
Nubecularia, 60
Nummulites, 2, 19, 65
Nuttalinella, 61

Odontochitina, 297
O. operculata, 310
Oepikium, 145
O. tenerum, 133
Oertli, H.J., 110, 116, 142, 147, 149
Oertliella, 145, 147
O. reticulata, 137
Ogmoconcha, 145, 146
O. amalthei, 134
Okada, H., 79, 87, 88, 95, 106, 269
Oligocythereis, 146
Oligosphaeridium, 297
O. complex, 294, 304
O. pulcherrimum, 306
Ollachitina ingens, 356
Ommatartus, 224
O. antepenultimus, 237
O. avitus, 237
O. hughesi, 237
O. penultimus, 237
O. tetrathalamus, 236, 237
Onychocella, 194, 197, 201
Opdyke, N.D., 231, 234, 266
Operculodinium israelianum, 309
Opthalmidium, 60, 64
Orbitoides, 61, 66
Orbulina, 28, 56
O. universa, 74
Orbulinoides, 51
O. beckmanni, 72
Oridorsalis, 47
Orionina, 147
Orthoteca, 157
Ortonella, 174, 185
Osangularia, 45, 46, 47
Ostracodes,
— biology 110—121
— carapace, 112, 113
— classification, 128
— ecology, 121—123
— geologic distribution, 144—147
— history of research, 109, 110
— major morphological groups, 128—140
— paleoecology, 123—128
— paleogeography, 140—142
— phylogenetic trends, 142, 143, 144
Ovulites, 185

Paalzowella, 67
Paasche, E., 84, 106
Pachysphaera, 323, 324
P. marshalliae, 322, 323
P. pelagica, 322
Pachysphaera sp., 322
Packham, G.H., 15, 17
Palaeoniscus, 279
Paleobiogeography, 14—15
Palmatolepis, 282, 287, 289
P. hassi, 287
Palmula, 60
Pander, C.H., 278, 290
Panderodus, 286, 289
P. simplex, 286
Papp, A., 89, 107
Parachaetetes, 174, 178, 184, 185
Parathurammina, 59
Pareodina, 310, 313
P. ceratophora, 310
Parhabdolithus angustus, 103

P. liasicus, 103
Parke, M., 80, 83, 106, 322, 323, 326
Parker, W.K., 3
Paroistodus, 286, 289
P. parallelus, 286
Partridge, A.D., 335, 339
Partridge, T.M., 243
Patelloides, 164
Patrick, R., 266
Pautard, F.G.E., 173, 186
Peck, R., 332, 339
Pemma, 94
P. papillatum, 94
Peneroplis, 29, 40
Penicillus, 174, 179, 180
Pentagona pentagona, 121
Peraclis, 157
P. apicifulva, 154
P. bispinosa, 154
P. depressa, 154
P. moluccensis, 154
P. reticulata, 154
Peraclis sp., 153
P. triacantha, 154
Perch-Nielsen, K., 98, 106, 270, 271, 272, 273, 274, 275
Peridinium, 296, 297, 300, 301, 306, 314
P. grande, 299
P. leonis, 297, 305
P. limbatum, 313
Perisseiasphaeridium sp., 307
Peritrachelina joidesa, 85
Pessagno, E.A., 76, 219, 223, 231, 244
Petrushevskaya, M.G., 206, 211, 213, 214, 242, 244
Phacorhabdotus, 147
Phillips, J.D., 76
Phleger, F., 20
Phoberocysta neocomica, 310
Phormocyrtis striata, 232
Phylloporina, 194, 196, 197
Pia, J., 172, 287
Picea, 336, 337
P. sitchensis, 335, 337
Pickard, G., 76
Pinus, 333, 337
P. contorta, 337
P. monticola, 337
Pitrat, C.W., 147
Planicircus, 272
Planoglobulina, 66
P. glabrata, 70
Planomalina buxtorfi, 69
Planorbulina, 40
Platycythereis, 147
Plectochitina carminae, 357
Pleurocythere, 145
P. impar, 136
Pliny the Elder, 2
Plymouth Biology Laboratory, 80
Poag, C.W., 42
Poelchau, H.S., 272, 275
Podocyrtis, 224
P. ampla, 232, 233
P. aphorma, 233
P. chalara, 233
P. diamesa, 233
P. goetheana, 233
P. mitra, 233
P. papalis, 232, 233
P. sinuosa, 233
P. trachodes, 233
Podorhabdus, 91, 92
P. granulatus, 92
Pokornyella, 145, 147
P. limbata, 137

Pokornyopsis, 145
P. feifeli, 126
Pokorný, V., 31, 76, 109, 114, 115, 116, 117, 120, 127, 133, 134, 135, 137, 139, 141, 147, 149, 213, 244
Pollen, 327
— distinguishing criteria, 328, 329
— factors affecting distribution in marine environment, 330—335
Poloniella symmetrica, 139
Polyancistrodorus, 317
Polycladolithus operosus, 95
Polycope, 145
P. punctata, 132
Polydeunffia eisenacki, 319
Polyedryxium trifissilis, 319
Polygnathus, 280, 287, 289
P. asymmetricus, 287
P. tuberculatus, 287
Polypora, 194, 197
Polysphaeridium pastielsi, 308
Pontocypris mytiloides, 138
Pontosphaera, 91, 92
P. japonica, 92
Poseidonamicus pintoi, 126
Posidonia, 125
Postuma, J., 76
Power, D.M., 332, 339
Praetintinnopsella, 165
P. andrusovi, 166, 168
Prediscosphaera, 91, 92
P. cretacea, 92
Premoli-Silva, I., 16, 55
Price, I., 228, 244
Primitiopsis, 145
P. planifrons, 133
Prins, B., 97, 105, 107
Prinsius, 91
P. bisulcus, 101
P. martinii, 91, 101
Procytheridea, 146
Progonocythere, 146
Protocythere, 145, 147
P. triplicata, 136
Pullenia, 47, 65
Pulleniatina, 40
P. obliquiloculata, 75
Puncia, 145
P. novozealandica, 139
Puri, H.S., 110, 123, 147, 148, 149
Pustulopora, 194, 197, 198, 199
P. virgula, 199
Psammosphaera, 27
Pseudoammodochium, 275
P. dictyoides, 274
Pseudobolivina, 65
Pseudoemiliania lacunosa, 105
Pseudoeunotia doliolus, 247, 257, 259, 260, 264
Pseudoguembelina excolata, 70
Pseudohastigerina, 68, 72
P. barbadiensis, 73
Pseudomesocena, 272
Pseudopolygnathus, 287, 289
Pseudopolygnathus sp., 282
Pseudorocella, 272
P. barbadiensis, 272
P. corona, 272
Pseudotsuga menziesii, 337
Pterocanium praetextum, 241
P. prismatium, 241
Pterochitina perivelata, 357
Pteropods,
— ecology, 152—154
— evolutionary trends, 156—157
— major morphological groups, 154—156

— morphology of soft parts, 151, 152
— paleoecology, 157, 158
Pterospathodus, 286, 289
P. amorphognathoides, 286
Pterospermopsis, 317
P. helios, 319
Pterygocythereis, 147
Ptilodictya, 194, 197
Pygodus, 286, 289
P. anserinus, 286
Pyrgo, 30
Pyrophacus, 306, 314
Pyxilla, 256

Quadrijugator, 145
Q. permarginatus, 133
Querus, 335, 336
Quinqueloculina, 29, 60

Rabenhorst, L., 80, 106
Rabien, A., 132, 146, 149
Racemiguembelina fructicosa, 70
Radimella, 141, 145
Radiolaria,
— biogeography, 213, 214
— biology, 206—212
— biostratigraphy, 230—231
— ecology, 212—214
— evolution and geologic history, 224—225
— history of study, 203, 205, 206
— major morphological groups, 214—224
— nassellarians, 220—224
— nutrition, 209
— paleoecology and paleooceanography, 225—230
— reproduction, 208, 209
— skeleton, 209—212
— spumellarians, 216—219
— symbiotic algae, 212, 213
Radoičić, R., 165, 170
Ramochitina magnifica, 357
Ramon, G., 159
Raup, D., 16, 76, 172, 187
Recurvoides, 27
Reimer, C., 266
Reinhardt, P., 89, 106
Rehacythereis, 147
R. (?) kodymi, 120
Remane, J., 161, 162, 166, 167, 170
Remanellina, 164
R. cadischiana, 161, 166
R. ferasini, 166
Renalcis, 184, 185
Renard, A.F., 80
Renz, G.W., 213, 244
Reschetnjak, V.V., 212, 244
Reticulofenestra, 91
R. dictyoda, 91
R. pseudoumbilica, 105
R. umbilica, 104
Reubebella, 145
R. amnekhoroshevi, 144
R. kramtchanini, 119, 144
Reuss, 3
Rexroad, C.B., 286, 290
Rezak, R., 186
Rhabdochitina hedlundi, 357
R. magna, 357
Rhabdoporella, 185
Rhabdosphaera, 92
R. clavigera, 93
R. procera, 93
Rhaphidodiscus marylandicus, 256
Rhaphoneis sachalinensis, 258
Rhizammina, 27

Rhizosolenia, 251, 259
R. barboi, 258, 264
R. bergonii, 247, 264
R. curvirostris, 258, 260, 264
R. miocenica, 258, 265
R. praealata, 258, 265
R. praebarboi, 258
R. praebergonii, 257, 259, 260, 265
Rhodes, F.H.T., 289, 290
Rhombodinium glabrum, 303, 310
Rhombopora, 194, 197, 198
Rhumbler, 21
Richterina, 145
R. zimmermanni, 132
Riculasphaera, 317
Riding, R., 183, 187
Riedel, W.R., 80, 206, 213, 214, 216, 217, 222, 224, 226, 230, 231, 242, 243, 244
Rinaldi, R.A., 22, 77
Rocella, 272
R. gemma, 272
Roperia tesselata, 247
Ropolonellus, 145
R. kettneri, 135
Rosalina, 24, 25, 26
Rose, M., 152, 159
Rossignol-Strick, M., 335, 339
Rotalipora, 32, 67, 69
Roth, P.H., 15, 16, 76, 93, 100, 107
Rouxia californica, 260
Rowe, G.T., 149
Ruddiman, W.F., 106
Ruggieri, G., 149
Rugoglobigerina, 12
R. rugosa, 70
Rüst, 162
Ryland, J.S., 190, 201

Sachs, H.M., 230—244
Sagenachitina, 352
S. striata, 357
Saito, T., 56, 266
Sanfilippo, A., 214, 224, 231, 242, 244
Sangster, A.G., 333, 339
Saracenaria, 64
Sarjeant, W.A.S., 311, 319, 323, 325, 326
Sars, G.O., 3, 109, 132, 138, 149
Savin, S.M., 77
Scaliognathus, 287, 289
S. anchoralis, 288
Schaudin, 21
Schenck, H.G., 5
Schiller, J., 80, 81, 106
Schizoporella, 192, 194, 197, 201
Schlanger, S.O., 15, 17
Schmidt, H., 279, 284, 291
Schneider, G.F., 135, 149
Schnitker, D., 22, 25, 26
Schopf, J.W., 319, 326
Schornikov, E.I., 121, 149
Schott, H., 20, 77
Schrader, H.-J., 256, 266
Schubert, R.J., 20
Schumann, 267
Schwalb, H., 347, 355
Sclater, J.G., 48, 77
Sclerochilus levis, 135
Scolopodus, 285, 289
S. rex, 286
Scott, H.W., 279, 291
Scott, R.A., 339
Scriniodinium crystallinum, 306
Scyphosphaera, 91, 92
S. pulcherrima, 92
Seddon, G., 290

Semicytherura angulata, 136
Semikhatov, M.A., 184, 186
Seneš, J., 122, 149
Senoniasphaera protrusa, 311
Sequoia, 335, 336
Sertella, 189
Sethochytris, 234
S. babylonis, 234
S. triconiscus, 234
Shackoina, 67
Shackleton, N.J., 339
Shannon-Wiener Diversity Index, 256
Sherborne, 3
Sherwood, R., 28
Shinn, E.A., 187
Shulz, P., 270, 274, 275
Siberiella, 185
Sibiritia ventriangularis, 129
Sigmobolbina, 145
S. variolaris, 121
Silén, L., 201
Silicoflagellates,
— biology, 267—269
— ecology, 269
— geologic distribution, 273
— history of study, 267
— major morphological groups, 269—272
— nutrition, 268
— paleoclimatology, 272, 273
— reproduction, 268, 269
— skeleton, 267, 268
Simonsen, R., 266
Simpson, G.G., 10, 11, 12, 17
Siphonochitina, 352
S. copulata, 357
S. formosa, 349, 357
S. veligera, 357
Siphonodella, 287, 289
S. duplicata, 288
Skipp, B., 55, 77
Skogsberg, T., 110, 129, 149, 295, 325
Sliter, W.V., 38, 77
Sloan, J.R., 243
Smith, R.N., 115, 149
Smittipora, 194, 197, 201
Sohn, I.G., 124, 132, 138, 149
Sokal, R.A., 253, 266
Solenopora, 174, 185
Sollasites, 90
S. horticus, 90
Solnhofen Limestone, 229
Sorby, H.C., 79
Soutar, A., 229, 244
Spathognathodus, 286, 287, 288, 289
S. steinhornensis, 286
Sphaerocalyptra papillifera, 85
Sphaerochitina longicollis, 357
S. pistilliformis, 357
S. sphaerocephala, 352
S. vitrea, 357
Sphaerocodium, 184, 185
Sphaeroidinella dehiscens, 75
Sphaeroidinellopsis, 68
S. dehiscens, 68
S. seminulina, 74
Sphenolithus, 96
S. belemnos, 105
S. ciperoensis, 104
S. distentus, 95, 104
S. heteromorphus, 95, 105
S. moriformis, 95
S. pseudoradians, 104
S. radians, 95
Spinacopia sandersi, 129
Spiniferites, 297, 301, 305
S. bentori, 300

S. cingulatus, 307
S. pseudofurcatus, 309
S. ramosus, 304, 308
Spiniferites sp., 307
Spinoceberis, 147
Spirillina, 64
Spjeldnaes, N., 143, 149
Spongaster, 225
S. berminghami, 240
S. pentas, 240
S. tetras, 240
Spores, 328
— characteristics, 350
Spores and pollen,
— distribution in marine environment, 335—339
Stadum, G.J., 244
Stanley, E-A., 335—339
Staplin, F.L., 321—326
Starbo, 2
Stehli, F.G., 77
Steinmann, G., 228, 244
Stelidiella stelidium, 162
Stenosemella nivalis, 162
Stephanelytron redcliffense, 319
Stephanodiscus astraea, 255
Stephanolithion, 92
S. bigotii, 103
S. laffittei, 93
Stephanopyxis, 248, 256
S. californica, 258
S. dimorpha, 258
Stichocorys delmontensis, 240
S. peregrina, 240
S. wolffii, 240
Stilostomella, 47
Stockman, K.W., 179, 187
Stoppel, D. 288, 290
Stover, L.E., 335, 339
Stradner, H., 89, 107
Streeter, S., 29, 40, 42
Strelkov, A.A., 242
Streptochilus, 68
Streptognathodus, 290
Stromatolites, 181, 182
Sturmer, W., 229, 244
Styliola, 155, 157
S. subula, 153, 155
Styliolina, 157
Subbotina, 51, 68
S. frontosa, 72
Subbotina, N.N., 20, 77
"*Subbotina*" *pseudobulloides*, 71
S. triloculinoides, 71
Surdam, R.C., 183, 187
Surirella, 251
Svantovites, 145
S. primus, 135
Swain, F.M., 110, 149
Sweet, W.C., 283, 289, 290, 291
Sylvester-Bradley, P.C., 110, 113, 125, 129, 132, 138, 140, 142, 147, 149
Syracosphaera, 92
S. lamina, 93
S. pulchera, 93
Systematophora schindewolfi, 306

Tallinella, 145
T. dimorpha, 133
Tappan, H., 33, 77, 89, 106, 163, 164, 170, 252, 266, 275, 309, 311, 317, 320, 325, 326
Tappanina, 46
Tasmanites, 322, 323, 324
T. cf. *tardus*, 322
T. huronensis, 323
T. suevicus, 323

Tasmanitids, 293
— evolutionary trends, 324
— historical background, 322
— major morphological groups, 323, 324
— morphology, 323
— paleoecology, 324
— stratigraphic distribution, 324
Taugourdeau, P., 355
Tavener-Smith, R., 201
Tentaculites, 157
Terrestricythere, 121
Tesch, J.J., 154, 156, 159
Tetradella, 145
T. quadrilirata, 133
Tetralithus trifidus, 103
Tetrataxis, 59
Thalassiphora pelagica, 308
Thalassiosira, 248, 259
T. antiqua, 258
T. convexa, 257, 258, 265
T. gracilis, 260
T. gravida, 260
T. nativa, 257
T. nitzschioides, 260
T. praeconvexa, 257, 258, 265
T. punctata, 259
T. usatschevii, 257, 258, 265
Thalassocythere acanthoderma, 126
Thelodus, 279
Theocampe mongolfieri, 232
Theocorythium trachelium, 241
T. vetulum, 241
Theocyrtis annosa, 238
T. tuberosa, 238
Theonoa, 199
T. disticha, 199
Theyer, F., 231, 244
Thierstein, H.R., 93, 94, 100, 103, 105, 107, 168, 170
Thieuloy, J.P., 168, 170
Thlipsura, 145
Thlipsurella, 145
T. discreta, 114
Thompson, D., 332, 339
Thompson, G.C., 339
Thoracosphaera, 96
T. operculata, 96
Thoracosphaera sp., 96
Thyrsocyrtis bromia, 235
T. hirsuta, 234, 235
T. rhizodon, 235
T. tetracantha, 234
T. triacantha, 235
Ticinella, 67, 69
Time-scale,
— philosophy behind establishment, 5—8
Tintinnidium neapolitanum, 162
Tintinnopsis prawazeki, 162
Tintinnopsella carpathica, 161, 166, 167, 168, 170
T. longa, 161, 166
Tintinnopsella sp., 163, 166
T. remanei, 166
Tintinnus macilentus,
Tolypammina, 27, 57
Toweius, 91
T. craticulus, 91
Trachelomonas, 347
Trachyleberidea, 147
Transversopontis, 91, 92
T. pulcher, 92
Traverse, A., 330, 333, 334, 335, 339
Tribrachiatus, 94
T. orthostylus, 94, 104
Triceratium, 251, 256

T. cinnamomeum, 247
Tregouboff, G., 152, 159
Triebel, E., 110, 125, 136, 149
Trigonopyxidia ginella, 319
Triloculina, 24, 40, 60
Triloculinella, 29
Triquetrorhabdulus, 89, 97
T. carinatus, 96, 105
Trinocladus, 185
Tristylaspyris triceros, 239
Tritaxia, 46
Trochammina, 27, 47, 59, 65
Trochiliscus, 174
Trochoaster, 96
Truncorotaloides, 68
Tschudy, R.H., 339
Tsuga heterophylla, 335, 336, 337
Tuberculodinium, 306, 312
Tubucellaria, 194, 197, 201
Tubucellaria sp., 201
Tubulibairdia antecedens, 134
Tubulipora, 194, 197
"*Turborotalia*" *cerroazulensis*, 68, 72
Turitella, 57
Turrilina, 60
Tyler, S.A., 320
Tynan, E.J., 273, 275
Tytthocoris, 164
Tytthodiscus sp., 322, 324

Udden, J.A., 4
Ulrich, E.O., 134, 149, 279, 291
Umbilicosphaera, 90
U. sibogae, 91
Undellopsis entzi, 162
Unicospirillina, 67
Urey, H., 49
Urochitina simplex, 357
Urochitina sp., 354
Urocythereis, 147
Ussuricavina rakovkensis, 144
Uvigerina, 28, 38, 47

Vagalapilla, 92, 93
V. octoradiata, 93

Vaginella, 156, 157
V. bicostata, 156
V. chipolana, 156
V. clavata, 156
V. floridana, 156
Vaginulinopsis, 46
Vallacerta, 271, 273
V. hortonii, 272
Valentine, J.W., 38, 39, 77
Van den Bold, 140, 149
Van der Lingen, G.J., 15, 17
Van Hinte, J.E., 7, 8, 17, 56, 77
Van Leeuwenhoek, A., 2
Van Morkhoven, F.P.C.M., 116, 149
Van Valkenburg, S.D., 268, 275
Vávra, V., 111, 149
Vermiporella, 184, 185
Veryhachium, 320
V. calandrae, 319
V. europaeum, 316, 320
Vesper, B., 126, 149
Vespremeanu, E.E., 121, 148
Vestrogothia, 278
V. spinata, 278
Visscher, H., 162
Von Koenigswald, G.H.R., 76
Voss-Foucart, M.F., 348, 355
Vozzhennikova, T.F., 314, 326
Vulvulina, 27

Wagner, C.W., 76, 136, 149
Wall, D., 299, 301, 305, 306, 314, 319,
    321, 322, 325, 326
Wallich, G.C., 79
Walliser, O.H., 286, 290
Wallodinium lunum, 319
Walter, M.R., 186
Wass, R.E., 195, 201
Watabe, N., 81, 82, 83, 84, 85, 107
Watznauria, 90
W. barnesae, 90
W. communis, 103
Weaver, F., 272, 275
Webb, III, T., 331, 333, 339
Wehmiller, J., 57, 77

Westergaardodina, 285, 289
W. bicuspidata, 281
W. tricuspidata, 285
Wetzel, O., 301, 326
Wetzeliella, 296
W. condylos, 310
W. lunaris, 302
W. reticulata, 304
W. tenuivirgula, 310
Whitney, F.L., 4
Winterer, E.L., 51, 77, 228, 230, 242,
    243
Wilbur, K.M., 81, 82, 83, 84, 85, 107
Williams, A., 201
Williams, D.B., 314, 326
Williams, G.L., 301, 306, 325, 326, 335,
    339
Williamson, N.C., 3
Wise, S.W., 100, 107
Wollin, G., 20, 254, 266
Wornardt, W.W., 249, 266, 268, 270,
    275
Wray, J.L., 183, 184, 186, 187

Xanthidium, 301
Xenascus, 297
X. ceratioides, 310, 311
Xestoleberis, 145
X. aurantia, 136
Xystonellopsis inaequalis, 162

Yvonniellina, 164

Zaitzeff, J.B., 339
Zalányi, B., 110, 113, 149
Zarillo, G.A., 272, 275
Zaspelova, V.S., 134, 149
Zeuner, F.E., 309, 326
Ziegler, W., 290
Zygobolba, 145
Z. decora, 133
Zygodiscus, 92, 93
Z. sigmoides, 93
Zygrhablithus bijugatus, 85